Edited by
Jovitas Skucas, M.D.
Department of Diagnostic Radiology
School of Medicine and Dentistry
University of Rochester
Rochester, New York

RADIOGRAPHIC CONTRAST AGENTS

Second Edition

AN ASPEN PUBLICATION®
Aspen Publishers, Inc.
Rockville, Maryland
Royal Tunbridge Wells
1989

Library of Congress Cataloging-in-Publication Data

Radiographic contrast agents.
"An Aspen publication."
Includes bibliographies and index.
1. Contrast media. 2. Diagnosis, Radioscopic. I. Skucas, Jovitas.
[DNLM: 1. Contrast Media. WN 160 R129]
RC78.7.C65R34 1989 616.07'572 88-24185
ISBN: 0-8342-0006-6

The authors have made every effort to ensure the accuracy of the information herein, particularly with regard to drug selection and dose. However, appropriate information sources should be consulted, especially for new or unfamiliar procedures. It is the responsibility of every practitioner to evaluate the appropriateness of a particular opinion in the context of actual clinical situations and with due consideration to new developments. Authors, editors, and the publisher cannot be held responsible for any typographical or other errors found in this book.

Editorial Services: Jane Coyle Garwood

Library of Congress Catalog Card Number:88-24185
ISBN: 0-8342-0006-6

Printed in the United States of America

1 2 3 4 5

Dedicated
with gratitude to the memory of
Roscoe E. Miller, M.D.
who did so much to promote
excellence in radiology

Table of Contents

Contributors

Michael A. Bettmann, MD
Associate Professor of Radiology
Harvard Medical School
Boston, Massachusetts

Francis A. Burgener, MD
Professor of Radiology
School of Medicine and Dentistry
University of Rochester
Rochester, New York

H. Joachim Burhenne, MD
Professor and Head
Department of Radiology
University of British Columbia
Vancouver, B.C., Canada

Andrew B. Crummy, MD
Professor of Radiology
University of Wisconsin
Madison, Wisconsin

Peter Dawson, PhD, MRCP, FRCR
Senior Lecturer and Honorary Consultant in Diagnostic Radiology
Royal Postgraduate Medical School
Hammersmith Hospital
London, England

Sven E. Ekholm, MD
Assistant Professor of Radiology
School of Medicine and Dentistry
University of Rochester
Rochester, New York
Currently at
Department of Radiology
University of Gothenburg
Gothenburg, Sweden

Harry W. Fischer, MD
Professor of Radiology
Former Chairman, Department of Radiology
School of Medicine and Dentistry
University of Rochester
Rochester, New York

W. Dennis Foley, MD
Professor of Radiology
Medical College of Wisconsin
Milwaukee, Wisconsin

E.A. Franken, Jr., MD
Professor and Head
Department of Radiology
University of Iowa School of Medicine
Iowa City, Iowa

P. Gauthier
Chef de Clinique Assistant
Service Central de Radiologie
Groupe Hospitalier PITIE-SALPETRIERE
Paris, France

Robert G. Gibney, MB, MRCPI, FRCPC
Assistant Professor of Radiology
University of British Columbia
Vancouver, B.C., Canada

Raymond Gramiak, MD
Professor of Radiology
School of Medicine and Dentistry
University of Rochester
Rochester, New York

F. A. D. Heitz
Professeur de Radiologie
Service Central de Radiologie
Groupe Hospitalier PITIE-SALPETRIERE
Paris, France

Anthony House, MB, ChB, FRACR
Consultant Radiologist
Mater Misericordiae Hospital
Epsom
Auckland, New Zealand

Richard W. Katzberg, MD
Associate Professor of Radiology
School of Medicine and Dentistry
University of Rochester
Rochester, New York

Martti Kormano, MD
Professor and Chairman
Department of Diagnostic Radiology
University of Turku
Turku, Finland

Robert M. Lerner, MD, PhD
Associate Professor of Clinical Radiology
School of Medicine and Dentistry
University of Rochester
Rochester, New York

Alexander R. Margulis, MD
Professor and Chairman
Department of Radiology
University of California
School of Medicine
San Francisco, California

William H. McAlister, MD
Professor of Radiology and Pediatrics in Radiology
Mallinckrodt Institute of Radiology
Washington University Medical Center
St. Louis, Missouri

F.P. Meyer
Ancien Chef de Clinique Assistant
Service Central de Radiologie
Groupe Hospitalier PITIE-SALPETRIERE
Paris, France

Thomas W. Morris, PhD
Associate Professor of Radiology and Physiology
School of Medicine and Dentistry
University of Rochester
Rochester, New York

Hannu Paajanen, MD
Department of Diagnostic Radiology
University of Turku
Turku, Finland

Kevin J. Parker, PhD
Associate Professor of Electrical Engineering
Department of Electrical Engineering
University of Rochester
Rochester, New York

Michael R. Sage, MD, FRACR, FRCR, FRCP
Professor and Chairman
Department of Radiology
Flinders Medical Center
Bedford Park, South Australia

Yutaka Sato, MD
Associate Professor of Radiology
University of Iowa School of Medicine
Iowa City, Iowa

Jack H. Simon, MD, PhD
Assistant Professor of Radiology
School of Medicine and Dentistry
University of Rochester
Rochester, New York

Jovitas Skucas, MD
Professor of Radiology
School of Medicine and Dentistry
University of Rochester
Rochester, New York

Wilbur L. Smith, MD
Professor of Radiology and Pediatrics
University of Iowa School of Medicine
Iowa City, Iowa

Robert F. Spataro, MD
Associate Professor of Radiology
School of Medicine and Dentistry
University of Rochester
Rochester, New York

Ruedi F. Thoeni, MD
Associate Professor of Radiology
University of California
School of Medicine
San Francisco, California

Kenneth R. Thomson, MB, ChB, FRACR, FRCR
Associate Radiologist
Department of Radiology
University of Melbourne
Parkville, Victoria, Australia

Michael R. Violante, PhD
Associate Professor of Radiology
School of Medicine and Dentistry
University of Rochester
Rochester, New York

Heun Y. Yune, MD, FACR
Professor of Radiology
Indiana University Medical Center
Indianapolis, Indiana

Preface

Major changes have occurred in the availability and application of radiographic contrast agents since the publication of the first edition of this work in 1977. Not only have such new modalities as magnetic resonance imaging and digital angiography appeared on the horizon but the introduction of nonionic preparations has also changed the basic approach to the use of intravascular contrast agents. The application of contrast agents in ultrasonography and magnetic resonance imaging, although currently still in the research arena, undoubtedly will continue to expand.

The material presented in this text is geared toward the practicing radiologist. The purpose is to provide an understanding of (1) the basic concepts involved in the application of various contrast agents available for different examinations, (2) a rationale for their use, (3) their advantages, and (4) their limitations. A number of publications are available to the scientist and researcher involved in the forefront of new intravascular contrast media development and testing. Pertinent references are provided in Chapters 7 and 8 on intravascular contrast agents. Discussion of theoretical and research topics involving the intravascular contrast agents has thus been limited intentionally. This work is meant to impart a practical knowledge in the clinical selection of contrast agents.

The barium sulfate preparations are discussed in considerably more detail than the other contrast agents because there is no other publication in the English language dealing primarily with the medical applications of barium sulfate. In contrast, a number of recent studies have discussed the intravascular contrast agents.

Since 1977, the volume of barium gastrointestinal tract examinations has decreased in most institutions, with various endoscopic procedures, newer imaging modalities, and newer therapies for peptic ulcer disease accounting for some of this decrease. Nevertheless, commercial barium sulfate preparations have continued to evolve in sophistication to the point where today a number of commercial products are available for the study of each segment of the gastrointestinal tract. Not only have the individual products been further refined but new packaging and new formulations have also been introduced. In addition, newer methods have been developed for adding a second contrast agent in order to achieve double-contrast examinations.

Most mathematical formulas were omitted from this book. The reason was not to underestimate their importance but to make the work more readable and readily available for reference in a clinical setting. Wherever necessary, references to the pertinent literature are included.

A number of individuals aided in the preparation of this work. Especially sincere thanks go to those radiology residents, fellows, and staff at the University of Rochester Medical Center and other institutions who asked pertinent questions about the choice of contrast materials. Likewise, a number of the contributors to this second edition provided stimulating ideas as it evolved.

Unusual bibliographic requests and the wide range of references involving a number of disciplines were cheerfully handled by Ms. Iona Mackey. Secretarial assistance and other special services were provided by Ms. Judy Olevnik and Ms. Alyce Norder. Anne S. Patterson, editorial director of clinical medicine at Aspen Publishers, gave encouragement, assistance, and expert advice as the project evolved.

Part I

Gastrointestinal Agents

Chapter 1

Introduction

Jovitas Skucas

Radiopaque contrast media were used to study the gastrointestinal (GI) tract almost from the days of Roentgen, while in other structures the application of contrast agents developed only later. Two of the reasons for the early use of contrast agents may have been the need for better clinical delineation of the GI tract and the ready availability of bismuth salts.

EARLY STUDIES

The first known study of the GI tract was published in 1896.[1] It involved the stomach and bowel of a guinea pig, with lead subacetate used as the contrast agent. The next year, Rumpel studied the esophagus with bismuth subnitrate.[2] Also in 1897, two French investigators mixed food and bismuth subnitrate to study gastric motility in a frog.[3] Within several months, the same investigators published their results on the contraction of the stomach in humans, also using bismuth subnitrate.[4] A year later, in 1898, Walter Cannon published his classic studies on the movement of the stomach in a cat using bismuth subnitrate mixed with food.[5]

The first known study comparing the relative radiopacity of several contrast agents for use within the human GI tract was performed by Rumpel in 1897.[2] He studied potassium iodide, potassium bromide, bismuth subnitrate, and other substances and found that, among 5% solutions or suspensions of these agents, bismuth subnitrate cast the most dense radiographic shadow. Numerous studies using either solid or liquid suspensions of bismuth subnitrate followed. This contrast agent was mixed with various meals, allowing a fluoroscopic study of motility of the esophagus and stomach. Most investigators varied the amount of contrast agent used, whether it was liquid or solid, and the amount and type of accompanying meal.

Rieder, in 1904, is credited with establishing a standardized meal.[6] He gave clear and succinct descriptions of the usual gastric outline, as well as its numerous variations in position and size. He also described the position of the jejunum, ileum, and colon. He evaluated bowel transit time with his meal and found no difference in motility whether carbohydrates or proteins were ingested. The Rieder meal, consisting of approximately 40 g of bismuth nitrate added to food, yielded valuable initial data on the physiology of the alimentary tract. European investigators mixed the contrast medium with cooked cereal. American investigators found the Rieder meal made with cereal to be too coarse and preferred to substitute milk, buttermilk, or water.[7]

Schüle in 1904 published the first contrast enema results.[8] He used an oil suspension of bismuth subnitrate. In 1913, Haenisch described the horizontal fluoroscope,[9] and its use for a contrast enema became established.

The toxicity from the reduction of bismuth subnitrate to nitrite became apparent in 1907.[10] Bismuth subcarbonate was also used extensively for GI studies. Unfortunately, poisoning also occurred with this substance.[11]

During these early years of contrast media evolution, numerous other heavy metals were evaluated for use in the GI tract. Among others, cerium,[12] thorium,[12–14] mercury sulfide,[15] iron,[12,16] and zirconium[12,17–19] compounds were studied. Most of the early reports are suspect because of contamination of these compounds with impurities. There probably also was some cross-contamination among the heavy metals. With some of these products there was an initial enthusiasm, which was followed by disenchantment when toxicity became apparent. For example, Thorotrast, a suspension of thorium oxide, is chemically inert and without acute toxicity. It provided excellent opacification and was used for years in a number of applications before the grave late complications of its radioactivity became manifest.

NONBARIUM SULFATE MEDIA

The use of water-soluble organic iodine compounds in the GI tract was reported by Canada in 1955.[20] Shortly afterward several clinical reports appeared on the use of Gastrografin.[21,22] The application of water-soluble contrast media in the GI tract is discussed in Chapter 5.

An iodine substituted aromatic and benzene polymer that is insoluble in water and is useful for the GI tract was patented in 1975.[23] The polymer contains hydrophilic groups that swell in contact with water and form a gel. The future role of such polymeric agents is not known.

Tetraiodophthalimidoethanol, an organic iodine compound, has been ground to particle sizes as small as 1 to 2 micrometers.[24] This compound did not settle out as readily as the barium products then available in 1947; it adhered well to bowel mucosa. The major limitation was the complexity and expense involved in the preparation of this compound.

Tantalum powder has been proposed as a contrast agent for the upper GI tract.[25–27] The powder is insufflated through a catheter passed into the esophagus or stomach. Tantalum is essentially insoluble, nontoxic, and chemically inert. To my knowledge tantalum is not currently used as a bowel contrast agent. It is rather expensive, and the powder presents an explosion hazard.

The brominated fluorochemicals were initially used medically as possible synthetic oxygen carriers. One such chemical, perfluoroctylbromide (PFOB) ($C_8F_{17}Br$), is a stable, non-water-soluble compound with a sufficiently high boiling point so that it is a liquid at body temperature. Because of its bromide content, it has been considered for use as a contrast agent.[28,29] PFOB is an odorless and tasteless liquid with a molecular weight of 499. In animals, oral and rectal instillation of large doses has not produced any significant toxicity.[28,29] There is good mucosal coating because of its low surface tension. Only small amounts are absorbed systemically.

The future role of the fluorochemicals has not yet been established. They may have a role as (1) intravascular computed tomography (CT) contrast agents,[30–33] (2) as liver- and spleen-specific contrast agents in ultrasonography where they alter the acoustic properties of normal tissue and lesions,[34,35] or (3) as oral contrast agents in magnetic resonance where their lack of hydrogen results in a signal void in the bowel lumen both on T1- and T2-weighted images.[36]

Ethylenediaminetetraacetic acid, when combined with sodium hydroxide, chelates heavy metals. Chelates of some heavy metals have been proposed as contrast agents in the GI tract.[37,38] One of their limitations is dissociation with subsequent heavy metal absorption. Although theoretically feasible as contrast agents, so far these chelates have not been found to be clinically useful.

The lanthanide rare earths are good x-ray absorbers. Their oxides are water insoluble and relatively inert. Particulate contrast agents of such oxides have been proposed for visualization of the reticuloendothelial system[39]; their peroral use seems to be limited because of high cost.

Air and various gases have been used as bowel contrast agents since early in this century.[40–42] Currently, such negative contrast agents are combined with barium sulfate in double-contrast studies of the bowel. Used alone, the negative contrast agents are finding an increasing role in delineating bowel in CT examinations of the abdomen.

Barium titanate has been proposed and is used in several places in Europe.[43] It is discussed in more detail in Chapter 4. Its full role in the double-contrast examination of the GI tract remains to be determined.

New GI contrast media continue to evolve. They are often designed for specific indications and thus have limited application. One such agent was made up of air-containing polyiodostyrene spheres dispersed in a diluent and water. It has a specific gravity of one and was designed to detect gastroesophageal reflux; (it has proven to be no better than barium).[44]

A different method of improving contrast is to employ a relative monochromatic x-ray beam, with the x-ray photon energy being at the upper absorption edge of the contrast medium used.[45] A relative band-pass effect simulating a monochromatic x-ray beam can be achieved with various filters. As an example, a gadolinium filter results in relative filtration of both high and low energy x-ray photons, leading to contrast enhancement for both iodine and barium.[46]

BARIUM SULFATE

It is not known who first worked with barium sulfate in the GI tract. Cannon used primarily bismuth subnitrate, but did perform some experiments with barium sulfate, probably as early as 1896.[47,48] Because of the toxicity, expense, and scarcity of the bismuth salts and the publication in 1910 of the advantages of barium sulfate by Bachem and Günther,[49] and in 1911 by Günther,[50] the latter agent came into increasing use. However, as late as 1923 bismuth subcarbonate was still used to study bowel folds.[51]

The initial reluctance about using barium sulfate was gradually overcome by its ready availability, high density, relative insolubility, and resultant low toxicity. It has replaced other agents and currently remains the non-water-soluble agent of choice for evaluation of the GI tract. An early improvement in barium sulfate contrast media was made in 1913, when Mallinckrodt Chemical Works prepared finely precipitated barium sulfate by adding sulfuric acid to barium carbonate.[52] Further refinements have been made over the years. Currently, different products are available for each segment of the GI tract, with each product having specific properties best suited for its intended task.

The radiologist can use barium sulfate suspensions to much better advantage if the fundamental physical, chemical, and colloidal properties of these materials are understood. These properties influence processing, the end product, and eventual behavior of the contrast media. Chapters 2 and 3 describe in detail the clinical applications, toxicity, and complications of barium sulfate.

Although the importance of "impurities" has long been stressed, the incorrect notion persists that a chemical analysis will tell all one needs to know about a manufactured or natural product. Thus, dropping a watch on the floor may not change its chemical composition, but the shock may ruin its value as a time piece.[53] Similarly, the listing of sodium citrate or lactose among the ingredients in a barium sulfate preparation does not tell whether these substances were added in the precipitation process to control particle size and structure, used as a grinding or granulation agent, or added at the end as a suspending or flavoring agent. These and other additions in minute amounts can have a profound influence on subsequent performance.

Many barium sulfate additives are not listed on the container. Yet, additives are almost invariably present and form a coat that is adsorbed on the aggregate particles' surface. This coat consists of many substances in a variety of patterns with multiple degrees of ionic charge and hydration.

All publications involving barium sulfate preparations should include the brand name and quantity used. Where applicable, hydrometer readings or weight-to-volume or weight-to-weight percentages should be given. Although these terms are not synonymous, they do allow a comparison of results.

REFERENCES

1. Becher W: Zur Anwendung des röntgen'schen Verfahrens in der Medicin. *Deutsch Med Wschr* 1896;22:202–203.

2. Rumpel T: Die klinische Diagnose der spindelförmigen Speiseröhrenerweiterung. *Münch Med Wschr* 1897;44:420–421.

3. Roux J-C, Balthazard V: Sur l'emploi des rayons de Röntgen pour l'etude de la motricité stomacale. *C R Soc Biol (Paris)* 1897;49:567–569.

4. Roux J-C, Balthazard V: Étude des contractions de l'estomac chez l'homme à l'aide des rayons de roentgen. *C R Soc Biol (Paris)* 1897;49:785–787.

5. Cannon WB: The movements of the stomach studied by means of the roentgen rays. *Am J Physiol* 1898;1:359–382.

6. Rieder H: Radiologische Untersuchungen des Magens und Darmes beim lebenden Menschen. *Münch Med Wschr* 1904;51:1548–1551.

7. George AW, Leonard RD: *The Roentgen Diagnosis of Surgical Lesions of the Gastrointestinal Tract.* Boston, Colonial Medical Press, 1915.

8. Schüle A: Über die Sondierung und Radiographie des Dickdarms. *Arch Verdauungskr* 1904; 10:111–118.

9. Haenisch F: The Roentgen examination of the large intestines. *Arch Roentgen Ray* 1913; 17:208–215.

10. Böhme A: Über Nitritvergiftung nach interner Dareichung von Bismuthum subnitricum. *Naunyn Schmiedeberg's Arch Pharmakol* 1907;57:441–453.

11. Alexander W: Ueber Wismutvergiftungen und einen ungiftigen Ersatz des Wismut für Röntgenaufnahmen. *Deutsch Med Wschr* 1909;35:877–879.

12. Krause P, Schilling X: Die röntgenologischen Untersuchungsmethoden zur Darstellung des Magendarmkanales mit besonderer Berücksichtigung der Kontrastmittel. *Fortschr Roentgenstr* 1913; 20:455–505.

13. Kaestle C: Die Thorerde, thorium oxydatum anhydricum, in der Röntgenologie des menschlichen Magendarmkanals, ein Ergänzungsmittel und teilweiser Ersatz der Wismutpräparate. *Münch Med Wschr* 1908;55:2666–2667.

14. Kalkbrenner H: Über eine neue röntgenologische Untersuchungsmethode des Dickdarms: die Darstellung des Schleimhautreliefs mit Umbrathor. *Fortschr Röntgenstr* 1928;38:325–332.

15. Kopp JG: Cinneber als Contrastmiddel bij de Röntgendiagnostiek van Maag en Darmkanaal en Bloedvaatstelsel. *Ned Tijdschr Geneesk* 1920;64:2009–2014.

16. Taege K: Eisen als Ersatz des Wismut für Röntgenaufnahmen. *Münch Med Wschr* 1909; 56:758.

17. Kaestle C: Zirkonoxyd als kontrastbildendes Mittel in der Röntgenologie. *Münch Med Wschr* 1909;56: 2576–2578.

18. Shapiro R: An experimental study of zirconium compounds in contrast radiography. *Radiology* 1955;65:429–432.

19. von Elischer J: Über eine Methode zur Röntgenuntersuchung des Magens. *Fortschr Röntgenstr*. 1911;18:332–340.

20. Canada WJ: Use of Urokon (sodium-3-acetylamino-2,4,6-triiodobenzoate) in roentgen study of the gastrointestinal tract. *Radiology* 1955;64:867–873.

21. Jacobson HG, Shapiro JH, Poppel MH: Oral Renografin 76 percent: a contrast medium for examination of the gastrointestinal tract. *Am J Roentgenol* 1958;80:82–88.

22. Robinson D, Levene JM: Oral Renografin: a new contrast medium for gastrointestinal examination. *Am J Roentgenol* 1958;80:79–81.

23. Pharmacia Aktiebolag: X-ray contrast agent. 1975; British Patent 1400985.

24. Jones GE, Chalecke WE, Dec J, et al: Iodinated organic compounds as contrast media for radiographic diagnosis. Studies on tetraiodophthalimidoethanol as a medium for gastro-intestinal visualization. *Radiology* 1947;49:143–151.

25. Dodds WJ, Goldberg HI, Kohatsu S, et al: Insufflation of tantalum powder into the stomach. *Invest Radiol* 1970;5:30–34.

26. Nadel JA, Dodds WJ, Goldberg H, et al: Insufflation of powdered tantalum in the esophagus. *Invest Radiol* 1969;4:57–62.

27. Esguerra A, Segura J: Tantalum esophagography. *Radiology* 1970;97:181–182.

28. Long DM, Liu M-S, Szanto PS, et al: Efficacy and toxicity studies with radiopaque perfluorocarbon. *Radiology* 1972;105:323–332.

29. Liu M-S, Long DM: Perfluoroctylbromide as a diagnostic contrast medium in gastroenterography. *Radiology* 1977;122:71–76.

30. Mattrey RF, Long DM, Multer F, et al: Perfluoroctylbromide: a reticuloendothelial-specific and tumor-imaging agent for computed tomography. *Radiology* 1982;145:755–758.

31. Mattrey RF, Andrè M, Campbell J, et al: Specific enhancement of intra-abdominal abscesses with perfluoroctylbromide for CT imaging. *Invest Radiol* 1984;19:438–446.

32. Mattrey RF, Long DM, Peck WW, et al: Perfluoroctylbromide as a blood pool contrast agent for liver, spleen, and vascular imaging in computed tomography. *J Comput Assist Tomogr* 1984; 8:739–744.

33. Patronas N, Miller DL, Girton M: Experimental comparison of EOE-13 and perfluoroctylbromide for the CT detection of hepatic metastases. *Invest Radiol* 1984;19:570–573.

34. Mattrey RF, Leopold GR, vonSonnenberg E, et al: Perfluorochemicals as liver- and spleen-seeking ultrasound contrast agents. *J Ultrasound Med* 1983;2:173–176.

35. Mattrey RF, Strich G, Shelton RE, et al: Perfluorochemicals as ultrasound contrast agents for tumor imaging and hepatosplenomegaly: preliminary clinical results. *Radiology* 1987;163:339–343.

36. Mattrey RF, Hajek PC, Gylys-Morin VM, et al: Perfluorochemicals as gastrointestinal contrast agents for MR imaging: preliminary studies in rats and humans. *AJR* 1987;148:1259–1263.

37. Rubin M, DiChiro G: Chelates as possible contrast media. *Ann NY Acad Sci* 1959;78:764–778.

38. Shapiro R, Papa D: Heavy-metal chelates and cesium salts for contrast radiography. *Ann NY Acad Sci* 1959;78:756–763.

39. Seltzer SE: Rare earth contrast agents in hepatic computed tomography, in Felix R, Kazner E, Wegener OH (eds): *Contrast Media in Computed Tomography*. Amsterdam, Excerpta Medica, 1981, pp 76–84.

40. Pfahler GE: Physiologic and clinical observations on the alimentary canal by means of the roentgen rays. *JAMA* 1907;49:2069–2077.

41. Cole LG, Einhorn M: Radiograms of the digestive tract by inflation with air. *NY Med J* 1910; 92:705–708.

42. Williams FH: *The Roentgen Rays in Medicine and Surgery*. New York, Macmillan, 1901.

43. Heitz F, Heitz L: Presentation d'un nouveau produit de contraste en radiologie digestive: le titanate de baryum. *J Radiol Electrol Med Nucl* 1974;55:430–431.

44. Fransson S-G, Sökjer H, Johansson K-E, et al: Radiologic diagnosis of gastro-esophageal reflux; comparison of barium and low-density contrast medium. *Acta Radiol* 1987;28:295–298.

45. Ter-Pogossian M: Monochromatic roentgen rays in contrast media roentgenography. *Acta Radiol* 1956;45:313–322.

46. Atkins HL, Fairchild RG, Robertson JS, et al: Effect of absorption edge filters on diagnostic x-ray spectra. *Radiology* 1975;115:431–437.

47. Cannon WB: The passage of different foodstuffs from the stomach and through the small intestine. *Am J Physiol* 1904;12:387–418.

48. Cannon WB: Early use of the Röntgen ray in the study of the alimentary canal. *JAMA* 1914; 62:1–3.

49. Bachem C, Günther H: Bariumsulfat als schattenbildendes Kontrastmittel bei Röntgenuntersuchungen. *Z Röntgenk Rad Forschr* 1910;12:369–376.

50. Günther H: Bariumsulfat als schattenbildendes Kontrastmittel bei Röntgenuntersuchungen. *Deutsch Med Wschr* 1911;37:717.

51. Rendich RA: The roentgenographic study of the mucosa in normal and pathological states. *Am J Roentgenol* 1923;10:526–537.

52. Wallingford VH: General aspects of contrast media research. *Ann NY Acad Sci* 1959; 78:707–719.

53. Alexander J: *Colloid Chemistry*, vol V. New York, Reinhold Publishing, 1944, pp 10, 443.

Chapter 2

Barium Sulfate: Clinical Application

Jovitas Skucas

FINAL SUSPENSIONS

The ideal barium suspension must be dense, yet flow readily under the influence of gravity. It must not be too fluid and should not leave too thin a coat on the mucosa. The suspension must be slightly thixotropic so that the barium remains suspended throughout the examination. It should form a sediment slowly, and any cake at the bottom of the suspension should resuspend readily. The suspension should not foam excessively, and it should coat the mucosa evenly without artifacts. It must adhere and remain firmly attached to damp mucosa for many minutes without undue peeling. Any surface defects must level out. The suspension must not drag or flow too slowly. It must be palatable, inert, relatively inexpensive, and nontoxic. The suspension must do all this at varying bowel pH ranges; different degrees of patient hydration; in the presence of flocculating substances, such as sodium chloride, mucin, and blood; and in other extremes of a patient's internal environment.

The ideal gastrointestinal (GI) contrast agent is not yet available. For practical purposes, a compromise in terms of the various barium sulfate suspension characteristics is used. Most advances in barium mucosal coating properties have been made empirically.

Numerous terms have been used to describe commercial barium sulfate preparations. Many of these perpetuate confusion. The following phrases have been used in advertisements and in medical publications: "micronized," "stabilized," "specially micronized form," "microparticle barium," "microbarium," "colloidal-like," "extremely fine grain," "colloidal powder," and "microparticle size." The particle sizes are generally not specified by the manufacturers. "Micronized" can mean that the barium preparation has been processed through a micropulverized hammer mill, a micronizer, a reductionizer jet mill, or a colloidal mill. With none of these mechanical methods of size reduction are the barium sulfate particles primarily in the submicron range.

"Thick" and "thin," when applied to barium suspensions, are meaningless. These terms, which describe viscosity, are frequently misused to describe radiodensity. A thin suspension may be thin primarily because of its viscosity and yet be relatively radiodense. As an example, the double-contrast media designed to coat the stomach are very fluid but of high density; thus, they are dense but thin. Likewise a thick suspension may be radiolucent but highly viscous.

Previously, most commercial barium preparations were sold in bulk as a dry powder. The radiologist then prepared a suspension, generally using tap water, thus introducing another variable. Currently, commercial preparations are available as (1) dry powders in bulk or in unit doses, (2) as viscous suspensions requiring additional dilution, or (3) as ready-to-use liquid suspensions sold in bulk or unit doses.

Measurement

It is difficult to analyze the precise amount of barium present in diagnostic barium sulfate preparations. The *United States Pharmacopeia* (USP) specifies a complex assay involving conversion to barium chromate.[1] Atomic adsorption spectroscopy has also been proposed as a means of analysis.[2] This procedure is relatively accurate and specific, but is beyond the reach of most radiologists.

The accuracy of measurement is the most important factor in obtaining reproducible results. Some radiologists measure dry barium sulfate by volume. Although a volumetric measurement of water is accurate, a similar measurement of a dry powder is highly inaccurate. A given volume of one brand of barium sulfate may not weigh the same as another because different brands may have different densities. The large differences in weight of equal volumes of a number of brands were first shown by Miller in 1965 (Table 2-1).[3] Even the weight of the same brand may vary from day to day. A cup filled from the top of a container with loose powder will have a different weight than a cup filled from the bottom with packed powder.

The weight of a barium sulfate product depends upon the quantity of moisture and air adsorbed, the type and quantity of additives, the percentage of voids present, the amount of compaction during filling, the ease and degree of eventual compaction, handling during shipping, and whether or not the container was leveled when a measured amount was obtained. There is a considerable difference in the volume of a 200-ml leveled cup between "loose" and "packed" barium sulfate (Table 2-2).

Table 2-1 Comparison of 8-Ounce Level Cups of Dry, Loose Barium Compounds

Product	*Weight, <g>*	*Product*	*Weight <g>*
1. Baridol	198	12. Large-particle barium (average 4 μm)	396
2. Bari-O-Meal	230	13. Large-particle barium with additives	252
3. Barium sulfate compound	188	14. Liquibarine	255
4. Baroloid	301	15. Mallinckrodt USP barium	252
5. Barosperse	222	16. Micropaque	357
6. Barotrast	238	17. Stabarium	201
7. Basolac	247	18. Ultrapaque B	215
8. Gastriloid	194	19. Ultrapaque C	234
9. Gastropaque	441	20. Unibaryt C	211
10. Intropaque	233	21. Unibaryt rectal	208
11. 1-X Barium	421	22. Veri-O-Pake	190

Source: Reprinted with permission from "Barium Sulfate Suspensions" by RE Miller in *Radiology* (1965;84:241–251), Copyright © 1965, Radiological Society of North America Inc.

In general, the larger particle barium products have less bulk, "pack" more, and weigh more per unit volume. An exception is Micropaque, which is a granulated material with many small particles packed tightly together to form larger particles. Dry Micropaque acts like and has a weight similar to large-particle barium, but after suspension in water, the granulated particles break up. The smaller particle barium sulfate suspensions have more adsorbed air cushions that decrease weight. The larger particle suspensions have a wider range of particle sizes that increase weight.

It should be obvious that it is inaccurate to use measurements based on "packed" or "loose" barium sulfate. These terms are not precise enough for present-day formulations. In general, if volumetric measurements are used, level measurements of loose, dry barium are more precise. To sift the barium powder helps even more.

Table 2-2 Weight of 200-ml Cup of Barium Sulfate

Medium	*Loose Cup, <g>*	*Packed Cup, <g>*
Barosperse	212	267
Intropaque	184	308
Unibaryt C	140	235
USP (Mallinckrodt)	187	341

Mattsson pointed out that, even if the contrast medium is accurately mixed, the percentage of barium sulfate present in different samples used throughout the day may vary.[4] Sedimentation in storage containers can lead to a variation in concentration of the suspension. Likewise, it is not possible to judge the concentration of a suspension by its external appearance. Miller showed that even an accurately prepared suspension may superficially look well suspended; a horizontal x-ray beam radiograph is needed to show how much barium has settled out (Fig. 2-1).[3]

Testing

Standardization of a system of measurement is necessary to achieve reproducible results. The three standardization systems currently used are specific gravity,

Figure 2-1 Settling of various commercial barium products. All suspensions were 20% w/w. This radiograph was obtained after 30 minutes of settling. The numbers correspond to the barium products listed in Table 2-1. Several of the suspensions appeared to be well suspended visually, with the true degree of settling being apparent only on a radiograph. *Source*: Reprinted with permission from "Barium Sulfate Suspensions" by RE Miller in *Radiology* (1965;84:241–251), Copyright © 1965, Radiological Society of North America Inc.

weight-to-volume (w/v), and weight-to-weight (w/w), with the latter two expressed on a percentage basis.

In the weight-to-volume system, a certain *weight* of barium sulfate is added to enough water to obtain a predetermined *total volume*. For example, a 15% w/v suspension can be prepared by adding 15 g of barium sulfate to enough water to obtain a total volume of 100 ml.

In the weight-to-weight system, a certain *weight* of barium sulfate is added to enough water to obtain a predetermined *total weight*. As an example, a 15% w/w suspension can be prepared by adding 15 g of barium sulfate to 85 g (85 ml) of water to obtain a total weight of 100 g. Because barium sulfate also has some volume, the total volume in the example would be slightly greater than 85 ml.

Either of the above systems can be used; however, they are not readily interchangeable. The barium density in the two examples is not the same. This difference is accentuated at the higher densities (Fig. 2-2). A 100% w/v suspension contains 100 g of barium sulfate with sufficient water for a total volume of 100 ml. Several commercial preparations having even greater densities are avail-

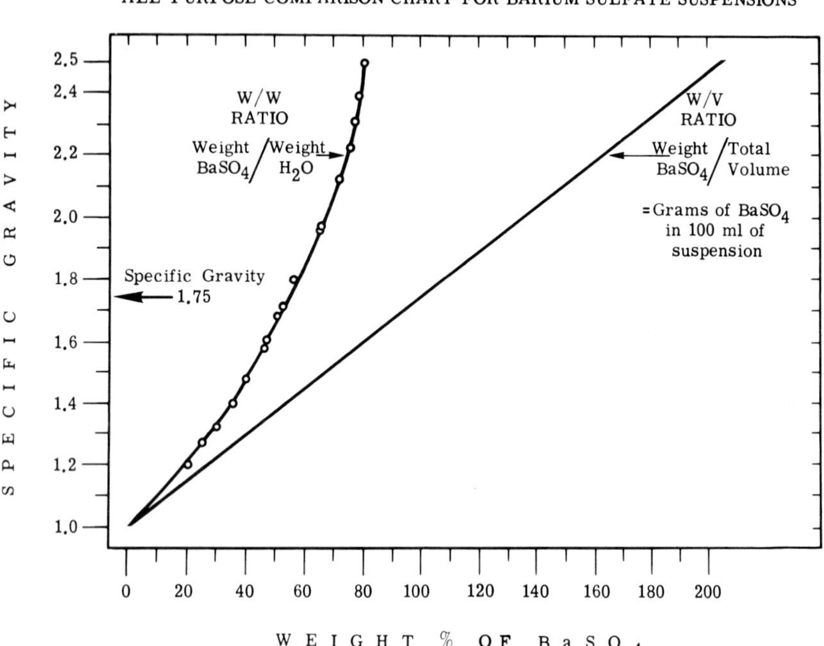

Figure 2-2 Measurement conversion chart for one particular product. The conversion factors vary depending upon the amount of additives present and other factors.

able and are used for stomach double-contrast examinations. On the other hand, a 100% w/w suspension represents the dry powder.

For either of the two above systems to be accurate, the *weight* of pure barium sulfate should be used because it is this compound that is associated with x-ray photon absorption. Practically, however, some investigators use the total weight of the dry powder and ignore the error introduced by the non-barium sulfate additives present in most commercial products.

Conversion factors exist for converting measurements from one system to the other (Table 2-3, Fig. 2-3). Because the amount of additives differs in different preparations, some error is introduced with such an all-purpose conversion system. The error is least at low specific gravities, but it can be significant at higher levels.

The specific gravity of various barium preparations can be used to compare density. An accurate instrument for the measurement of specific gravity is a special weight cup designed by the Institute for Paint and Varnish Research. This cup holds 83.2 g of water at 77°F. A sample of a well-stirred barium sulfate mixture can be weighed in this cup. The weight of the suspension, multiplied by 0.012, gives the specific gravity.

Within gross limits, a hydrometer may be used for standardization. However, such properties of barium sulfate as viscosity, foaming, and rate of settling also influence the results. The readings change as the barium sulfate settles out.[3] If the suspended material is in a stable equilibrium, the density, as determined by a hydrometer, will be the same as is obtained by weighing a given volume. If, however, the suspended barium is settling with sufficient rapidity, the density determined by a hydrometer will be less and may approach that of water. In dense suspensions, such as 200–250% w/v as used for upper GI examinations, the hydrometer simply gives no reading unless held vertically. This, in turn, can lead to further errors.

Although a hydrometer measurement is related primarily to the specific gravity of a suspension, for a given brand it also indirectly helps control viscosity, because viscosity varies directly with specific gravity, and subsequent mucosal coating. Because the coating is determined to a large degree by the type and amount of additives, consistent hydrometer readings from different batches of the same brand ensure that the same amount of additives is present in each batch.

Specific gravity readings do not necessarily correspond to the resultant mucosal coating; they do not show how thick the mucosal coating will be.[4] A less dense product may adhere better to mucosa, resulting in a thicker coating. It may thus absorb more x-ray photons and result in greater contrast. Also, an identical hydrometer reading for different commercial preparations does not imply the presence of the same amount of barium sulfate. Different amounts of additives and adsorbed air will produce different results.

Table 2-3 All-Purpose Conversion Table for Barium Sulfate Suspension*

Specific Gravity	% *w/v*	% *w/w*
2.500	200.0	80.0
2.388	185.0	77.4
2.313	175.0	75.6
2.24	156.0	75.0
2.125	150.0	70.5
1.97	131.8	65.0
1.96	125.0	64.4
1.8	100.0	56.0
1.75	96.3	55.0
1.72	90.0	52.3
1.68	85.0	50.6
1.64	80.0	48.8
1.60	75.0	46.9
1.58	72.0	45.5
1.48	60.0	40.0
1.40	50.0	35.6
1.39	48.9	35.0
1.32	40.0	30.0
1.30	37.5	29.0
1.27	33.3	26.3
1.24	30.0	25.0
1.23	28.5	23.2
1.20	25.0	20.8
1.18	22.2	18.8
1.16	20.0	17.3
1.14	18.3	15.8
1.13	16.6	14.7
1.12	15.4	13.7
1.11	14.3	12.8
1.104	13.0	11.78
1.096	12.0	10.95
1.080	10.0	9.26
1.072	9.0	8.40
1.056	7.0	6.63

*Weight-to-weight and weight-to-volume figures are approximate, but should fall within a plus or minus range of 10%. The variation occurs because the amount of additives differs. With the high specific gravity suspensions used for double-contrast gastric examinations, the error introduced by such an all-purpose conversion table may exceed 10%.

The most practical way of measuring viscosity is to use a no. 4 Ford cup viscometer. The procedure is simple and can be performed in several minutes whenever a new batch or product is introduced (Table 2-4).[5] The viscosity of a no. 4 Ford cup is the time required for 100 cc of fluid to flow through an orifice

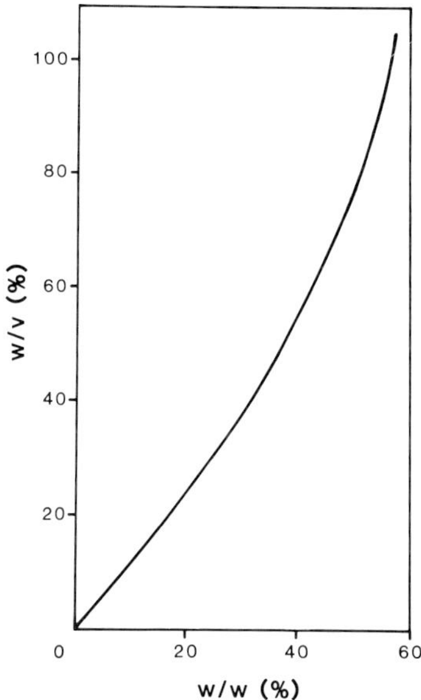

Figure 2-3 Conversion factors between weight-to-weight system and weight-to-volume system for USP barium sulfate. *Source*: Based on data in "Barium Sulfate Suspensions" by RE Miller in *Radiology* (1965;84:241–251), Copyright © 1965, Radiological Society of North America Inc.

that is 4.1 mm in diameter. Because the barium suspension flows by gravity, unnecessarily high forces are not produced as with a Brookfield viscometer.

The adhesion of a barium sulfate product to colon mucosa may be judged by dipping a clean microscope slide into a freshly stirred mixture. The slide should be held vertically and allowed to drain for 30 seconds. The gross appearance of the slide is a useful guide to subsequent coating. In addition to measuring thickness, it also shows any irregular coating caused by foam, flocculation, and poor suspension (Fig. 2-4). The observed defects compare very roughly to colonic mucosal images seen on radiography; colonic strips from a dog have also been used for similar purposes.[6] To date, no completely satisfactory model of mucosal coating exists, and of necessity, the actual examination must be the final arbiter (Fig. 2-5).

A similar in vitro model of gastric coating is not readily available,[7] although some investigators have used gastric strips from animals,[8–11] the surface of a football,[12] and other materials.[5,13,14]

One cannot use taste to distinguish between uniform and wide particle size ranges. All have a "nongritty" texture, even when some of the particle sizes range up to 11 μm in diameter. Only when the majority of particles are 4μm or larger is there a minor difference in feel, taste, or texture. Even then, the suspensions have a nongritty texture.

Table 2-4 Viscosity of Selected Barium Sulfate Products

Contrast Agent	*Density, % w/v*	*Viscosity, sec**
Intropaque	40	41.0
Solopake	60	9.7
Barosperse	60	10.0
	135	11.5
E-Z Pake	60	10.1
Micropaque	60	11.3
HD-85	60	17.1
	85	86.6
Novopaque	60	20.5
Variopake	60	30.4
Baroloid	60	62.8
Barotrast	60	161.1
Liquid Solopake	72	39.9
Liquid Polybar	100	54.9

*Viscosity was measured with a no. 4 Ford cup.

Source: Reprinted with permission from "Barium Sulfate Suspensions: An Evaluation of Available Products" by DW Gelfand and DJ Ott in *American Journal of Roentgenology* (1982;138:935–941), Copyright © 1982, American Roentgen Ray Society.

Aging

One characteristic of most colloidal adsorption complexes is that they are not stable. Many colloid-coated materials, including barium sulfate preparations, change their properties with prolonged standing. Such aging may be studied through changes in the rheological properties of the substance, such as viscosity, dilatancy, surface tension, and conductivity. Syneresis in some suspensions is a consequence of aging. Because the amount of colloidal coating material varies among different barium sulfate preparations, the effect of aging also varies. Continuous transformations are induced in these systems by hydration, heat, light, electrical fields, extrinsic ions, and molecules of the dispersion medium. There are thus variations in the physical properties of any one system. Colloids can coagulate and cause extensive changes in the suspension's physical properties. Aside from sedimentation and cake formation, the viscosity changes are usually the most noticeable. They can influence markedly the barium sulfate suspension's flow and coating behavior.

Viscosity and dilatancy of colloidal systems often change with aging. The viscosity of linear polymer additives may either increase or decrease with time.[15] A decrease in viscosity usually occurs with long-chain linear colloids, such as cellulose. Oxygen can rupture long-chain molecules. The chief cause of long-

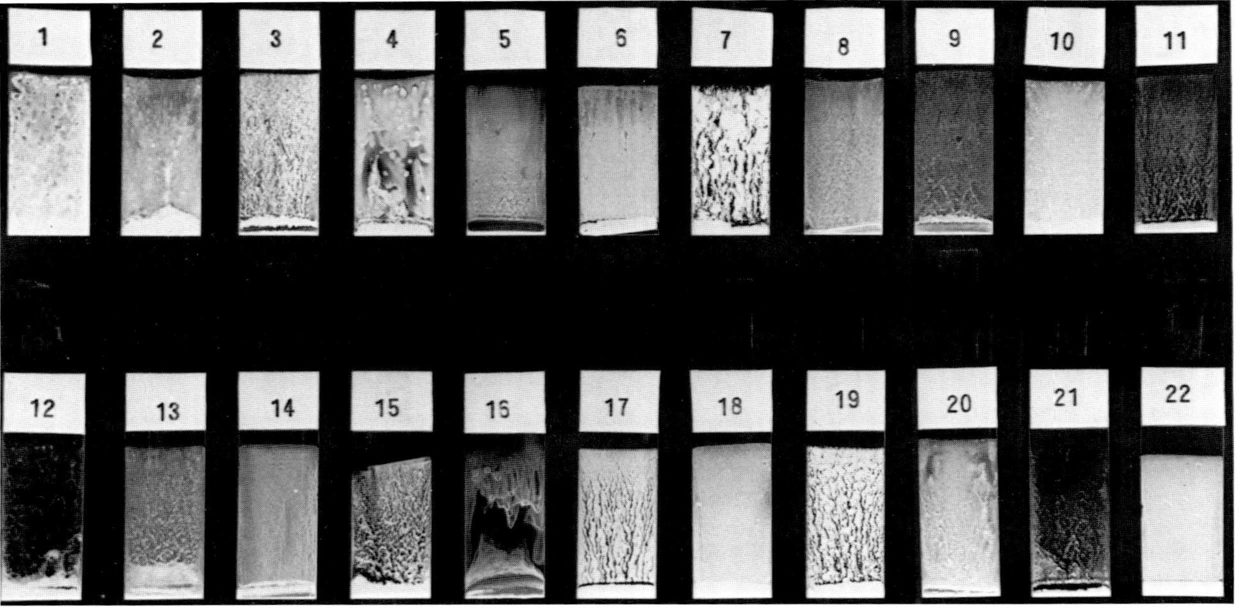

Figure 2-4 Barium coating of glass slides. Such coating is useful as a rough test of subsequent colonic mucosal coating. The artifacts and irregular coating by some products are readily apparent. The numbers correspond to the barium products listed in Table 2-1. *Source*: Reprinted with permission from "Barium Sulfate Suspensions" by RE Miller in *Radiology* (1965;84:241–251), Copyright © 1965, Radiological Society of North America Inc.

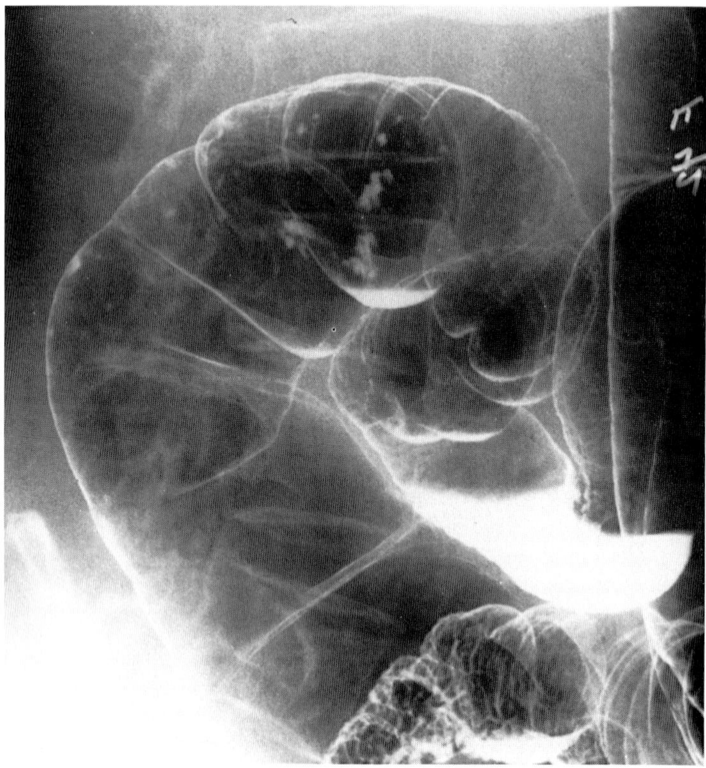

Figure 2-5 Inadequate colonic mucosal coating. Overall, the coating is too thin; both ulcers and neoplasms can escape detection with such a thin coating. In addition, some of the barium has flocculated, leading to artifact formation and further confusion. In this particular example, the barium suspension is excessively dilute.

---

chain degradation is oxidation, and even small traces of oxygen are enough to produce substantial changes. In contrast, an increase in viscosity occurs if there is linear aggregation of the polymers.

Some suspensions prepared from the dry powder in one institution's radiology department coated the stomach significantly better than commercially bought suspensions.[16] These authors postulated that some of the additives used to prevent excessive sedimentation in the commercial suspension may be responsible for their inferior coating. A number of other factors, such as different water supply and degradation of some of the additives, may also be responsible.

The effect of aging on barium sulfate contrast media has not been studied adequately. It is probably of little significance if the media are stored for short time periods. The radiologist can control the effects of aging by letting the preparations stand for a uniform length of time and by using relatively fresh suspensions.

Cost

In 1969, the cost of commercial barium sulfate products ranged from 13 cents/kg (6 cents/lb) to $4.96/kg ($2.25/lb).[17] Currently, the cost is considerably greater.

Determining the cost of barium sulfate is not easy. Undoubtedly, the extent of processing and quality control during manufacture influences the purchase price. As an example, one commercial barium sulfate preparation undergoes over 200 separate tests on the individual components and the finished product.

Because a radiologist uses barium sulfate on a volume basis, the cost should also be based on volume. Costs based on the price per kilogram are misleading. The weight of a certain volume of one brand may be considerably different from an identical volume of another brand. A more viscous barium product may produce better mucosal coating, yet its density may be lower than another brand. As a result, less barium may be required with the first brand, possibly resulting in a lower cost. The final estimate of cost should thus be the price per examination. Such factors as the type of examination performed, the specific gravity of the barium sulfate used, the mixing process, and the method of measuring barium sulfate all influence cost. As an initial step, the price per liter of the final product should be determined. An approximate estimate of the amount of suspension used per patient can be obtained by dividing the total volume of suspension used by the number of examinations performed. If the same barium sulfate preparation is used for different examinations, the relative amount for each can be estimated. The amounts discarded should be included in the total.

Sometimes, the overall cost can be reduced by using a mixture of two barium sulfate products with different flow and coating characteristics and different costs. When such products are mixed at optimal proportions, the quality of the resultant double-contrast examinations may be improved, yet at the same time, the overall cost may be lower.

As a rule, the prepackaged liquid preparations cost more per volume than the corresponding dry bulk powder. Thus, barium bought as bulk powder and mixed locally may appear to be cheapest; yet, if one includes the time of the technician for mixing, the amounts discarded, and other losses, the powder may end up being more expensive than the liquid formulations.

The most important factor in the choice of a commercial brand should be the quality of the resultant examination, not the cost. The radiologist should, however, calculate a relative cost per examination and use this information in the selection of similar but competing brands. Certainly, the cost per kilogram should not be looked upon as the true cost.

MUCOSAL COATING

The amount of mucosal coating during an examination is based on a subjective evaluation. Although most radiologists would agree on good and bad coating, small differences tend to reflect personal preferences.

The variables in commercial barium sulfate products include particle size, shape, surface area, effect of packing, type of additives present, and the presence of ions, mucin, blood, and other organic substances found within bowel. These variables influence mucosal coating, flocculation, and flow properties and account for many of the anomalous characteristics of barium sulfate media. The ions and complex molecules found within bowel lumen introduce additional characteristics and influence markedly the resultant suspension's physical properties. The type of additives and their subsequent modification by various physiologic and pathologic secretions and their ions must also be taken into consideration.

A coating thick enough to serve its intended purpose is needed. With commercial products the barium sulfate mucosal coating is thousands to millions of molecules thick. A chemist generally refers to a film as being only several molecules thick. Such a thin film in radiology is not useful because the resultant radiographic contrast would be too low.

There are forces that work against adhesion and serve to detach the barium sulfate film from the mucosa. Because of differences in expansion that occur with stretching and contraction of mucosa, the barium sulfate coating has a tendency to crack and become detached in places. A poor preparation will start cracking soon after it is applied. However, a short delay during the examination should not result in failure.[18] Highly pliable mucosal films that can accommodate themselves well to dimensional changes without cracking are therefore desirable. These films should be flexible, highly extensible, soft, plastic, yet firmly attached to the mucosa.

Breakdown in mucosal films starts at certain weak spots (Fig. 2-6). The film will crack if the force applied is greater than the ability of the film to stretch. Weak spots can be produced by mechanical means, or they can result from the action of mucin or air bubbles. Weak spots are also produced if the barium suspension flocculates. Flocculation is not the same as mucosal film breakdown. However, flocculation can lead to eventual film breakdown.

Flow lines, called "silking," must be leveled out by the action of surface tension, which tends to reduce the surface area to a minimum. Crater-like pockmarks from intact or broken bubbles must also be leveled out. Defects caused by intermittent contact of one mucosal surface with another, which occur with evacuation, compression, or peristalsis, must also be eliminated by surface tension forces. However, surface tension should not be so strong that it results in other defects. High surface tension with failure to wet and adhere to the mucosa can cause streaking and tree-like branching blemishes in the film's surface.

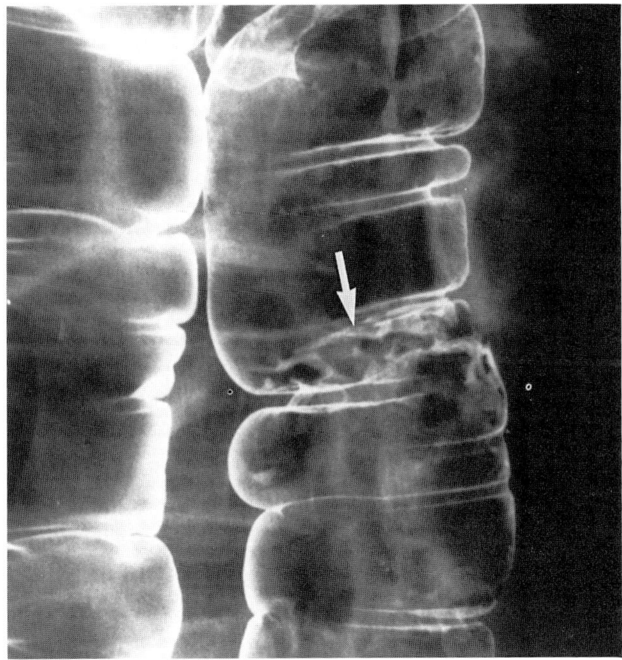

Figure 2-6 Breakdown in mucosal coating (*arrow*) presumably because of differences in elasticity; the barium coating has cracked at this particular site.

"Mottling," "crow's feet," and "drying" are terms given to various imperfections in the mucosal film seen on radiographs.

The radiologist cannot yet diagnose small lesions that protrude from the mucosal surface or that differ in size by only a few micrometers. A few micrometers' difference in the thickness of the barium film, however, has a marked effect on its ability to absorb x-rays and record a lesion on the radiograph. When measured by a wet film thickness gauge, barium suspensions vary more than 10 to 250 μm in thickness. A suspension with high fluidity with its thin coating can impair visualization of mucosa. Too little material may be left on the mucosal surface. A high density, low viscosity suspension can produce a thinner coating, yet still have excellent radiographic contrast.[19] For some barium suspensions, mucosal coating increases as the barium content is increased, up to a limit, beyond which coating starts to decrease.

Even the same suspension at two different film thicknesses may exhibit different artifacts. If the coating is too thin, local weakness and rapid breakdown occur more frequently. Too thick a coating may cause "sagging" and "rippling."

Film coatings have different characteristics at different thicknesses. The elasticity of the coating before cracking varies with the film thickness. The thicker films generally result in better coatings. However, excessively thick coatings introduce their own artifacts.

The rate of "setting" should be fast enough to prevent sagging or uneven drainage, but slow enough for proper leveling. Initial leveling is a function of the maximum fluidity of any given suspension. Subsequent slower leveling depends upon a change in fluidity that occurs with time when the patient is not moving and the barium suspension is not flowing. Rapid flow from high fluidity promotes streaks and uneven drainage.

Because leveling and sagging are related, it is necessary to compromise on these properties in good barium sulfate suspensions. Viscosity is less important than thixotropy; but, if the suspension is at all thixotropic, resistance to flow will affect the rate of both leveling and sagging. Therefore, slightly retarded flow will reduce the extent to which leveling and sagging progress before they are arrested by the setting of the suspension's thixotropic structure. For some suspensions this means that a moderate viscosity results in a smoother mucosal coating than a highly fluid one.

Different areas of the gut have different mucosa and different electrical charges. The pH may range from 1.0 to 8.4. A barium sulfate medium that is ideal for one area may fail completely in another part of the bowel. The type of suspension that gives excellent coating in the stomach is quite different from one that coats the colon mucosa well.[20] The clinical application is thus best studied by considering each structure separately.

ESOPHAGUS

Quite often the esophagus is studied without much thought being given to the type of barium sulfate used. It is not uncommon to show only the esophageal outline with such a single-contrast study. As a result, multiple views in various obliquities are required to cover the esophageal circumference adequately. Even such an examination may not demonstrate small ulcers or tumors. Some authors therefore advocate a multiphasic study consisting of single- and double-contrast studies, a mucosal relief study, and fluoroscopic evaluation of motility.[21]

Esophageal distention can be obtained with a dilute barium sulfate suspension.[22] Unfortunately, such a dilute suspension can prevent a subsequent double-contrast examination of the stomach.

The ideal double-contrast examination of the esophagus should meet the following criteria.[23]

1. It should be well tolerated by the patient.
2. It should not interfere with a subsequent double-contrast examination of the stomach.[24] Ideally, it should be integrated into the upper GI examination.
3. It should have few artifacts, such as gas bubbles (Fig. 2-7).

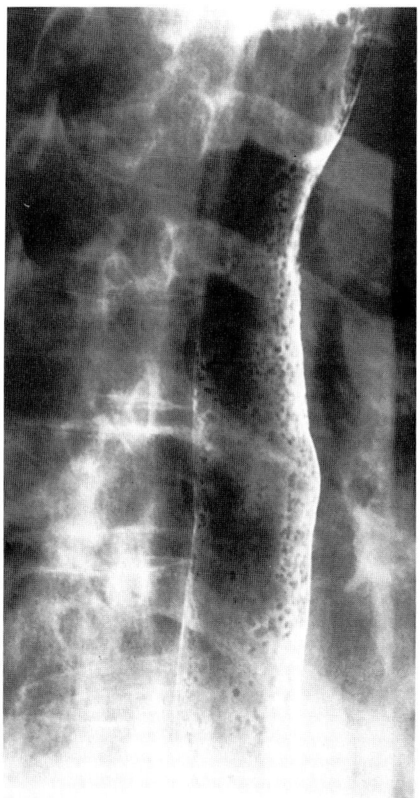

Figure 2-7 Numerous air bubbles scattered throughout the esophagus prevent adequate study. The examination is not acceptable.

4. It should be simple to perform even by the radiologist who does only an occasional examination.

Normal esophageal tonicity prevents prolonged distention; this intrinsic tonicity cannot be abolished by glucagon or anticholinergic agents, a condition not found in the rest of the bowel.[25] Some degree of esophageal dilation can be achieved with propantheline bromide (Pro-Banthine).[26,27]

Many radiologists prefer a cold contrast medium; although a cold suspension is more viscous than one at body temperature, there is better patient acceptance of a cold fluid.

A number of esophageal double-contrast techniques have been proposed; most consist of combining barium with a low-contrast agent.

Barium-Gas Techniques

Initially, an attempt was made to have the patient swallow air together with the barium suspension.[28] Or, the patient could attempt to swallow air spontaneously after drinking the barium suspension.[29] A drawback of both techniques is that some patients simply cannot swallow sufficient air. In addition, initially a low specific gravity barium suspension was used, resulting in poor mucosal coating. It has been found that a high density barium suspension, when used with such a simple technique, can be adequate for a double-contrast examination.[20] A refinement of this technique is to have the patient drink the barium suspension through a large-bore straw that has a number of side holes to help draw in air.[30] Another method is to inject the barium suspension through a nasogastric tube positioned with the tip in the esophagus, followed by air insufflation.[31,32] Air can be insufflated through the tube while the patient drinks the barium suspension. Such a double-contrast esophagram can be performed at the conclusion of an upper GI examination or enteroclysis if an indication for such a tube study exists. The major disadvantages of such a study are intubation and the extra time required. It is thus not routinely performed by most radiologists, but is reserved for specific indications.

At times, reflux of gas from the stomach can be used to achieve a double-contrast esophagram.[33] Small, rapid swallows of barium tend to allow reflux of gas into the esophagus.

Carbon dioxide can be incorporated directly into the barium suspension to produce "bubbly barium."[34–36] Such barium can be used for double-contrast studies not only of the esophagus but also of the stomach and duodenum. A commercial bottled barium product containing carbon dioxide is available (Baritop, Concept Pharmaceuticals). When the container is opened, CO_2 is slowly released, as when a bottle of club soda is opened. The intrinsic gas can be used to obtain a double-contrast esophagram in some patients. I prefer faster and greater esophageal and gastric distention than is available with this product; others have also found the evolution of gas from this product to be slower than with effervescent tablets.[37] The gas volume varies with temperature, extent of shaking, and time interval after opening the can and drinking.

Barium-Water Techniques

If the patient swallows a barium sulfate suspension followed by water, a double-contrast effect can be achieved.[38,39] A 2.5% suspension of sodium methylcellulose has also been found effective in achieving a double-contrast study.[40] If these techniques are combined with a biphasic upper GI examination, they should be performed at the end; otherwise, the extra water necessary for the esophageal views tends to dilute the barium used for the gastric study.

Barium-Effervescent Liquid or Solid Techniques

A double-contrast esophagram should be an integral part of a biphasic study of the upper GI tract. Such a study is best achieved by combining a barium sulfate suspension with either liquid or solid effervescent compounds. A number of carbon-dioxide-producing effervescent tablets, granules, and powders are commercially available. Although all are satisfactory in producing sufficient gastric distention, most effervescent tablets dissolve too slowly to produce adequate distention in the esophagus. At times, tablets can produce artifacts in the stomach. Considerable variation in the dissolution time of tablets from different batches has been reported.[41]

Commercial effervescent powders and granules come in single-dose packages. The patient places the granules or powder in the mouth, uses a small amount of water to wash it down, and immediately drinks the barium suspension. Double-contrast views of the esophagus are then obtained. Some radiologists have designed their own flavored effervescent granules and prepare them in the hospital pharmacy.[41]

I prefer liquid effervescent solutions.[24] The acid can consist of citric or tartaric acid, and the base is sodium bicarbonate. The dose per patient is 12 ml of each element. Generally, 1.5 ml of simethicone, which is equivalent to 100 mg, is added to one of the effervescent agents to prevent gas bubbles from forming. The acid and base solutions are prepared separately by the hospital pharmacy. Generally, enough of the solution is prepared at one time for several weeks or a month's use. Individual portions are then poured into small cups for each patient.

One method to obtain optimal double-contrast views of the esophagus is to place the patient upright in a slight left posterior oblique (LPO) position (with respect to the table). In quick succession, the patient drinks first one and then the other of the effervescent solutions (the sequence of effervescent solutions does not matter). The effervescent solutions begin to react in the esophagus and produce carbon dioxide. They are immediately followed by 60–120 ml of a high density, low viscosity barium sulfate suspension. The gas-distended esophagus is thus coated by the barium suspension; appropriate double-contrast views can then be obtained (Fig. 2-8).

There is better visualization of the esophagus if the liquid effervescent agents are given first, followed by the barium suspension. This sequence, however, leads to inferior gastric mucosal coating. If the effervescent agents are given first, they coat the gastric mucosa and continue to release carbon dioxide. The subsequently ingested barium suspension does not adhere as readily to the gastric mucosa,

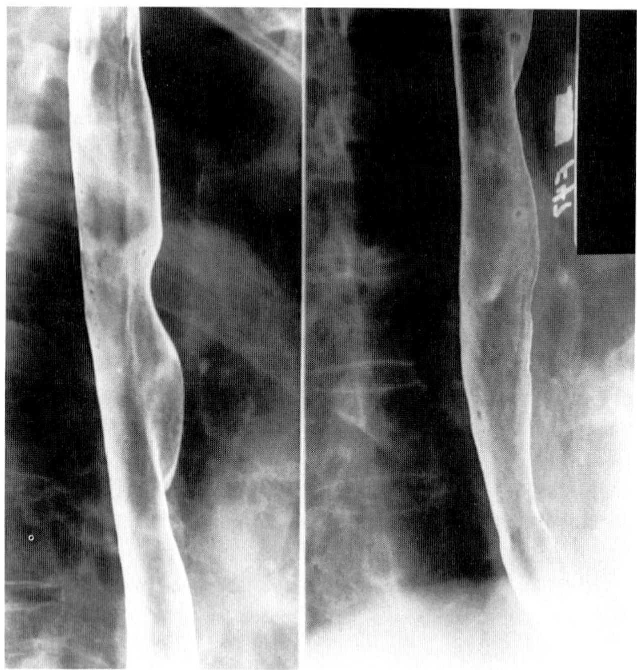

Figure 2-8 Normal esophagram. The liquid effervescent solutions were ingested first, followed by barium. Such coating can be obtained with most of the high density, low viscosity barium preparations.

presumably because of the continued carbon dioxide bubble formation. Barium coating of the gastric mucosa is improved if the barium suspension is given first, followed by the effervescents. The optimal coating in the esophagus and in the stomach thus depend on different sequences of ingestion. One compromise found useful in routine practice is to tailor the sequence of ingestion to the patient's presenting complaint. If an esophageal abnormality is suspected, the effervescent agents are given first; if a gastroduodenal abnormality is suspected, the barium sulfate suspension is given first. A similar effect is not seen when air is used as the second contrast medium, such as through a nasogastric tube.

An obstruction, gross stricture, fistula, or a large esophageal tumor can be detected with both a single-contrast and a double-contrast esophagram. Early lesions, whether inflammatory or neoplastic in nature, are best studied with a double-contrast technique.[30,31,33,42]

Other Techniques

A study of varices requires a different barium sulfate product that coats and adheres to the mucosa for extended periods of time. This requirement is met by high viscosity media, yet the viscosity should not be so high that the barium sulfate flows in a localized bolus. The commercial barium pastes are somewhat too viscous for adequate coating of the mucosa and can be difficult for the patient to swallow. Mixtures of high and low viscosity media are best. The paste should approach the consistency of honey at room temperature. Some of the low viscosity, high density barium products are also suitable for variceal detection.

If an impacted foreign body is encountered in the esophagus, some authors advocate glucagon to relieve underlying spasm.[43] Although glucagon occasionally will relieve spasm,[44] its current use for this condition is questionable. A greater success rate in relieving an obstruction can probably be obtained with judicious use of liquid effervescent agents.[45]

Reflux esophagitis has been studied using acid barium sulfate preparations. Acidification can be obtained with hydrochloric acid to a pH of 1.7, which may cause an abnormal motor response in the presence of esophagitis.[46,47] Most radiologists no longer use this technique, however.

Barium sulfate marshmallows and fudge were developed for study of the swallowing mechanism and possible localization of foreign bodies. They have had limited success.

Barium sulfate tablets, 12.5 mm in diameter, have been recommended for detection of minimal esophageal stenosis.[48,49] These tablets contain 650 mg of barium sulfate plus sugar, cornstarch, sorbitol, cellulose, and magnesium stearate. Although fresh tablets dissolve readily within 30 minutes,[49] tablets ''aged'' within a radiology department for several years may not dissolve for up to 6 hours, producing prolonged esophageal obstruction.[50]

STOMACH

Gastric pH can vary over a wide range. The barium sulfate product must adhere to the mucosa readily, be of low viscosity yet of sufficient density to produce adequate radiographic contrast, and not flocculate in the presence of gastric juice. Although there are reports of an association between barium flocculation and hyperacidity,[51] the correlation probably depends on the brand of barium product used. With some suspensions, agglomeration into clumps and settling of the barium sulfate may occur in the presence of hypersecretion and hyperacidity.[52,53] In the stomach, the barium sulfate particles adhere to the mucosa not solely because of electrostatic attraction but probably also because of direct ion bridging.[54]

The rate of gastric emptying varies with the viscosity of the test meal. Low viscosity fluids empty considerably faster than a medium or high viscosity meal.[55] Thus, the rate of barium emptying can vary with the amount of gastric secretions and retained food.

Both acid and alkaline barium suspensions have been used in the stomach.[56] An acid medium results in delayed pyloric opening and gastric hypoperistalsis, whereas an alkaline medium increases gastric emptying but does not produce any other significant radiographic changes. An action similar to acid is seen with fats, degraded protein products, and hyper- and hypotonic solutions.[57] These products may stimulate enterogastrone release in the duodenum.

Pretreatment of the gastric mucosa by a mucolytic agent is theoretically useful. An in vitro study of rat gastric mucosa found that pretreatment with the mucolytic agent, N-acetyl-L-cysteine, before adding the barium suspension improved adhesion.[58] Adhesion was improved further by adding sodium bicarbonate to the pretreatment regimen. A recent in vivo study found more frequent areae gastricae detectibility following oral pretreatment with N-acetylcysteine.[59]

In 1953, Shufflebarger et al. concluded that radiographic visualization did not correlate with the amount of secretions.[60] James et al., in 1977, used cimetidine to suppress gastric secretions and found no difference in mucosal coating[61]; in 1981, another study found that inhibition of gastric secretions before a barium examination did improve coating.[62] Currently, an antisecretory agent before a barium study is used only by a minority of radiologists.

Pharmacologic hypoperistalsis of the stomach can be readily achieved. In the United States, generally glucagon is used, whereas in some other countries Buscopan is preferred. Gas is then introduced, the patient rotated several times, and up to 50 radiographs obtained.

A barium spray nebulizer technique has been proposed, but it adds complexity to the examination and has not been widely adopted.[63]

Single-Contrast Examination

For years a single-contrast upper GI examination was performed by most radiologists. The barium density varied from 35% w/v[64] to about 80% w/v. Swallowed air is invariably present, and as usually performed, this examination generally includes some double-contrast views. Use of a compression paddle allows a "mucosal" outline to be obtained. The barium sulfate should not settle out readily; any significant settling seen with a horizontal x-ray beam should be sufficient justification to discard that particular mixture from routine use.

Double-Contrast and Biphasic Examination

A double-contrast gastric study was already described by von Elischer in 1911.[65] He used zirconium oxide (mixed with gum arabic) and air as the two contrast agents. A number of double-contrast techniques were reported in the late

1920s and early 1930s.[66–68] In the United States, the value of double-contrast studies in detecting ulcers was suggested by Hampton in 1937.[69] Occasional papers in the 1940s and 1950s continued to emphasize the usefulness of this technique.[70,71] In Japan, a further impetus to development of double-contrast gastric studies was the high incidence of gastric cancer.

At times a lesion can be missed with a double-contrast study, but will be seen by single contrast.[72] To overcome this pitfall, a combined study involving both modalities has been developed, called the biphasic examination.[73] This examination consists of barium-filled, compression, mucosal relief, and double-contrast views.

Negative Contrast Agents

A double-contrast examination can be obtained by adding air through a nasogastric tube as the second contrast agent.[32,74–76] Gas-forming tablets, granules, dry effervescent powder, swallowed air,[71] and liquid effervescent solutions have also been used.[77] I favor liquid effervescent agents for reasons discussed in the esophagus section. The main requirements of a gas-forming agent are generation of up to 400–500 ml of gas, noninterference with the barium coating, rapid disintegration, and ready patient acceptance.[41] Increasing the amounts of gas-forming agents above what is recommended does not result in significantly greater distention, presumably because of the patient's inability to retain the extra volume in the stomach.[78] Rapid addition of air through a tube can, of course, eventually lead to massive distention.

The effervescent agent should require little additional fluid; a large volume inhibits gastric mucosal coating by the barium suspension. One study suggested that the quality of mucosal coating was relatively independent of whether a liquid or solid effervescent was used.[79] A comparison of effervescent powder (E-Z-Gas powder, E-Z-EM) versus granules (Baros, Mallinckrodt, and E-Z-Gas granules, E-Z-EM), however, found that there was better patient acceptance of the granular products.[80] Patients had difficulty swallowing powder with small amounts of water. Another study compared two granular products—Baros effervescent granules and E-Z-Gas II—and found better patient acceptance, better visualization of the areae gastricae, and fewer gas bubble artifacts with Baros.[81] The major difference in ingredients between these two products is that Baros contains tartaric acid, whereas E-Z-Gas II contains citric acid (both contain sodium bicarbonate as the other active ingredient). The difference in results should not be ascribed only to the difference in acid composition; the manufacturing steps in effervescent granulation and the possible presence of other ingredients may also play a role. Likewise, a particular barium sulfate product may lead to better results with a particular effervescent agent.

Formation of bubbles can be prevented by adding an antifoam agent, such as simethicone, to reduce surface tension. Although some barium sulfate media and many commercial effervescent agents contain simethicone, in some localities additional amounts must be added to control foam adequately.

Barium Sulfate

Some radiologists prepare their own contrast media.[73] Some of these media include gas-forming agents, which are also often prepared locally.[82]

An unrefined barium powder, "barytes," mined in South Australia and mixed with Micropaque, was found to result in good mucosal coating.[83] This particular barium product was not sold as fit for human consumption, and the author cautioned that it should first be analyzed for impurities.[83] To my knowledge, this product is not used in the United States. Other investigators have also found that better gastric mucosal coating is obtained with crushed natural barium sulfate particles that have jagged edges than with the precipitated variety.[84] Some barium contrast manufacturers are using such crushed mined particles in their barium products designed for double-contrast stomach examinations. A wide range in particle size is believed to be beneficial.[84]

A good double-contrast barium suspension should result in routine identification of the areae gastricae,[20,82,85–88] although the relationship between visualization of the areae gastricae and mucosal coating is not straightforward.[79] High viscosity in these products is detrimental because small anatomic details can be obscured. Therefore, most contrast media designed for double-contrast gastric studies have a low viscosity.[10,12,89] High density has also been found to be desirable; densities up to 250% w/v and volumes of 70-100 ml are commonly used.[84] One such product achieves its high density by including heterogeneous barium particles ranging up to 12 μm[90] or 18.8 μm[10] in diameter, with some particles being even larger. Although in this product the number of such large particles is small, they contain a disproportionately large amount of barium sulfate. Such large barium particles may result in rapid sedimentation, with the larger particles settling in the mucosal grooves.[10] Because of the rapid sedimentation, these products are generally sold as a dry powder; once a liquid suspension is formed and allowed to stand for some time, the thick cake at the bottom resists resuspension, and at times the entire jug must be discarded.

If a biphasic examination is being performed, some investigators prefer a lower density barium suspension, such as 82.5% w/v.[73] Others use two contrast media; a high density product is given for the initial double-contrast phase, followed by a lower density product for the subsequent single-contrast phase. For the latter, a density as low as 15% w/v is used.[64]

The relative mucosal coating depends, in part, on the technique used and whether the barium product is given before or after the effervescent agent. The speed of patient rotation also affects the thickness of the layer deposited on the mucosa; rapid rotation reduces the time of contact between a point on the mucosa and the suspension, thus decreasing the amount of suspension adhering to the surface at any point. In general, by rotating the patient several times a better coating can be achieved. One study found improved visualization of the areae gastricae with age.[91] It is thus necessary, when comparing different contrast agents, that an identical technique and patient population be adopted.

Although the double-contrast examination has been described many times, it is unfortunate that often the type of barium sulfate used and its concentration are not mentioned.[92,93] The amount of barium sulfate given should be measured. In one study, when patients were asked to drink one mouthful from a full cup, the amount of contrast varied from 6–45 ml.[94]

Comparison of Techniques

The relative accuracy of a radiographic examination of the stomach compared to endoscopy is still not settled. Using endoscopy as a gold standard, one study found that a single-contrast barium examination is limited in detecting gastric and duodenal ulcers.[95] Several studies have questioned whether the double-contrast method is truly superior to the single-contrast examination.[72,96–98] In Japan, however, the double-contrast examination has become indispensable in detecting small gastric cancers.[99,100] In fact, mass screening for gastric cancer and the subsequent detection of early gastric cancer have resulted in a significantly higher cure rate in Japan than in the Western world.[101] The double-contrast and biphasic studies have also gained acceptance in the West and are widely advocated. In addition to detecting small gastric cancers, the double-contrast examination is accurate in detecting linear ulcers[102] and gastritis.[103,104]

A number of comparison studies of double-contrast techniques have been published.[10,34,78,79,105–107] In general, the high density, low viscosity products improved visualization.[10,12,86,108]

SMALL BOWEL

The study of mucus coating the small bowel mucosa is still in its infancy. Among other ingredients, mucus contains electrolytes, cellular content, and mucin, a high molecular weight glycoprotein. From a barium coating viewpoint, the most important property of mucin is its ability to form a gel that adheres to the mucosa. It is the interaction of a barium product with mucin that determines the radiographic appearance of a barium coating. One early report described better visualization of the intestinal pattern when mucin was added to the barium product.[109]

A complex interplay of several intricate systems occurs in small bowel contrast examinations. When barium media are added to the acid and mucus present in the stomach, the resultant pattern can be quite complex. The small bowel in turn adds secretions and changes the overall pH. As a result, various bizarre patterns can be produced. Some of the earlier radiologists found that the onset of strong emotions could be associated with the appearance of flocculation.[110–112] These results are not surprising considering that vagal stimulation leads to increased mucus secretion throughout the GI tract. It was not until 1949 that mucus was shown to produce flocculation with some barium media.[113] The value of a protective coating on the barium suspension was noted shortly afterward.[114] Because of the reports by Kirsh,[115] Golden,[110] and others,[116] a "flocculation pattern" has lost meaning and is de-emphasized by most radiologists as a sign of organic disease.

There is a lack of correlation between malabsorption disorders and flocculation. There is also poor correlation between flocculation and jejunal biopsy results. Flocculation is more prominent with barium sulfate USP and least apparent with barium media containing suspending agents.[56] A number of commercial barium products are available that were designed to withstand flocculation; yet, in an occasional patient such flocculation can be identified in the absence of any known disease.

In the small bowel, flocculation is seen as a coarse, irregular outline. The barium suspension tends to form and travel in clumps. The fine, feathery pattern seen in the jejunum after the main barium bolus has passed is also the result of flocculation.[117] The clumping that is sometimes formed at the head of a barium column in the small bowel likewise is caused by flocculation; it disappears when additional barium sulfate reaches this segment.[118]

An acid barium medium produces a variety of abnormal patterns in the duodenum and jejunum, ranging from spasm to enlarged folds and dilation.[56] There may be intestinal hypersecretion. An alkaline barium medium at times improves coating of the valvulae conniventes and may aid in diagnosis.

Currently, there are four radiographic techniques of studying the small bowel, with each having advantages and disadvantages: (1) conventional antegrade small bowel examination, (2) enteroclysis, (3) retrograde small bowel examination, and (4) peroral pneumocolon. All require a barium suspension that does not flocculate nor precipitate.

Antegrade Small Bowel Examination

The simplest small bowel study is performed by having the patient drink a barium suspension and then obtaining serial radiographs as the suspension gradually travels through the small bowel. This examination was the first small bowel study described and, to a large degree, continues to be the most commonly performed study. Most radiologists prefer a slightly concentrated barium suspension, such as 40–60% w/v. Such a concentration allows for subsequent dilution by small bowel secretions.

Generally, only a single-contrast effect can be achieved with a conventional antegrade study; the exception is with distal small bowel obstruction. In this setting intestinal secretions tend to distend the small bowel, while eventually the ingested barium partly coats the mucosa and there is an apparent double-contrast effect (Fig. 2-9). Yet even here, the results are unpredictable. Some authors achieve a "semitransparent" effect by adding a cellulose base as a thickening agent to the barium suspension.[119]

The amount of barium sulfate used for an antegrade small bowel study remains controversial. In 1959, Golden[110] believed that 240 ml of barium sulfate was sufficient, whereas others used as little as 100 ml.[120] The lack of overlapping loops was believed to outweigh the advantage of distention when larger volumes

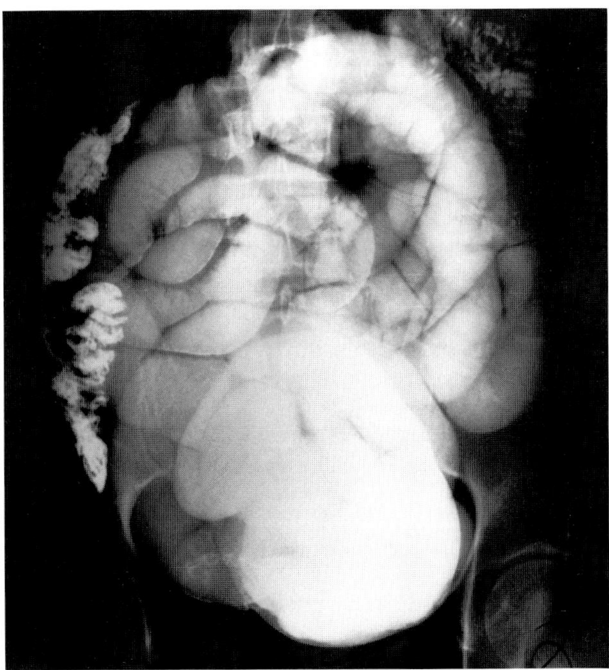

Figure 2-9 Partial distal small bowel obstruction. A single-contrast antegrade small bowel examination was performed. The apparent double contrast is produced by intestinal secretions retained proximal to the obstruction.

of contrast were used. Marshak, in 1976, recommended larger volumes (500 ml).[121] Currently, the trend is toward using even larger volumes, possibly augmented by pharmacologic agents. Often, 900 ml of barium is used.[122]

Although inspissation of barium proximal to the site of a small bowel obstruction has been feared, this complication does not appear to exist.[123] Rather, the contrast medium tends to become more dilute, serving as an excellent marker for the degree of obstruction and bowel dilation present.[124] A large volume of contrast media in such a clinical setting is safe.[125]

Once the proximal small bowel pattern is established and there is slow transit, several methods of stimulating transit are available. Ingestion of cold contrast media, water, or food generally promotes peristalsis and rapid small bowel transit.[118,126] Small amounts of a water-soluble, iodinated medium likewise increase peristalsis. The drawback of the latter is that large amounts of iodinated media may result in a bizarre small bowel pattern that has little relation to any underlying disorder.

Lactase deficiency has been studied by adding lactose to the barium medium. Some media contain lactose; the Micropaque powder sold in the United States has not contained lactose since the early to mid-1960s. The Micropaque powder (but not the liquid) distributed in Europe contained 5% lactose. Oratrast produced before 1970 contained 3% lactose; sucrose has subsequently been substituted. To study lactase deficiency, at least 25 g of lactose should be added. Although the lactose-barium small bowel study is believed to be a relatively sensitive diagnostic test,[127] it has been almost universally superseded by other laboratory tests.

Enteroclysis

One of the first to investigate the possibilities of duodenal intubation was Cole in 1911.[128] He produced intermittent duodenal obstruction by a balloon at the end of an Einhorn tube. The contrast agents used, bismuth and buttermilk, accumulated in the dilated duodenum proximal to the balloon. Thorium oxide has also been used as the contrast agent.[129]

A double-contrast hypotonic duodenogram was described in 1927.[130] The two contrast agents used were a barium suspension and air, and hypotonia was achieved with atropine. The hypotonic duodenogram was subsequently refined and was used for years to study the pancreaticoduodenal interface; its usefulness decreased markedly after the introduction of computed tomography in the 1970s, when direct visualization of the pancreas became possible.

Pesquera in 1929 intubated the duodenum, added a barium suspension through the tube, and performed a small bowel study.[131] The limitation of duodenal intubation is that the flow rate is limited by gastric reflux; in approximately half of the patients, however, the entire small bowel can be filled in 15 minutes.[132]

In enteroclysis, a tube is passed either into the duodenum or the jejunum, followed by instillation of a barium suspension through the tube. A second contrast agent can also be instilled. The term "barium enteroclysis" was first used by Gershon-Cohen and Shay in 1939.[133] Schatzki introduced the term "small bowel enema" in 1943.[132] Enteroclysis was investigated in detail by Sellink in the early 1970s.[117,134,135] In the past decade it has been popularized by Sellink and Miller[136,137] and others.

Technique

A flexible nasogastric tube can be difficult to maneuver through the pylorus and into the duodenum. Initially, an attempt was made to pass two catheters; a rigid, large-bore catheter was passed into the stomach, and a small-bore catheter was then maneuvered through the larger catheter into the duodenum.[138] The adoption of a steerable guide wire enabled relatively easy intubation in most patients.[139] Currently, several catheter assemblies up to 130 cm in length are available to aid placement of the catheter tip in the jejunum. Some catheters have an inflatable balloon proximal to the side holes; by inflating the balloon, the barium suspension flow rate can be increased without significant reflux into the stomach. Fournier et al. described a two-balloon catheter;[140] a larger proximal balloon is positioned in the gastric antrum while a smaller distal balloon is in the duodenal bulb. The tip of the catheter is just distal to the bulb. The advantage claimed for this catheter is that the entire duodenum need not be intubated because the two balloons prevent gastric reflux.

A tubeless, double-contrast small bowel examination has also been described.[141] The patient drinks a barium suspension and then swallows capsules that release gas once they reach the small bowel.

Initially, contrast infusion was achieved by gravity from a plastic bag. The flow rate was varied simply by changing the bag height above the table. More viscous suspensions required infusion with hand-held syringes. Since then, a number of infusion systems have been described.[127,142–144] One study found an electric pump more efficient than a hand-operated pump.[145] For the radiologist who performs only an occasional enteroclysis examination, I maintain that infusion with syringes is still the best compromise.

Controversy exists whether enteroclysis should be performed as a single-contrast examination,[136,137,146–148] or whether a double-contrast study using water[149,150] or methylcellulose[151–153] yields better results (Figs. 2-10 and 2-11). One non-crossover study of 200 patients found similar image quality in the single-contrast examination and a methylcellulose-enhanced study.[154]

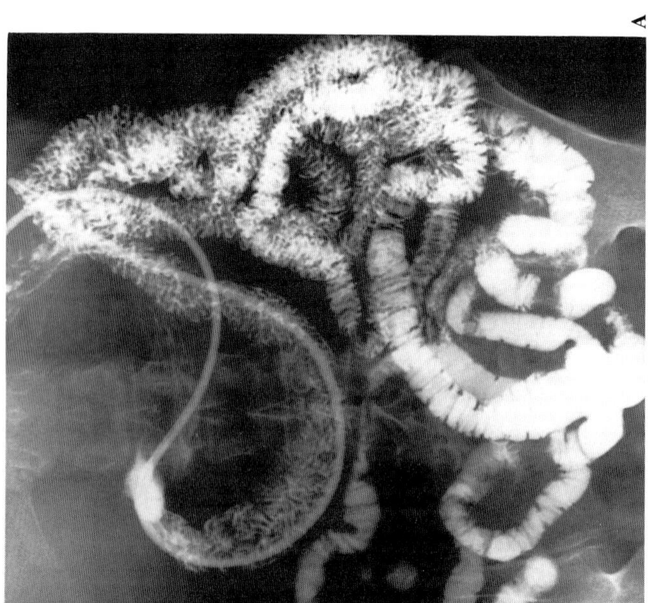

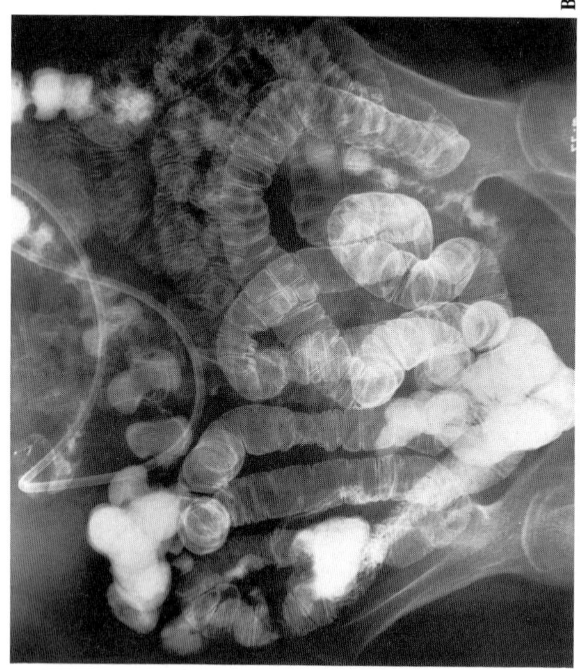

Figure 2-10 (*A*) Single-contrast enteroclysis. The infusion rate was rather slow, and the jejunum is not fully distended. (*B*) Double-contrast study using 1% methylcellulose as the second contrast agent. Same patient as in (*A*). The infusion rate has been increased and approximately 500 ml of methylcellulose infused.

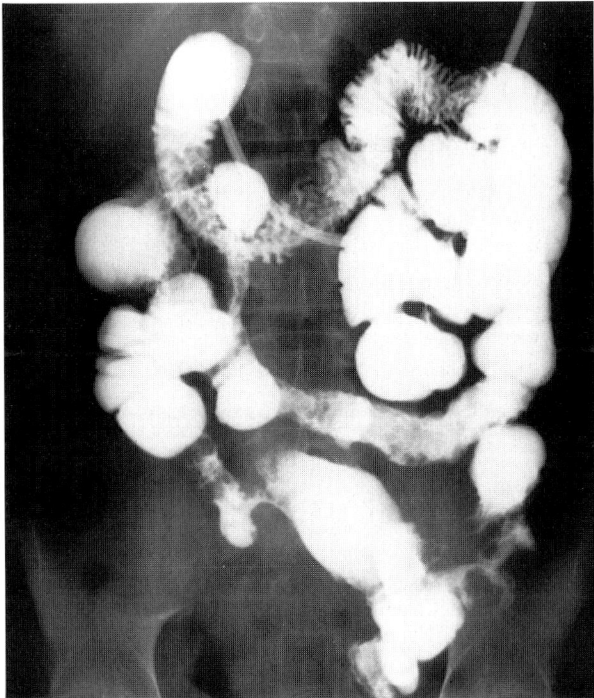

Figure 2-11 Single-contrast enteroclysis in a 16-year-old patient with extensive small bowel regional enteritis. The barium specific gravity was approximately 1.25. Although a double-contrast study would provide ''prettier'' views, it is questioned whether it would add any significant additional information.

Double-Contrast Agents

Some radiologists modify the single-contrast examination by instilling water once the barium column has reached the ileum.[155] The water tends to push the barium column ahead and helps distend the ileum. The use of water also enables a double-contrast study of the jejunum; a disadvantage is that water washes the barium away from the mucosa rather quickly, and the time available for double-contrast filming is limited.

Air as the second contrast is being used by some radiologists,[140,156–159] but it has not been accepted universally (Fig. 2-12). Air does not advance the column of barium; rather, it percolates through the intestine.[158] In addition, when the examination is performed in search of polyps, air bubbles can introduce another source of confusion. Sinuses and fistulas appear to be less well identified when air is used.[160] Infusion of air is helpful, however, in distending overlapping loops of bowel, especially in the pelvis. Injection of air is also useful after barium has

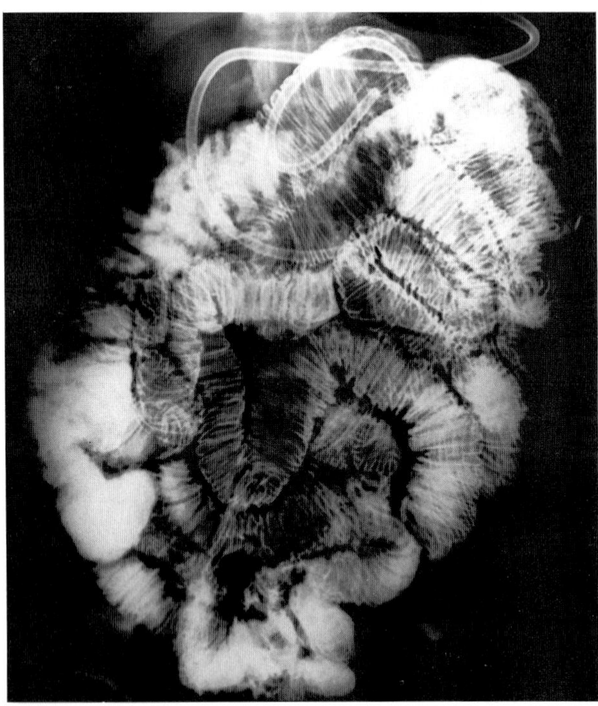

Figure 2-12 Double-contrast enteroclysis using air as the second contrast agent. Air was added after the barium suspension reached the ileum. *Source*: Reprinted from *Radiology of the Acute Abdomen* (p 137) by J Skucas and RF Spataro with permission of Churchill Livingstone Inc, © 1986.

reached the terminal ileum, especially when evaluating the full extent of inflammatory bowel disease.[157] The primary proponents of air are Japanese investigators, although air is also used in some centers in Europe.

A 1% suspension of methylcellulose had been used in the 1960s as the second contrast agent for barium enemas.[161] It was believed that the methylcellulose led to better adherence of barium to the colon mucosa. This observation was subsequently adapted to double-contrast study of the small bowel. Generally, a solution of 0.5% or more of methylcellulose is used.[162] If the methylcellulose solution is prepared from powder it should be refrigerated for 24–36 hours to allow adequate stabilization.[162] It should also be refrigerated until use to prevent bacterial and fungal growth. Commercial concentrated solutions of methylcellulose are available, and the only preparation is dilution to the desired concentration. Up to 2 liters may be required to distend the small bowel adequately.

Instead of methylcellulose, a high-molecular-weight guar preparation has been proposed.[163] Guar gum is a polysaccharide that has been used as a viscosity-enhancing agent.

Barium Sulfate

Through personal experience, some investigators have established optimal barium concentrations and flow rates (Table 2-5).[136,137] For a single-contrast study, Sellink and Miller preferred a specific gravity of 1.27 (approximately 34% w/v) for a normal size adult, 1.3 for an obese patient, and 1.2 for a thin patient. Examinations in children are performed with the barium at the lower specific gravity range. Their examinations were mainly single contrast. In general, a flow rate of approximately 100 ml/min is recommended for adults, 75 ml/min for infants, and 50 ml/min for babies,[136,137] although Sellink has since modified the flow rate to 75 ml/min as being the most favorable rate in adults.[164]

Those who perform double-contrast enteroclysis using methylcellulose as the second contrast agent tend to use a barium suspension at a higher specific gravity than recommended by Sellink and Miller. Thus, Herlinger[153] uses approximately a 95% w/v suspension with a high viscosity, whereas Thoeni[122] prefers a 50% w/v suspension. Several commercial suspensions are currently available that have been designed for either the single-contrast or double-contrast technique.

If the flow rate is too fast there is loss of peristalsis; as a result, larger amounts of contrast are required to study the entire small bowel. In addition, the incidence of reflux into the stomach, with its associated nausea and vomiting, is greater. However, if the contrast is instilled too slowly, the small bowel is incompletely filled, and the examination is prolonged. A slow infusion rate approaches that of a conventional antegrade small bowel examination. Overall, it is probably best to

Table 2-5 Enteroclysis Barium Concentration (Density)*

Abdomen Thickness, cm	*HD-85, <ml>*	*Water <ml>*	*Ratio*	*Weight/Volume, <%>*	*SPG*	*kV*
Obese (25+)	720	720	1:1	42	1.32	125
Normal (20–24)	600	900	1:1.5	34	1.27	125
Thin (15–19)	480	960	1:2	28	1.23	120
Child (9–14)	300	750	1:2.5	24	1.2	100
Infant (5–8)	200	600	1:3	21	1.17	80
Baby (<4)	100	350	1:3.5	19	1.15	60

*The above figures apply only to HD-85 (Lafayette Pharmacal), ⅜-inch internal diameter tubing and 130 cm Sellink-type enteroclysis duodenal intubation tube as supplied by Cook, Inc., Bloomington. Any other system must be tested to assure flow rates of 100 ml/min for adults.

Source: Adapted with permission from *Gastrointestinal Radiology* (1979;4:269–283), Copyright © 1979, Springer-Verlag.

adjust the flow rate depending upon the degree of peristalsis encountered during the examination.

Retrograde Small Bowel Examination

The small bowel can be studied by refluxing the ileum during a barium enema. This examination, also known as retrograde ileography, was popularized by Miller in the 1960s.[125,165,166] Although a number of previous indications for this examination have been replaced by enteroclysis, it is an excellent method of visualizing the distal small bowel.[167] Because the retrograde flow of the barium column can be controlled, the distal small bowel can be studied without superimposed loops of more proximal small bowel. In general, a somewhat greater barium concentration than that used in a single-contrast barium enema is optimal. A 20% w/v concentration has been satisfactory. The same barium suspension and same specific gravity as used in enteroclysis are also satisfactory in the retrograde small bowel examination.

Premedication with glucagon to induce bowel hypotonia is helpful. Glucagon also relaxes the ileocecal valve.[168,169] The barium sulfate suspension is instilled until the area in question is filled. Fluoroscopic monitoring ensures that there is little or no reflux into the stomach; fatal barium aspiration has been reported.[170] One modification is to instill the barium suspension until the distal ileum is filled and to follow with a saline solution. The saline pushes the barium more proximally and at the same time clears most of the barium from the colon so a ''see-through'' effect is achieved for the underlying loops of small bowel. This is especially helpful when a redundant sigmoid overlaps the ileum.

The distal small bowel can also be studied in combination with a double-contrast barium enema. Reflux of barium and air can provide either a single-contrast or double-contrast study. These approaches are especially useful with suspected gynecologic malignancies involving the distal small bowel or inflammatory disease involving the distal small bowel (Fig. 2-13).

Peroral Pneumocolon

The peroral pneumocolon contains elements of both antegrade and retrograde examinations. The patient ingests barium orally, and a conventional antegrade small bowel examination is then performed. Once the barium column reaches the right side of the colon, sufficient air is introduced rectally to produce a double-contrast examination of the area in question.[171]

Originally the peroral pneumocolon was used to study the colon,[172] although lately it is being applied to study the distal ileum.[173,174] A barium preparation

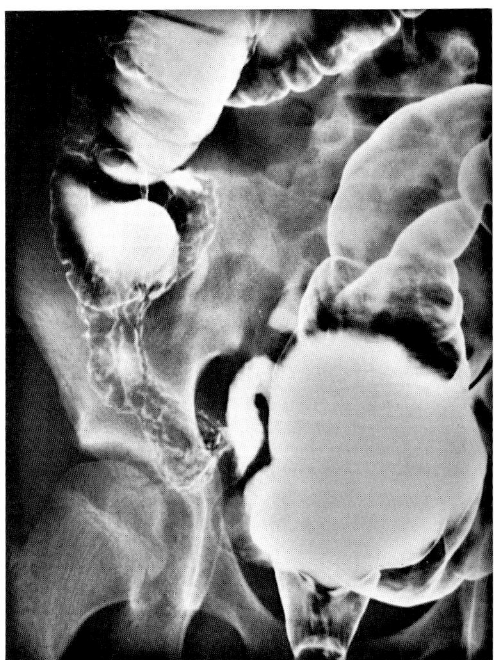

Figure 2-13A Distal ileal regional enteritis in a 12-year-old patient. A double-contrast retrograde study reveals extensive inflammatory changes, including ulcerations. *Source*: Reprinted from *The Radiological Examination of the Colon* (p 139) by RE Miller and J Skucas with permission of Martinus Nijhoff Publishers, © 1983.

similar to that used in a conventional antegrade study or enteroclysis is satisfactory. In general, this technique is reserved for the patient for whom more conventional studies, including enteroclysis, barium enema, or colonoscopy, have failed. Some use this technique as an adjunct to a conventional small bowel examination to study the distal ileum more effectively.[175] It can also be combined with enteroclysis.

Comparison of Techniques

The antegrade small bowel examination is easily performed and does not require intubation. Enteroclysis allows direct contrast media infusion into the small bowel at a controlled rate, and the resultant bowel distention aids detection of both intra- and extraluminal abnormalities.

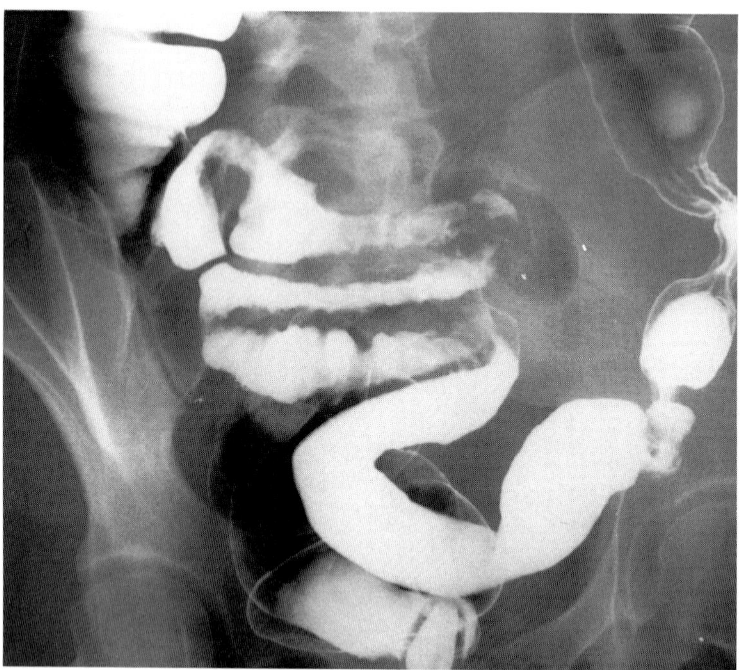

Figure 2-13B Ileal regional enteritis. A single-contrast retrograde study outlines the long segment involved.

The relative efficacy of the various small bowel examinations is difficult to gauge. One variable is the degree of technical expertise and experience of the examiner. Technical errors are the major cause of missed lesions with a conventional antegrade study.[176] A number of comparative studies through 1987 were summarized by Thoeni.[122] In one series where enteroclysis and a conventional examination were performed in the same patient, enteroclysis was superior in visualizing disorders in most of the small bowel, with no significant difference being found in the terminal ileum.[177] Others have found that enteroclysis was positive more than twice as often as a conventional study in the same group of patients[152] and that enteroclysis provided new and useful additional information in one-half of the patients when compared with a conventional large-volume antegrade study.[150] A study in patients with regional enteritis found that double-contrast enteroclysis, using air as the second contrast media, showed the diseased segments better than a conventional antegrade study;[160] however, the amount and brand of barium used for the enteroclysis and antegrade studies were different, and the findings are difficult to put in proper perspective.

COLON

Almost 145,000 patients develop colorectal cancer each year in the United States.[178] This incidence is more than in the rest of the digestive organs combined. If detected and treated early, a high cure rate is possible. The barium enema continues to be the major study in detecting colon cancer. Other techniques, such as sigmoidoscopy and colonoscopy, complement rather than supplement the barium enema.

Up to 10% of colon cancers are missed on the initial barium enema.[179] Most missed carcinomas result from poor patient preparation, poor technique during examination, and inadequate attention to detail.[180] For a good examination, the radiologist needs a barium suspension that achieves excellent mucosal coating. Various barium media should be tried until the most acceptable suspension is found. In all instances, accurate measurement of the amount and density of barium sulfate is essential.

Ideally, the barium suspension should be administered close to body temperature. A warm suspension is less viscous than a cold one, and especially if a double-contrast barium enema is being performed, the lower viscosity enables a faster examination. In addition, there may be less colon spasm, resulting in greater patient comfort, more uniform bowel distention, and a better examination.

Disposable enema bags, which became readily available in the early 1960s,[181] are generally made out of polyvinylchloride. Some are sold already containing a dry barium product to which one simply adds water. Because simple shaking does not result in thorough mixing and wetting of the barium particles, the bags must be shaken vigorously and for a considerable period of time. After shaking and before use, the bags should be stored on their side and not upright; otherwise, the barium sulfate gradually settles out, with the densest portion settling in the inferiorly located enema tubing.

Currently, there are three radiographic examinations of the colon: a single-contrast barium enema, a double-contrast examination using barium and air as the two contrast agents, and a biphasic examination that is primarily designed to detect polyps in a distorted sigmoid. The single-contrast barium enema was used initially and has achieved the greatest acceptance. Some radiologists routinely perform a single-contrast barium enema in elderly and debilitated patients.[182] A double-contrast study can, however, be performed in most patients over the age of 60.[183]

Single-Contrast Barium Enema

With a single-contrast barium enema, a relatively low concentration of barium is used—generally between 12% and 25% w/v. The main criterion is that the suspension not precipitate or settle during the time of the examination. The tendency for settling should be evaluated at the density used for the actual examination. Some products are well suspended at higher densities, but dilution of the barium sulfate also leads to dilution of the suspending additives, with resultant settling during the examination. In clinical practice, the settling properties of a commercial barium preparation can be evaluated with a horizontal x-ray beam; significant settling into the most dependent portion of bowel is readily apparent. Such horizontal radiography will also evaluate whether an adequate amount of barium sulfate is present in the uppermost part of the bowel. Most of the barium products satisfactory for a single-contrast barium enema contain relatively small barium particles. The overall aim of a single-contrast examination is to have "see-through" capability. Numerous products designed for single-contrast barium enema are commercially available.

Double-Contrast and Biphasic Barium Enema

The double-contrast examination was first mentioned by Hugo Laurell of Sweden in 1922.[184] It was popularized by Fischer of Frankfurt in 1923[185] and was introduced into the United States by Weber in 1931.[186] Its initial acceptance was rather limited. More recently, the double-contrast examination has been popularized by Welin worldwide[187] and by Miller in the United States.[188,189] Currently, in most academic centers the double-contrast barium enema is the procedure of choice in the radiographic study of the large bowel.

The choice of contrast medium for a double-contrast examination is more stringent than for a single-contrast study. The barium suspension should be relatively dense but also have sufficiently low viscosity to flow readily. The mucosa should be coated with a sufficiently thick coating that is visible on subsequent radiographs. The coating should be uniform in thickness without significant artifacts. The contrast medium should form a homogeneous suspension upon shaking, and the barium sulfate should not settle out. It is obvious that all of the above criteria cannot be met with a single product. Most commercial preparations have a density of 60%–120% w/v.

The double-contrast examination does take longer to perform, in part because of the higher viscosity barium sulfate suspension necessary for this examination. The examination can be speeded up somewhat by using relatively large internal diameter tubing.[190] The large-caliber tubing (12.5 mm) allows not only faster introduction but also aids in drainage of the barium suspension. The flow rate varies with the pressure, viscosity, and length of the tubing and is directly proportional to the fourth power of the internal radius of the tubing.[190] Thus, the single most important factor in determining the rate of flow is the internal diameter of the tubing.

The same barium concentration cannot optimally be used for both a single-contrast and a double-contrast examination. The low concentration needed for the single-contrast examination simply leads to poor coating if a double-contrast study is attempted. Likewise, the suspensions useful for a double-contrast study result in such a dense outline if a single-contrast study is attempted that no "see-through" effect is possible.

Currently, both powder and liquid commercial products specifically designed for double-contrast barium enemas are available. Dry barium powder in pre-packaged enema bags requires only the addition of a specified amount of water. The problems encountered with these dry powders are twofold. First, it takes time and vigorous shaking to ensure that the barium particles are in adequate suspension and that their surface is adequately wetted. Cursory mixing before an examination can lead to erratic results. Second, the barium powder comes in collapsible plastic bags; although a level marking is generally inscribed on the bag, the amount of water to be added can easily vary. Even small variations in the amount of water can influence the bowel-coating properties. The amount of water to be added should be established precisely with a graduated glass container, but this practice is not uniformly followed.

Because of the variability in (1) mixing, (2) water hardness in different localities, and (3) the amount of water added, several barium manufacturers have formulated premixed liquid preparations. The required amount of contrast is simply poured into an enema bag without any further dilution. Yet, even here a variable exists. With prolonged storage the barium suspension tends to settle out; therefore, before use the containers should be shaken vigorously. The amount of settling, resultant barium "cake" formation, and subsequent readiness to resuspend depend considerably upon the barium particle sizes and additives used by the manufacturer. Some of these products are sold in tall plastic jugs; subsequent resuspension is easier if these jugs are stored on their side, rather than in an upright position.

Even with all mixing factors standardized, there can be variation from one batch to another. The length of time in storage varies. One batch was found to have completely different coating characteristics from subsequent batches; investigation revealed that that particular batch may have been allowed to freeze while stored in a warehouse.

In order to speed up the examination, some manufacturers have introduced "low viscosity" barium preparations. In general, this lower viscosity is achieved by using less barium sulfate and changes in additives. Although the barium does flow more readily, the resultant barium coating on the mucosa can be compromised. If the thickness of the barium coating is believed to be too thin, coating can be improved by using a higher specific gravity barium product. However, an increase in specific gravity beyond a certain point is counterproductive; the resultant highly viscous contrast medium flows poorly through plastic tubing and results in an uneven mucosal coating.

Some radiologists perform colonic lavage 30 minutes or more before the barium enema, whereas others rely on the previously given laxatives. With colonic lavage, invariably there is retained water and subsequent dilution of the barium suspension. Thus, with any barium preparation, the resultant mucosal coating will vary depending on whether colonic lavage has been performed. Some barium manufacturers recognize this fact and market two similar preparations. The preparation designed to be used with colonic lavage generally has a slightly greater specific gravity.

Some barium products are simply not designed for double-contrast use. Others precipitate and flake if excess mucus is present. In general, the presence of mucus leads to an inferior barium coating. Thus, in patients with inflammatory bowel disease, the barium coating tends to be thinner and more intermittent than in patients with a normal bowel.

A delay between barium addition and subsequent radiography may lead to the drying out of the suspension. Initially, there are small filling defects that eventually coalesce into a coarse pattern. The better commercial products do not form significant artifacts in the time required to complete an average barium enema.

A gas other than air has been used as the second contrast agent in double-contrast studies.[18,191,192] The type of gas used probably does not influence significantly the quality of the examination, although a gas that is absorbed faster than air may provide greater patient comfort.[193] One such gas is carbon dioxide; several delivery systems for its use have been described.[194,195]

At times, marked sigmoid diverticulosis precludes adequate evaluation of this segment by a double-contrast technique; there simply is too much radiographic contrast, and among the diverticula a polyp can be missed. The detection rate can be improved by following the double-contrast study with a limited single-contrast study of the sigmoid. Such a combination is known as a biphasic examination.[196] The single-contrast component should be performed with a low specific gravity suspension as is typically used in single-contrast barium enemas. A recent suggestion has been to use a 1.5% CT barium suspension for this part of the examination;[197] the CT barium contains a large amount of additives for suspension purposes. The authors found an improvement in interpretation when using such 1.5% CT barium compared to distention by tap water.[197]

The radiologist beginning to perform double-contrast barium enemas should start with one of the commercial liquid preparations specifically designed for double-contrast examinations. Several such products can then be compared and the optimal one chosen. It is important that during the comparison no change either in technique or exposure factors be made. Because of the variability of coating, a number of examinations must be performed in order to gauge the relative coating properties of different products. If significant air bubbles are encountered, adding simethicone to the barium suspension should eliminate the problem.[198] Many commercial brands already contain an antifoam agent.

Comparison of Techniques

Over the years, the issue of single- versus double-contrast barium enema has been hotly debated, often at an emotional level. A number of studies have compared the relative accuracy of a single-contrast versus double-contrast barium enema. In general, these studies have concentrated on the detection of inflammatory bowel disease or the detection of polyps and cancers. Many of these studies were summarized by Miller and Skucas in 1983.[199] Several recent studies have concluded that the single-contrast examination is accurate in the detection of colon neoplasms.[200–202] Most studies, however, have upheld the double-contrast examination as being more accurate both in detecting the extent and incidence of inflammatory bowel disease and in detecting polyps.[96,199,203] Because the techniques of the two types of examinations differ, many of the previous reports are biased. The accuracy of each examination in the best of hands is difficult to gauge. One recent study comparing a single- versus double-contrast barium enema found that the double-contrast examination was better in detecting polyps under 1 cm in size.[182] Both techniques detected 94%–96% of polyps larger than 1 cm. Yet, an indirect bias exists in this study; all of the studies were performed with remote control equipment and decubitus radiographs were not obtained.

THE BLEEDING PATIENT

Citric acid, sodium citrate, potassium citrate, hexametaphosphate, sodium and potassium oxalates, sodium fluoride, heparin, and certain azo, amine, and benzidine dyes precipitate or bind calcium.[204] Compounds that bind calcium are effective anticoagulants and are used in the laboratory to prevent coagulation.[205] Some of these agents are used in commercial barium preparations as suspending, deflocculating, preservative, and flavoring agents.

Barium sulfate has been used to separate some of the blood coagulation factors.[206] Barium sulfate forms metalloprotein complexes with the prothrombin complex. The result is a barium glycoprotein.[207] Similarly, barium citrate can be used to adsorb certain glycoproteins secreted by platelets.[208]

Although the additives do not dissolve blood clots, they will delay clotting when given internally with barium sulfate contrast media. The combination of blood and barium sulfate occurs when patients with GI bleeding are examined with contrast materials.

A series of experiments was performed with several brands and concentrations of barium products, some of which contained a sodium citrate additive. Two milliliters of several barium media were added to an equal volume of freshly drawn human blood. The resulting flocculation, rate of blood clot formation, and degree of hemolysis were then evaluated (Table 2-6).[209]

Some of the contrast media showed no clotting, whereas others clotted faster than blood alone. At the present time, it is not possible to predict in advance the coagulation properties of a barium preparation because the additives are generally not known.

The blood-barium flocculation was accentuated as the barium specific gravity was increased. The type of flocculation observed varied with the medium used; with some a flocculated homogeneous mass was produced, whereas with others there were gross clumps present.

The resultant hemolysis probably results from the suspension's hypo-osmolality. The effect of additives on hemolysis is not known. The osmolality of a barium suspension is primarily the result of the additives, because barium sulfate is only minimally soluble. As a result, USP barium sulfate, which has few additives, results in severe hemolysis; for some others, a true concentration dependence is seen.

With a bleeding patient, the radiologist should use a nonflocculating barium product. In experimental studies of animals, the presence of intraluminal blood decreased the mucosal adsorption of barium considerably,[9] and a similar effect may occur in humans. If an upper GI examination is being performed, any significant residual contrast in the stomach should probably be aspirated so that any anticoagulation agents present in the barium medium do not reach the bleeding site. A contrast medium that flocculates in the presence of blood may result in large blood-barium masses in the stomach or bowel. These can eventually result in stercoral ulcers or obstruction.

A therapeutic barium enema for major colonic bleeding has been described in the surgical literature.[210] Because different barium sulfate preparations have different effects when mixed with blood, the results can be quite variable. As seen from Table 2-6, tap water is very effective in clotting blood. Whether a tap water enema is effective is not known. The role of hypotonic agents is also not known.[25]

SUMMARY

Current knowledge suggests that nonflocculating, nonclot-forming barium sulfate suspensions should be used for the examination of bleeding patients. These media should be aspirated from the stomach of patients with pyloric or duodenal obstruction or other causes of significant gastric retention.

Table 2-6 Compatibility of Blood and $BaSO_4$

Media	*% w/v*	*Rate of Clotting Relative to Blood**	*Degree of Flocculation***					*Hemolysis in 2 Hours*
			5	*20*	*30*	*60*	*120*	
Tap water		F						++
Saline		E						−
USP $BaSO_4$	20	F	−	+	+	+	+	++
	40	F	−	−	+	++	++	++
	60	F	++	++	++	++	++	++
USP $BaSO_4$: 0.5% Tannic acid	20	N	+	+	++	++	++	++
	60	F	++	++	++	++	++	++
Barosperse	20	N	−	−	−	−	−	+
	40	N	−	−	−	−	−	+
	60	N	−	−	−	+	+	+
	80	N	−	+	+	+	+	+
	100	N	−	+	+	+	+	+
	120	N	++	++	++	++	++	+
Intropaque	10	N	−	−	+	++	++	++
	20	N	+	+	+	++	++	++
	40	N	+	+	++	++	++	+
	60	F	++	++	++	++	++	−
Intropaque: Barosperse (5:2)	10	F	−	−	−	−	−	++
	20	N	−	+	+	++	++	++
	40	N	+	+	+	++	++	+
	60	N	++	++	++	++	++	+
Barosperse: Seidlitz (8.1:1)	80	N	−	−	−	−	+	+
	100	N	+	+	+	+	+	+
	120	N	++	++	++	++	++	+
Novopaque	60	N	+	++	++	++	++	++
Liquipake	20	F	−	−	+	++	++	++
	50	F	+	+	+	+	+	+
	67	S	+	++	++	++	++	+
Micropaque	20	F	−	−	−	−	−	++
	30	F	−	+	+	++	++	++
	50	S	−	+	+	++	++	+
	80	N	−	+	+	++	++	+
Redi-Flow	38	F	+	++	++	++	++	−
Ultrapaque	25	N	−	+	+	+	+	++
	45	N	−	++	++	++	++	+
Barodense	33	N	−	−	−	−	−	++
	50	N	−	−	−	−	−	+
	67	N	−	−	−	−	−	+

Table 2-6 continued

Media	% w/v	Rate of Clotting Relative to Blood*	Degree of Flocculation** 5	20	30	60	120	Hemolysis in 2 Hours
Stabarium	25	N	−	−	−	+	+	++
	33	N	−	+	+	+	+	++

*Rate of clotting: F—faster than blood; E—same rate as blood; S—slower than blood but did clot within 2 hours; N—no clotting within 2 hours.
**Flocculation (minutes after mixing) and hemolysis: −:—none; +:—moderate; ++:—severe.

Source: Adapted with permission from "The Effect of Barium on Blood in the Gastrointestinal Tract" by RE Miller et al in *Radiology* (1975;117:527–530), Copyright © 1975, Radiological Society of North America Inc.

The radiologist has a simple method for improving suspensions. Various barium contrast media can be tested clinically, noting the good and bad qualities of each. The practicing radiologist can then select the product best suited to a particular practice and locality.

The radiologist, knowingly or unknowingly, purposefully or haphazardly, changes the viscosity, density, surface tension, and mucosal coating qualities by selecting different brands of barium sulfate and by choosing the concentration and mixing methods in their preparation.

REFERENCES

1. *The United States Pharmacopeia*, rev 21. Rockville, MD, United States Pharmacopeial Convention, 1985, p 90.

2. Sharp RA, Knevel AM: Analysis of barium in barium sulfate and diagnostic meals containing barium sulfate using atomic absorption spectroscopy. *J Pharm Sci* 1971;60:458–460.

3. Miller RE: Barium sulfate suspensions. *Radiology* 1965;84:241–251.

4. Mattsson O: A simple method of ensuring correct concentration of barium contrast media. *Acta Radiol* 1953;39:501–506.

5. Gelfand DW, Ott DJ: Barium sulfate suspensions: an evaluation of available products. *AJR* 1982;138:935–941.

6. Schwartz SE, Fischer HW, House AJS: Studies in adherence of contrast media to mucosal surfaces. *Radiology* 1974;112:727–731.

7. Roberts GM, Roberts EE, Davies RL, et al: *In vivo* and *in vitro* assessment of barium sulphate suspensions. *Br J Radiol* 1977;50:541–545.

8. Dietze R: Strahlenphysik, Strahlenschutz, Röntgentechnik. *Radiol Diagn* 1986;27:143–149.

9. Virkkunen P, Retulainen M, Keto P: Observations on the behavior of barium sulfate contrast media in vitro. *Invest Radiol* 1980;15:346–349.

10. Anderson W, Harthill JE, James WB, et al: Barium sulphate preparations for use in double contrast examination of the upper gastrointestinal tract. *Br J Radiol* 1980;53:1150–1159.

11. Virkkunen P, Retulainen M: A new method for studying barium sulphate contrast media in vitro. Some factors contributing to the visualization of areae gastricae. *Br J Radiol* 1980;53:765–769.

12. Gelfand DW: High density, low viscosity barium for fine mucosal detail on double-contrast upper gastrointestinal examinations. *AJR* 1978;130:831–833.

13. Vlasov PV, Jakimenko VF: Verbesserung der physikalisch-chemischen Eigenschaften von Bariumsuspensionen zur Untersuchung des oberen Verdauungstraktes. *Radiol Diagn* 1983;24:35–41.

14. Treugut H, Hübener KH: In-Vitro-Test von Bariumsulfatsuspensionen zur Magendoppelkontrastuntersuchung. *Fortschr Röntgenstr* 1980;132:504–508.

15. Jirgensons B, Straumanis ME: *A Short Textbook of Colloid Chemistry*, ed 2. Oxford, Pergamon Press, 1962, p 174.

16. Kormano M, Mäkelä P, Rossi I: Visualization of the areae gastricae in a double contrast examination—dependence on the contrast medium. *ROEFO* 1978;128:52–56.

17. Kunz AL: Enema materials, in Miller RE (ed): *Detection of Colon Lesions*. Chicago, American College of Radiology, First Standardization Conference, 1969, p 5.

18. Pochaczevsky R, Sherman RS: A new technique for the roentgenologic examination of the colon. *Am J Roentgenol* 1963;89:787–796.

19. Brown GR: High density barium sulfate suspension. *Radiology* 1963;81:839–846.

20. James WB: Double contrast radiology in the gastrointestinal tract. *Clin Gastroenterol* 1978;7:397–430.

21. Maglinte DDT, Schultheis TE, Krol KL, et al: Survey of the esophagus during the upper gastrointestinal examination in 500 patients. *Radiology* 1983;147:65–70.

22. Brombart M: Fluoroscopically controlled single contrast examination of the gastro-intestinal tract, in Margulis AR, Burhenne HJ (eds): *Alimentary Tract Radiology*, ed 3. St. Louis, CV Mosby, 1983, p 195.

23. Skucas J: The routine double-contrast examination of the esophagus. *Crit Rev Diag Imag* 1978;11:121–143.

24. Skucas J, Schrank WW: The routine air-contrast examination of the esophagus. *Radiology* 1975;115:482–484.

25. Miller RE, Chernish SM, Skucas J, et al: Hypotonic colon examination with glucagon. *Radiology* 1974;113:555–562.

26. Ghahremani GG, Heck LL, Williams JR: A pharmacologic aid in the radiographic diagnosis of obstructive esophageal lesions. *Radiology* 1972;103:289–293.

27. Novak D: Die hypotone Ösophagographie mit Propanthelinbromid (Pro-Banthine). *ROEFO* 1975;123:409–414.

28. Palugyay J: Die Luftreliefdarstellung der Speiseröhre im Röntgenbild. *Fortschr Geb Roentgenstr* 1933;47:579–596.

29. Jacobs P: Carcinoma of the oesophagus presenting unusual features, with a note on technique of examination. *Br J Radiol* 1955;28:317–319.

30. Koehler RE, Moss AA, Margulis AR: Early radiographic manifestations of carcinoma of the esophagus. *Radiology* 1976;119:1–5.

31. Suzuki H, Kobayashi S, Endo M, et al: Diagnosis of early esophageal cancer. *Surgery* 1972;71:99–103.

32. Tréheux A, Bigard M-A, Plouvier B: La Radiographie oeso-gastro-duodénale en double contraste expérience personnelle après un an d'utilisation. *Ann Med (Nancy)* 1974;13:1797–1810.

33. Cassel DM, Anderson MF, Zboralske FF: Double-contrast esophagram. The prone technique. *Radiology* 1981;139:737–739.

34. O'Reilly GVA, Bryan G: The double-contrast barium meal—a simplification. *Br J Radiol* 1974;47:482–483.

35. Pochaczevsky R: ''Bubbly barium.'' A carbonated cocktail for double-contrast examination of the stomach. *Radiology* 1973;107:461–462.

36. Eisenscher A: Une technique d'insufflation endogène pour l'examen gastrique et duodénal en double contraste: la baryte gazéifiée. *J Radiol* 1980;61:139–140.

37. Bagnall RD, Galloway RW, Annis JAD: Double contrast preparations: An in vitro study of some antifoaming agents. *Br J Radiol* 1977;50:546–550.

38. Brombart M: *Clinical Radiology of the Oesophagus*. Bristol, John Wright and Sons, 1961, p 3.

39. Goldstein HM, Dodd GD: Double-contrast examination of the esophagus. *Gastrointest Radiol* 1976;1:3–6.

40. Wiljasalo M, Rissanen P: A new double-contrast method in esophageal roentgenology. *Ann Med Int Fenn* 1966;55:77–80.

41. de Lacey GJ, Wignall BK, Bray C: Effervescent granules for the barium meal. *Br J Radiol* 1979;52:405–408.

42. Moss AA, Koehler RE, Margulis AR: Initial accuracy of esophagograms in detection of small esophageal carcinoma. *AJR* 1976;127:909–913.

43. Ferrucci JT Jr, Long JA Jr: Radiologic treatment of esophageal food impaction using intravenous glucagon. *Radiology* 1977;125:25–28.

44. Trenkner SW, Maglinte DDT, Lehman GA, et al: Esophageal food impaction: treatment with glucagon. *Radiology* 1983;149:401–403.

45. Rice BT, Spiegel PK, Dombrowski PJ: Acute esophageal food impaction treated by gas-forming agents. *Radiology* 1983;146:299–301.

46. Donner MW, Silbiger ML, Hookman P, et al: Acid-barium swallows in the radiographic evaluation of clinical esophagitis. *Radiology* 1966;87:220–225.

47. McCall IW, Davies ER, Delahunty JE: The acid-barium test as an index of intermittent gastroesophageal reflux. *Br J Radiol* 1973;46:578–584.

48. Wolfe BS: ''Contraction rings'' associated with gross hiatal herniation: a roentgen method for the detection and measurement of minimal esophageal stricture. *J Mt Sinai Hosp* 1956;23:735–738.

49. Wolfe BS: Use of a half-inch barium tablet to detect minimal esophageal strictures. *J Mt Sinai Hosp* 1961;28:80–82.

50. Schabel SI, Skucas J: Esophageal obstruction following administration of ''aged'' barium sulfate tablets. A warning. *Radiology* 1977;122:835–836.

51. Chang CH, Carroll RM: Roentgenographic barium patterns and gastric secretion. *Am J Dig Dis* 1967;12:614–618.

52. Bryk D, Robinson KB: Roentgen evaluation of the dilated stomach and its contents in the differential diagnosis of pyloric obstruction. *Radiology* 1967;89:893–895.

53. Roberts GM, Roberts EE, Davies RL, et al: Observations on the behaviour of barium sulphate suspensions in gastric secretion. *Br J Radiol* 1977;50:468–472.

54. James AM, Goddard GH: A study of barium sulphate preparations used as x-ray opaque media. I. Particle size and particle charge. *Pharm Acta Helv* 1971;46:708–720.

55. Ehrlein HJ, Pröve J: Effect of viscosity of test meals on gastric emptying in dogs. *Quart J Experim Physiol* 1982;67:419–425.

56. Bryk D, Roska JC: Upper gastrointestinal studies with acid and alkaline barium sulfate suspensions. *Radiology* 1969;92:832–837.

57. Margieson GR, Williams HBL: Radiopharmacology of the stomach. *Aust Radiol* 1968; 12:239–244.

58. Lindgren I, Nevalainen T, Mäki J, et al: Effect of mucolytic pretreatment on gastric mucosal coating with barium sulfate in the rat. *Acta Radiol (Diagn)* 1980;21:443–446.

59. Persigehl M: Influence of N-acetylcysteine on visualization of the areae gastricae (abstract no. 87). *Radiology* 1987;165(suppl):57.

60. Shufflebarger HE, Knoefel PK, Telford J, et al: Some factors influencing the Roentgen visualization of the mucosal pattern of the gastrointestinal tract. *Radiology* 1953;61:801–805.

61. James WB, Stanton B, Teasdale E, et al: The effect of cimetidine on barium coating of the gastric mucosa. *Br J Radiol* 1977;50:445–456.

62. Brühlmann W, von Büren U, Zollikofer C, et al: Die Verbesserung der radiologischen Schleimhautdarstellung durch Magensekretions hemmer. *ROEFO* 1981;134:681–684.

63. Buonocore E, Meaney TF: Barium spray examination of the stomach—preliminary report of a new roentgenographic technic. *Cleveland Clin Q* 1965;32:133–141.

64. Moss AA, Beneventano T, Gohel V, et al: The current status of upper gastrointestinal radiology. *Invest Radiol* 1980;15:92–102.

65. von Elischer J: Über eine Methode zur Röntgenuntersuchung des Magens. *Fortschr Röntgenstr* 1911;18:332–340.

66. Feissly R: Le poudrage baryté combiné à l'insufflation, pour la mise en évidence, par voie radiographique, du relief de la muqueuse gastrique. *Arch d. mal. de l'app. digestif* 1930;20:76–81.

67. Vallebona A: Nuovo methodo di esame radiologico del tubo digerente. *Radiol Med* 1926;13:241–248.

68. Hilpert F: Das Pneumo-Relief des Magens. *Fortschr Roentgenstr Nuclearmed* 1928;38: 80–87.

69. Hampton AO: A safe method for the roentgen demonstration of bleeding duodenal ulcers. *AJR* 1937;38:565–570.

70. Ruzicka FF Jr, Rigler LG: Inflation of the stomach with double contrast. A roentgen study. *JAMA* 1951;145:696–702.

71. Amplatz K: A new and simple approach to air-contrast studies of the stomach and duodenum. *Radiology* 1958;70:392–394.

72. Montagne J-P, Moss AA, Margulis AR: Double-blind study of single and double contrast upper gastrointestinal examinations using endoscopy as a control. *AJR* 1978;130:1041–1045.

73. Op Den Orth JO: *The Standard Biphasic-Contrast Examination of the Stomach and Duodenum*. The Hague, Martinus Nijhoff, 1979.

74. Solanke TF, Kumakura K, Maruyama M, et al: Double-contrast method for the evaluation of gastric lesions. *Gut* 1969;10:436–442.

75. Keto P, Suoranta H, Ihamäki T, et al: Double contrast examination of the stomach compared with endoscopy. *Acta Radiol Diagn* 1979;20:762–768.

76. Suteanu S, Ionescu P, Stoichita S, et al: Our experience in double contrast x-ray examination in the diagnosis of more difficult cases of diseases of the esocardiotuberosity region (translation). *Med Intern (Bucharest)* 1976;28:63–72.

77. James WB, McCreath G, Sutherland GR, et al: Double contrast barium meal examination—a comparison of techniques for introducing gas. *Clin Radiol* 1976;27:99–101.

78. Montgomery DP, Clamp SE, de Dombal FT, et al: A comparison of barium sulphate preparations used for the double contrast barium meal. *Clin Radiol* 1982;33:265–269.

79. Lintott DJ, Simpkins KC, de Dombal FT, et al: Assessment of the double contrast barium meal: method and application. *Clin Radiol* 1978;29:313–321.

80. Koehler RE, Weyman PJ, Stanley RJ, et al: Evaluation of three effervescent agents for double-contrast upper gastrointestinal radiography. *Gastrointest Radiol* 1981;6:111–114.

81. Agha FP, Trenkner SW, Woolsey EJ, et al: Comparison of two effervescent agents for double-contrast upper gastrointestinal tract radiography. *Radiology* 1985;157:533–534.

82. Lotz W, Liebenow S: Erweiterte diagnostische Möglichkeiten der röntgenologischen Magenuntersuchung mit verbessertem Kontrastmittel. *ROEFO* 1979;131:157–165.

83. Kalokerinos J: Barium for double-contrast examinations. *Aust Radiol* 1973;17:155–156.

84. James WB: The double contrast meal—New high density barium sulphate powders. *Br J Radiol* 1978;51:1020–1022.

85. Kreel L, Herlinger H, Glanville J: Technique of the double contrast barium meal with examples of correlation with endoscopy. *Clin Radiol* 1973;24:307–314.

86. Kormano M, Mäkelä P, Rossi I: Visualization of the areae gastricae in a double contrast examination—dependence on the contrast medium. *ROEFO* 1978;128:52–56.

87. Lotz W, Liebenow S: Areae gastricae und varioliforme Erosionen—Qualitätskriterien der röntgenologischen Magenuntersuchung. *ROEFO* 1980;132:491–495.

88. Seaman WB: The areae gastricae. *AJR* 1978;131:554.

89. Miller RE: The air-contrast stomach examination: an overview. *Radiology* 1975; 117:743–744.

90. Virkkunen P, Lounatmaa K: On the differences between the $BaSO_4$ particles and additives in media for the double contrast examination of the stomach. *ROEFO* 1980;133:542–545.

91. Hirai K, Suezawa Y, Sugawara T: High-density barium sulfate as contrast media for mucosal examination of the stomach. *Sakura X-ray Photograph Studies* 1975;26:1–18.

92. Gelfand DW, Hachiya J: The double-contrast examination of the stomach using gas-producing granules and tablets. *Radiology* 1969;93:1381–1382.

93. Quattromani F, Finby N: Roentgenographic double-contrast examination of stomach and duodenum. *NY State J Med* 1972;72:1140–1143.

94. Kalokerinos J: Double-contrast barium meal technique. *Aust Radiol* 1967;11:246–249.

95. Tedesco FJ, Griffin JW Jr, Crisp WL, et al: "Skinny" upper gastrointestinal endoscopy—the initial diagnostic tool: a prospective comparison of upper gastrointestinal endoscopy and radiology. *J Clin Gastroenterol* 1980;2:27–30.

96. Gelfand DW, Ott DJ: Single- vs. double contrast gastrointestinal studies: critical analysis of reported statistics. *AJR* 1981;137:523–528.

97. Pyhtinen J, Päivänsalo M, Myllylä V, et al: Accuracy of single and double contrast barium meal studies. *Ann Clin Res* 1982;14:177–180.

98. Hedemand N, Kruse A, Madsen EH, et al: X-ray examination or endoscopy? A blind prospective study including barium meal, double contrast examination, and endoscopy of esophagus, stomach, and duodenum. *Gastrointest Radiol* 1977;1:331–334.

99. Koga M, Nakata H, Kiyonari H, et al: Roentgen features of the superficial depressed type of early gastric carcinoma. *Radiology* 1975;115:289–292.

100. Shirakabe H: *Atlas of X-Ray Diagnosis of Early Gastric Cancer*. Philadelphia, JB Lippincott, 1966.

101. Yamada E, Nakazato H, Koike A, et al: Surgical results for early gastric cancer. *Int Surg* 1974;59:7–14.

102. Kawai K, Takada H, Takekoshi T, et al: Double-contrast radiograph on routine examination of the stomach. *Am J Gastroenterol* 1970;53:147–153.

103. Poplack W, Paul RE, Goldsmith M, et al: Demonstration of erosive gastritis by the double-contrast technique. *Radiology* 1975;117:519–521.

104. Rienmüller R: Die Bedeutung der Areae gastricae bei der hypotonen Doppelkontrastuntersuchung des Magens. *ROEFO* 1980;132:485–490.

105. Obata WG: A double-contrast technique for examination of the stomach using barium sulfate with simethicone. *Am J Roentgenol* 1972;115:275–280.

106. Escobar-Billing R, Kallman R, Slezak P: Ett nytt dubbelkontrastmedel för röntgenundersökning av ventrikel och esofagus. *Läkartidningen* 1983;80:2880–2882.

107. Hyslop JS, Mitchelmore AE, Cox RR, et al: Double contrast barium meal examination: a comparison of two high density barium preparations, E-Z-HD and X-Opaque. *Clin Radiol* 1982; 33:83–85.

108. Toischer HP: Verbesserung der Magenfeindiagnostik durch ''high-density''—Kontrastmittel niedriger Viskosität. *ROEFO* 1983;139:21–24.

109. Alexander GH, Alexander RE: The use of gastric mucin as a barium suspension medium. *Radiology* 1950;54:875–877.

110. Golden R: Technical factors in the roentgen examination of the small intestine. *Am J Roentgenol* 1959;82:965–972.

111. Friedman J: Roentgen studies of the effects on the small intestine from emotional disturbances. *Am J Roentgenol* 1954;72:367–379.

112. Goin LS: Some obscure factors in the production of unusual small bowel patterns. *Radiology* 1952;59:177–184.

113. Frazer AC, French JM, Thompson MD: Radiographic studies showing the induction of a segmentation pattern in the small intestine in normal human subjects. *Br J Radiol* 1949;22:123–136.

114. Ardran GM, French JM, Mucklow EH: Relationship of the nature of the opaque medium to small intestine radiographic pattern. *Br J Radiol* 1950;23:697–702.

115. Kirsh IE: Motility of the small intestine with nonflocculating medium: a review of 173 roentgen examinations. *Gastroenterology* 1956;31:251–259.

116. Stacy GS, Loop JW: Unusual small bowel diseases. Methods and observations. *Am J Roentgenol* 1964;92:1072–1079.

117. Sellink JL: *Radiological Atlas of Common Diseases of the Small Bowel*. Leiden, HE Stenfert Kroese, 1976.

118. Brun B, Hegedüs V: Radiography of the small intestine with large amounts of cold contrast medium. *Acta Radiol (Diagn)* 1980;21:65–70.

119. Emons D: Semitransparente Dünndarmdarstellung per os. *ROEFO* 1981;135:446–452.

120. Laws JW, Shawdon H, Booth CC, et al: Correlation of radiological and histological findings in idiopathic steatorrhoea. *Br Med J* 1963;1:1311–1314.

121. Marshak RH: *Radiology of the Small Intestine*. Philadelphia, WB Saunders, 1976, p 2.

122. Thoeni RF: Radiography of the small bowel and enteroclysis. A perspective. *Invest Radiol* 1987;22:930–936.

123. Nelson SW: Facts versus folklore. *Am J Surg* 1965;109:543–545.

124. Miller RE, Brahme F: Large amounts of orally administered barium for obstruction of the small intestine. *Surg Gynec Obstet* 1969;129:1185–1188.

125. Miller RE: Reflux examination of the small bowel. *Radiol Clin N Am* 1969;7:175–184.

126. Weintraub S, Williams RG: A rapid method of roentgenologic examination of the small intestine. *Am J Roentgenol* 1949;61:45–55.

127. Morrison WJ, Christopher NL, Bayless TM, et al: Low lactose levels: evaluation of the radiologic diagnosis. *Radiology* 1974;111:513–518.

128. Cole LG: Artificial dilation of the duodenum for radiographic examination. *Am Quart Roentgenol* 1911;3:204–205.

129. Ghélew B, Mengis O: Mise en évidence de l'intestin grêle par une nouvelle technique radiologique. *Presse Med* 1938;46:444–445.

130. Pribram BO, Kleiber N: Ein neuer Weg zur röntgenologischen Darstellung des Duodenums (Pneumo-Duodenum). *Fortschr Geb Roentgenstr* 1927;36:739–744.

131. Pesquera GS: A method for direct visualization of lesions in the small intestines. *Am J Roentgenol* 1929;22:254–257.

132. Schatzki R: Small intestinal enema. *Am J Roentgenol* 1943;50:743–751.

133. Gershon-Cohen J, Shay H: Barium enteroclysis. A method for the direct immediate examination of the small intestine by single and double contrast techniques. *Am J Roentgenol* 1939; 42:456–458.

134. Sellink JL: Radiological examination of the small intestine by duodenal intubation. *Acta Radiol (Diagn)* 1974;15:318–332.

135. Sellink JL: *Examination of the Small Intestine by Means of Duodenal Intubation*. HE Stenfert Kroese, Leiden, 1971.

136. Miller RE, Sellink JL: Enteroclysis: the small bowel enema. *Gastrointest Radiol* 1979;4: 269–283.

137. Sellink JL, Miller RE: *Radiology of the Small Bowel. Modern Enteroclysis Technique and Atlas*. The Hague, Martinus Nijhoff, 1982.

138. Pygott F, Street DF, Shellshear MF, et al: Radiological investigation of the small intestine by small bowel enema technique. *Gut* 1960;1:366–370.

139. Bilbao MK, Frische LH, Dotter CT, et al: Hypotonic duodenography. *Radiology* 1967;89:438–443.

140. Fournier AM, Cave P, Duval J, et al: Pour un grêle en double contraste . . . et en 6 minutes. *J Radiol* 1979;60:71–74.

141. Novak D: Doppelkontrastuntersuchung des Dünndarms ohne intubation. *ROEFO* 1976; 125:38–41.

142. Salomonowitz E, Czembirek H: Dynamische Doppelkontrastuntersuchung des Dünndarmes. *ROEFO* 1980;133:274–278.

143. Abu-Yousef MM, Benson CA, Lu CH, et al: Enteroclysis aided by an electric pump. *Radiology* 1983;147:268–269.

144. Maglinte DDT, Burney BT, Miller RE: Technical factors for a more rapid enteroclysis. *AJR* 1982;138:588–591.

145. Maglinte DDT, Miller RE: A comparison of pumps used for enteroclysis. *Radiology* 1984;152:815.

146. Sellink JL, Rosenbusch G: Moderne Untersuchungstechnik des Dünndarms oder Die zehn Gebote des Enteroklysmas. *Radiologe* 1981;21:366–376.

147. Hippéli R, Grehn S: Untersuchungen zur Intensivdiagnostik des Dünndarmes mit der Sondenmethode. *ROEFO* 1978;129:713–723.

148. Nolan DJ, Piris J: Crohn's disease of the small intestine: a comparative study of the radiological and pathological appearances. *Clin Radiol* 1980;31:591–596.

149. Scott-Harden WG, Hamilton HAR, McCall Smith S: Radiological investigation of the small intestine. *Gut* 1961;2:316–322.

150. Sanders DE, Ho CS: The small bowel enema. Experience with 150 examinations. *AJR* 1976;127:743–751.

151. Antes G, Lissner J: Double-contrast small-bowel examination with barium and methylcellulose. *Radiology* 1983;148:37–40.

152. Vallance R: An evaluation of the small bowel enema based on an analysis of 350 consecutive examinations. *Clin Radiol* 1980;31:227–232.

153. Herlinger H: Double contrast enteroclysis, in Margulis AR, Burhenne HJ (eds): *Alimentary Tract Radiology*, ed 3., St. Louis, CV Mosby, 1983, p 892.

154. Taverne PP, van der Jagt EJ: Small-bowel radiography. *ROEFO* 1985;143:293–297.

155. Skjennald A, Samset JH: Duodeno-jejunal intubation in examination of the small intestine. *Clin Radiol* 1980;31:221–224.

156. Pajewski M, Eshchar J, Manor A: Visualization of the small intestine by double contrast. *Clin Radiol* 1975;26:491–493.

157. Dyet JF, Pratt AE, Flouty G: The small bowel enema: description and experience of a technique. *Br J Radiol* 1976;49:1039–1044.

158. Ekberg O: Double contrast examination of the small bowel. *Gastrointest Radiol* 1977;1: 349–353.

159. Loubière M, Grimaud A, Coussement A, et al: L'étude radiologique en double contraste de l'intestin grêle sous intubation duodéno-jéjunale. *J Radiol* 1977;58:75–79.

160. Ekberg O: Crohn's disease of the small bowel examined by double contrast technique: a comparison with oral technique. *Gastrointest Radiol* 1977;1:355–359.

161. Sinclair DJ, Buist TAS: Water contrast barium enema technique using methylcellulose. *Br J Radiol* 1966;39:228–230.

162. Herlinger H: Small bowel, in Laufer I (ed): *Double Contrast Gastrointestinal Radiology*. Philadelphia, WB Saunders, 1979, pp 423–494.

163. Desaga JF: Röntgenologischer Nachweis pathologischer Veränderungen der Dünndarmschleimhaut. *Fortschr Röntgenstr* 1987;146:689–694.

164. Sellink J: Enteroclysis, in Margulis AR, Burhenne HJ (eds): *Alimentary Tract Radiology*, ed 3., St. Louis, CV Mosby, 1983, p 875.

165. Miller RE: Complete reflux small bowel examination. *Radiology* 1965;84:457–463.

166. Miller RE, Miller WJ: Inflammatory lesions of the small bowel. Complete reflux small bowel examination. *Am J Gastroenterol* 1966;45:40–49.

167. Miller RE, Lehman G: Localization of small bowel hemorrhage. *Am J Dig Dis* 1972;17:1019–1023.

168. Violon D, Steppe R, Potvliege R: Improved retrograde ileography with glucagon. *AJR* 1981;136:833–839.

169. Monsein LH, Halpert RD, Harris ED, et al. Retrograde ileography: value of glucagon. *Radiology* 1986;161:558–559.

170. Castellino RE, Verby HD, Friedland GW, et al: Delayed barium aspiration following complete reflux small bowel enema. *Br J Radiol* 1968;41:937–939.

171. Fitzgerald EJ, Thompson GT, Somers SS, et al: Pneumocolon as an aid to small-bowel studies. *Clin Radiol* 1985;36:633–637.

172. Pochaczevsky R: Oral examination of the colon. "The colonic cocktail." *Am J Roentgenol* 1974;121:318–325.

173. Kressel HY, Evers KA, Glick SN, et al: The peroral pneumocolon examination: technique and indications. *Radiology* 1982;144:414–416.

174. Kelvin FM, Gedgaudas RK, Thompson WM, et al: The peroral pneumocolon: its role in evaluating the terminal ileum. *AJR* 1982;139:115–121.

175. Wolf K-J, Goldberg HI, Wall SD, et al: Feasibility of the peroral pneumocolon in evaluating the ileocecal region. *AJR* 1985;145:1019–1024.

176. Maglinte DDT, Burney BT, Miller RE: Lesions missed on small-bowel follow-through: analysis and recommendations. *Radiology* 1982;144:737–739.

177. Fleckenstein P, Pedersen G: The value of the duodenal intubation method (Sellink modification) for the radiological visualization of the small bowel. *Scand J Gastroenterol* 1975;10:423–425.

178. Silverberg E, Lubera JA: Cancer statistics, 1988. *Cancer J Clin* 1988;38:5–22.

179. Saunders CG, MacEwen DW: Delay in diagnosis of colonic cancer: a continuing challenge. *Radiology* 1971;101:207–208.

180. Miller RE: Detection of colon carcinoma and the barium enema. *JAMA* 1974;230:1195–1198.

181. Steinbach HL, Burhenne HJ: Performing the barium enema: equipment, preparation and contrast medium. *Am J Roentgenol* 1962;87:644–654.

182. Ott DJ, Chen YM, Gelfand DW, et al: Single-contrast vs double-contrast barium enema in the detection of colonic polyps. *AJR* 1986;146:993–996.

183. Wolf EL, Frager D, Beneventano TC: Feasibility of double-contrast barium enema in the elderly. *AJR* 1985;145:47–48.

184. Laurell H: Discussion of S. Ström's "On the Roentgen diagnostics of changes in the appendix and caecum." *Acta Radiol* 1922;1:491–492.

185. Fischer AW: Über eine neue röntgenologische Untersuchungsmethode des Dickdarms: Kombination von Kontrasteinlauf und Luftaufblähung. *Klin Wochenschr* 1923;2:1595–1598.

186. Weber HM: The roentgenologic demonstration of polypoid lesions and polyposis of the large intestine. *Am J Roentgenol* 1931;25:577–589.

187. Welin S, Welin G: *The Double Contrast Examination of the Colon. Experiences with the Welin Modification*. Stuttgart, Georg Thieme, 1976.

188. Miller RE, Maglinte DDT: Barium pneumocolon: technologist-performed '7 pump' method. *AJR* 1982;131:1230–1232.

189. Miller RE: Examination of the colon. *Curr Prob Radiol* 1975;5:3–40.

190. Miller RE: Faster-flow enema equipment. *Radiology* 1977;123:229–230.

191. Levene G: Rates of venous absorption of carbon dioxide and air used in double-contrast examination of the colon. *Radiology* 1957;69:571–575.

192. Levene G, Kaufman SA: An improved technic for double contrast examination of the colon by the use of compressed carbon dioxide. *Radiology* 1957;68:83–85.

193. Coblentz CL, Frost RA, Molinaro V, et al: Pain after barium enema: effect of CO_2 and air on double-contrast study. *Radiology* 1985;157:35–36.

194. Bassette JR, Maglinte DDT: Double-contrast barium enema study: simple conversion to CO_2. *Radiology* 1987;162:274–275.

195. Bernier P, Coblentz C: CO_2 delivery system for double-contrast barium enema examinations. *Radiology* 1986;159:264.

196. de Roos A, Hermans J, Op den Orth JO: Polypoid lesions of the sigmoid colon: a comparison of single-contrast, double-contrast, and biphasic examinations. *Radiology* 1984;151:597–599.

197. Lappas JC, Maglinte DDT, Kopecky KK, et al: Post double-contrast sigmoid flush; an adjuvant technique in imaging diverticular disease (abstract #89). *Radiology* 1987;165(suppl):57.

198. Virkki R, Mäkelä P, Kormano M: Dimethylpolysiloxane as an adjuvant in double-contrast barium enema. *Europ J Radiol* 1981;1:134–136.

199. Miller RE, Skucas J: *Radiological Examination of the Colon—Practical Diagnosis.* The Hague, Martinus Nijhoff, 1983.

200. Kaude JV, Harty RF: Sensitivity of single contrast barium enema with regard to colorectal disease as diagnosed by colonoscopy. *Europ J Radiol* 1982;2:290–292.

201. Teefey SA, Carlson HC: The fluoroscopic barium enema in colonic polyp detection. *AJR* 1983;141:1279–1281.

202. Johnson CD, Carlson HC, Taylor WF, et al: Barium enemas of carcinoma of the colon: sensitivity of double- and single-contrast studies. *AJR* 1983;140:1143–1149.

203. Winthrop JD, Balfe DM, Shackelford GD, et al: Ulcerative and granulomatous colitis in children. *Radiology* 1985;154:657–660.

204. Prandoni A, Wright I: The anti-coagulants. *Bull NY Acad Med* 1942;18:433–458.

205. Conley CL: Hemostasis, in Mountcastle VB (ed): *Medical Physiology*, ed 13., St. Louis, CV Mosby, 1974, p 1043.

206. Duckert F, Yin ET, Straub W: Separation and purification of the blood clotting factors by means of chromatography and electrophoresis, in Preters H (ed): *Prolides of the Biological Fluids. Proceedings of the 8th Colloquium, Bruges, 1960.* Amsterdam, Elsevier, 1961, p 41.

207. Tishkoff GH, Williams LC, Brown DM: Preparation of highly purified prothrombin complex. *J Biol Chem* 1968;243:4151–4167.

208. Alexander RJ, Detwiler TC: Quantitative adsorption of platelet glycoprotein G (thrombin sensitive protein thrombospondin) to barium citrate. *Biochem J* 1984;217:67–71.

209. Miller RE, Skucas J, Violante MR, et al: The effect of barium on blood in the gastrointestinal tract. *Radiology* 1975;117:527–530.

210. Adams JT: The barium enema as treatment for massive diverticular bleeding. *Dis Colon Rect* 1974;17:439–441.

Chapter 3

Barium Sulfate: Toxicity and Complications

Jovitas Skucas

Barium is widely distributed in the earth's crust and in sea water. Life on earth evolved in the presence of barium, and trace amounts are found in many organisms. Some marine plants concentrate barium in their tissues. The exoskeleton of the rhizopod *Xenophyofera* is composed primarily of barium sulfate.[1] Some body tissues concentrate barium.[2] In humans, most of the barium is found in bone and connective tissue;[1] barium can cross the placental barrier, and trace amounts have been reported in normal amniotic fluid.[3] A barium deficiency or mild excess does not appear to be associated with any disease. Little is known about barium metabolism.

SOLUBILITY

Before the 20th century, various soluble barium salts were used for a number of medical conditions. Even early in the 20th century the soluble barium chloride was employed orally in patients with heart block until such therapy was supplanted by better drugs.

Barium sulfate is poorly soluble in water. The resultant trace concentrations of barium can be measured by a number of spectroscopic or spectrometric techniques.[4,5] Both a laser-enhanced ionization technique and use of inductively coupled plasma emission allow the detection of barium as low as 0.1–0.2 ng/ml,[6,7] with greater sensitivity possible in water than in blood.[8]

Extremely fine grinding of barium sulfate particles or their formation by precipitation increases the solubility in water. Hulett[9] observed that relatively coarse crystalline barium sulfate particles, either 3.6 μm or 1.8 μm, were soluble at 2.29 mg/liter at 25°C. When ground to a diameter of 0.2 μm, the solubility rose to 4.15 mg/liter, and when ground to 0.1 μm the solubility rose slightly further to 4.5 mg/liter. The corresponding figures for natural barytes were 2.38 mg and

6.18 mg/liter.[9] The solubility of barium sulfate does not decrease appreciably after the crystals reach 2 μm in diameter.[10] Barium sulfate particles in the 0.04–0.1 μm range can be absorbed from bowel. Eventually they end up in the lymphatics.[11]

Is there absorption of barium sulfate following a typical radiologic barium examination? Mauras et al. used emission spectrometry to measure the plasma and urine barium levels in ten patients before ingestion of a commercial barium product (Barytgen) and at serial times after the study.[12] The barium blood levels before the study were less than one μg/liter, with subsequent statistically significant elevations occurring at 4 hours (2.5 μg/liter) and 8 hours (1.3 μg/liter) after ingestion. The urinary elimination of barium was 4.2 μg/24 hours before the study and 13.4 μg/24 hours after ingestion, also a statistically significant elevation. The increased plasma levels after oral ingestion of barium, although statistically significant, are extremely small. Likewise, the rise in 24-hour urinary excretion after ingestion of the 350-g dose was less than 10 μg.

Clavel et al. used atomic absorption spectrometry to measure the urinary excretion of barium after both oral ingestion and rectal administration of several commercial barium sulfate preparations.[13] Following oral ingestion of varying amounts of barium sulfate, they found the total urinary excretion of barium was a function of the amount ingested initially, regardless of the preparation used (Fig. 3-1). For all preparations the total urinary excretion was $0.16–0.26 \times 10^{-6}$ of the ingested dose. Excretion during the first 24 hours was 71–98% of the total,

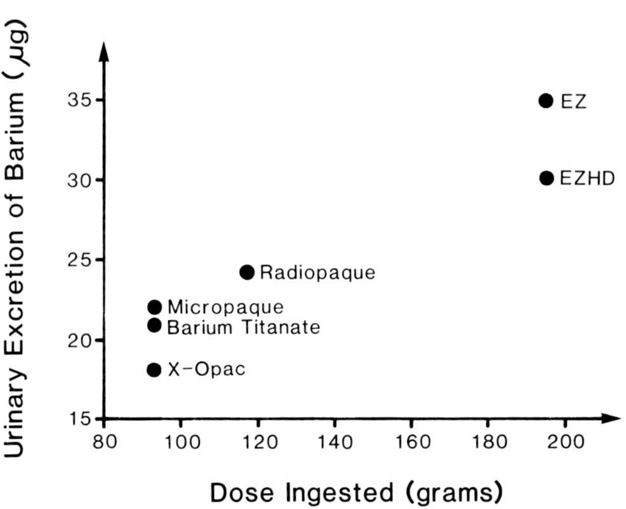

Figure 3-1 Urinary excretion of several barium preparations. In general, the excretion varies with the dose ingested. *Source*: Adapted with permission from *Therapie* (1987;42:239–243), Copyright © 1987, Doin Editeurs.

depending upon the preparation. Administration as a barium enema led to considerably lower subsequent urinary excretion;[13] different concentrations of four barium sulfate preparations resulted in urinary excretion of 0.023–0.096 $\times$ 10^{-6} of the administered dose. In both of these studies, the measured changes in barium concentration should be viewed as an indication of the increased sophistication and sensitivity of the recording apparatus available, rather than as an indictment of any possible side effects of the contrast agent.

In a rat, barium instillation of up to 40% of body weight is needed for a toxic dose.[14] Hecht[15] claimed that the oral intake by adults of 3 mg/kg of soluble barium salts may be considered harmless. Therefore, even with the smaller barium sulfate particle sizes, a 70-kg adult could ingest over 40 liters of a 50% barium sulfate suspension before exceeding this margin of safety. This does not mean that ingested barium sulfate is without gross effect. The constipating tendency of earlier commercial preparations is well known among radiologists. Less well known is that intravenously the barium ion acts as a purgative. Although the amount absorbed from oral ingestion of barium sulfate is considerably less than that required for a purgative effect, it is not unreasonable to expect some effect upon bowel motility, especially if sensitive recording devices are used.

In one reported instance, plaster of Paris was being stored in a barium sulfate container and was administered to four patients.[16] The plaster formed hard casts in the stomachs, although eventually the casts disintegrated.

POISONING

Contamination of barium sulfate media with soluble barium salts is rare, although inadvertent substitution of soluble barium salts for the insoluble barium sulfate has been reported.[17] Poisoning by ingestion of soluble barium salts, either accidentally or with suicidal intent, does occur. Some of these soluble salts are used as rat poisons, pesticides, or depilatories. These salts include barium sulfide,[18] barium carbonate,[19–21] barium polysulfide,[22] and barium chloride.[23–25] Barium chloride was found as a contaminant of table salt in parts of China and led to endemic muscle paralysis that mimicked familial periodic paralysis.[24,26] Barium carbonate resembles flour in appearance; mass poisonings have occurred when it was added to various foods.[19,27]

Patients with barium toxicity can develop hypertension that is unresponsive to phentolamine.[28] A profound acute muscle paralysis may evolve; involvement of the respiratory muscles results in a respiratory acidosis, and mechanical ventilation may be necessary. Electromyography during the paralysis phase shows the muscles to be electrically silent and not excitable. Sensory function is not involved.

Hypokalemia is common in patients with barium toxicity. The erythrocyte potassium level is increased because of intracellular migration of potassium.[28] Sulfhemoglobin may be present.[22] Although some authors have ascribed the muscle paralysis to hypokalemia,[29] high plasma barium ion levels may be its primary cause.[21] The barium ion results in vascular smooth muscle contraction by a direct intracellular effect.[30] The ion enters the cell by the same channels as calcium.

Therapy consists of intravenous potassium administration, marked diuresis to help clear the barium, and if necessary, ventilation assistance. Intravascular barium is partially cleared by the kidneys. Intravenous sulfates, such as magnesium sulfate, should be avoided; the resultant barium sulfate can precipitate in the renal tubules.[25]

BACTERIAL CONTAMINATION

Bacterial contamination of barium sulfate products is not common.[31] Unsanitary mixing on a day-to-day basis or storing the liquid suspensions in opened containers without refrigeration may lead to such contamination. Cross-contamination between different examinations has been virtually eliminated by use of disposable equipment.

Bacteremia may occur in up to 11% of patients after a barium enema.[32,33] It can be associated with underlying disease.[34] Such bacteremia should not be ascribed to bacterial contamination of the barium preparation.

BARIUM PNEUMOCONIOSIS

Barium normally accumulates with age in lung tissue, probably as a contaminant of dust.[1] The gradual increase has not been associated with any specific disorder. Larger amounts of barium dust, however, have been associated with a pneumoconiosis.[35] The pulmonary pattern in barium sulfate mill workers consists of reticular or reticulonodular densities bilaterally with sparing of the apices and bases. The extent of change correlates with the duration of exposure. Generally there are few clinical symptoms, and the condition appears to be benign. Coexisting silica dust may be present, which could account for most of the pulmonary changes.

BARIUM ASPIRATION

Barium sulfate particles deposited on ciliated portions of the trachea and bronchi are generally cleared rapidly. Experimental work in rats has shown, however, that a small fraction of the dose is retained in macrophages for days or even months.[36] This small residue is cleared exponentially, possibly by lymphatic drainage. No adverse effects have been reported for this residue (assuming the barium is not radioactive, which may not be true in certain mines). Barium entering alveoli may persist for a considerable time (Fig. 3-2).

The aspiration of modest amounts of commercial barium sulfate preparations is believed to be of little harm; in fact, barium sulfate has been used as a bronchographic contrast agent in patients allergic to iodine.[37] However, aspiration of large amounts can compromise pulmonary function in a manner similar to other inert materials. If the aspiration includes gastric juice, a chemical pneumonitis can ensue. In general, if aspiration or a tracheoesophageal fistula is suspected, barium contrast media are preferred over hyperosmolar iodinated water-soluble media.[38,39] The role of the nonionic media in such a situation is not currently established.

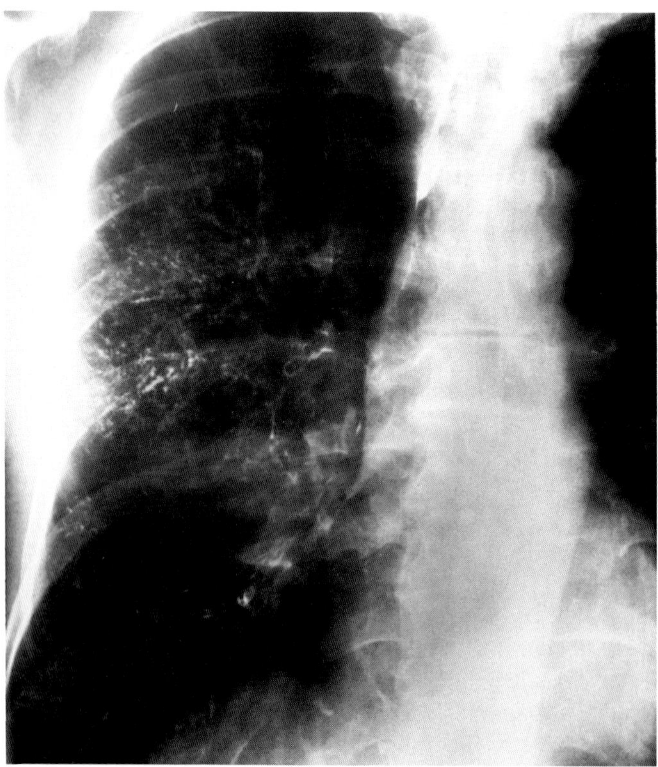

Figure 3-2 Barium aspiration. Barium is present in the trachea, bronchi, and alveoli. It took almost a week for the more peripherally located barium to clear.

Aspiration of any highly viscous material can obstruct the pharynx, trachea, or bronchi with fatal results.[40] Especially in debilitated individuals, introduction of viscous contrast media into the oropharynx should be performed with care. Fluoroscopy aids in evaluating the swallowing function of these patients.

HYPERSENSITIVITY REACTIONS

Hypersensitivity to commercial barium sulfate products is rare, but has been reported. The most common reaction involves the skin.[41] Respiratory complications, loss of consciousness, and anaphylaxis have also been reported.[42] The reactions occur about equally following upper GI examinations and barium enemas.[41] Because commercial products contain a number of additives, the obvious question is whether the hypersensitivity is to the barium sulfate or to one of the additives. Among several patients who developed urticaria following a barium enema, one patient had a positive skin test to methylparaben, a preservative that has been used in some commercial products.[43] Sensitivity reactions have also been reported following use of barium suspensions after the injection of glucagon to induce bowel hypotonia.[44] In general, the incriminating agent has not been sought nor detected.

BARIUM APPENDICITIS

Barium retention in the appendix is not uncommon and is generally not associated with any sequelae. A number of authors have suggested that prolonged barium retention in the appendix can eventually result in acute appendicitis,[45–49] although others disagree.[50] One literature search up to 1987 yielded only 25 patients with so-called barium appendicitis reported in the English literature.[51]

Whether such retention is indeed a causative factor is difficult to prove. At times retained barium will coat an appendiceal calculus. If appendicitis does develop in the future, it is tempting to associate it with the retained and highly visible barium. In some patients, undoubtedly the initial barium study is performed because of vague symptomatology that is a precursor to the eventual acute appendicitis.

It has been suggested that appendicitis in the face of retained barium in the appendix is associated with a high risk of being complicated and that a long interval between the barium study and subsequent appendicitis results in increased complications.[51] Whether such barium retention is indeed responsible for the increased complications or whether it is simply a reflection of an underlying appendiceal abnormality is conjecture at present.

BOWEL PERFORATION

Of greater clinical concern are bowel perforation, barium peritonitis, and barium intravasation. A number of peptic ulcer perforations during an upper GI examination have been reported.[52] Most of these reports date from the 1940s or earlier; vigorous compression and palpation were given as the reasons for the perforation. The first report of such gastric perforation and contrast extravasation is believed to have been in 1916.[53] A more recent case report of gastric outlet obstruction and a gastric ulcer described eventual perforation 72 hours after the barium study.[54] Surgery within 1.5 hours of the perforation revealed 3 liters of serosanguineous fluid mixed with barium in the peritoneal cavity. The report did not mention any gastric decompression between the time the obstruction was detected by the barium study and the perforation.

Many of the barium enema perforations published before 1980 have been summarized by Gelfand.[55] The true incidence of colorectal perforation during a barium enema is difficult to determine. Undoubtedly, some minor, localized extraperitoneal perforations are undetected. The published incidence of perforation has ranged from one in 2,250[56] to one in 12,000 patients.[57]

A perforation may be secondary to an error in technique. Rectal laceration can occur during introduction of the enema tip or during inflation of a balloon. Perforation may be associated with diverticulitis, inflammatory bowel disease, radiation colitis, disuse atrophy if a colostomy is present proximally, recent rectal biopsy, ulcer, necrotic tumor, or unsuspected bowel damage as a result of sigmoidoscopy or colonoscopy.[58,59] A perforation can occur in normal bowel[60,61] and in infants and children,[62] a population that may be particularly at risk.[58] Insertion of a balloon catheter into the vagina can lacerate the vaginal wall. Subsequent introduction of barium can propel the contrast media into the paravaginal venous plexus and then into the systemic circulation.[63]

A perforation can be difficult to detect during fluoroscopy. Contrast media can spread throughout the peritoneal cavity or be localized in the extraperitoneal tissues. Initially, some of these patients have few symptoms.

The relative incidence of intra- versus extraperitoneal perforation is difficult to gauge from the literature. Some patients described as having "barium peritonitis" appear to have an extraperitoneal perforation when their clinical course and published radiographs are analyzed in retrospect.[54]

It has been suggested that a double-contrast barium enema is more dangerous than a single-contrast enema,[64] yet no convincing proof is available that this is indeed so. A perforation is probably directly related to the intraluminal pressure. At times, perforation during a double-contrast study is associated with spillage of air only;[65] a similar situation during a single-contrast study would result in leakage of barium sulfate, clearly a more dangerous situation.

A unique finding after a colon perforation was development of a high-pitched voice.[66] The patient had intraperitoneal and extraperitoneal gas following a double-contrast barium enema, with spread of the gas superiorly into the mediastinum and soft tissues of the neck.

Extraperitoneal Perforation

One survey reported one *extraperitoneal* perforation for every 40,000 barium enema examinations.[57] Some of these extraperitoneal perforations are not associated with immediate symptoms.[67] Many are along the anterior rectal wall. Barium in the perirectal soft tissues results in an inflammatory reaction that evolves into fibroblastic proliferation, multinucleated giant cells with intracytoplasmic barium sulfate crystals, and eventual extensive fibrosis.[67,68] The barium sulfate particles can enter adjacent lymphatics.[61] Such extraperitoneal barium collections can eventually lead to indurated masses that mimic a carcinoma,[52,69] although there is no convincing evidence so far that barium deposits are carcinogenic.

Barium spillage into the mediastinum, such as with an esophageal perforation, likewise results in inflammation and eventual granuloma formation (Fig. 3-3).[70,71] The greater radiographic density of barium can enable detection of a perforation that is missed by water-soluble agents. Although such mediastinal barium can persist, there is little evidence that it is associated with any sequelae more severe than with the water soluble agents (Fig. 3-4).

It has been thought that barium in the mediastinum may obscure the operative field,[72] although this viewpoint has been refuted[73] and is not generally held today.

An esophago-subarachnoidal fistula has been studied with barium esophagography.[74] The patient had an unresectable squamous cell carcinoma, was treated by radiotherapy, and eventually developed meningitis with spinal fluid infection. The subsequent barium study showed extravasation into the mediastinum, destroyed vertebral bodies, and extravasation into the spinal canal. It is speculative whether the subarachnoidal barium played any significant role in the demise of this patient with extensive tumor and an esophagosubarachnoid fistula.

The choice of contrast agents in the study of bowel perforation is still controversial. Each patient should be evaluated individually. If a recent perforation is suspected, if the site of perforation is in doubt, or if communication with the peritoneal cavity is a possibility, most radiologists would use one of the water-soluble agents. With a chronic or loculated perforation, barium sulfate will, at times, provide more information. If intestinal contents have been leaking into a walled-off cavity for some time, the introduction of a barium preparation is not associated with any additional morbidity. In fact, the better visualization of bari-

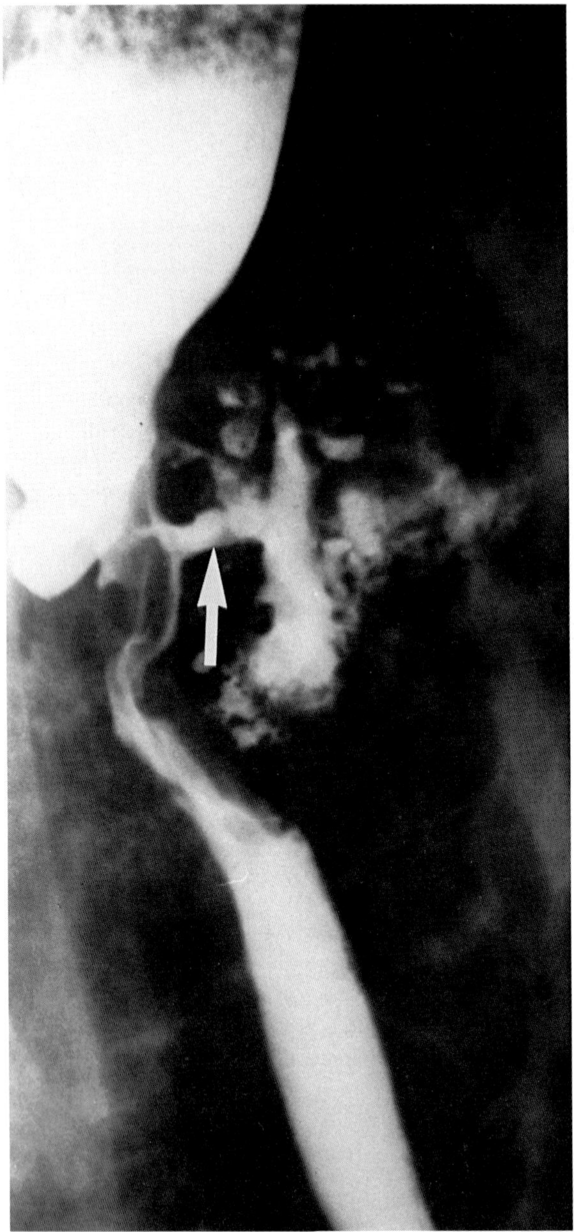

Figure 3-3 Squamous carcinoma of the esophagus with a perforation into adjacent carinal lympn nodes (*arrow*). Barium sulfate was the contrast agent in this examination. The spill of barium is not believed to have added to any increased morbidity.

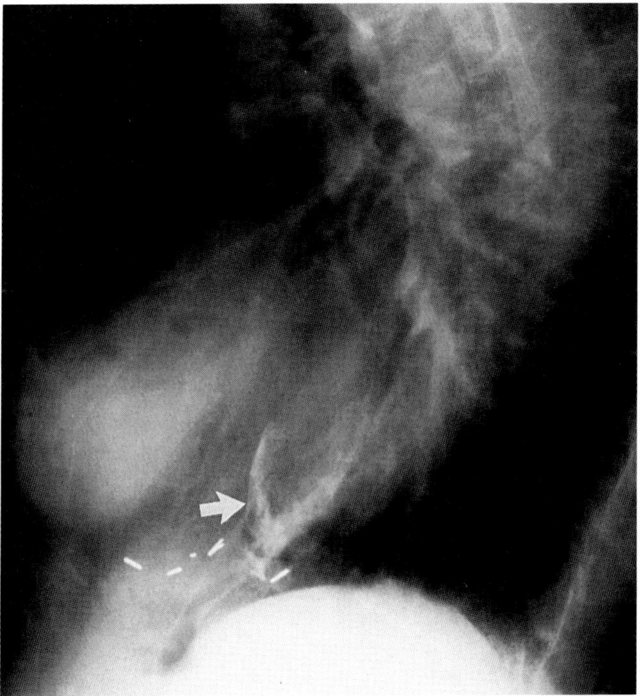

Figure 3-4 Residual barium in mediastinal soft tissues (*arrow*). A suspected esophageal perforation in this patient with achalasia was studied with barium years previously. The residual barium should be innocuous.

---

um aids in a full evaluation of such a cavity and its communication with bowel (Fig. 3-5a and 3-5b).

Prior extravasation of barium sulfate into the soft tissues can be recognized radiographically (Fig. 3-6). The dense, linear collections are generally distinguishable from soft tissue calcifications, although in an occasional patient the appearance can be confusing.[75] A biopsy is usually diagnostic, with barium sulfate crystals being identified histologically; a number of other exogenous materials may, however, mimic the histologic appearance of barium crystals.[76]

Some barium crystals are birefringent. The birefringence of barium deposits may not be due to the barium but rather to accompanying magnesium and silicon impurities; these ingredients are found in talc that has been used in some commercial barium sulfate preparations.[77] More recently, using energy dispersive x-ray analysis, it was shown that birefringence is due to the barium sulfate crystals themselves and not to any accompanying magnesium or silicon impurities.[78] The author postulated that barium sulfate particles prepared from crushed ore are

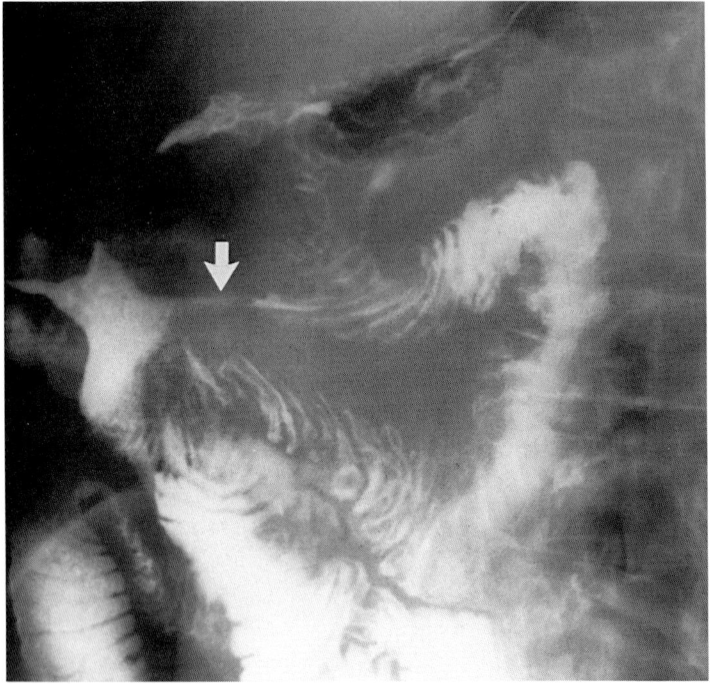

Figure 3-5A Duodenal perforation following pancreatitis and surgical drainage of an abscess. Barium sulfate was used to show the site of communication (*arrow*) into an adjacent cavity.

birefringent, whereas particles prepared by a precipitation process are not.[78] This issue is still confusing because colloidal dispersions, in general, exhibit birefringence. Although the commercial barium suspensions are not colloidal in size, they do contain colloidal additives, and some birefringence is to be expected from them.

Another hypothesis has been that barium sulfate crystals prepared by precipitation appear as green, granular particles, whereas the crushed variety is seen as rhomboid crystals.[78] The published data, however, are also confusing. Barium sulfate crystals found in some parenteral solutions, presumably formed by precipitation, have a rhombohedral outline and smooth surfaces.[79]

If barium deposits are suspected clinically, sodium rhodizonate is a useful histochemical reagent, although there is variation in the results obtained with different batches of rhodizonate.[76]

Barium Peritonitis

There have been a number of papers, often single case reports, describing barium peritonitis. The incidence of *intraperitoneal* perforation has been reported

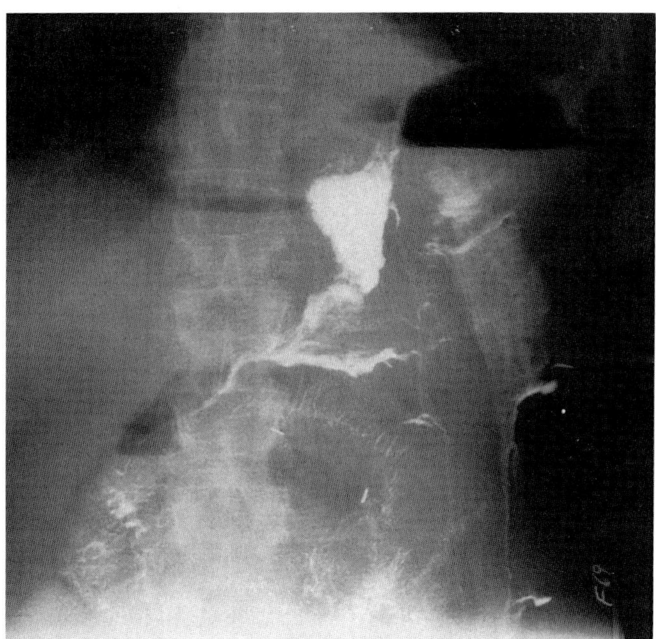

Figure 3-5B Another patient developed a large abdominal abscess following a hemigastrectomy. With the patient upright, barium can be seen streaming from the stomach into the mostly fluid-filled cavity. The left upper quadrant gas is in the most superior portion of the abscess. The use of barium was not associated with any undue sequelae in this patient; in fact, barium allowed a better definition of the full extent of the abscess that was subsequently drained.

as one in 17,000 barium enema examinations.[57] In a review in 1952, Zheutlin et al. reported a mortality of 50% from perforation into the peritoneal cavity during a barium enema.[80] Since then, the mortality does not appear to have improved significantly. The morbidity and mortality depend upon the amount of contrast and stool spilled into the peritoneal cavity.[81] Barium sulfate and stool are probably more toxic in combination than either alone.[82,83] However, even sterile barium sulfate in the peritoneal cavity of dogs produces hemorrhagic peritonitis.[82] Perforation during a barium enema seems to be associated with a higher mortality than a perforation during endoscopy.[84]

Barium peritonitis has been studied by injecting the contrast agent into the peritoneal cavities of dogs.[85] In this study, most deaths occurred within 24 hours; surgery was not a major factor in the resultant mortality, and mortality was lowest when vigorous IV fluid therapy was used. In dogs, the commercial preparations had a more adverse effect than USP barium sulfate.[82] This may be due, in part, to the additives in the commercial products.

In rats, intraperitoneal sterile barium sulfate results in suppression of delayed cutaneous hypersensitivity.[83] Such barium affects the inflammatory response even

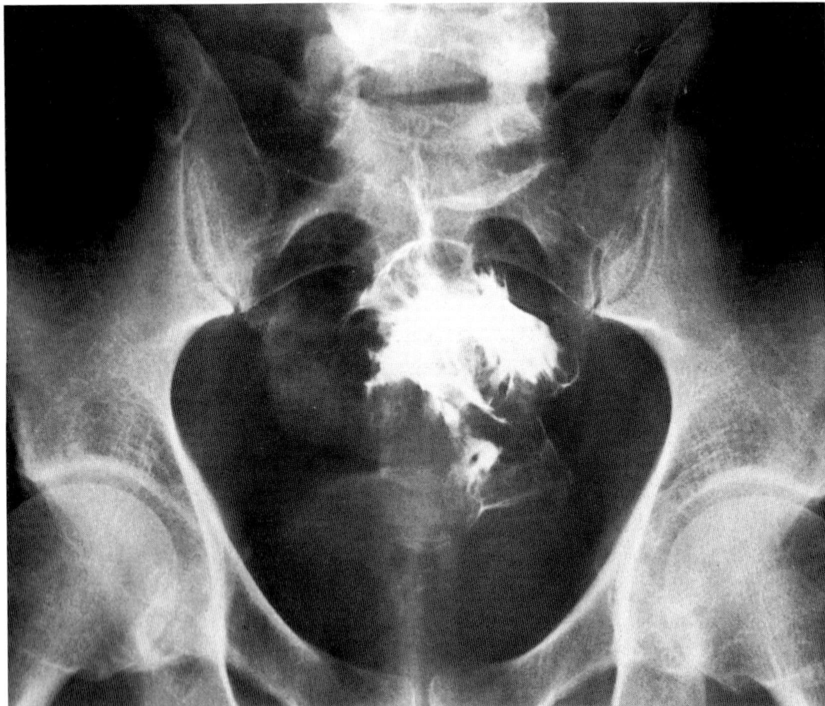

Figure 3-6 Prior colon perforation. Residual barium in the soft tissues will continue to be readily identified on subsequent radiographs. *Source*: Reprinted from *The Radiological Examination of the Colon* (p 90) by RE Miller and J Skucas with permission of Martinus Nijhoff Publishers, © 1983.

in sites remote from the peritoneal cavity. The hypersensitivity suppression was associated with reduced barium particle clearance from the peritoneal cavity.

Following barium spill, there is a migration of leukocytes into the peritoneal cavity.[80] The barium sulfate crystals become coated by a fibrin membrane,[86] and there is an inpouring of fluid into the peritoneal cavity. Eventually, signs and symptoms of acute peritonitis develop. The intraperitoneal sequestration of fluid can lead to a profound hypovolemia. Invariably, a septic component is present, and shock can ensue, at times manifesting several hours later.

The management of barium sulfate peritonitis is generally surgical, but the care of each patient should be individualized.[87,88] In addition to antibiotic coverage, these patients require large amounts of IV fluids well beyond the usual maintenance levels.[54,56,88] At surgery, the involved bowel feels sandy and grating when it is handled.[86] Irrigation of the peritoneal cavity can remove most of the barium suspension, but invariably significant amounts remain. Wiping with gauze increases the underlying hemorrhage.

Long-term sequelae of barium sulfate peritonitis include extensive bowel adhesions and barium granulomas (Fig. 3-7). Ureteral obstruction has been reported.[89,90] Occasionally the barium is not detected by radiographic examination.[91] There is dense fibrosis around the barium sulfate particles. Two children with massive barium peritonitis had opacification of retrosternal lymph nodes days to weeks later,[92] confirming earlier experimental work of the lymphatic drainage of the peritoneum.

Deixonne et al. reported a patient with barium peritonitis who subsequently developed barium poisoning, manifested primarily by neurological abnormalities.[93] Assay using emission spectrometry revealed elevated barium levels in the cerebrospinal fluid. The type of barium sulfate product used in the barium enema was not stated. Such a complication of barium peritonitis must indeed be rare.

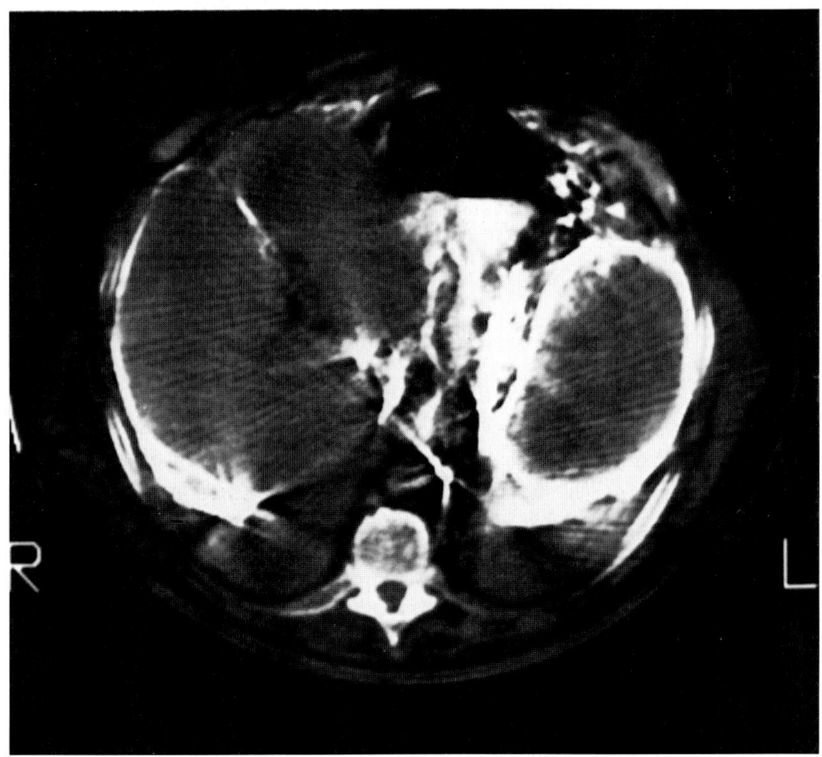

Figure 3-7 Barium peritonitis occurred during a barium enema in this patient with ulcerative colitis. At subsequent surgery he underwent a hemicolectomy, and as much of the intraperitoneal barium as possible was removed. This CT scan, obtained 3 weeks after the perforation, reveals the extensive residual barium scattered throughout the abdomen.

BARIUM INTRAVASATION

A relatively lethal complication is barium sulfate intravasation.[63,94–96] It can occur not only during a barium enema but has also been reported during an upper GI examination, presumably from mucosal ulceration.[97] The lethal dose has not been established and probably varies between species. In rabbits, the IV injection of a suspension of barium sulfate in a dose of 0.5 g/kg of body weight was not associated with any observable toxic reaction.[11]

Not all barium intravasation comes from radiological examinations. Several parenteral solutions have been found to contain barium sulfate crystals.[79,98] It has been postulated that in these solutions barium sulfate formed in situ, with the barium ion originating from the glass vial (which contains barium oxide) and the sulfate ion originating either from the drug or associated antioxidant.

The published reports of visible barium intravasation in humans reveal that many of these patients died, generally within minutes of the extravasation.[99] At times, no obvious site of perforation is found, even at postmortem.[94] The contrast media can pass through the lung vessels into the systemic circulation.[94,95] Barium sulfate in the lung vessels can be seen as a transient phenomenon in survivors.[100]

Prolonged hypotension and disseminated intravascular coagulopathy may occur in these patients, possibly secondary to the release of bradykinin.[95,101] Treatment should be directed against the underlying endotoxic shock. In survivors, barium intravasation can result in increased radiographic opacification of the liver, spleen, and bones. In one such patient, CT liver attenuation of 240 HU and splenic attenuation of 880 HU were found.[95]

Barium intravasation into the portal venous system can also occur;[96,102–104] one sequelae is an eventual liver abscess.[105] There are numerous reports of intravasation of air or gas. In the presence of inflammatory bowel disease, portal venous gas during a double-contrast barium enema can be a benign condition,[64,106–112] although barium can also intravasate with fatal results.[104] Barium sulfate or air intravasation may be more common than is reported.

Not all sudden deaths at the time of barium enema are caused by the examination. One of my patients died while waiting for the examination (the waiting time was not unduly prolonged). One death shortly after a barium enema was secondary to a massive pulmonary embolus.[113] In some patients who die during a barium enema no specific cause is found; cardiac electrical irregularities have been postulated.[114,115]

RELATED COMPLICATIONS

Enteric-coated aspirin is often used by patients requiring long-term aspirin administration. These tablets do not begin to dissolve until the pH is greater than

6 to 7. Thus, in patients with gastric outlet obstruction the tablets accumulate intact in the stomach for a prolonged period of time. One such patient had a rise in serum salicylate level following a barium upper GI examination, implying that there was intragastric dissolution of these tablets.[116] Subsequent in vitro tests showed that enteric-coated aspirin tablets could dissolve in the presence of some commercial barium preparations. The addition of effervescent granules to the barium suspensions resulted in faster and more complete tablet dissolution.[116] In general, whenever enteric-coated aspirin tablets are detected by an upper GI examination in a patient with gastric outlet obstruction, the tablets should be evacuated.

REFERENCES

1. Schroeder HA, Tipton IH, Nason AP: Trace metals in man: strontium and barium. *J Chronic Dis* 1972;25:491–517.

2. McCauley PT, Washington IS: Barium bioavailability as the chloride, sulfate, or carbonate salt in the rat. *Drug Chem Toxicol* 1983;6:209–217.

3. Hall GS, Carr MJ, Cummings E, et al: Aluminum, barium, silicon, and strontium in amniotic fluid by emission spectrometry. *Clin Chemistry* 1983;29:1318.

4. Hovis FE, Gelbwachs JA: Determination of barium at trace levels by laser induced ionic fluorescence spectrometry. *Anal Chem* 1984;56:1392–1394.

5. Sharp RA, Knevel AM: Analysis of barium in barium sulfate and diagnostic meals containing barium sulfate using atomic absorption spectroscopy. *J Pharm Sci* 1971;60:458–560.

6. Turk GC, Travis JC, DeVoe JR: Laser enhanced ionization spectrometry in analytical flames. *Anal Chem* 1979;51:1890–1896.

7. Fassel VA, Kniseley RN: Inductively coupled plasma—optical emission spectroscopy. *Anal Chem* 1974;46:1110A–1120A.

8. Mauras Y, Allain P: Dosage du baryum dans l'eau et les liquides biologiques par spectrometrie d'emission avec source plasma haute frequence. *Anal Chem Acta* 1979;110:271–277.

9. Hulett GA: Beziehungen Zwischen Oberflächenspannung und löslichkeit. *Z Physik Chem* 1901;37:385–406.

10. Trimble HM: The coalescence of an unfilterable precipitate of barium sulfate. *J Phys Chem* 1927;31:601–606.

11. Adolph W, Taplin GV: Use of micropulverized barium sulfate in x-ray diagnosis. A preliminary report. *Radiology* 1950;54:878–883.

12. Mauras Y, Allain P, Roques MA, et al: Étude de l'absorption digestive du baryum après l'administration orale du sulfate de baryum pour exploration radiologique. *Thérapie* 1983;38:109–110.

13. Clavel JP, Lorillot ML, Buthiau D, et al: Absorption intestinale du baryum lors d'explorations radiologiques. *Thérapie* 1987;42:239–243.

14. Boyd EM, Abel M: The acute toxicity of barium sulfate administered intragastrically. *Can Med Assoc J* 1966;94:849–853.

15. Hecht G: Röntgenkontrastmittel, in Heubner W, Schüller J (eds): *Handbuch der Experimentellen Pharmakologie*, Erg—Werk VIII. Berlin, Springer-Verlag, 1939, p 97.

16. Felson B: Radiologist on the rocks. *Semin Roentgenol* 1973;4:361–363.

17. Govindiah D, Bhaskar GR: An unusual case of barium poisoning. *Antiseptic* 1972; 69:675–677.

18. Gould DB, Sorrell MR, Lupariello HD: Barium sulfide poisoning. *Arch Intern Med* 1973;132:891–894.

19. Lewi Z, Bar-Khayim Y: Food poisoning from barium carbonate. *Lancet* 1964;2:342–343.

20. Maretic Z, Homadovski J, Razbojnikov S, et al: Ein beitrag zur Kenntnis von Vergiftung mit Barium. *Med Klin* 1957;52:1950–1953.

21. Phelan DM, Hagley SR, Guerin MD: Is hypokalaemia the cause of paralysis in barium poisoning? *Br Med J* 1984;289:882.

22. Jobba G, Renge B: Über die Neopol-Vergiftung. *Arch Toxikol* 1971;27:106–110.

23. Graham CF: Barium chloride poisoning. *JAMA* 1934;102:1471.

24. Ku D, Yen CK, Li CC: Acute poisoning by common salt containing barium chloride. *Chin Med J* 1943;61:303–304.

25. Wetherill SF, Guarino MJ, Cox RW: Acute renal failure associated with barium chloride poisoning. *Ann Intern Med* 1981;95:187–188.

26. Huang K-W: Pa ping. *Chin Med J* 1943;61:305–312.

27. Schott GD: Some observations on the history of the use of barium salts in medicine. *Med Hist* 1974;18:9–21.

28. Roza O, Berman LB: The pathophysiology of barium: hypokalemic and cardiovascular effects. *J Pharmacol Exp Ther* 1971;177:433–439.

29. Layzer RB: Periodic paralysis and the sodium-potassium pump. *Ann Neurol* 1982; 11:547–552.

30. Hansen TR, Dineen DX, Petrak R: Mechanism of action of barium ion on rat aortic smooth muscle. *Am J Physiol* 1984;246:c235–c241.

31. Amberg JR, Unger JD: Contamination of barium sulfate suspension. *Radiology* 1970; 97:182–183.

32. Le Frock J, Ellis CA, Klainer AS, et al: Transient bacteremia associated with barium enema. *Arch Intern Med* 1975;135:835–837.

33. Butt J, Hentges D, Pelican G, et al: Bacteremia during barium enema study. *Am J Roentgenol* 1978;130:715–718.

34. Richman LS, Short WF, Cooper WM: Barium enema septicemia: occurrence in a patient with leukemia. *JAMA* 1973;226:62–63.

35. Levi-Valensi P, Drif M, Dat A, et al: A propos de 57 observations de barytose pulmonaire. Resultats d'une enquete systematique dans une usine de baryte. *J Franç Med Chir Thorac* 1966;20:443–455.

36. Takahashi S, Patrick G: Long-term retention of ^{133}Ba in the rat trachea following local administration of barium sulfate particles. *Radiat Res* 1987;110:321–328.

37. Nelson SW, Christoforidis A, Pratt PC: Further experiments with barium sulfate as a bronchographic contrast medium. *Am J Roentgenol* 1964;92:595–614.

38. Frech RS, Davie JM, Adatepe M, et al: Comparison of barium sulfate and oral 40% diatrizoate injected into the trachea of dogs. *Radiology* 1970;95:299–303.

39. Chiu CL, Gambach RR: Hypaque pulmonary edema. *Radiology* 1974;111:91–92.

40. Lareau DG, Berta JW: Fatal aspiration of thick barium. *Radiology* 1976;120:317.

41. Janower ML: Hypersensitivity reactions after barium studies of the upper and lower gastrointestinal tract. *Radiology* 1986;161:139–140.

42. McAvoy M, Young JWR, Keramati B: Hypersensitivity reactions to barium suspension (letter). *AJR* 1985;144:1316.

43. Schwartz EE, Glick SN, Foggs MB, et al: Hypersensitivity reactions after barium enema examination. *AJR* 1984;143:103–104.

44. Gelfand DW, Sowers JC, DePonte KA, et al: Anaphylactic and allergic reactions during double-contrast studies: is glucagon or barium suspension the allergen. *AJR* 1985;144:405–406.

45. Bergman JJ, Rosen GD, Moeller DA: Appendicitis associated with recent barium study. *J Fam Pract* 1979;8:931–935.

46. Merten DF, Lebowitz ME: Acute appendicitis in a child associated with prolonged appendiceal retention of barium (barium appendicitis). *South Med J* 1978;71:81–82.

47. Totty WG, Koehler RE, Cheung LY: Significance of retained barium in the appendix. *AJR* 1980;135:753–756.

48. Wobbes T: Appendicitis acuta na röntgenonderzoek van het colon met bariumsulfaat als contraststof. *Ned T Geneesk* 1981;125:10–12.

49. Sisley JF, Wagner CW: Barium appendicitis. *South Med J* 1982;75:498–499.

50. Maglinte DDT, Bush ML, Aruta EV, et al: Retained barium in the appendix: diagnostic and clinical significance. *AJR* 1981;137:529–533.

51. Cohen N, Modai D, Rosen A, et al: Barium appendicitis; fact or fancy. *J Clin Gastroenterol* 1987;9:447–451.

52. Puylaert CBAJ: Barium perforation of a peptic ulcer and barium granuloma of the colon. *Radiol Clin Biol* 1969;38:84–95.

53. Rosenthal E: Röntgenologisch beobachtete Magenperforation. *Berl Klin Wochnschr* 1916;53:945–947.

54. Sung JP, O'Hara VS, Lee C-Y: Barium peritonitis. *West J Med* 1977;127:172–176.

55. Gelfand DW: Complications of gastrointestinal radiologic procedures. I. Complications of routine fluoroscopic studies. *Gastrointest Radiol* 1980;5:293–315.

56. Gardiner H, Miller RE: Barium peritonitis. A new therapeutic approach. *Am J Surg* 1973;125:350–352.

57. Masel H, Masel JP, Casey KV: A survey of colon examination techniques in Australia and New Zealand, with a review of complications. *Austral Radiol* 1971;15:140–147.

58. Vogel H, Steinkamp U, Grabbe E: Perforationen beim Kontrasteinlauf. *Röntgen-Bl* 1983;36:271–278.

59. Butson ARC, D'Souza TJ, Thomson JG: Perforation of simple ulcer of the colon. *Br J Radiol* 1980;53:723–725.

60. Kahn SP, Lindenauer SM, Wojtalik RS: Perforation of the normal colon during barium contrast examination. *Am Surg* 1976;42:789–792.

61. Reitamo J, Häyry P, Nordling S: Nonperforating extrarectal escape of barium complicating barium enema. *Dis Colon Rect* 1971;14:381–385.

62. Eklöf O, Hald J, Thomasson B: Barium peritonitis. Experience in five pediatric cases. *Pediatr Radiol* 1983;13:5–9.

63. David R, Berezesky IK, Bohlman M, et al: Fatal barium embolization due to incorrect vaginal rather than colonic insertion: an ultrastructural and x-ray microanalysis study. *Arch Pathol Lab Med* 1983;107:548–551.

64. Lazar HP: Survival following portal venous air embolization. *Am J Dig Dis* 1965;10:259–264.

65. Gelfand DW, Ott DJ, Ramquist NA: Pneumoperitoneum occurring during double-contrast enema. *Gastrointest Radiol* 1979;4:307–308.

66. Rabin DN, Smith C, Witt TR, Holinger LD: Voice change after barium enema: a clinical sign of extraperitoneal colon perforation. *AJR* 1987;148:145–146.

67. Lewis JW Jr, Kerstein MD, Koss N: Barium granuloma of the rectum: an uncommon complication of barium enema. *Ann Surg* 1975;181:418–423.

68. Röckert H, Zettergren L: Tissue reaction to barium sulphate contrast medium. *Acta Pathol Microbiol Scand* 1963;58:445–450.

69. Phelps JE, Sanowski RA, Kozarek RA: Intramural extravasation of barium simulating carcinoma of the rectum. *Dis Colon Rect* 1981;24:388–390.

70. Ginai AZ, ten Kate FJW, ten Berg RGM, et al: Experimental evaluation of various available contrast agents for use in the upper gastrointestinal tract in case of suspected leakage. Effects on mediastinum. *Br J Radiol* 1985;58:585–592.

71. Vessal K, Montali RJ, Larson SM, et al: Evaluation of barium and Gastrografin as contrast media for the diagnosis of esophageal ruptures or perforations. *AJR* 1975;123:307–319.

72. Schwartz SS: Barium or Gastrografin: which contrast media for diagnosis of esophageal tears? (letter). *Gastroenterology* 1975;69:1377.

73. James AE: Barium or Gastrografin: which contrast media for diagnosis of esophageal tears? (response). *Gastroenterology* 1975;69:1377.

74. Cornwell J, Walden C, Ghahremani GG: CT demonstration of fistula between esophageal carcinoma and spinal canal. *J Comp Assist Tomogr* 1986;10:871–873.

75. Broadfoot E, Martin G: Barium granuloma of the rectum. *Austral Radiol* 1977;21:50–52.

76. Chaplin AJ, Turner ELT: Observations on the histochemistry of barium. *Histochemistry* 1983;79:111–116.

77. Marek J, Jurek K: Comparative light microscopical and x-ray microanalysis study of barium granuloma. *Pathol Res Pract* 1981;171:293–302.

78. Womack C: Unusual histological appearances of barium sulphate—a case report with scanning electron microscopy and energy dispersive x-ray analysis. *J Clin Pathol* 1984;37:488–493.

79. Boddapati S, Butler LD, Im S, et al: Identification of subvisible barium sulfate crystals in parenteral solutions. *J Pharm Sci* 1980;69:608–610.

80. Zheutlin N, Lasser EC, Rigler LG: Clinical studies on effect of barium in the peritoneal cavity following rupture of the colon. *Surgery* 1952;32:967–979.

81. Sisel RJ, Donovan AJ, Yellin AE: Experimental fecal peritonitis. *Arch Surg* 1972; 104:765–768.

82. Cochran DQ, Almond CH, Shucart WA: An experimental study of the effects of barium and intestinal contents on the peritoneal cavity. *Am J Roentgenol* 1963;89:883–887.

83. Bohnen JMA, Christou NV, Meakins J: Suppression of delayed cutaneous hypersensitivity and inflammatory cell delivery by sterile barium peritonitis. *J Surg Res* 1987;43:430–435.

84. Nelson RL, Abcarian H, Prasad ML: Iatrogenic perforation of the colon and rectum. *Dis Colon Rect* 1982;25:305–308.

85. Nahrwold DL, Isch JH, Benner DA, et al: Effect of fluid administration and operation on the mortality rate in barium peritonitis. *Surgery* 1971;70:778–781.

86. Westfall RH, Nelson RH, Musselman MM: Barium peritonitis. *Am J Surg* 1966;112:760–763.

87. Hardy TG Jr, Hartmann RF, Aguilar PS, et al: Survival after colonic perforation during barium-enema examination. *Dis Colon Rect* 1983;26:116–118.

88. Grobmyer AJ, Kerlan RA, Peterson CM, et al: Barium peritonitis. *Am Surg* 1984;50:116–120.

89. Herrington JL Jr: Barium granuloma within the peritoneal cavity: ureteral obstruction 7 years after barium enema and colonic perforation. *Ann Surg* 1966;164:162–166.

90. Elliot JS, Rosenberg ML: Ureteral occlusion by barium granuloma. *J Urol* 1954;71:692–694.

91. Walther JM, Romas NA, Lowe FC: Barium granuloma: an unusual cause of unilateral ureteral obstruction. *J Urol* 1987;138:614–616.

92. Berdon WE, Baker DH, Poznanski A: Opacification of retrosternal lymph nodes following barium peritonitis. *Radiology* 1973;106:171–173.

93. Deixonne B, Baumel H, Mauras Y, et al: Un cas de baryto-péritoine avec atteinte neurologique. Intérêt du dosage du baryum dans les liquides biologiques. *J Chir (Paris)* 1983;120:611–613.

94. Cove JKJ, Snyder RN: Fatal barium intravasation during barium enema. *Radiology* 1974;112:9–10.

95. Chan FL, Tso WK, Wong LC, et al: Barium intravasation: radiographic and CT findings in a nonfatal case. *Radiology* 1987;163:311–312.

96. Juler GL, Dietrick WR, Eisenman JI: Intramesenteric perforation of sigmoid diverticulitis with nonfatal venous intravasation. *Am J Surg* 1976;132:653–656.

97. Mahboubi S, Gohel VK, Dalinka MK, et al: Barium embolization following upper gastrointestinal examination. *Radiology* 1974;111:301–302.

98. Aoyama T, Horioka M: Barium sulfate crystals in parenteral solutions of aminoglycoside antibiotics. *Chem Pharm Bull* 1987;35:1223–1227.

99. Rosenberg LS, Fine A: Fatal venous intravasation of barium during a barium enema. *Radiology* 1959;73:771–773.

100. Zatzkin HR, Irwin GAL: Nonfatal intravasation of barium. *Am J Roentgenol* 1974; 92:1169–1172.

101. Blom H, Nauta EH, van Rosevelt RF, et al: Disseminated intravascular coagulation and hypotension after intravasation of barium. *Arch Intern Med* 1983;143:1253–1255.

102. Schumacher F: $BaSO_4$—Übertritt in die Vena mesenterica inferior bei Kolonkontrastuntersuchung—eine seltene Komplikation. *ROEFO* 1980;133:99–100.

103. Archer FH, Freeman AH: A case of non-fatal intravasation of barium during barium enema. *Br J Radiol* 1981;54:69–72.

104. Salvo AF, Capron CW, Leigh KE, et al: Barium intravasation into portal venous system during barium enema examination. *JAMA* 1976;235:749–751.

105. Isaacs I, Nissen R, Epstein BS: Liver abscess resulting from barium enema in a case of chronic ulcerative colitis. *NY State J Med* 1950;50:332–334.

106. Kees CJ, Hester CL Jr: Portal vein gas following barium enema examination. *Radiology* 1972;102:525–526.

107. Weinstein GE, Weiner M, Schwartz M: Portal vein gas. *Am J Gastroenterol* 1968; 49:425–429.

108. Sadhu VK, Brennan RE, Madan V: Portal vein gas following air-contrast barium enema in granulomatous colitis: report of a case. *Gastrointest Radiol* 1979;4:163–164.

109. Christensen MA, Lu CH: Gas in the portal vein after air-contrast barium enema in a patient with inflammatory colitis. *South Med J* 1982;75:1291–1292.

110. Pappas D, Romeu J, Tarkin N, et al: Portal vein gas in a patient with Crohn's colitis. *Am J Gastroenterol* 1984;79:728–730.

111. Birnberg FA, Gore RM, Shragg B, et al: Hepatic portal venous gas: a benign finding in a patient with ulcerative colitis. *J Clin Gastroenterol* 1983;5:89–91.

112. Katz BH, Schwartz SS, Vender RJ: Portal venous gas following a barium enema in a patient with Crohn's colitis. *Dis Colon Rect* 1986;29:49–51.

113. Irshad M: Association of sudden death with barium enema examination. *JAMA* 1986; 256:2264.

114. DeLeonardis EA, Pearce JG: Death during routine outpatient barium enema examination. *Applied Radiol* 1987;16:44a–44b.

115. Harrington RA, Kaul AF: Cardiopulmonary arrest following barium enema examination with Glucagon. *Drug Intel Clin Pharm* 1987;21:721–722.

116. Bogacz K, Caldron P: Enteric-coated aspirin bezoar: elevation of serum salicylate level by barium study. *Am J Med* 1987;83:783–786.

Chapter 4

Barium Titanate

F.A.D. Heitz, F.P. Meyer, and P. Gauthier

The addition of double-contrast techniques in gastrointestinal (GI) radiology led to an investigation of contrast preparations other than barium sulfate. One product found satisfactory in some centers is barium titanate.[1–3] Table 4-1 compares the physical and chemical properties of barium sulfate and barium titanate. Neither their close molecular weights nor crystalline shapes accounts for the differences in mucosal coating between those two contrast agents.

Because of the fluidity and relatively poor mucosal coating of earlier barium sulfate preparations, a number of methods were employed to increase their vis-

Table 4-1 Comparison of Physical and Chemical Properties of Barium Sulfate and Barium Titanate

Properties	*Barium Sulfate* ($BaSO_4$)	*Barium Titanate or Barium Metatitanate* ($BaTiO_3$)
Molecular weight	233.43	233.26
% Ba	58.84	58.89
BaO	65.70	—
SO_4	41.15	—
Ti	—	20.54
Density	4.5	6.0
Melting temperature	1580°C	1625°C
Crystalline forms	Rhombic	Tetragonal, hexagonal, cubic, orthorhombic
Crystal sizes — commercial	0.07–1.50 μm	0.07–1.50 μm

Sources: Rhône-Poulenc, France, and Prof. Pradeau, Pharmacie Centrale de l'Assistance Publique, Paris.

Table 4-2 Viscosity for Some Gums at a Concentration of 20%

Type of Gum	*Viscosity* (cP)
Arabic	50
Tragacanth	200
Stercula	1,500
Guar	3,300

Source: Personal research by F Heitz and A Thuillier, Hôpital de la Pitié-Salpêtrière, Paris.

cosity and mucosal coating. One approach was to use various additions, such as gums. Table 4-2 shows the viscosity values for some gums at a concentration of 20%. The values vary over a large range, increasing from gum arabic to guar gum. Most natural gums and some artificial polymers have been studied as viscosity-enhancing agents. In general, their viscosities change not only with concentration (Fig. 4-1), but also with time (Fig. 4-2), temperature (Fig. 4-3), and pH (Fig. 4-4).

Table 4-3 shows the settling values of barium titanate with sorbitol, saccharose, and bentonite (sodium montmorillonite) as added ingredients. Thus, small amounts of additives profoundly effect settling.

Viscosity of a suspension can be increased by adding methylcellulose (Table 4-4). Such increase in viscosity does not necessarily lead to an increase in

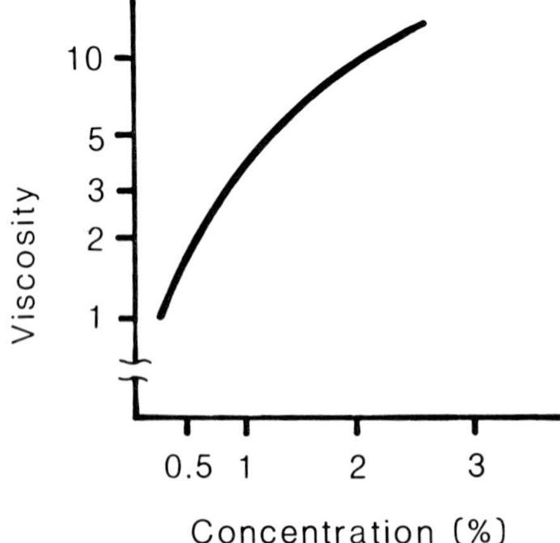

Figure 4-1 Viscosity of XB-23 sol at different concentrations. Measurements were made at 25°C. Sol XB-23 is an anionic heteropolysaccharide obtained by fermenting sugars with *Xanthomonas campestris*. Viscosity was measured with a Brookfield viscometer at 30 reus/mn and is expressed relative to the viscosity of water. *Source:* Rhône—Poulenc, France, 1978.

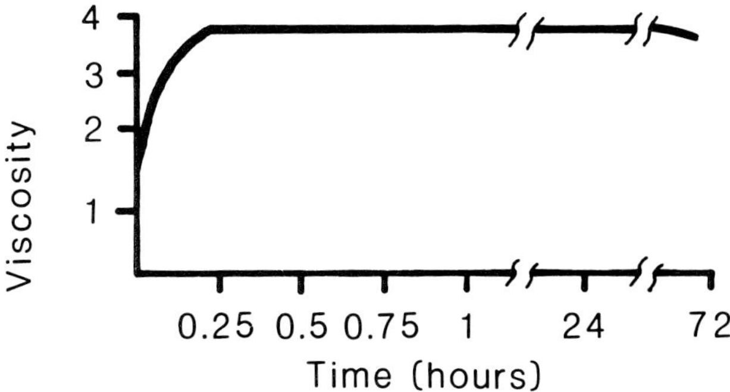

Figure 4-2 Change in viscosity with time of a 1% concentration of XB-23 sol. Measurement conditions are the same as for Figure 4-1. *Source:* Rhône—Poulenc, France, 1978.

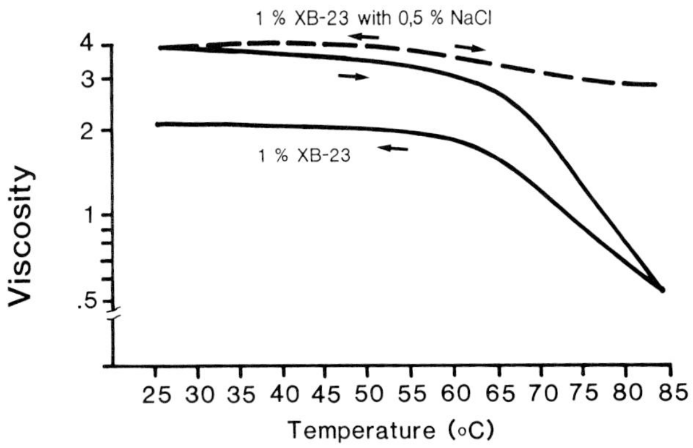

Figure 4-3 Viscosity dependence on temperature of a 1% concentration of XB-23 sol. Measurement conditions are the same as for Figure 4-1. *Source:* Rhône—Poulenc, France, 1978.

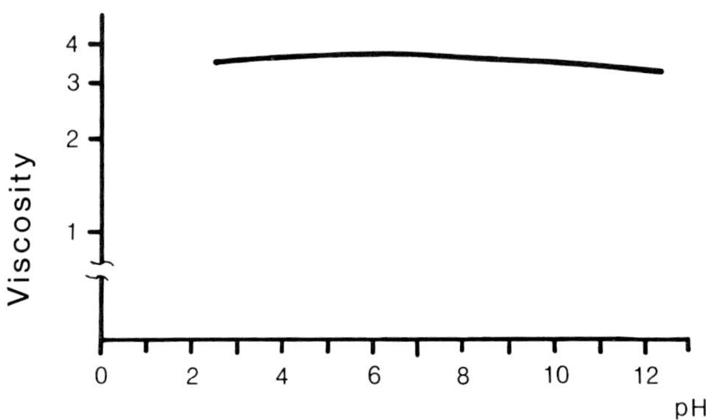

Figure 4-4 Viscosity variation with pH of a 1% concentration of XB-23 sol. Measurement conditions are the same as for Figure 4-1. *Source:* Rhône—Poulenc, France, 1978.

coating. In fact, adhesion is even decreased with barium titanate. Such experimental studies, summarized in Tables 4-3 and 4-4, have led to the use of pure barium titanate preparations in clinical radiology. These preparations have two advantages: (1) The suspension is easy to prepare as the water-powder blend is nearly instantaneous (10 seconds), and (2) the mixture can be used immediately. One has to wait when a gum is added to barium sulfate in order to benefit from the maximal viscosity.

Coating ability seems to be related to surface tension, the values of which are numerically identical whether expressed in dyne/cm or in erg/cm^2. The surface

Table 4-3 Influence of Additives on Barium Titanate Settling

Barium Titanate (g)	*Additives* *Sorbitol* (g)	*Saccharose* (g)	*Bentonite* (g)	*Settling** *After 1h*	*After 3h*
200	20			18	28
200		20		19	28
200			70	0	1
200				5	8
200	40		50	4	8
200		60	50	7	7
200	20	30		20	26
200	20	30	20	13	17
200	20	30	10	14	21
200	20	30	40	5	6
200	5	35	50	5	6

*Settling is expressed in millimeters of supernatant liquid in the tube after a specified settling time.

Source: Personal research by F Heitz and A Thuillier, Hôpital de la Pitié-Salpêtrière, Paris.

Table 4-4 The Effect of Methylcellulose on the Viscosity and Adhesion of Barium Sulfate and Barium Titanate*

Concentration of Methylcellulose	*Barium Sulfate* *Viscosity* (cP)	*Adhesion* (sec)	*Barium Titanate* *Viscosity* (cP)	*Adhesion* (min)
0.2%	6	10	5	120
0.55%	80	10	10	20

*Adhesion is the length of time that the product adhered to the wall of a glass tube.

Source: Personal research by F Heitz and A Thuillier, Hôpital de la Pitié-Salpêtrière, Paris.

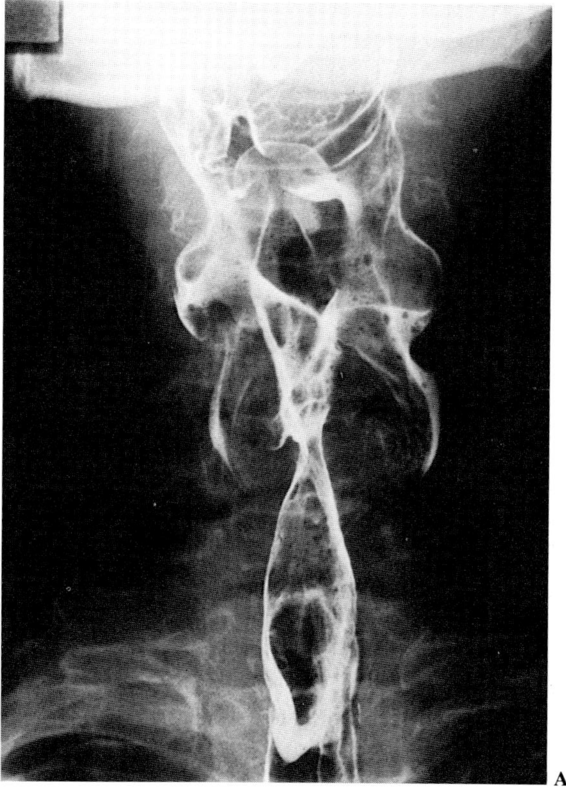

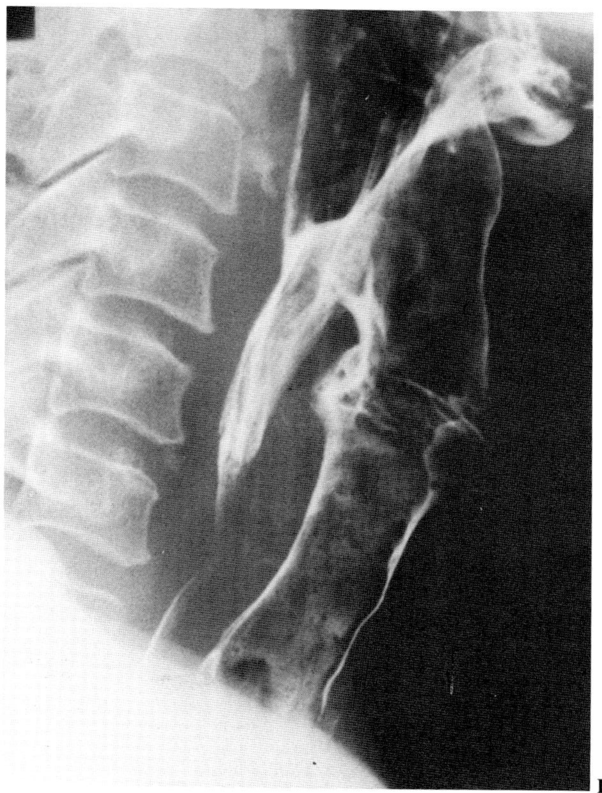

Figure 4-5 Pharyngogram with barium titanate: (*A*) frontal and (*B*) lateral views. Opacification of the pharyngeal structures persists for 10 to 15 minutes.

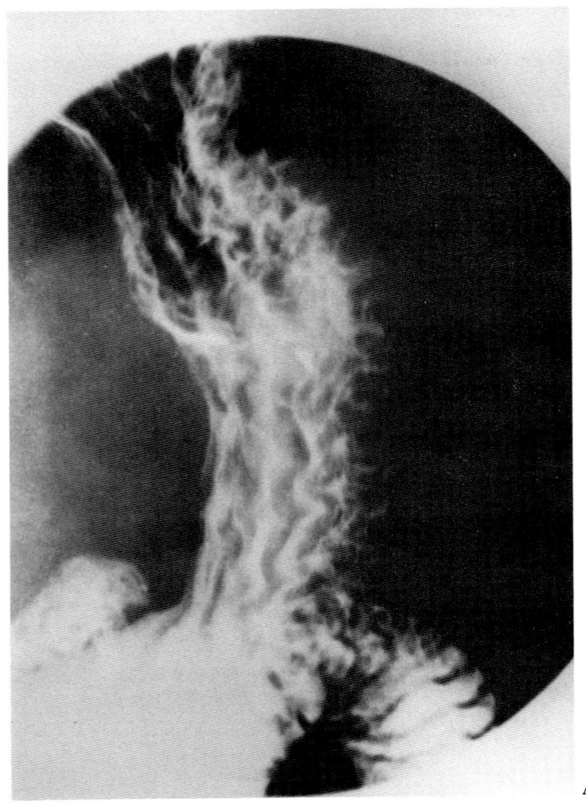

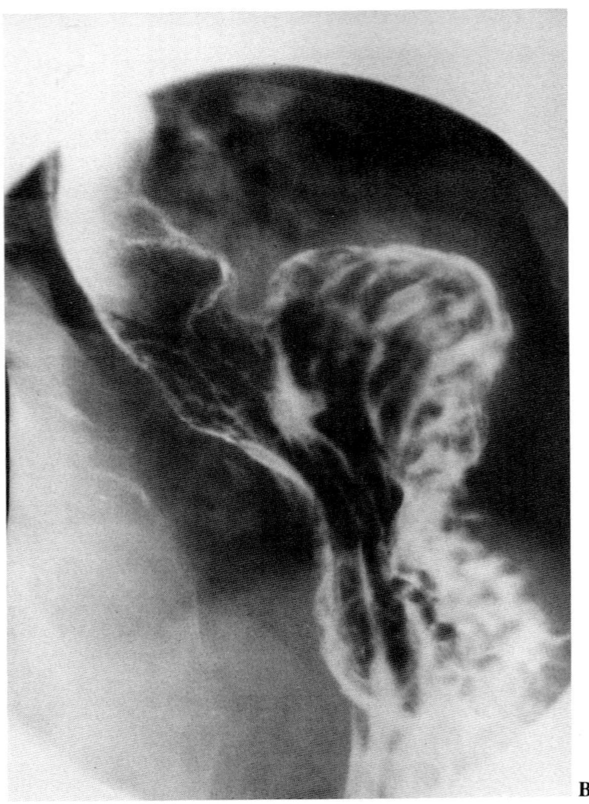

Figure 4-6 Gastric coating with barium sulfate (*A*) and barium titanate (*B*).

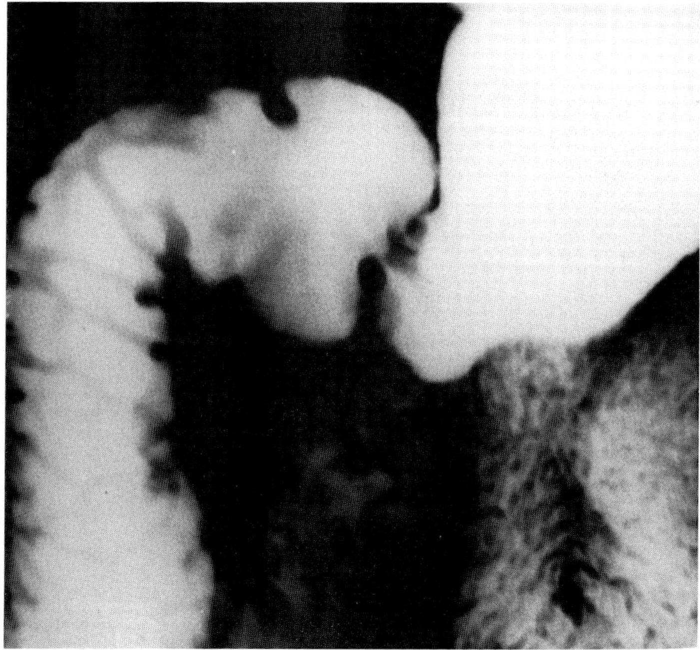

Figure 4-7A Duodenal views with barium titanate: single contrast.

tension of different liquids varies with their molar volumes, the electrical charge of their molecules, and the ability of molecules to enter into hydrogen bridgings. The surface tension of water is greater than that of most other liquids and salts.

A rise in temperature of a liquid, leading to an increase in volume, decreases the mutual attraction forces of the inner and superficial molecules. Thus, surface tension decreases with a rise of temperature.

Based on the above observations, we prepare barium titanate suspensions by not using any additives, keeping the water at room temperature, and using the suspension immediately after vigorous shaking for 10 seconds. For most clinical applications, water amounting to two-thirds of the weight of barium titanate powder is used.

The use of barium titanate in double-contrast examination of the upper digestive tract is straightforward. With the patient in an upright position, effervescent tablets in a quantity sufficient to induce gastric distention are swallowed. The barium titanate suspension is then immediately ingested. The patient is premedicated with IV Buscopan. The patient is then placed supine and rotated three times in order to coat the gastric mucosa. Appropriate radiographs are then obtained.

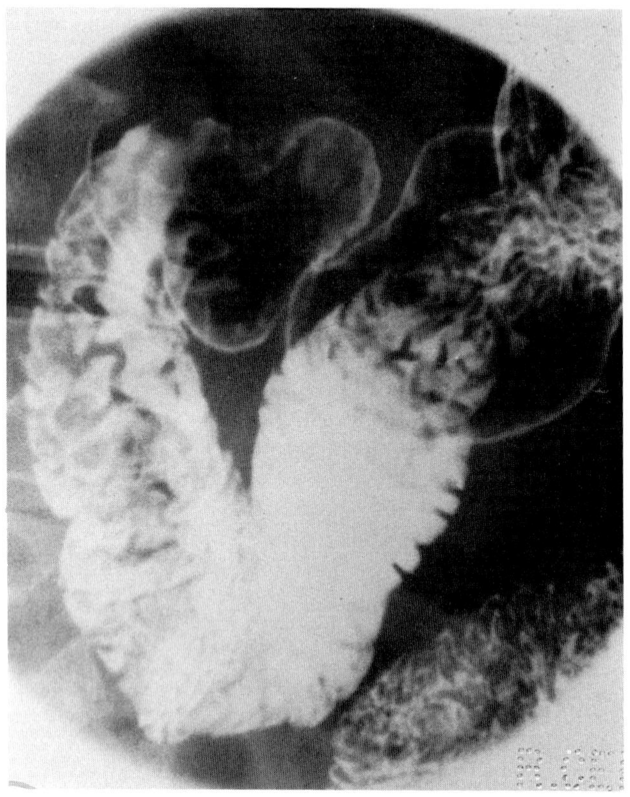

Figure 4-7B Duodenal views with barium titanate: double contrast.

Figures 4-5 through 4-7 illustrate the relative coating properties of barium titanate and barium sulfate in examinations of the upper GI tract.

REFERENCES

1. Heitz F: Utilisation du titanate de baryum comme moyen de contraste en radiologie digestive, abstracted. *Bull l'Acad Natl Méd* 1973;157:3.

2. Heitz F, Heitz L: Présentation d'un nouveau produit de contraste en radiologie digestive: le titanate de baryum. *J Radiol Electrol Med Nucl* 1974;55:430–431.

3. Heitz F: Utilisation du baryum pour le double contraste: ses indications, la toxicité. *J Pharm Clin* 1983;2:387–389.

Chapter 5

Water-Soluble Agents

A.R. Margulis and R.F. Thoeni

Water-soluble contrast media were introduced into the examination of the gastrointestinal (GI) tract in 1955 by Canada.[1] After a short rash of enthusiastic articles,[2–4] the use of these media in the GI tract became controversial. Already in 1956 it was found that these contrast materials act like saline laxatives because of their hyperosmolarity.[1] Many authors[3,4] reported the side effect of diarrhea, as well as the lack of good radiographic detail of the small bowel after the use of iodine-containing water-soluble contrast agents. It was not until Nelson and associates[5] and Harris and co-workers[6] reported on the dangers to dehydrated patients, especially infants, that the use of water-soluble, iodine-containing contrast agents became unpopular and many radiologists stopped using them.

Yet, Vest and Margulis[7] reported in 1962 on the advantages of water-soluble contrast agents in differentiating between adynamic ileus and mechanical small bowel obstruction, particularly in the postoperative patient. Even though some distinguished GI radiologists publicly make statements that they see no reason whatsoever for using water-soluble contrast materials in the GI tract, most radiologists agree that water-soluble contrast materials are the contrast media of choice for demonstrating the site of perforation in the GI tract[8,9] and for injecting sinus tracts that may connect with the general peritoneal cavity. Because so many contradictory statements have been made in the literature and in lectures by well-known radiologists, and in view of the fact that many surgeons strongly encourage radiologists into the use of these contrast media, a review of the advantages, disadvantages, indications, and contraindications for their use is in order.

METHOD OF ADMINISTRATION

About 100 ml of undiluted Gastrografin is used if the patient has to swallow the contrast material. The objectionable taste of the contrast material is somewhat improved if a flavoring agent is added, such as is present in the commercially available products Gastrografin and Gastroview. If a nasogastric tube is in place, it is advisable to inject sodium diatrizoate (Hypaque) through it; this solution is mixed from powder to a concentration of 40%. An adult receives 100 ml of this solution. Hypaque is less costly than Gastrografin and, as it does not contain a flavoring agent, is less likely to produce a local tissue reaction.[10] Thus far, however, no complications resulting from the flavoring agents have been shown.

For a water-soluble enema in an adult, a 20% to 25% solution of diatrizoate should be employed, and the amount of contrast material administered should be carefully observed during fluoroscopy. Instillation of water-soluble contrast material should be discontinued if the colonic lesion is defined.

For fecal impaction and in the absence of an obstructing lesion, a larger amount of contrast medium needs to be administered. The infusion is discontinued when the cecal tip is reached. For meconium ileus, usually a 40% solution of Gastrografin as an enema is employed. The additive Tween 80 (polysorbate 80) has been implicated as a cause of severe topical toxicity produced by Gastrografin enemas combined with Tween 80 in rats[11]; however, another study refuted this conclusion and attributed the toxic effect in the colon of these rats to an excessive volume of undiluted contrast material.[12] The use of dilute diatrizoate preparations avoids both fluid depletion of the patient and uncontrolled dilution that leads to further distention and compromise, particularly of the abnormal bowel wall by osmotic fluid transfer.

For injection of sinuses, fistulas, and stab wounds, no more than 60% concentration of diatrizoate should be used. In cases of perforation, the examination is monitored under fluoroscopy to prevent massive filling of the peritoneal cavity with the hyperosmolar contrast medium.

In recent years, metrizamide (Amipaque) has been recommended for high-risk patients, particularly for neonates or infants. It is not generally indicated in the GI tract if a water-soluble contrast material is needed because of its high cost. It is used as an isotonic solution at a concentration of 180 mg/ml of iodine. Metrizamide offers considerable advantages over other existing agents in that it is nontoxic to the peritoneum, is absorbed less than 1% from the adult gut, does not dilute in the gut, does not damage the bowel mucosa, and is stable in the presence of gastric acids and other secretions.[13]

INDICATIONS

Under certain conditions, water-soluble, iodine-containing contrast media have advantages over barium sulfate because barium sulfate is used as a suspension and is therefore osmotically inert, whereas water-soluble contrast media are hyperosmolar solutions. Barium permits excellent radiographic detail (Fig. 5-1) and, in

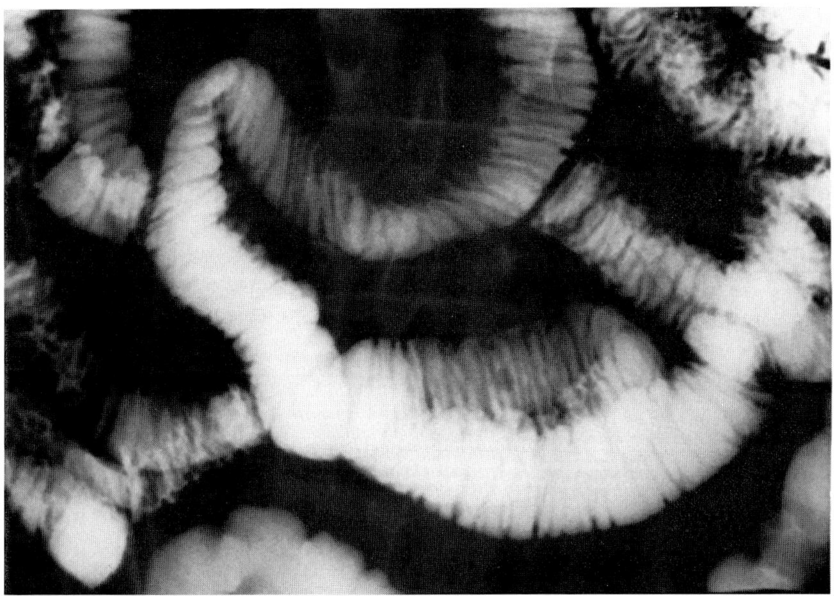

Figure 5-1 In this normal patient, a coned-down view during a small bowel follow-through with barium shows good mucosal detail.

the presence of normal small bowel motility, enables both unequivocal and specific diagnoses. However, if there is an adynamic ileus present, the progress of barium is very slow. Furthermore, barium sulfate spilled into the peritoneal cavity produces a fibrotic reaction, resulting in avascular adhesions that may envelop and shield bacteria in small abscesses from contact with blood-borne antibiotics. In patients with postoperative complications of an ischemic or obstructive nature, it is imperative to outline rapidly the GI tract and determine whether immediate surgical intervention is necessary or whether there is time to observe the patient without endangering his or her life. Surgeons usually feel that patients with postsurgical obstructive complications can be safely observed for 3 to 4 hours. The oral use of barium sulfate in these patients usually requires a protracted procedure far beyond the acceptable time period.

Water-soluble contrast media are hyperosmolar and draw interstitial fluid across the intestinal mucosa into the lumen. This process increases the interstitial volume, inducing the Bayliss-Starling reflex, which stimulates peristaltic activity.[14] Water-soluble contrast media thus act as saline cathartics by speeding up passage but, in the course, are diluted, thereby precluding adequate radiographic detail. This latter phenomenon prevents specific diagnoses beyond the mere statement of whether there is or is not a complete or partial bowel obstruction.

Mechanical Obstruction

Small Bowel

The only two indications for the use of water-soluble contrast media in studying possible small bowel obstructions are (1) in very ill patients where it makes a significant difference if the diagnosis is made early and (2) in patients who have had postoperative adynamic ileus with a partial small bowel obstruction. The administration of water-soluble contrast media in these patients may actually be therapeutic; by dilating the bowel proximally to a partial obstruction there is relief of the kinking and obstruction by the induction of peristalsis.

The best diagnostic results are obtained with complete and partial obstructions in the jejunum (Fig. 5-2).[15] For examinations in patients with partial small bowel obstruction high in the jejunum, films should be obtained at short and regular intervals; preferably, these films should be combined with multiple compression

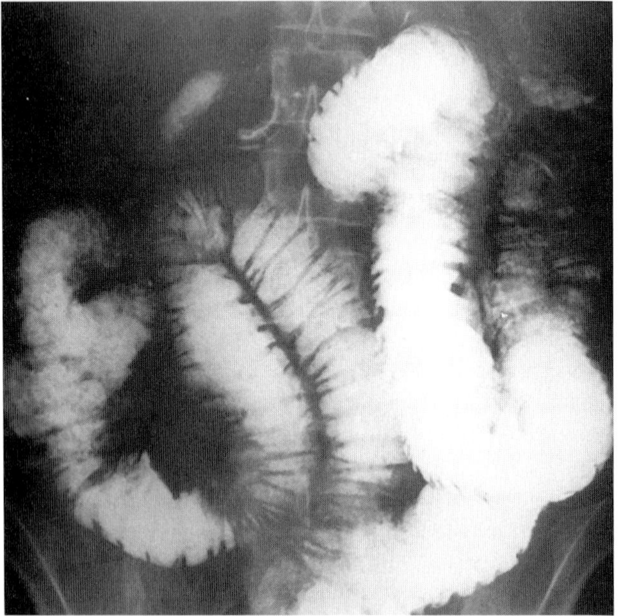

Figure 5-2 High-grade obstruction in jejunum due to adhesions is visualized by water-soluble contrast material.

spot films. Partial obstruction may be missed if early films are not observed as the contrast medium may move very rapidly.

In the proximal and mid-ileum, results are less specific but still conclusive as to the presence or absence of partial or total obstruction.[16] If specific diagnoses are needed, enteroclysis or small bowel enema is the preferred method. At the end of this examination, the barium may be aspirated through the nasogastric tube that had been placed into the stomach and via the guidewire manipulated into the proximal jejunum. For the enteroclysis examination, barium and methylcellulose are administered by an infusion pump.

For distal ileal obstructions, it is recommended that water-soluble, iodine-containing contrast materials be used only if retrograde small bowel enema via the colon and ileocecal valve is unsuccessful. The retrograde small bowel filling, as introduced by Miller,[17] is the fastest and most accurate method for delineation of obstructions in the last 4 to 5 feet of small bowel.

Closed Loop Small Bowel Obstruction

This condition combines obstruction and impairment of blood supply of the involved segment and is a surgical emergency. It usually produces a profound adynamic ileus. If the involved loop becomes nonviable, contrast material will not move out of the stomach or duodenum.[18] If the loop is still viable, barium sulfate given orally will progress extremely slowly, whereas water-soluble contrast media may rapidly reach the point of obstruction and outline the fluid-filled closed loop. Water-soluble contrast media are ideal for the study of closed loop small bowel obstruction, particularly in the jejunum or proximal ileum.[19]

Small Bowel Study in the Presence of Large Bowel Obstruction

If an obstruction in the colon is demonstrated by barium enema in a patient with advanced cancer involving the large and small bowel, it may be important for the surgeon to determine whether there is a small bowel obstruction in addition to the large bowel obstruction. This information is necessary in order to place a colostomy. In those instances, ingestion of water-soluble contrast medium with multiple films and detailed spot films is indicated. The delineation of abnormalities is not as good as with barium sulfate, but the patient avoids the danger of inspissation of barium proximally to a large bowel obstruction (Fig. 5-3). However, caution is advised because water-soluble iodine-containing contrast media irritate the bowel mucosa and, if left in contact with them for a long time because of complete distal obstruction, will produce a severe inflammatory reaction.[20]

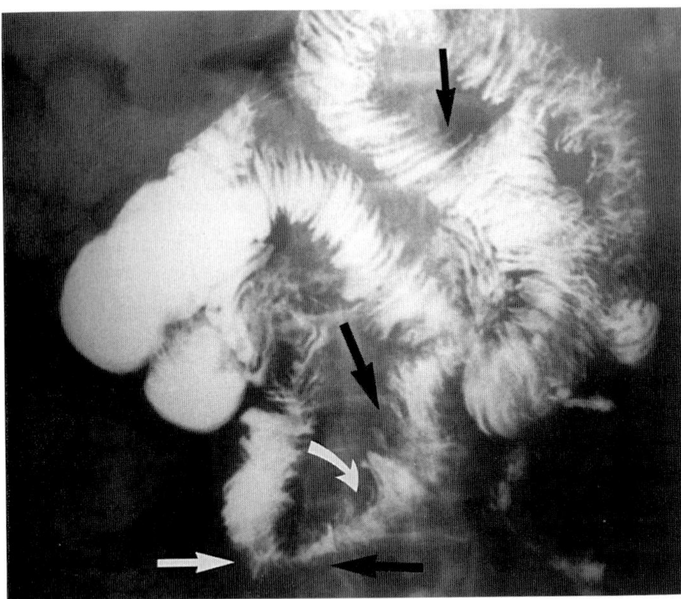

Figure 5-3 Water-soluble contrast material was used in this patient with near-complete obstruction at splenic flexure (see Fig. 5-4) to avoid inspissation of barium proximal to obstruction. Multiple areas with spiculation (*arrows*) are also seen due to peritoneal seeding of tumor.

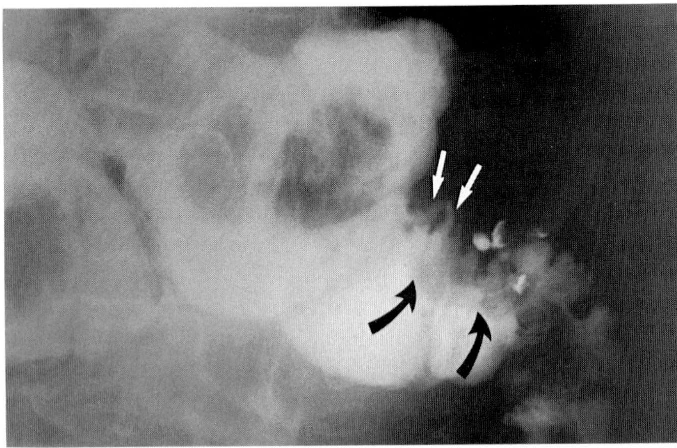

Figure 5-4 A Hypaque enema was performed in this patient with suspected near-complete obstruction of the splenic flexure due to metastatic disease from a gastric carcinoma. Spiculation (*arrows*) and mass impression (*curved arrows*) are clearly outlined. Use of Hypaque was necessary to prevent barium inspissation proximal to the obstruction.

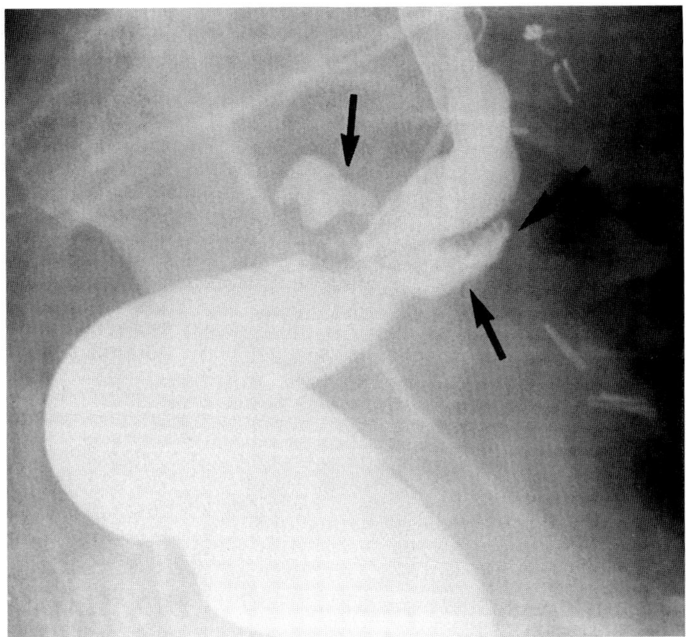

Figure 5-5 A Hypaque enema shows two small walled-off abscesses (*arrows*) at the site of a recto-sigmoid anastomosis in a patient with resection for rectosigmoid carcinoma.

Large Bowel

Water-soluble contrast media enema should be reserved for suspected perforations, such as with diverticulitis, perforated carcinoma, leaking anastomosis, an abdominal stab wound communicating with the colon, or for fecal impaction above the rectum and chronic obstipation (pseudo-obstruction). If a near-complete obstruction is present and the colon proximal to the abnormal area needs to be demonstrated, Hypaque should be used to avoid barium inspissation (Fig. 5-4). For demonstration of leaks from colonic diverticulitis, it is advisable to initiate the enema study with a 20% solution of Hypaque (Fig. 5-5). If it is ascertained that there is no communication between the abscess and the free peritoneal cavity, the use of barium is advocated in order to show anatomical detail better. Barium in a walled-off abscess communicating with the colon is not harmful. For treatment of fecal impaction or obstipation, it is important to fill the entire colon with a similar Hypaque solution, delay evacuation, and obtain postevaluation films to assess the degree of colonic emptying.

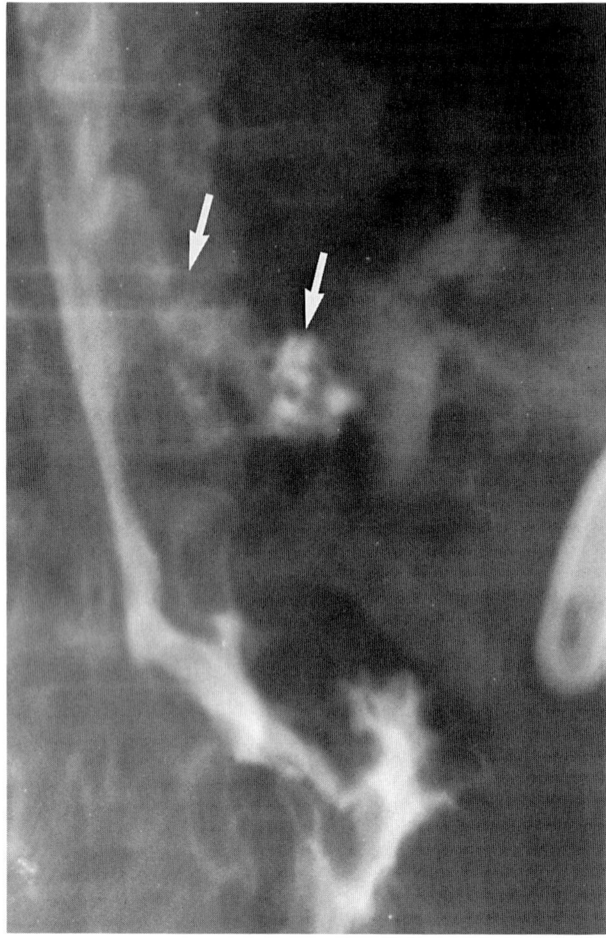

Figure 5-6 Extravasation of water-soluble contrast material from an esophageal perforation due to bouginage. The dilation was performed for a lye stricture. The leak into the mediastinum is well shown (*arrows*).

Leakage from Gastrointestinal Tract

Water-soluble contrast media must be used whenever a leak from the GI tract into the peritoneal cavity is suspected,[8,21] as barium sulfate produces adhesions and aggravates the peritonitis. For suspected esophageal perforation, a small amount of water-soluble contrast material is administered and more added, if

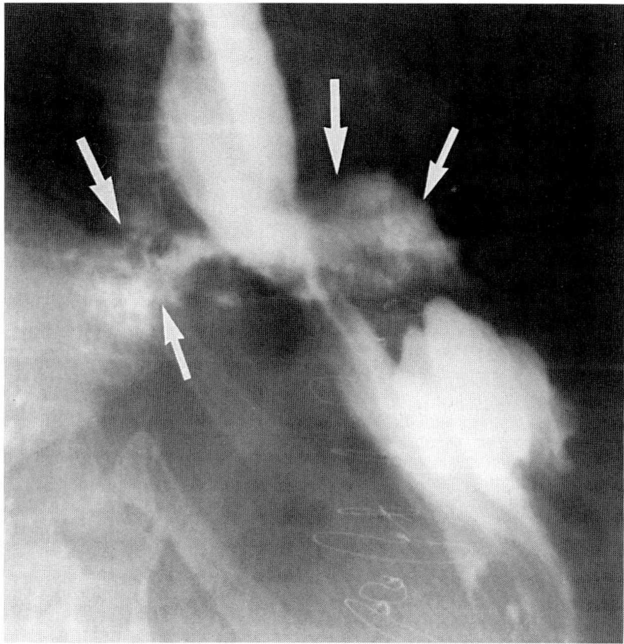

Figure 5-7 Spontaneous leakage of contrast material caused by perforation in the area of a stricture near the gastroesophageal junction. Water-soluble contrast material was used to demonstrate the extravasation (*arrows*) in the absence of a communication between esophagus and lung.

necessary, after leakage into the mediastinum alone is demonstrated (Figs. 5-6 and 5-7). If leakage into the lungs occurs, administration of water-soluble contrast material should be discontinued and barium cautiously used if needed for further evaluation because the hyperosmolarity of the water-soluble medium produces lung edema. Demonstration of perforated ulcers (Figs. 5-8, 5-9, and 5-10), anastomotic leaks, internal fistulas (Fig. 5-11*A-C*), and perforations caused by trauma are part of everyday practice in departments of radiology throughout the Western world.

Stab Wounds, Sinuses, and Fistulas

The extent of stab wounds, sinuses, and fistulas can be shown by introducing a soft rubber catheter into the skin defect and injecting contrast media under fluoroscopic control. Demonstration of penetration of the peritoneal cavity after

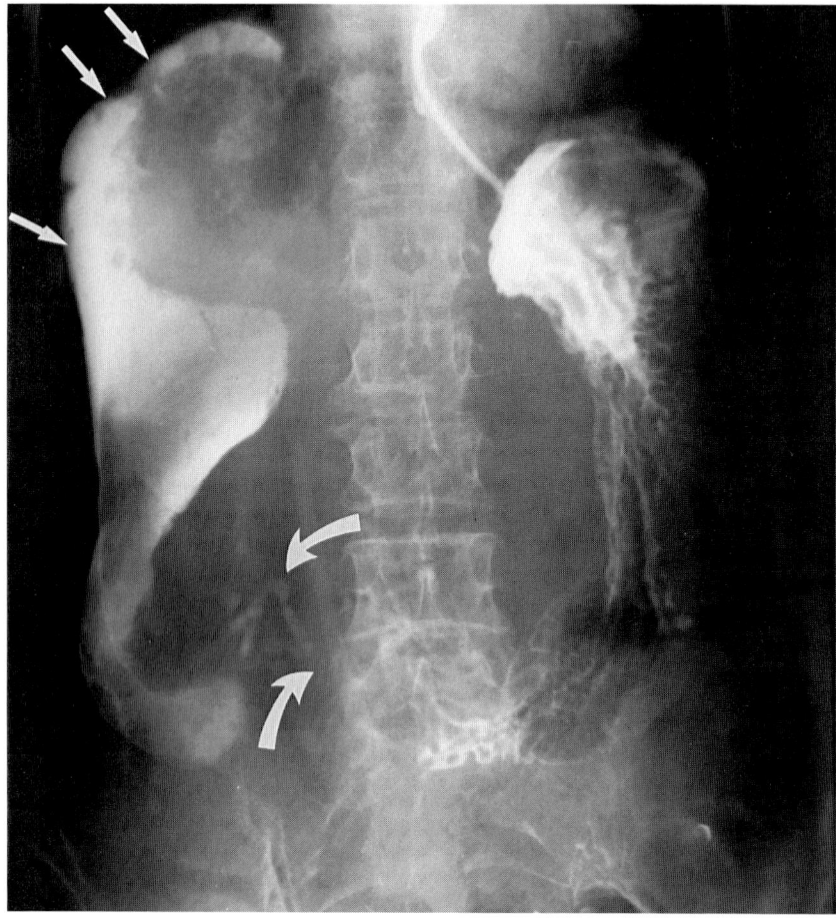

Figure 5-8 Perforated duodenal ulcer. Water-soluble contrast material is noted in the right flank and surrounding the liver (*straight arrows*). Vicarious excretion of contrast material through the kidneys (*curved arrows*) is also identified.

stabbing has important surgical implications.[22] Lack of communication with the peritoneal cavity may save unnecessary laparotomies.

Sinuses and fistulas should be demonstrated by injection through catheters. As the contrast medium is absorbed from the soft tissues, no residual opacities remain to confuse peroral studies. Such studies must be performed under fluoroscopic control in order to obtain the best projections and prevent injection of too much contrast material. (Fistulography is covered in Chapter 29.)

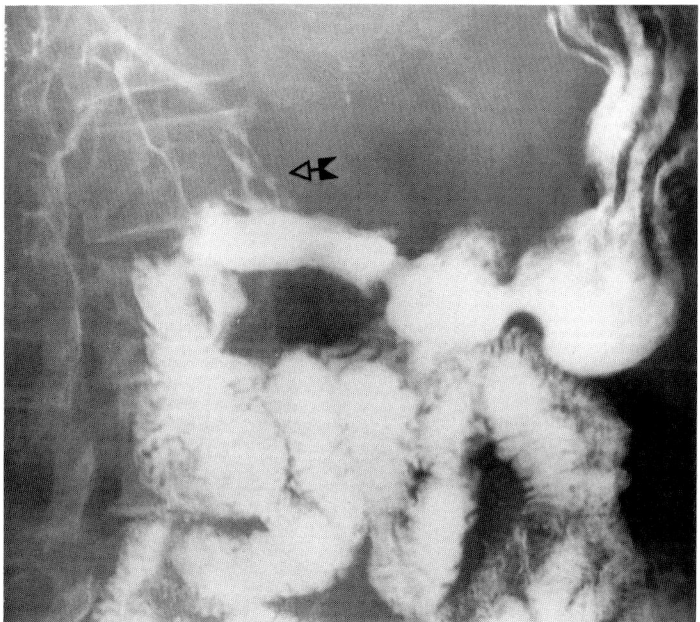

Figure 5-9 Water-soluble contrast material shows a fistula (*arrow*) between duodenum and common bile duct caused by a perforated duodenal ulcer.

Visualization of Gastrointestinal Tract for Computed Tomography (CT)

Optimal opacification of the bowel on CT examinations of the abdomen and pelvis is very important in order to avoid confusion of fluid-filled bowel loops with intraabdominal masses or adenopathy (Fig. 5-12). Furthermore, distention of the bowel wall with air and/or water-soluble contrast material enables correct assessment of bowel wall thickness caused by neoplastic, inflammatory, or ischemic processes. Although diluted barium mixtures are advocated by some radiologists for CT demonstration of the GI tract, we find that a dilute 1%–2% solution of Hypaque or Gastrografin gives the best results. In our CT experience, the water-soluble contrast material shows homogeneous opacification of the intestinal tract, whereas the barium products available for CT frequently flocculate or become too dense in the colon due to water absorption.

Barium sulfate in the small and large bowel renders subsequent imaging with CT difficult or impossible because of streak artifacts. Whenever a contrast study of the small bowel or colon is needed before a CT examination, a water-soluble

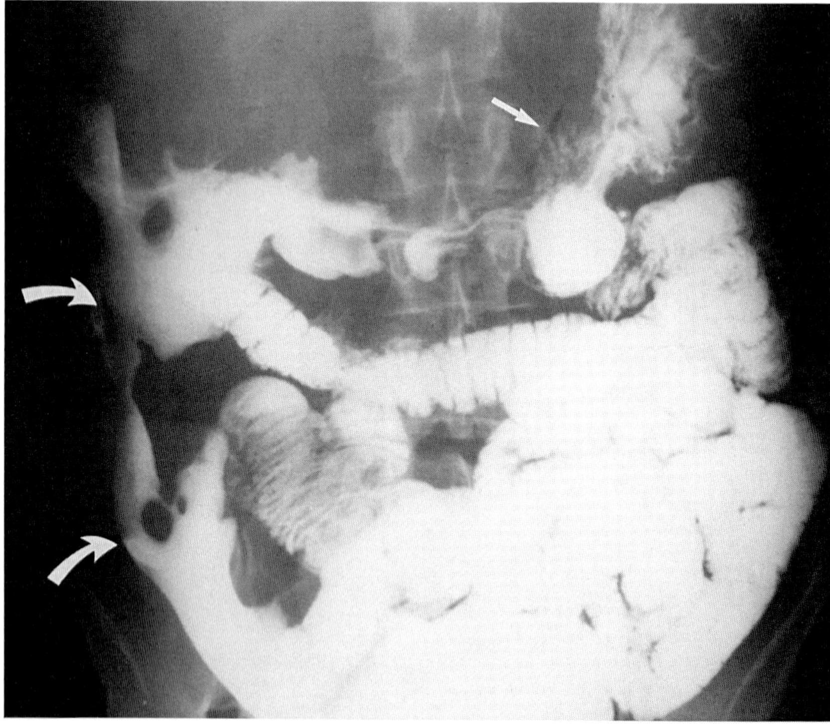

Figure 5-10 Perforated gastric ulcer shows extravasation of water-soluble material into the lesser sac (*straight arrows*) and into the left flank (*curved arrows*).

iodine-containing contrast medium should be used. These water-soluble media administered orally for conventional radiographic examinations are still too concentrated for CT, but are less disturbing than barium, which becomes progressively more concentrated. Usually, a single water enema sufficiently clears and dilutes the dense contrast material so that a CT examination becomes possible on the same day (see Chapter 6 for a detailed discussion on the use of oral contrast material for CT).

DANGERS AND CONTRAINDICATIONS

As mentioned earlier, water-soluble contrast media draw interstitial fluid into the intestinal lumen, thus speeding up the examination but diminishing demonstration of detail because of dilution. Progressive dilution of the contrast medium as it approaches the site of obstruction, particularly in the distal small bowel, further diminishes detail. If hypovolemia is present, oral administration of water-soluble media may result in shock and possibly death, particularly in infants, children, the aged, and the very ill. The electrolyte status of all patients who are hypovolemic, particularly those enumerated above, should be followed carefully. Should an oral study with water-soluble contrast be necessary in this type of patient, fluid and electrolyte replacement therapy must be instituted. Rarely, such systemic reactions as urticaria may be seen in hypersensitive individuals.[21]

Water-soluble contrast media can be used in the study of possible esophageal perforations. However, if an esophagobronchial fistula exists, serious complications may result from aspiration of the water-soluble hyperosmolar contrast medium.[23] Pulmonary edema occurs, and if both lungs are drowned the results may be fatal. Because such fistulas usually present clinical manifestations of cough with ingestion of liquids, this clinical symptom should preclude the use of water-soluble contrast media. Small amounts of barium sulfate carefully monitored by fluoroscopy would be the approach of choice. For the same reasons,

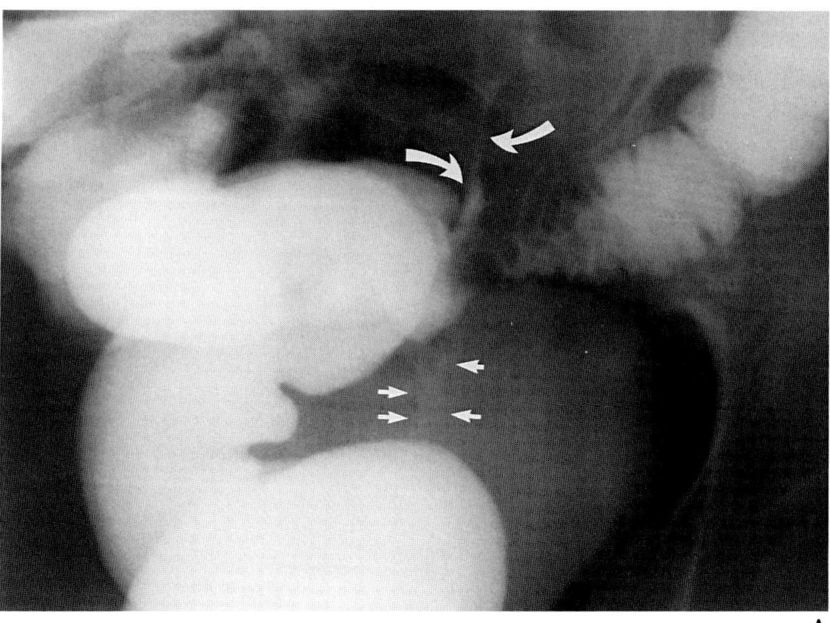

Figure 5-11 *(A)* Hypaque enema demonstrates colovaginal fistula (*straight arrows*) and sinus tract (*curved arrows*) originating from sigmoid anastomosis in a patient following surgery for diverticulitis.

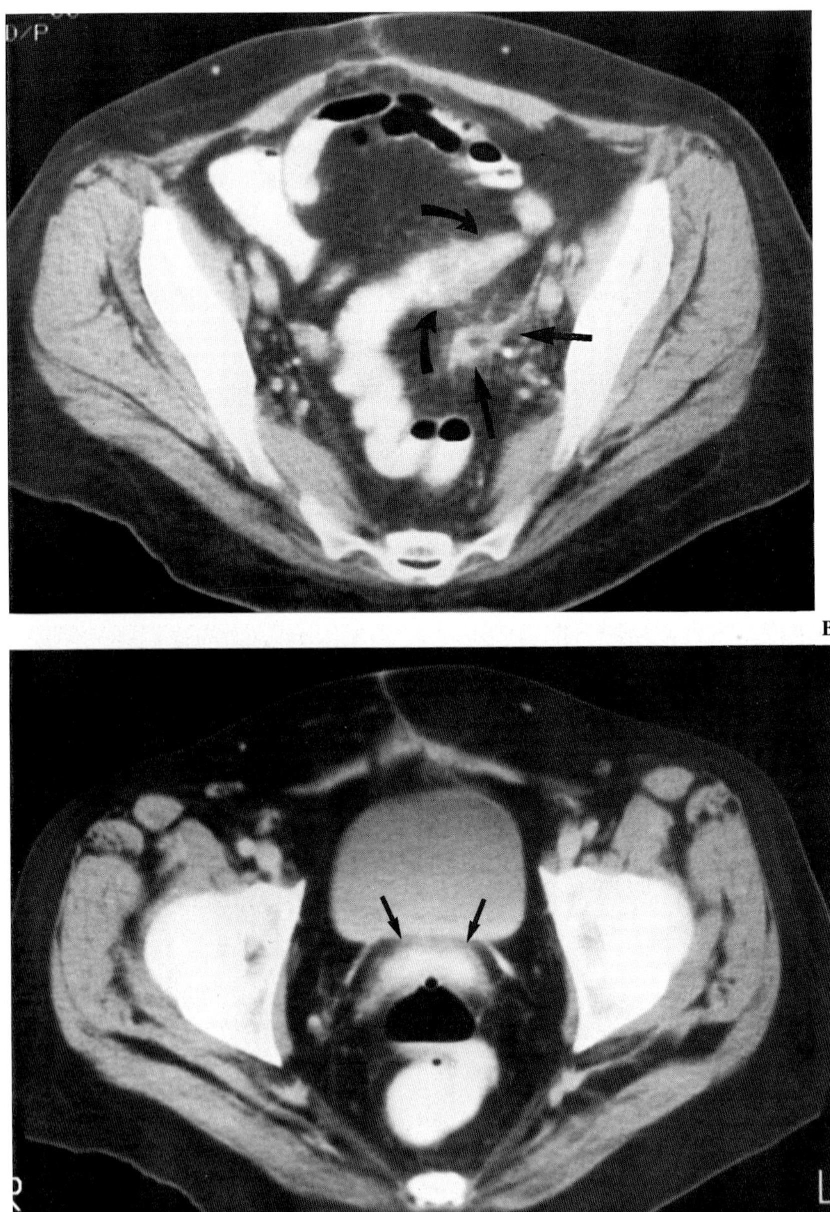

Figure 5-11 *(B)* Same patient as in Figure 5-11*A*. Irregular inflammatory reaction (*straight arrows*) is noted in the pericolonic areas, and an irregular and thickened sigmoid (*curved arrows*) is identified. *(C)* Same patient as in Fig. 5-11*A*. Water-soluble contrast material is present in the vagina (*arrows*).

it is advisable not to use water-soluble contrast media in the study of esophageal atresia, with or without tracheoesophageal fistulas.

THERAPEUTIC USES

The use of water-soluble contrast media in the presence of mild partial obstruction and postoperative adynamic ileus has been mentioned earlier. This approach is used by many surgeons, as it is believed to relieve the obstruction by stimulating peristalsis and freeing distended adynamic small bowel loops from adhesions. Another therapeutic method is the use of water-soluble contrast media by enema for relieving fecal impaction in severely constipated patients. This method is often used at the University of California, San Francisco, in patients on the renal transplant service. It has also been described for the treatment of meconium ileus and meconium plug syndrome.[24] For these latter cases, careful monitoring of the patient's electrolytes is necessary.

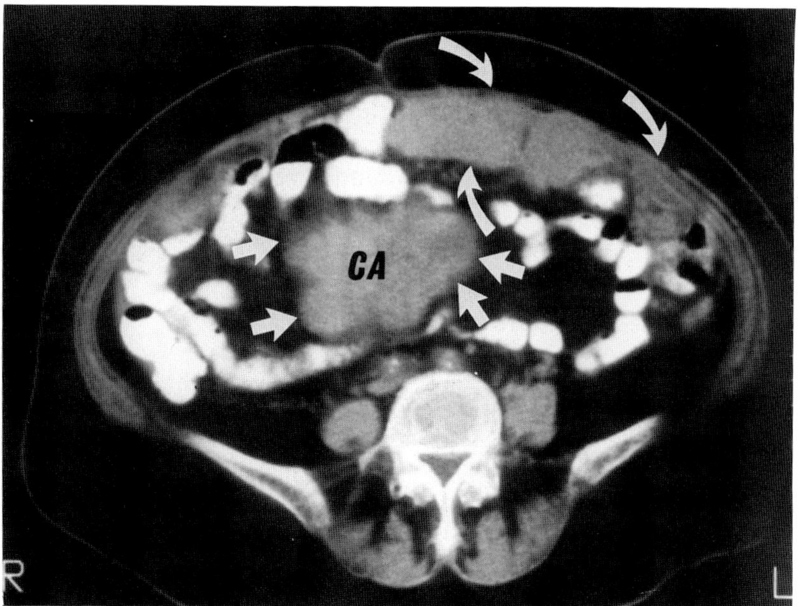

Figure 5-12A Optimal opacification of large and small bowel by a 1% solution of Gastrografin permits a large soft tissue mass in the midabdomen (*straight arrows*) and multiple small soft tissue masses adherent to the anterior abdominal wall (*curved arrows*) to be distinguished from fluid-filled bowel loops. This patient had a large cystadenocarcinoma of the ovary with multiple serosal implants throughout the abdomen.

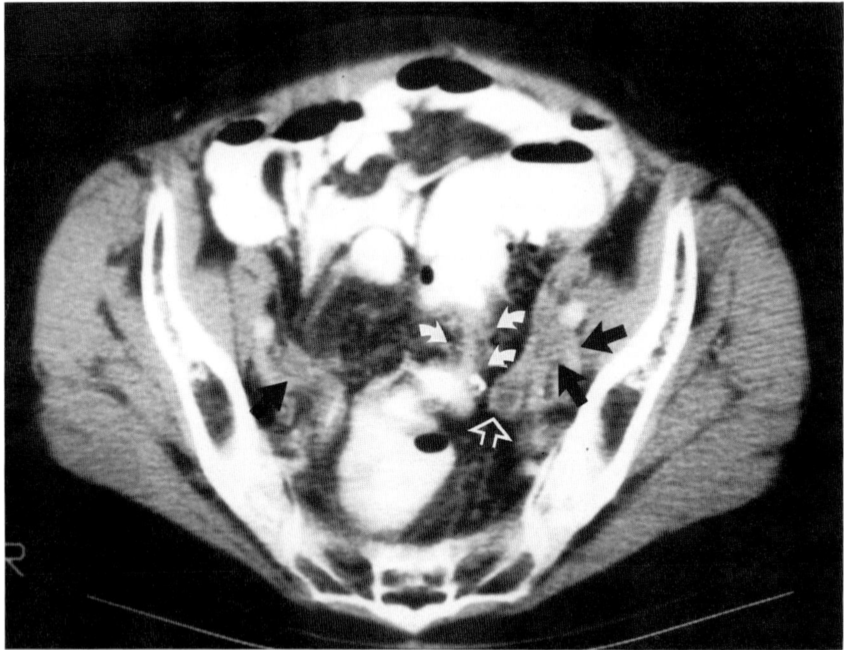

Figure 5-12B Optimal distention of the colon and small bowel due to rectal and oral administration of water-soluble contrast material is demonstrated in this patient. A stenosed area in the midsigmoid colon (*curved arrows*) is clearly outlined, and bilateral adenopathy in the iliac chains (*straight arrows*) and left hydronephrosis (*open arrow*) are identified. This patient suffered from a recurrent fallopian tube carcinoma with encasement of the midsigmoid colon.

URINARY TRACT OPACIFICATION

Although the urinary bladder will opacify regularly in the presence of perforation,[25] it will also opacify frequently from absorption of the iodinated water-soluble contrast media from the GI tract. In our experience, the bladders of about 10% of adults ingesting water-soluble contrast media opacify in the absence of perforation. In children, urinary bladder opacification without perforation of the gut is three to five times more common than in adults. However, opacification of renal calices and ureters is generally a reliable sign of perforation of viscus and is not seen otherwise.[21]

USE OF MIXTURES

Goldstein and associates[26] have reported that the transit time of a mixture of barium sulfate and Gastrografin through the small bowel is faster than that of barium alone. In their experience, 16 fluid ounces of barium mixed with 10 ml of Gastrografin produced the most efficacious mixture. It was reported that such a contrast medium gave excellent radiographic detail. These authors recommend the use of this mixture, as it combines the advantages of both media and overcomes the objections to either. Although the transit time of this mixture[27] is slower than that of iodinated contrast media alone, it is superior to barium sulfate. These mixtures should not be used in patients with perforation, as the contraindications to barium remain the same.

In our clinical experience at the University of California, San Francisco, and in experimental animals,[27] the combination of barium sulfate and the water-soluble contrast media disassociate in the presence of distal small bowel obstruction; the water-soluble media pass more rapidly and become diluted, whereas barium sulfate remains behind and precipitates out of the suspension, progressing only very slowly.

REFERENCES

1. Canada WJ: Use of Urokon (sodium-3-acetylamino-2,4,6,-triiodobenzoate) in roentgen study of the gastrointestinal tract. *Radiology* 1955;64:867–873.

2. Davis LA, Huang KC, Pirkey EL: Water-soluble nonabsorbable radiopaque mediums in gastrointestinal examination. *JAMA* 1956;160:373–375.

3. Jacobson HG, Shapiro JH, Poppel MH: Oral Renografin 76 percent: a contrast medium for examination of the gastrointestinal tract. *Am J Roentgenol* 1958;80:82–88.

4. Robinson D, Levene JM: Oral Renografin: a new contrast medium for gastrointestinal examinations. *Am J Roentgenol* 1958;80:79–81.

5. Nelson SW, Christoforidis AJ, Roenigk WJ: Dangers and fallibilities of iodinated radiopaque media in obstruction of the small bowel. *Am J Surg* 1965;109:546–559.

6. Harris PD, Neuhauser EBD, Gerth R: The osmotic effect of water-soluble contrast media on circulating plasma volume. *Am J Roentgenol* 1964;91:694–698.

7. Vest B, Margulis AR: Roentgen diagnosis of postoperative ileus-obstruction. *Surg Gynecol Obstet* 1962;115:421–427.

8. Jacobson G, Berne CJ, Meyers HI, et al: Examination of patients with suspected perforated ulcer using a water-soluble contrast medium. *Am J Roentgenol* 1961;86:37–49.

9. Meyers HI, Jacobson G: The use of water-soluble contrast medium in suspected perforated peptic ulcer. *Radiol Clin N Am* 1964;2:55–69.

10. McAlister WH, Shackleford GD, Kissane J: The histologic effects of some iodine-containing contrast media on the rat peritoneal cavity. *Radiology* 1972;105:581–582.

11. Lutzger LG, Factor SM: Effects of some water-soluble contrast media on the colonic mucosa. *Radiology* 1976;118:545–548.

12. Wood BP, Katzberg RW, Ryan DH, et al: Diatrizoate enemas: facts and fallacies of colonic toxicity. *Radiology* 1978;126:441–444.

13. Cohen M, Wilbur LS, Smith JA, et al: The use of metrizamide to visualize the gastrointestinal tract in children: a preliminary report. *Clin Radiol* 1980;31:635–641.

14. Bayliss WM, Starling EH: The movements and innervation of the small intestine. *J Physiol* 1899;24:99–143.

15. Shehadi WH: Orally administered water-soluble iodinated contrast media. *Am J Roentgenol* 1960;83:933–941.

16. Nelson SW, Christoforidis AJ, Roenigk WJ: Barium suspensions vs. water-soluble iodine compounds in the study of obstruction of the small bowel: an experimental study of physiologic characteristics and radiographic value. *Radiology* 1963;80:252–254.

17. Miller RE: Complete reflux small bowel examination. *Radiology* 1965;84:457–463.

18. Vest B: Roentgenographic diagnosis of strangulating closed-loop obstruction of the small intestine. *Surg Gynecol Obstet* 1962;115:561–567.

19. Margulis AR: Use of iodinated water-soluble contrast agents in acute gastrointestinal disease, in Margulis AR, Burhenne JH (eds): *Alimentary Tract Roentgenology*, ed 2, vol 1. St. Louis, CV Mosby, 1973, pp 271–280.

20. Creteur V, Douglas D, Galante M, et al: Inflammatory colonic changes produced by contrast material. *Radiology* 1983;147:77–78.

21. Margulis AR: Contrast media: the present status of water-soluble iodine containing material in the examination of acute abdominal disease. *Calif Med* 1969;110:193–199.

22. Steichen FM, Pearlman DM, Dargan EL, et al: Wounds of the abdomen: radiographic diagnosis of intraperitoneal penetration. *Ann Surg* 1967;165:77–82.

23. Reich SB: Production of pulmonary edema by aspiration of water-soluble nonabsorbable contrast media. *Radiology* 1969;92:367–370.

24. Frech RS, McAlister WH, Ternberg J, et al: Meconium ileus relieved by 40 percent water-soluble contrast enemas. *Radiology* 1970;94:341–342.

25. Mori PA, Barrett HA: A sign of intestinal perforation. *Radiology* 1962;79:401–407.

26. Goldstein HM, Poole GJ, Rosenquist CJ, et al: Comparison of methods for acceleration of small intestinal radiographic examination. *Radiology* 1971;98:519–523.

27. Noonan CD, Margulis AR: Small bowel transit time of water-soluble iodinated contrast medium and barium sulfate in cats with simulated surgical acute abdomen. *Am J Roentgenol* 1970; 110:334–337.

Chapter 6

Agents in Computed Tomography

Jovitas Skucas

Computed tomography (CT), which was introduced in the 1970s, provided the ability to differentiate fluid and various soft tissue densities from each other and was a significant advance over previous radiographic techniques. However, the variable location of bowel, the degree of bowel distention, and the amount of fluid within the lumen needed to be identified because loops of bowel can mimic an adjacent abscess, hematoma, fluid-filled cyst, or neoplasm. Especially when dilated, a loop of bowel can mimic a cyst; when collapsed, that same loop has an attenuation similar to liver. Therefore, early in the development of body CT it became apparent that a major factor in the accurate evaluation of various soft tissue structures of the abdomen is the ability to identify bowel clearly.[1–5] In patients with little intra-abdominal fat, the tissue planes between normal structures may be difficult to identify; in these patients, contrast within bowel aids identification of adjacent lymph nodes, blood vessels, and other structures.

Motion artifacts are considerably reduced with present-day scanners. However, contrast artifacts generated at the interface between two structures with large contrast differences are still encountered. These artifacts are most prominent at the interface between low-contrast bowel gas and any positive contrast media within bowel. With the patient supine, the most common site for these artifacts is in the stomach.

NEGATIVE CONTRAST AGENTS

A number of substances can be used to opacify the bowel. The cheapest and quite often the easiest method is to use residual gas within bowel as a marker. Under some conditions, air can be introduced just before scanning.[6] Excessive amounts of gas, when used with early generation scanners, can result in streak artifacts because of the large contrast differences. When significant amounts of gas are present, wide window settings are needed to ensure optimal soft tissue visualization.

Rather than using the highly negative contrast air, a mildly negative contrast, such as fat, can be introduced into the bowel lumen.[7] A number of substances high in fat content, such as mineral oil, have been tried but with limited success. The low contrast differential between fat and adjacent soft tissues essentially eliminates streak artifacts. In dogs, a corn oil emulsion administered in divided doses led to consistent identification of bowel lumen and visualization of the bowel wall.[8] Subsequent use in humans of such a flavored 12.5% corn oil emulsion resulted in consistent discrimination of bowel wall.[9] In this study, the lumen density ranged between −20 and +15 CT units, whereas the wall density was +30 to +55 CT units before intravenous contrast and +60 to +100 CT units after contrast injection. The authors concluded that the difference between the bowel wall density and the intraluminal corn oil emulsion densities was optimal and superior to the results obtained with more conventional high contrast agents.[9] One potential source of confusion would be with structures having a similar CT density, such as cysts and abscesses.

NEUTRAL CONTRAST AGENTS

Water is readily available and can be used as a contrast agent. It is of most use in the stomach. Distention of the stomach with approximately 400 ml of water can improve visualization of the gastric wall.[10] This technique is useful in patients where the initial scans are inconclusive. Water is least useful in the small bowel where a distinction between a water-filled loop of bowel and a cyst or abscess is difficult.

POSITIVE CONTRAST AGENTS

Among positive contrast agents both the iodinated water-soluble materials and various barium sulfate suspensions have been used. Ideally, sufficient x-ray absorption should occur with the contrast medium so that it can be readily differentiated from surrounding structures, but the absorption should not be so great that streak artifacts are produced. Full-strength suspensions as used in conventional radiology cannot be readily applied; the resultant high contrast differentials lead to unacceptable streak artifacts in most patients. These agents must be diluted.

Calcium phosphate is inert, is not absorbed from the gastrointestinal (GI) tract, and can be readily identified by CT. A dose of 1 g orally 3 times per day for the

2 days before the examination has been proposed.[11] Such dosage is believed to label the colon in most patients.

Water-Soluble Contrast Agents

Theoretically, any of the ionic and nonionic contrast media, if diluted, can be used to opacify bowel. The bad taste of these media, however, has limited their acceptance by some patients. If instilled through a nasogastric tube or by enema they can produce adequate bowel opacification.

Although these contrast media are dilute solutions, precipitation can occur at the low pH found in the stomach and result in artifacts.[12] Adding sodium bicarbonate to the Gastrografin solutions can raise the pH sufficiently so that precipitation is essentially eliminated. One study found that adding 9 g of sodium bicarbonate to a 2% Gastrografin-flavoring mixture produced sufficient buffering to eliminate precipitation in most patients.[12]

Water-soluble contrast media have been available for examination of the GI tract since 1955.[13] If the patient is to drink the contrast, those media containing flavoring are preferred. In the United States currently two such products are available: Gastrografin, a sodium and meglumine diatrizoate solution containing various flavoring agents, and Oral Hypaque, which is a flavored sodium diatrizoate solution. Both can be diluted with water. A fruit juice can be added for extra flavoring.[12,14]

There has been controversy about the optimum amount of dilution; initially 10% solutions were proposed,[2] although currently more dilute solutions are generally used. Some prefer a 5% solution of Gastrografin,[14] whereas others use a 2.2%[15] to 2.8%[16] solution. With such dilutions, the solutions are hypo-osmotic. Especially with the earlier CT equipment, many investigators also injected an intravascular bowel hypoperistaltic agent to decrease artifacts from bowel peristalsis.

The nonionic contrast agents do not have any theoretical advantage over the ionic agents at the low dilutions that are used. In addition, no flavored nonionic contrast agents especially applicable to CT are currently available on the market.

Even when diluted, there is poor patient acceptance of these solutions. Many children and patients undergoing cancer therapy cannot tolerate these contrast media because of nausea and vomiting. One group in Canada found that Telebrix-38 (Andre Guerbet Laboratories), when diluted to a 2% solution, was better accepted by patients, especially children.[14] Telebrix has also been used for CT scanning in Holland.[1,3] One side effect of Gastrografin is that a significant number of patients develop diarrhea.[15]

Barium Sulfate

Because of the relatively poor patient acceptance of dilute iodinated solutions, several manufacturers introduced dilute barium sulfate suspension. Full-strength barium sulfate suspensions obviously cannot be used with CT because of the high contrast differentials and resultant streak artifacts.[17] Initially, an attempt was made to dilute the barium suspensions with water or saline; however, once the patient ingested these contrast media, the barium sulfate settled out into the most dependent portions of bowel. As a result, the nondependent loops of bowel tended to lack sufficient contrast for visualization, and the barium in dependent loops was so dense that streak artifacts occurred.

The ideal barium CT contrast agent should contain sufficient barium to be readily identified but not so much that streak artifacts are produced. It should contain significant amounts of suspending and antiflocculating agents so that the dilute barium preparation will stay in suspension. Simple dilution of conventional barium sulfate contrast agents thus is not applicable; rather, a low concentration of barium sulfate and high concentration of additives must be achieved during the manufacturing process.

In 1980, an Australian group developed a 6% weight-to-weight (w/w) suspension of barium sulfate to be used in CT examinations.[17] Because of additives, it had an osmotic pressure equivalent to 2.0% of a sodium chloride solution and provided satisfactory filling of the small bowel. Since then, the emphasis has been on using even lower concentrations of barium. In 1984, a Japanese study found that a preparation containing 2–3 μm barium sulfate particles (in a 1% suspension), 1.5% sodium carboxymethylcellulose, and 0.5% polyethylene glycol was useful for bowel opacification during CT scanning.[18] Similar products containing 1–2% W/W barium sulfate are currently available in most developed countries.

Clinical trials have shown better patient acceptance and fewer side effects with the barium products than with Gastrografin.[15,16,19] Both groups of contrast agents have similar bowel opacification and result in a similar degree of artifacts.[15,19–21] One study in England, however, found that the barium contrast was better than Gastrografin in labeling the duodenum.[16] As expected with the low contrast media concentrations used in CT, the barium preparations do not coat the intestinal mucosa[20]; rather, their function as a bowel marker is achieved simply by passive filling of the bowel lumen.

Although the water-soluble iodinated contrast media cost more than commercial barium preparations, at the low concentrations used in identifying bowel in CT examinations, the relative cost per examination for the two types of contrast media is comparable.[20]

Currently, most investigators prefer barium sulfate suspensions whenever a positive contrast medium is desired. The exception is in patients with suspected GI perforation, such as recent abdominal trauma, acute peritoneal irritation, or recent surgery where either the water-soluble contrast media, air, or no contrast at all is used.

CLINICAL APPLICATION

Esophagus

Identifying the lumen of the esophagus is useful in thoracic CT scanning. At times, air trapped in the esophagus will identify this structure. Conventional dilute water-soluble or barium contrast agents are not satisfactory for use in the esophagus. Invariably, the contrast agent drains promptly into the stomach, leaving little if any residue in the esophagus. The patient can be instructed to drink small amounts just before each scan, a procedure that is technically not satisfactory. A barium coating of the esophagus can be obtained with full-strength barium products but at the expense of producing possible artifacts.

A high-viscosity, low-barium-sulfate-concentration paste, specifically developed for CT, is available for use in the esophagus. It can outline the esophageal lumen in most patients.[22] The prolonged barium residue within the esophageal lumen presumably is caused by the high viscosity of the paste.

Stomach and Small Bowel

Various contrast ingestion schemes have been tried. Having the patient drink contrast from 1 hour to just before the examination results in opacification of the stomach and duodenum, but in some patients this leads to poor or no filling of the distal small bowel and colon. If the examination is being performed primarily to study the stomach, 400–500 ml of a more dilute contrast than used for the small bowel may be adequate, such as a 0.5%–1.0% barium suspension or slightly higher concentration iodine solution.

For a screening examination of the abdomen, the entire small bowel should be opacified. Because different parts of the bowel have different transit times and degrees of peristaltic activity, only varying segments of bowel tend to fill at any one time. Duodenal intubation and infusion of contrast directly into the small bowel (enteroclysis) have been proposed,[1] but this technique is not currently widely practiced. In general, the small bowel transit time is shorter in ambulatory patients.

Better distal bowel filling can be obtained if the contrast is given several hours before the scan. One useful sequence is to have the patient drink approximately 500 cc of diluted contrast several hours before scanning and a similar or larger amount immediately before the scan. The initial contrast should outline the small bowel and the proximal part of the colon, whereas the contrast ingested just before the scan outlines the stomach and duodenum. If there is a delay in scanning the upper abdomen, such as when scanning of the chest or pelvis is performed first, it is helpful if additional contrast is ingested just before the upper abdominal scans.

Some investigators distend the stomach with air or gas.[6] An effervescent agent, similar to that used in double-contrast upper GI examinations, can be administered to the patient just before scanning the upper abdomen. This method enables good visualization of the gastric wall and delineation of the abdominal viscera.[6] The patient is positioned in appropriate positions to evaluate suspicious regions. Generally, the first part of the duodenum is outlined by gas when the patient is supine. If visualization of the duodenal sweep is desired, the patient can be turned to a left-side-down decubitus position.

Colon

If scanning of the pelvic structures is to be performed, some radiologists have the patient drink the contrast agent the evening before the scan; residual contrast in the colon the next day can provide adequate visualization. Such schemes include having the patient drink 20–30 ml of full-strength Gastrografin[23] or 500–600 ml of a dilute barium suspension[24] the evening before the examination. Mitchell et al. compared having the patient drink 20 ml of full-strength Gastrografin, 30 ml of full-strength Gastrografin, and 600 ml of dilute barium the evening before the study; they found greatest opacification of the rectosigmoid in those patients who drank 30 ml of full-strength Gastrografin.[24] Only 46% of the patients who drank the dilute barium had opacification of the rectosigmoid. The better opacification with the full-strength Gastrografin may be due, in part, to its hypertonic state and to its ability to stimulate peristalsis.

The peroral techniques can opacify, but not distend the colon. Distention can be achieved by administering an enema of a dilute positive contrast medium[25] or insufflation of the colon with air.[26] Such colonic distention allows better control of the amount of contrast used and the degree of colon filling. Air has a higher contrast than the positive contrast media and may aid in evaluating bowel wall thickness.[27] Whether air or a dilute positive contrast medium is superior is still controversial. To opacify the rectosigmoid, 150–250 ml of positive contrast enema is generally sufficient. The amount of air insufflated is individualized and is generally based on patient discomfort.[26] Regardless of whether a positive or negative contrast agent is used, better results are achieved if the patient undergoes a colon cleansing regimen beforehand.

One possible use of air as a contrast agent is in the CT detection of colonic polyps.[28] A colon cleansing regimen was administered before the scan. The examination was performed by rectal insufflation of air, and a digital CT scan of the abdomen was obtained, followed by sequential contiguous thick sections. The authors then magnified selected CT sections and reconstructed them for maximum spatial resolution. An ability to resolve polyps 1–10 mm in size was claimed.[28]

REFERENCES

1. Angenent JFC, Schönfeld DHW, Mali WPT, et al: Potential pitfalls in computer tomography. *Diagn Imaging* 1979;48:326–335.

2. Kirkpatrick RH, Wittenberg J, Schaffer DL, et al: Scanning techniques in computed body tomography. *AJR* 1978;130:1069–1075.

3. Ruijs SH: A simple procedure for patient preparation in abdominal CT. *AJR* 1979;133:551–552.

4. Stanley RJ, Sagel SS, Levitt RG: Computed tomography of the body: early trends in application and accuracy of the method. *AJR* 1976;127:53–67.

5. Korobkin M: The use of contrast material in body CT (abstracted). *J Comput Assist Tomogr* 1979;3:556.

6. Megibow AJ, Zerhouni EA: Air contrast techniques, in Megibow AJ, Balthazar EJ (eds): *Computed Tomography of the Gastrointestinal Tract*. St. Louis, CV Mosby, 1986, pp 14–31.

7. Baldwin GN: Computed tomography of the pancreas. Negative contrast medium. *Radiology* 1978;128:827–828.

8. Raptopoulos V, Davis MA, Smith EH: Imaging of the bowel wall. Computed tomography and fat density oral-contrast agent in an animal model. *Invest Radiol* 1986;21:847–850.

9. Raptopoulos V, Davis MA, Davidoff A, et al: Fat-density oral contrast agent for abdominal CT. *Radiology* 1987;164:653–656.

10. Angelelli G, Macarini L, Fratello A: Use of water as an oral contrast agent for CT study of the stomach. *AJR* 1987;149:1084.

11. Kreel L: Contrast media for gastrointestinal examinations with computed tomography, in Felix R, Kazner E, Wegener OH (eds): *Contrast Media in Computed Tomography*. Amsterdam, Excerpta Medica, 1981, pp 271–274.

12. Ball DS, Radecki PD, Friedman AC, et al: Contrast medium precipitation during abdominal CT. *Radiology* 1986;158:258–260.

13. Canada WJ: Use of Urokon (sodium-3-acetylamino-2, 4, 6-triiodobenzoate) in roentgen study of the gastrointestinal tract. *Radiology* 1955;64:867–873.

14. Azouz EM, Hassell P, Nogrady MB, et al: Bowel opacification using "Telebrix 38" for CT scanning. *J Can Assoc Radiol* 1982;33:233–235.

15. Nymen U, Dinnetz G, Andersson I: E-Z-CAT. An oral contrast medium for use in computed tomography of the abdomen. *Acta Radiol Diagn* 1984;25:121–124.

16. Carr DH, Banks LM: Comparison of barium and diatrizoate bowel labelling agents in computed tomography. *Br J Radiol* 1985;58:393–394.

17. Hatfield KD, Segal SD, Tait K: Barium sulfate for abdominal computer assisted tomography. *J Comput Assist Tomogr* 1980;4:570–575.

18. Sako M, Hasegawa M, Watanabe H: A new contrast medium for bowel opacification in abdominal CT scans. *Nippon Igaku Hoshasen Gakkai Zasshi* 1984;44:93–95.

19. Kivisaari L, Kormano M: Comparison of diatrizoate and barium sulfate bowel markers in clinical CT. *Europ J Radiol* 1982;2:33–34.

20. Megibow AJ, Bosniak MA: Dilute barium as a contrast agent for abdominal CT. *AJR* 1980; 134:1273–1274.

21. Chambers SE, Best JJK: A comparison of dilute barium and dilute water-soluble contrast in opacification of the bowel for abdominal computed tomography. *Clin Radiol* 1984;35:463–464.

22. Cayea PD, Seltzer SE: A new barium paste for computed tomography of the esophagus. *J Comput Assist Tomogr* 1985;9:214–216.

23. Cranston PE: Colon opacification by oral water-soluble contrast medium administration the night prior to CT examination. *J Comput Assist Tomogr* 1982;6:413–415.

24. Mitchell DG, Bjorgvinsson E, terMeulen D, et al: Gastrografin versus dilute barium for colonic CT examination: a blind, randomized study. *J Comput Assist Tomogr* 1985;9:451–453.

25. Aronberg DJ: Techniques: Oral contrast material, in Lee JKT, Sagel SS, Stanley RJ (eds): *Computed Body Tomography*. New York, Raven Press, 1983, pp 16–17.

26. Megibow AJ, Zerhouni EA, Hulnick DH, et al: Air insufflation of the colon as an adjunct to computed tomography of the pelvis. *J Comput Assist Tomogr* 1984;8:797–800.

27. Hamlin DJ, Burgener FA, Sischy B: New technique to stage early rectal carcinoma by computed tomography. *Radiology* 1981;141:539–540.

28. Coin CG, Wollett FC, Coin JT, et al: Computerized radiology of the colon: a potential screening technique. *Comput Radiol* 1983;7:215–221.

Part II

Angiographic Agents

Chapter 7

Intravascular Contrast Media and Their Properties

Thomas W. Morris

Angiographic water-soluble contrast media are essential for radiologic diagnosis. In recent years several books have been published covering the chemical development, use, effects, and toxicity of these drugs.[1–7] The commercial development of new contrast media molecules is currently very active. The objective of this chapter is to provide a brief general description of the structure and properties of molecules that are currently being used.

PHYSICAL BASIS FOR CONTRAST MEDIA

The attenuation of x-ray energy is described by the exponential relationship in Fig. 7-1. For a single material, the exponential term is the negative product of the linear attenuation coefficient and the thickness of the material. The linear attenuation coefficient is dependent on the concentrations of elements in the material, the mass attenuation coefficients of the different elements, and the energy of the x-ray beam. In the energy range of most diagnostic x-ray systems, the mass attenuation coefficient of iodine is much greater than that for soft tissues and even for bone. As illustrated in Fig. 7-2, however, the difference in mass attenuation coefficients for soft tissue and iodine decreases with increasing x-ray energy.

Although iodine-containing contrast media have much higher attenuation coefficients than soft tissue, they are generally contained within a much smaller "thickness," such as a 3 mm coronary artery. Thus, a relatively large concentration of the contrast media molecules must be introduced into the structure. The concentration required is very dependent on the imaging modality used. For systems recording images on film/screen combinations or for fluoroscopy, concentrations of greater than 180 to 290 mgI/ml are generally needed. When using mask-mode digital subtraction angiography (DSA), concentrations as low as 15 to 65 mgI/ml can produce excellent contrast. In computed tomography (CT), con-

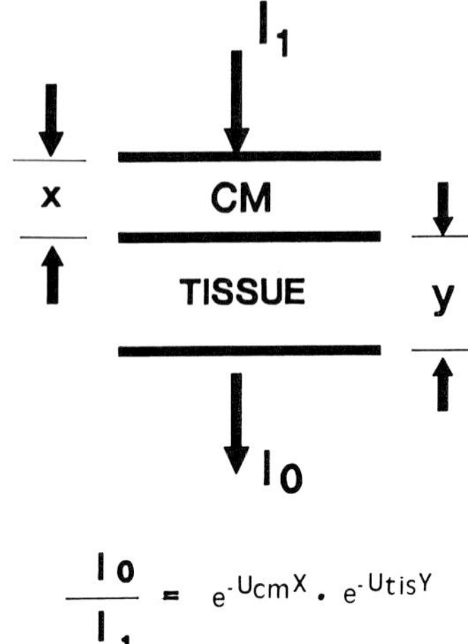

Figure 7-1 The attenuation of a monoenergetic x-ray beam through a two-component object is described in this figure. The linear attenuation coefficients, U1 and U2, are dependent on the concentrations of elements in each of the respective components and on the energy of the x-ray beam.

centrations of 3–5 mgI/ml are easily distinguished and quantified. Thus the imaging system used determines not only the method and site of injection but also the contrast media concentrations.

PROPERTIES OF CONTRAST MEDIA SOLUTIONS

Contrast media must be injected at high molar concentrations to achieve adequate opacification for clinical imaging. For example, concentrations of approximately 1 mole/liter (1 molar) are typically used for most cardiac angiography. At these high concentrations, the density, viscosity, and osmolality of contrast media are much greater than those of body fluids. Plasma, for example, has a density of approximately 1.05 g/ml, a viscosity of 1.2 cP, and an osmolality of 0.3 osmoles per kilogram of water, whereas angiographic contrast media at a concentration of 370 mgI/ml would have a density of 1.41 g/ml, a viscosity of 9 cP, and an osmolality of 2.0 osmoles per kilogram of water. Density represents the mass per unit volume of solution, viscosity represents the "resistance" the solution offers to deformation or flow, and osmolality represents the number of individual particles in solution. The density and viscosity are both inversely related to temperature. Viscosity is very strongly dependent on temperature. A decrease in temperature from 37°C to 20°C will double the viscosity of most contrast media.

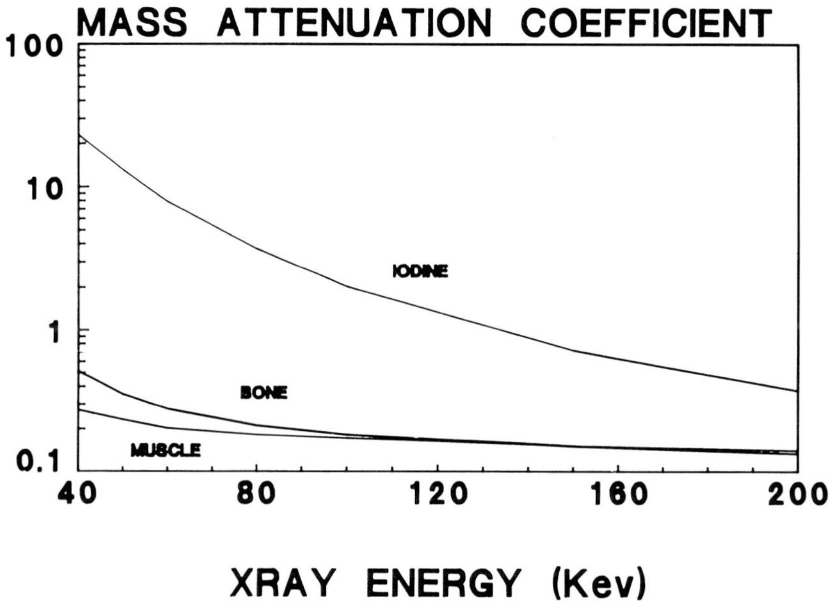

Figure 7-2 The mass attenuation coefficients for iodine, bone, and muscle are illustrated over a wide x-ray energy range. The differences between the attenuation of iodine and that of either bone or muscle decrease with increasing energy.

The colligative properties of solutions are related to their osmolality. These properties include the freezing point, boiling point, vapor pressure, and osmotic pressure. For contrast media, it is the osmotic pressure that is of importance. A .01 osmole per kilogram of water difference in osmolality produces approximately a 184 mm Hg pressure that moves water toward the more concentrated solution. The difference in osmolality between contrast media and body fluids is as large as 1.7 osm/Kg and produces osmotic pressure gradients measured in thousands of mm Hg.

Table 7-1 lists most of the important contrast media in commercial use at this time. Table 7-2 describes the properties of six contrast media that illustrate the differences and similarities of media in the four groups—ionic monomer, ionic dimer, nonionic monomer, and nonionic dimer—to be discussed.

Ionic Contrast Media

The ionic contrast media in current use can be characterized by the two generalized structures in Fig. 7-3. The first structure is described as an *ionic*

Table 7-1 Contrast Media

Generic Name	*Type*	*R**	*Patent Holder*
	CURRENTLY AVAILABLE		
Diatrizoate	Ionic	1.5	Shering AG, Winthrop
Metrizoate	Ionic	1.5	Nycomed
Iothalamate	Ionic	1.5	Mallinckrodt
Iodamide	Ionic	1.5	Bracco
Ioxithalamate	Ionic	1.5	Guerbet
Ioglicate	Ionic	1.5	Schering AG
Ioxaglate	Ionic	3.0	Guerbet
Metrizamide	Nonionic	3.0	Nycomed
Iohexol	Nonionic	3.0	Nycomed
Iopamidol	Nonionic	3.0	Bracco
Iopromide	Nonionic	3.0	Schering AG
Iotrol	Nonionic	6.0	Schering AG
	IN TRIALS		
Iodixanol	Nonionic	6.0	Nycomed
Ioversol	Nonionic	3.0	Mallinckrodt
Ioxilan	Nonionic	3.0	Biophysica, Inc.

*R is the ratio of the number of iodine atoms divided by the number of particles in solution.

Table 7-2 Properties of Selected Contrast Media

				Properties at 300 mgI/ml		
Generic Name	*Molecular Weight*	*Acute Lethal Dose in Mice*	*PC**	*Density 20°C, g/ml*	*Viscosity 37°C, mPa.s*	*Osmolality, Osm/kg*
Ionic monomer						
Diatrizoate	613	7.5	0.045	1.34	4.2	1.57
Ionic dimer						
Ioxaglate	1269	13.4	0.104	1.32	6.2	0.56
Nonionic monomer						
Iohexol	821	24.2	0.070	1.35	6.3	0.67
Iopamidol	777	22.1	0.113	1.33	4.7	0.62
Iopromide	791	16.5	0.069	1.33	4.6	0.61
Nonionic dimer						
Iotrol	1626	26.0	0.005	1.35	9.1	0.36

*PC = partition coefficient in butanol and water at pH 7.6 and 0.01 mg/ml of iodine.

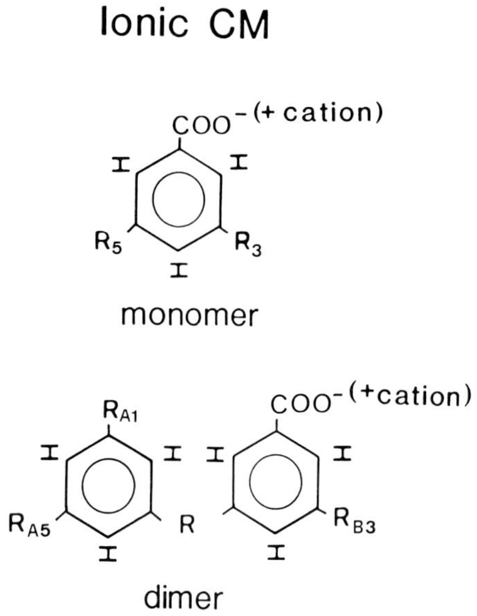

Ionic CM

COO⁻ (+ cation)
I I
R5 R3
I
monomer

RA1 COO⁻ (+cation)
I I I I
RA5 R RB3
I I
dimer

Figure 7-3 The two general forms of the ionic contrast media in use today. Changes in the R groups of the molecules greatly affect solubility, viscosity, and toxicity.

---

monomer or a ratio 1.5 molecule. These molecules are tri-iodinated derivatives of benzoic acid and are formulated as salts with cations of sodium or meglumine, an organic molecule. All of the ionic monomer media have a carboxyl group on the ring at the 1 position, which gives the molecule a valence of minus 1. The side groups labeled R3 and R5 vary among the different compounds. The ratio 1.5 description comes from the fact that the cation and anion dissociate in solution, leaving three iodine molecules (attached to the benzene ring) for two particles in solution. The ionic monomer molecules are very soluble and stable in water and have very little interaction with organic molecules. Because of their low molecular weights of between 600 and 700 and their extremely low lipid solubility, these compounds distribute in the extracellular space and are eliminated primarily by filtration at the renal glomerular capillaries.[8,9]

The second structure in Fig. 7-3 is an ionic dimer known as a *monoacidic dimer* and as a ratio 3 molecule. At the present time ioxaglate is the only molecule of this type in commercial use—ioxaglate (Table 7-1 and Table 7-2). Ioxaglate is formulated as a solution of sodium and meglumine salts.[10] This molecule is formed by joining two ionic monomer molecules at two R side groups. In addition, a carboxyl group on one of the ionic monomers is replaced by an organic group (shown as R1 in Fig. 7-3). This leaves the dimer molecule with a valence of minus 1 and a ratio of six iodine atoms for two particles in solution (or ratio 3).

The dimer molecule in commercial use is also very water soluble with extremely low lipid solubility. Its molecular weight of approximately twice the ionic monomer is still small enough that it has essentially the same distribution and elimination as the ionic monomers.

Nonionic Contrast Media

Since Torsten Almén developed the first commercial nonionic contrast media, metrizamide,[11] there has been an unprecedented effort to develop new nonionic contrast media molecules.[12–14] As of January 1988 there are five nonionic compounds in commercial use worldwide. At least three others are being evaluated in clinical trials. The nonionic molecules are desirable because they do not dissociate in solution as do the ionic salts. As a result, the number of particles in solution per atom of iodine is cut in half (compared to the ionics), and osmolality is also approximately reduced by half. The two types of nonionic molecule structures that parallel the two ionic structures are shown in Fig. 7-4. The *nonionic monomer* is a ratio 3 molecule, whereas the *nonionic dimer* is a ratio 6 molecule. Because the nonionic molecules require one additional side group to replace the carboxyl

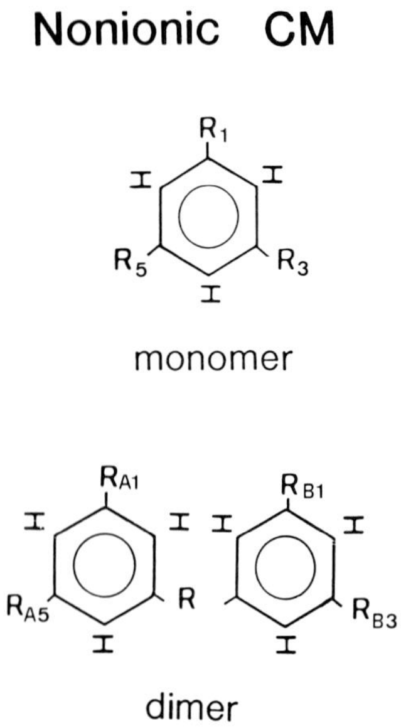

Nonionic CM

monomer

dimer

Figure 7-4 The structures of the nonionic media are essentially the same as the ionic media except that the COO- portion of the ionic molecule is replaced by an organic R group. As with the ionic molecules the choice of R groups is very important.

portion of the ionic molecules, they tend to have slightly larger molecular weights. However, the nonionic molecules are also distributed and eliminated essentially the same as the ionic molecules.

The properties of four commercially available nonionic media are listed in Table 7-2. At equal iodine concentrations, the densities, viscosities, and osmolalities of the nonionic monomers are relatively similar but not identical. Differences in the side groups affect both the size of the molecules and the molecule-to-molecule interactions. Increases in size tend to increase viscosity. Increases in the number of molecular interactions also tend to increase viscosity by increasing effective particle size and to decrease osmolality by decreasing the number of free particles in solution. The nonionic dimers are considerably larger than any of the other three groups of media and seem to have much greater viscosities, as well as the lowest osmolality at equal iodine concentration.

Comparison of Properties

Solution density is a linear function of iodine concentration and provides a very convenient method for accurately determining iodine content in commercial preparations.

The iodine forms such a large portion of the molecule that there is actually very little difference in the densities of the monomer and dimer compounds at equal iodine content (Table 7-2). However, if we were to plot density versus molar concentration, the dimers would be considerably denser at equal molar concentration.

Viscosity has a strong nonlinear dependence on iodine concentration, as shown in Fig. 7-5. At equal iodine concentrations there is a tendency for viscosity to increase with increasing molecular weight, although this is not always the case. For example, iohexol (MW 821) is slightly more viscous than the larger ionic dimer, ioxaglate (MW 1269). Viscosity is more closely related to molecular size and shape than weight, which may account for iohexol's relatively high viscosity. All contrast media have viscosities that exceed those predicted by the Einstein equation for a solution of rigid spheres.[12]

Osmolality is the property that seems to correlate most closely to the general effects of contrast media. As illustrated in Fig. 7-6, osmolality is strongly dependent on both concentration and molecular type. The nonionic dimer iotrol has the lowest osmolality and the ionic monomers the highest. The ionic dimer ioxaglate, and the nonionic monomers iohexol, iopamidol, iopromide, etc., have similar osmolalities, but they are not identical. As mentioned earlier, these differences are probably related to interactions between the molecules. It is not yet known whether these interactions are significantly affected by the presence of other solute molecules, such as sodium chloride.

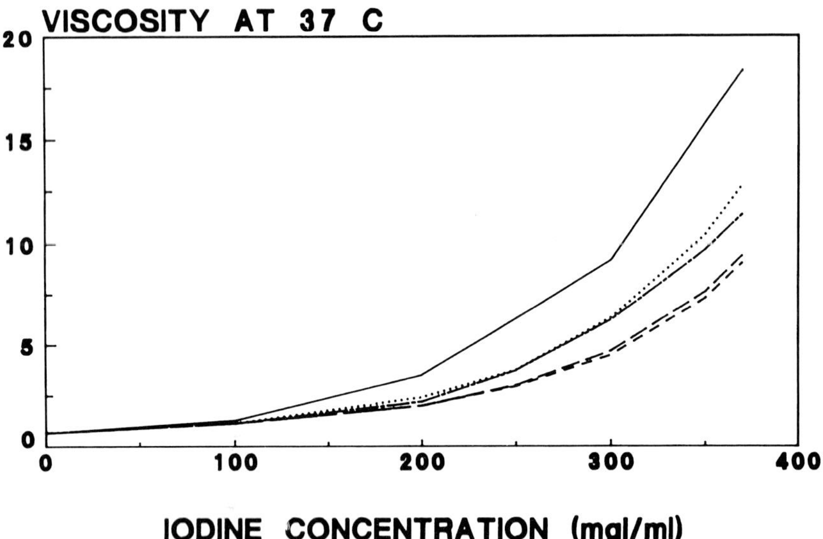

Figure 7-5 The viscosity of all contrast media increases with increasing iodine concentration. The ionic monomer media tend to have the lowest viscosities and the nonionic dimers the highest, but the differences are generally small and of little consequence. The specific R groups in the molecules control this property. The data illustrated are for iotrol, diatrizoate, iopamidol, iohexol, and ioxaglate.

Nonionic and ionic contrast media also interact differently with calcium and other divalent cations. All of the ionic media have been shown to be weak binders of ionic calcium.[15] Nonionic media do not seem to bind ionic calcium and do not affect calcium activity.

RATIONALE FOR COMMERCIAL FORMULATIONS

With the introduction of the nonionic and ionic dimer contrast media it has become clear that there are many factors that the manufacturers must consider in their final formulations. The highest concentrations of the new media range from 320–370 mgI/ml. Why were these concentrations chosen? One of the factors considered is solubility. Solutions that crystallize in the vial are of no use clinically. Manufacturers may also reduce concentrations to reduce viscosity and make the agents easier to inject. Fig. 7-5 illustrates that viscosity increases rapidly at the higher concentrations, and even a change of 50 mgI/ml makes a large

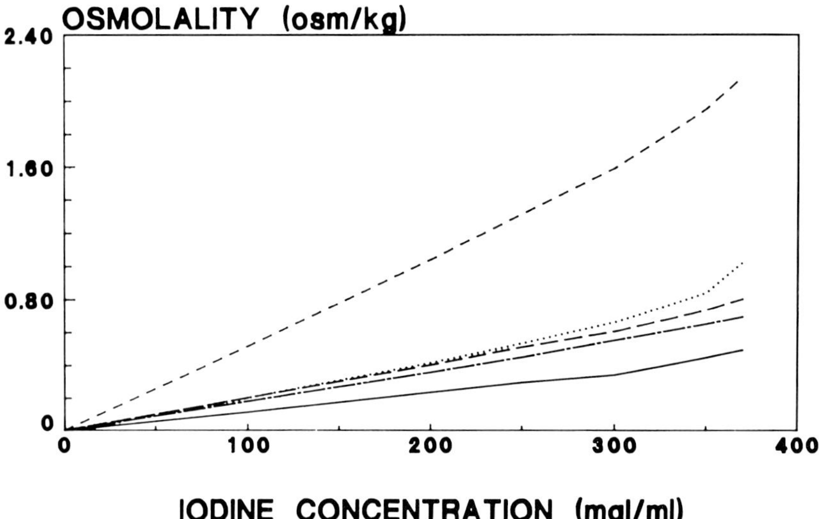

Figure 7-6 The osmolality of all contrast media increases with iodine concentration. The ionic monomers have the highest osmolality and nonionic dimers the lowest. Differences in the osmolality of the different types of molecules are large and of great physiologic consequence. The specific R groups exert relatively little control over this property.

---

difference. The pH of injected solutions must also be controlled using such buffers as sodium citrate or Tris or by titration with sodium hydroxide or hydrochloric acid. Virtually all media formulations contain a form of the chelator EDTA. Most include only a small amount of calcium disodium EDTA. This compound is included to bind potentially toxic heavy metal contaminates that might appear in the manufacturing process. Once bound, the heavy metals are excreted without problem by the kidney. The electrolyte balance of ionic contrast media has long been considered of importance,[16,17] and recently there has been some indication that it may also be of importance for nonionic media.[18]

FUTURE OF X-RAY CONTRAST MEDIA

It seems likely that x-ray contrast media will be used extensively for at least the next decade and probably much longer. However, the molecules and the procedures to be used are likely to change. Improvements in equipment should further reduce the concentrations required for imaging, thus increasing safety and decreasing cost. New molecules and manufacturing processes will probably lower the cost of the nonionic and ionic dimer media without reducing either their efficacy or safety. The major manufacturers realize that there are large financial rewards associated with these developments.

REFERENCES

1. Knoefel PK (ed): *Radiocontrast Agents,* vol 2. Oxford, Pergamon, 1971.
2. Amiel M (ed): *Contrast Media in Radiology*. New York, Springer-Verlag, 1982.
3. Taenzer V, Zeitler E (eds): *Contrast Media in Urography, Angiography and Computerized Tomography*. New York, Georg Thieme Verlag, 1983.
4. Sovak M (ed): *Radiocontrast Agents*. Berlin/New York, Springer-Verlag, 1984.
5. Granger RG: Intravascular contrast media; the past, the present and the future. *Br J Radiol* 1982;55:1–18.
6. Felix R, Fischer HW, Kormano M, et al (eds): *Contrast Media from the Past to the Future*. New York, Georg Thieme Verlag, 1987.
7. Parvez Z (ed): *Contrast Media: Biologic Effects and Clinical Application,* vols 1–3. Boca Raton, FL, CRC Press, 1987.
8. Dean PB, Kivisaari L, Kormano M: The diagnostic potential of contrast enhancement pharmacokinetics. *Invest Radiol* 1978;13:533–540.
9. Donaldson ML: Comparison of renal clearance of insulin and radioactive diatrizoate as measures of glomerular filtration rate in man. *Clin Sci* 1968;35:513–524.
10. Spataro RF, Fischer HW, Boylan L: Urography with low osmolality contrast media; comparative urinary excretion of iopamidol, Hexabrix and diatrizoate. *Invest Radiol* 1982;17:494–500.
11. McClennan BL (ed): Ioxaglic acid: A new low-osmolality contrast medium. *Invest Radiol* 1984; 19(suppl):S289–S392.
12. Almén T: Contrast agent design. Some aspects of the synthesis of water soluble contrast agents of low osmolality. *J Theor Biol* 1969;24:216–226.
13. Hoey GB, Smith KR: Chemistry of x-ray contrast media, in Sovak M (ed): *Radiocontrast Agents*. Berlin/New York, Springer-Verlag, 1984, pp 25–125.
14. Drayer BP (ed): Iopamidol: Intravascular and intrathecal applications. *Invest Radiol* 1984; 19(suppl):S157–S287.
15. Morris TW, Sahler LG, Violante MR et al: Reduction in calcium activity by radiopaque contrast media. *Radiology* 1983;148:55–59.
16. Potts DG, Higgins CB (eds): A worldwide clinical assessment of a new nonionic contrast medium: iohexol. *Invest Radiol* 1985;20(suppl):S1–S121.
17. Paulin S, Adams DF: Increased ventricular fibrillation during coronary arteriography with a new contrast medium preparation. *Radiology* 1971;101:45–50.
18. Morris TW, Hayakawa K, Sahler LG et al: Incidence of fibrillation with isotonic contrast media for intra-arterial DSA. *Diagn Imaging Clin Med* 1986;55:109–113.

Chapter 8

General Effects of Intravascular Contrast Media

Thomas W. Morris

Literally thousands of articles and book chapters have been written on the use and effects of intravascular x-ray contrast media.[1–6] Most of the literature has focused on the differences among the commercially available formulations. The objective of this chapter is to describe the events that occur with all contrast media. Later chapters discuss the specific use and effects of individual media in greater detail.

INTRA-ARTERIAL INJECTIONS

Intra-arterial injection or arteriography is performed by inserting a catheter into an artery and advancing the tip of the catheter to the desired vascular site of the injection. Injections are then made at a rate sufficient to opacify the vessels of interest. If the rate of injection is slow compared to the flow rate in the vessel, the contrast media will not mix with the blood uniformly and the images will be inadequate.[7] This is often described as ''streaming'' of the media, and the resulting image can be erroneously interpreted as a narrowed vessel. Adequate mixing is generally achieved at a rate that is close to the vessel flow (Table 8-1).

If we assume that the injection rate is equal to the vessel flow, then the average velocity (flow/cross-sectional area) in the catheter, which has a much smaller diameter, is much greater than the velocity in the vessel. As a result, a high-velocity injection jet of contrast medium enters the vessel. This type of concentric flow is unstable and breaks down into a turbulent or swirling flow pattern, which produces excellent mixing of blood and medium.[7]

Rapid arterial injection produces transient changes in vessel flow.[8,9] The mechanics of the injection jet produce an increase in distal pressure and flow and a decrease in proximal flow, as illustrated in Figure 8-1. These effects disappear

Table 8-1 Acute Effects of Arterial Injections

1. Injection jet causes increased distal pressure and flow and decreased proximal flow.
2. High media viscosity increases resistance and decreases flow.
3. High media osmolality causes water flux from red blood cells and tissue to plasma, and contrast media molecules enter interstitial fluid.
4. Vasodilation occurs, resistance decreases, and a sensation of heat or pain may occur. Systemic changes may occur depending on the vascular bed (heart, brain).
5. Contrast media are washed out of the local vascular bed by isotonic blood.
6. Water and contrast media fluxes are reversed.
7. Physiologic parameters return to preinjection levels.

---

within milliseconds after the injection stops; however, additional changes occur that are related to the properties of the contrast medium.

Blood flow through a tube is approximated very closely by the Poiseuille equation (1) for laminar flow.[10]

$$Q = \text{pi} \times d^4 \times P/128 \times u \times l \qquad (1)$$

where Q = flow, P = pressure gradient, pi = 3.1416, d^4 = diameter^4, u = viscosity, and l = length. In the mammalian vascular system the arterioles—arterial vessels from 300 to 30 microns in diameter—control local flow.[11] Blood

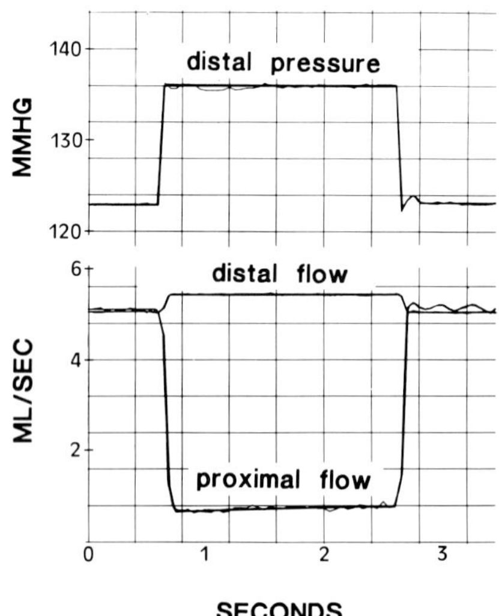

Figure 8-1 The effects of a catheter injection into a tube connected to a constant pressure source as a model of in vivo events. During the injection the distal pressure and flow in the tube are slightly elevated, and proximal flow is greatly reduced.

in large arteries at a hematocrit of 40 has a viscosity close to 4 cP.[12] As the blood reaches the arterioles the hematocrit drops and the effective viscosity begins to approach the viscosity of plasma (1.2 cP). In most cases the contrast medium injected has a viscosity that is greater than the viscosity of either blood (4 cP) or plasma (1.2 cP). As the blood-contrast media (CM) mixture reaches the arterioles, the overall resistance of the vascular bed changes, and flow decreases by a factor proportional to the viscosity of the blood previously in the vessel divided by the viscosity of the blood-CM mixture as it reaches the arteriole.[13] This decrease lasts only until the blood-CM mixture is washed out of the arterioles.

The high osmolality of contrast media also affects blood flow. The blood-CM mixture entering the vascular bed is very hypertonic. The red blood cell membrane is readily permeable to water. A high osmolality outside the red cell creates an enormous osmotic pressure that ''pulls'' the water out of the cells in a few milliseconds.[14] The red cells thus shrink by an amount that can be accurately predicted by the ratio of mixture osmolality divided by normal plasma osmolality.[15] These shrunken or crenated red cells are stiffer and less deformable than red cells at physiologic osmolality. In vitro this loss in deformability makes it more difficult for these cells to pass through small pores,[16] and it has been suggested that in vivo this loss contributes to the transient decrease in flow produced by contrast media.

As the blood-CM mixture passes through the vascular bed, the high osmolality also pulls water from the tissue interstitial and cellular space into the plasma, as illustrated in Figure 8-2.[15,17] Almost immediately after this outward flux of water from the tissue, a local vasodilation is observed. The mechanism for this dilation is not known; however, similar effects have been observed with many hypertonic solutions.[17–19] The amount of water shift and vasodilation seems to be closely related to the osmolality of the medium and the duration and rate of the injection. The vasodilation, although transient, is sustained for from several seconds to a few minutes, depending on the injection parameters and the physiologic state of the vascular bed. As is explained in a later chapter, the kidney responds with a delayed decrease in flow that also seems to be related to osmolality.[20,21]

After the hypertonic mixture of blood and contrast medium has washed through the vascular bed, the tissue is slightly more hypertonic than the incoming isotonic blood. The osmotic pressure is reversed, and water leaves blood and plasma to enter the tissue. It is this inward flux of water that most closely follows the delayed increase in flow observed in the kidney. Water may enter the tissue for an additional reason. Normally, plasma proteins are excluded from the interstitial space by the endothelial cells lining the capillary wall. Hypertonic contrast media solutions cause shrinkage and disruption of this lining, which may allow proteins to reach the interstitial fluid, thereby forcing more inward water flux.[22] There is also evidence in the literature that some contrast media, such as metrizamide, can alter endothelial cell shape even at isotonic concentrations.[23]

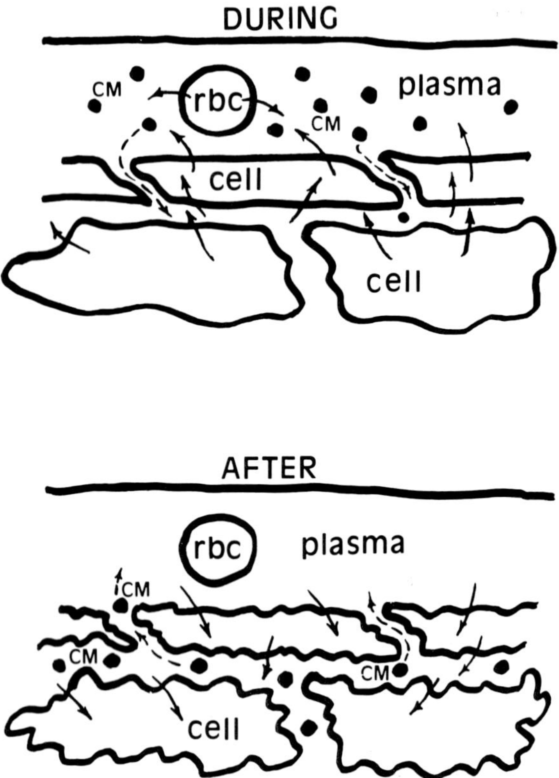

Figure 8-2 During an arterial injection the osmolality of the plasma is greatly elevated. In the capillaries there is a large rapid flux of water (*solid arrows*) from red blood cells (rbc), endothelial cells, interstitial fluid, and parenchymal cells into the plasma. At the same time the contrast media molecules (*CM, small dots*) enter the interstitial fluid. After the injection the isotonic blood entering the capillary is hypotonic compared to the tissue, and the fluid flux is from the blood back to the tissue. The contrast media molecules also reverse their movement and re-enter the plasma.

The molecular size of all contrast media is small enough so that these substances can pass through the endothelial cell junctions in all vascular beds except the brain and testes.[24–26] These two vascular beds have endothelial structures that greatly limit solute movement into tissue.[27] In most vascular beds there is an immediate and rapid flux of solute molecules into the interstitial space. This inward flux continues until the plasma concentration drops below the interstitial fluid concentration. The time it takes to reach this point is determined by the rate of flux into tissue and by the rate of clearance of the contrast media through the renal glomerulus.

The preceding paragraphs discussed the local effects of intra-arterial injections. It should be clear that each of these local effects can alter overall cardiovascular function. Some vascular beds, however, have greater effects on overall cardiovascular function than do others. Local changes in the heart and brain, for example, cause large alterations in cardiovascular function. In the heart the arterial injection of contrast media affects both contractility and electrophysiologic state.[28–31] Similarly, injections in the brain affect centers that control cardiovascular as well as other functions.[32–34] These effects seem to be partially related to osmolality, but they vary among the different commercial formulations and are discussed in later chapters.

INTRAVENOUS INJECTION

Intravenous (IV) injections are generally made at much higher doses than arterial injections; however, the contrast medium mixes with blood before it reaches the lung microcirculation. The iodine concentration of the mixture reaching the lung can be approximated by dividing the injection rate of the iodine—milligrams of iodine per second—by the cardiac output in milliliters per second. Because the average cardiac output is approximately 100 ml/sec, contrast media (CM) are quite dilute before they reach the lungs for all but the most rapid IV injections. IV injections cause many of the same effects as intra-arterial injections, but they are usually diminished by this large dilution with blood (Table 8-2).

The blood-CM mixture reaching the lungs contains shrunken or crenated red blood cells and has a higher "plasma" viscosity because of the CM molecules. The same fluid fluxes that occur in other tissues occur in the lung capillary endothelial cells and in the lung parenchyma.[35] It has also been suggested that

---

Table 8-2 Acute Effects of Intravenous Injections

1. High osmolality of media causes water flux from red blood cells and endothelial cells to plasma.
2. Hypertonic, hyperviscous mixture reaches lungs where further water shifts occur and contrast media enter lung parenchyma. Plasma protein flux to the lung interstitium may increase.
3. Pulmonary artery pressure and cardiac output increase while pulmonary resistance, systemic pressure, and hematocrit decrease.
4. The mixture enters the systemic vascular beds where further water shifts occur and media molecules enter the interstitial space.
5. Water moves from isotonic blood in the lung capillaries back into the hypertonic tissue, and the same process occurs in the systemic tissues.
6. Physiologic parameters return to preinjection values.

protein loss through the shrunken endothelial cells can lead to excess fluid retention after rapid IV injections.[36] The pulmonary endothelium contains neurotransmitters. Shrinkage and other CM effects may cause the lung endothelium to release neurotransmitters and may initiate some of the events in severe contrast media reactions. After the blood-CM mixture exits the lung the contrast media are even more dilute because of the water that is removed from the lung. This dilute blood-CM mixture is subsequently pumped to the entire body.

Rapid IV injections have been shown to produce an initial increase in pulmonary artery pressure and cardiac output and a decrease in pulmonary and systemic resistances.[37–39] Systemic arterial pressure usually falls slightly, but the increase in cardiac output and decrease in resistance tend to balance each other. These effects are greatly reduced when low osmolality agents or dilute solutions of ionic monomers are used.[38,40] The detailed mechanisms for these actions of contrast media are not known.

The contrast media molecules rapidly enter the interstitial space after an IV injection just as they do after arterial injections; however, their concentration is much lower and more uniform throughout the body than for the local arterial injection. In CT we actually make use of this fairly uniform distribution to differentiate the filtration of contrast media into tumors or other lesions. The concentration differences are usually greatest during or immediately after an injection.[41] After 30 minutes the concentration is quite uniform over the entire extracellular space.

It is quite easy to make rough approximations of the contrast media in plasma and the extracellular space. Plasma volume can be estimated as 45 ml per kilogram of body weight. In the first minute after a rapid IV injection, the contrast medium is primarily contained in the plasma, and the concentration in mg I/ml is approximately the dose injected (mg I) divided by (45 ml/kg times body weight in kilograms). The extracellular volume is approximately 200 ml per kg of body weight. Therefore, after 30 minutes the concentration is approximately the dose injected divided by 200 ml/kg times body weight in kg. More accurate estimates are possible if one can predict the time constants for distribution into the extracellular space and for excretions.[42]

All currently used intravascular agents are cleared by filtration through the glomerular capillaries in the kidneys. Because there is no active reabsorption or secretion and little binding to proteins or membranes, the pharmacokinetics of contrast media are closely approximated by a two-compartment model with excretion (Fig. 8-3). In patients with nonfunctioning kidneys there is a much slower clearance of contrast media, probably through the liver and gastrointestinal system.

FACTORS CONTROLLING RESPONSES TO CONTRAST MEDIA

The acute responses to contrast media are generally related to the nonspecific properties of the media formulations and the fact that they are used in large

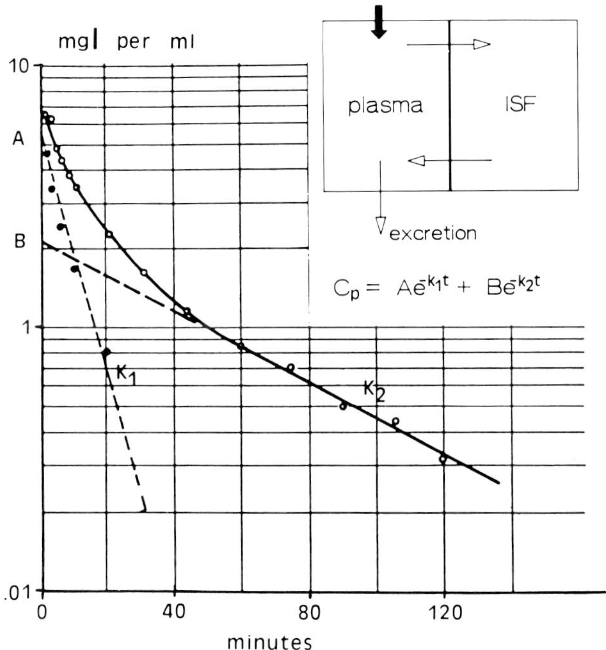

Figure 8-3 The plasma clearance of water-soluble contrast media is approximated by a two-compartment model with an injection (*bold arrow*) into the plasma. The coefficient A is the amount of iodine injected divided by the plasma volume. The coefficient B is the amount of iodine injected divided by the extracellular volume. The coefficient k1 is the slope (on the log plot) of the rapid component, and K2 is the slope of the slow component.

---

volumes and at high concentrations. Even injections of equal volumes of isotonic saline usually produce small responses. The osmolality of contrast media still seems to be the major factor involved in physiologic responses. The development of ratio 3 and ratio 6 compounds has greatly reduced the osmolality of solutions with iodine concentrations suitable for film and fluoroscopic imaging. The ratio 6 compounds (nonionic dimers) are actually isotonic at about 300 mg I/ml. The development of mask-mode digital subtraction angiography and computed tomography makes it possible to perform vascular imaging with isotonic solutions of even ratio 1.5 and ratio 3 media.

Solution viscosity produces a transient decrease in flow during arterial injections. All of the media have high viscosities, and the ratio 3 and 6 molecules tend to have higher viscosities in solution than the ratio 1.5 molecules. Fortunately, this high viscosity has not been associated with any real clinical risk or with any decrement in vascular imaging. Viscosity may, however, be a factor to consider if one is trying to measure flow or physiologic function quantitatively.

One fundamental difference between ionic and nonionic contrast media is their ability to bind calcium.[43] All of the ionic molecules are weak binders of calcium (and most likely other divalent cations), whereas all the nonionic media tested do not seem to bind calcium at all. A second difference, which is probably related, is that the ionic media have strong anticoagulant properties when compared to the nonionic media.[44]

Are there real differences in the molecule-specific "chemotoxicity" of the contrast media molecules? The answer to that question will continue to be debated, although there do seem to be differences in many different test systems. The problem is that the ordering of the molecules from least to most toxic is not always the same, and there is no consensus about which test system is thc bcst "model" for the patient. All of the molecules probably have some advantages and disadvantages, and all will cause physiologic responses.

REFERENCES

1. Knoefel PK (ed): *Radiocontrast Agents*, vol 2. Oxford, Pergamon, 1971.

2. Amiel M (ed): *Contrast Media in Radiology*. New York, Springer-Verlag, 1982.

3. Taenzer V, Zeitler E (eds): *Contrast Media in Urography, Angiography and Computerized Tomography*. New York, Thieme Verlag, 1983.

4. Sovak M (ed): *Radiocontrast Agents*. Berlin/New York, Springer-Verlag, 1984.

5. Felix R, Fischer HW, Kormano M, et al (eds): *Contrast Media from the Past to the Future*. New York, Thieme Verlag, 1987.

6. Parvez Z (ed): *Contrast Media: Biologic Effects and Clinical Application,* vols 1–3. Boca Raton, FL, CRC Press, 1987.

7. Mabon RF, Soder PD, Carpenter WA et al: Fluid dynamics in cerebral angiography. *Radiology* 1978;128:669–676.

8. Wolf GL, Shaw DD, Baltaxe HA: A proposed mechanism for transient increases in arterial pressure and flow during angiographic injections. *Invest Radiol* 1978;13:195–199.

9. Morris TW, Katzberg RW: A comparison of the hemodynamic responses to metrizamide and meglumine/sodium diatrizoate in canine renal angiography. *Invest Radiol* 1978;13:74–78.

10. Caro CJ, Pedley TJ, Schroter RC et al: *The Mechanics of the Circulation*. Oxford, Oxford University Press, 1978, pp 44–78.

11. Zweifach BW: Quantitative studies of microcirculatory structure and function. *Circ Res* 1974;34:843–857.

12. Chien S: Biophysical behavior of red cells in suspensions, in Surgenor DM (ed): *The Red Blood Cell*, vol 2. New York, Academic Press, 1975, pp 1031–1133.

13. Morris TW, Kern MA, Katzberg RW: The effects of media viscosity on hemodynamics in selective arteriography. *Invest Radiol* 1982;17:70–76.

14. Seidel VW, Solomon AK: Entrance of water into human red cells under an osmotic pressure gradient. *J Gen Physiol* 1957;41:243–257.

15. Morris TW, Harnish PP, Reece K et al: Tissue fluid shifts during renal arteriography with conventional and low osmolality agents. *Invest Radiol* 1983;18:335–340.

16. Aspelin P: Effect of ionic and nonionic contrast media on red cell deformability in vitro. *Acta Radiol Diagn* 1979;20:1–12.

17. Gazitua S, Scott S, Wubdakk B et al: Resistance response to local changes in plasma osmolality in three vascular beds. *Am J Physiol* 1971;220:384–396.

18. Hilal SK: Hemodynamic changes associated with the intra-arterial injection of contrast media. New toxicity tests and a new experimental contrast medium. *Radiology* 1966;86:615–633.

19. Morris TW, Francis M, Fischer HW: A comparison of the cardiovascular responses to ionic and nonionic contrast media. *Invest Radiol* 1980;15:248–259.

20. Katzberg RW, Morris TW, Burgener FA et al: Renal renin and hemodynamic responses to selective renal artery catheterization and angiography. *Invest Radiol* 1977;12: 381–388.

21. Katzberg RW, Schulman G, Meggs LG, et al: Mechanism of the renal response to contrast medium in dogs: decrease in renal function due to hypertonicity. *Invest Radiol* 1983;18:74–80.

22. Nyman U, Almén T: Effects of contrast media on aortic endothelium. Experiments in the rat with nonionic and ionic monomer and monoacidic dimeric contrast media. *Acta Radiol* 1980; 362(suppl):65–72.

23. Gospos C, Freuenberg N, Staubsand J et al: The effects of contrast media on the aortic endothelium of rats. *Radiology* 1983;147:685–688.

24. Dean PB, Kivisaari L, Kormano M: The diagnostic potential of contrast enhancement pharmacokinetics. *Invest Radiol* 1978;13:533–540.

25. Newhouse JH: Fluid compartment distribution of intravenous iothalamate in the dog. *Invest Radiol* 1977;12:364–367.

26. Caro CJ, Pedley TJ, Schroter RC et al: *The Mechanics of the Circulation*. Oxford, Oxford University Press, 1978, pp 350–433.

27. Sage MR: Kinetics of water soluble contrast media in the central nervous system. *Am J Roentgenol* 1983;141:815–824.

28. Fischer HW, Thomson KB: Contrast media in coronary arteriography. *Invest Radiol* 1978;13:450–459.

29. Gerber KH, Higgins CB, Yuh YS, et al: Regional myocardial hemodynamics and metabolic effects of ionic and nonionic contrast media in normal and ischemic states. *Circulation* 1982; 65:1307–1314.

30. Higgins CB: Contrast media in the cardiovascular system, in Sovak M (ed): *Radiocontrast Agents*. New York, Springer-Verlag, 1984, pp 193–251.

31. Morris TW, Ventura J: Incidence of fibrillation with dilute contrast media for intra-arterial coronary digital subtraction angiography. *Invest Radiol* 1986;21:416–418.

32. Hayakawa K, Morris TW, Katzberg RW, et al: Cardiovascular response to the intravertebral injection of hypertonic contrast media in the dog. *Invest Radiol* 1985;20:217–221.

33. Hayakawa K, Morris TW, Katzberg RW et al: Cardiovascular responses to the intracarotid injections of ionic contrast media and iohexol in the dog. *Acta Radiol Diagn* 1986; 27:729–733.

34. Hayakawa K, Nishimura Y, Yoshida M, et al: ECG changes during cerebral angiography: an analysis of 334 patients, 942 cerebral angiographies. *Neuroradiol* 1984;26:369–373.

35. Morris TW: Properties and pharmacology of intravascular radiocontrast media. *Can J Cardiol* 1987;3(suppl A):6A–10A.

36. Slutsky RA, Hackney DB, Peck WW, et al: Extravascular lung water: effect of ionic and nonionic contrast media. *Radiology* 1983;149:375–378.

37. Read RC, Johnson JA, Vick JA et al: Vascular effects of hypertonic solutions. *Circ Res* 1960;8:538–548.

38. Peck WW, Slutsky RA, Hackney DB, et al: Effects of contrast media on pulmonary hemodynamics: comparison of ionic and nonionic agents. *Radiology* 1983;149:371–374.

39. Harnish PP, Morris TW, Fischer HW: Drug actions and interactions of intravascular contrast media, in Parvez Z (ed): *Contrast Media: Biologic Effects and Clinical Application*, vol 2. Boca Raton, FL, CRC Press, 1987, pp 28–47.

40. Thomson WM, Mills SR, Bates M, et al: Pulmonary angiography with iopamidol and Renografin 76 in normal and pulmonary hypertensive dogs. *Acta Radiol Diagn* 1983;24:425–431.

41. Burgener FA, Hamlin DJ: Contrast enhancement in abdominal CT: bolus versus infusion. *Am J Roentgenol* 1981;137:351–358.

42. Hall JE, Guyton AC, Farr BM: A single injection method for measuring glomerular filtration rate. *Am J Physiol* 1977;232:F72–F76.

43. Morris TW, Sahler LG, Violante MR et al: Reduction in calcium activity by radiopaque contrast media. *Radiology* 1983;148:55–59.

44. Stormorken H, Skalpe IO, Testart MC: Effect of various contrast media on coagulation, fibrinolysis, and platelet function: an in vitro and in vivo study. *Invest Radiol* 1986;21:348–354.

Chapter 9

Idiosyncratic Reactions

H.W. Fischer

Patients undergoing an examination with contrast media are subject to adverse reactions. Excluding the sensation of warmth that patients feel to some degree as the contrast medium courses through the circulation, about 5% of all patients receiving intravascular media experience some kind of reaction requiring treatment, .05% have a reaction requiring hospitalization, and about .0025% have a fatal reaction.

Since adverse reactions first were observed, physicians have attempted to learn why they occur, but as yet, we do not have certain knowledge of why they do. Over the years a number of mechanisms have been proposed and supportive evidence presented. The list (Table 9-1) of proposed mechanisms is in alphabetical order, not in order of importance or chronology.

I do not believe there is one single mechanism underlying all adverse reactions to radiographic contrast media. Occasionally one of the proposed mechanisms may explain a certain observed reaction, but more likely multiple mechanisms

Table 9-1 Mechanisms That May Cause Adverse Reactions

Antigen-antibody
Anxiety and emotions
Blood-brain barrier penetration
Blood changes
Bronchospasm and pulmonary edema
Calcium-sodium changes
Cholinesterase inhibition
Complement activation
Drug synergism
Hemodynamic and heart changes
Histamine release
Injection of impurities
Protein binding and enzyme interface

may be involved in most reactions. Each of these is now considered separately, although often these mechanisms overlap.

PROPOSED ADVERSE REACTION MECHANISMS

Antigen-Antibody

Despite the fact the symptoms of some adverse reactions to contrast media closely resemble true allergic reactions, there is insufficient evidence to support this mechanism of reaction,[1,2] except in extremely rare circumstances.[3–5] The molecular structures of the currently used urographic and angiographic media do not have an antigenic potential.[6] The work supporting an antigen-antibody mechanism for reactions is that of Brasch and Sweeney,[7–9] and the interested reader may wish to consult these references.

Anxiety and Emotions

The idea that a person's emotions may be involved in a contrast media reaction is not new.[10] This continues to be a suspicion in the minds of some radiologists, but there has been little or no evidence of a scientific nature to support it. Tranquilizing and sedating drugs have been tried clinically, but the results are not impressive.[10–13] Occasionally a person responds to a needle puncture or some other anticipated unpleasant or painful event with a vagal reaction (see the section on hemodynamics). The needle puncture itself and the infusion of glucose or saline did not produce the decreases in pulmonary function or the ECG changes documented from the infusions of contrast media.[14–18] The perceived anxiety level of the patient was not found to be predictive of the chest pain or ECG changes caused by digital subtraction angiography.[19] Therefore, although the patient's emotional state cannot be excluded as a factor in reactions, the contrast medium itself seems necessary.

Blood-Brain Barrier Penetration

Injection of contrast media directly into the cerebral arterial circulation can produce blood-brain barrier (BBB) permeability changes; this phenomenon was first recognized by Olsson as long ago as 1948.[20] It seems reasonable that certain infrequent central nervous system (CNS) deficits that follow cerebral angiography may be initiated by damage to the BBB. However, for the adverse reactions usually termed idiosyncratic, damage to the BBB is not likely a first step because the blood concentration from the introduction of contrast media intravenously is not high enough to affect the BBB on a hyperosmolality basis, which is the major basis for action of the contrast media in cerebral angiography. Likewise, the blood concentration of contrast media flowing through the CNS after an arteriogram of some other part of the body is too low to similarly affect the BBB. However, despite the lack of definite evidence of an action by IV administered contrast media on the CNS that would cause idiosyncratic reactions, the possibility cannot be excluded.

Blood Changes

An adverse reaction to contrast media characterized by abnormal bleeding or clotting is rare. (See other sections for discussion of complement activation or histamine release.) Excessive bleeding is more likely due to the use of contrast media in patients with thrombocytopenia or sickle cell phenomena or patients on anticoagulant-antiplatelet drugs.[21] Patients with the sickling trait or sickle cell disease are more at risk for a thrombotic episode, because sickling is enhanced by contrast media.[22] The inhibition of platelet aggregation when tested by standard methods, is not long lasting.[21,23,24] Disseminated intravascular coagulation is also rare.[25]

The aggregation and deformation of the red cells by contrast media are well documented. This phenomenon is primarily related to the osmolality of the contrast medium, with the hyperosmolar media having much greater effects; however, the nature of the contrast media molecule is also a factor.[26,27] The interference with normal blood flow, particularly small vessel flow, may well play a role in initiating an adverse reaction.[28,29] What happens in the lungs just as likely happens in other capillary beds. For the heart, for example, cardiac dysfunction may be caused by disturbances in the microcirculation.

Bronchospasm and Pulmonary Edema

A number of adverse reactions to contrast media occur in which the bronchi and lungs are the primary reactive organs. Clinically, bronchospasm and/or pulmonary edema may be observed. Maximum and mean expiratory flow and forced expiratory volume are generally reduced during excretory urography,[14,15] with more pronounced decreases in allergic patients and tobacco smokers. The important question is whether the documented subclinical bronchospasm and the severe bronchospasm seen occasionally as a contrast media reaction are different degrees of the same phenomenon or whether they occur by different mechanisms.

Pulmonary edema may develop quickly or sometimes hours after the examination. A cardiac etiology has been suspected, but the edema also occurs in healthy young adults and children without heart disease.[30–32] In the dog, rapid IV injection of large doses of contrast media produce many hemodynamic alterations of the pulmonary circulation, respiration, heart function, and the general circulation; frequently, pulmonary edema and hemorrhage occur.[33] Many of these changes are thought to be due to the obstruction of pulmonary capillaries by aggregations of red cells caused by the hyperosmolality of the contrast agents and other toxic effects of the contrast media acting on the red cell membrane.[29]

A reflex beginning in the lung vessels may be the cause of these pulmonary phenomena, but serotonin release has also been suspected.[34] With the hyperosmolar agents the abnormalities observed in the dog can be markedly diminished or prevented by premedication with low molecular weight dextran, which prevents the red cell clumping and aggregation.[28,35] Both in humans and in dogs injection of ordinary clinical doses of contrast medium raises pulmonary artery pressure,[36–39] but there is disagreement over whether this is due to alteration of cardiac output or to an increase in pulmonary vascular resistance.[36,37] A clue to the pulmonary vascular resistance may lie in the inhibition of activation of the vasoactive prostaglandin E-2 by ionic media.[40]

Calcium-Sodium Changes

Ventricular fibrillation has been produced in dogs by the right coronary artery injection of a sodium methylglucamine diatrizoate containing a relatively high content of calcium chelators, specifically sodium citrate, disodium edetate, and the diatrizoate anion itself.[41] Addition of calcium or the use of a lower content of chelators in another diatrizoate contrast medium produced much less fibrillation.[42–44]

The lowering of ionic calcium values in the general circulation also has been demonstrated. When volumes of sodium methylglucamine diatrizoate contrast medium with strong chelators were injected into the general circulation, as in enhanced CT scanning, venography, or arteriography, the maximum fall of ionic calcium was 0.13–.20 mM, 8 to 10 minutes postinjection.[45,46] Ventricular fibrillation in clinical coronary arteriography is uncommon and how often the calcium chelation plays a role in causing it is unknown. It is not likely that lowered ionized calcium values from contrast media formulations are a frequent cause of adverse reactions. However, certain patients with severe cardiac dysfunction develop hypotension from an acute lowering of calcium levels, resulting in a clinical state indistinguishable from vascular collapse due to other mechanisms.

Another cause of cardiac dysfunction—a deficiency or excess of sodium ions in contrast media used for coronary arteriography—manifested as ECG changes and arrhythmias, most notably ventricular fibrillation, had previously been recognized. A number of studies, mainly in the laboratory but confirmed clinically, showed that a minimum incidence of fibrillation is associated with a sodium content of approximately 190 mEq/L for a 76% sodium methylglucamine diatrizoate medium and that lowering or raising the sodium content from this level results in a higher incidence of ventricular fibrillation.[1,47,48] A significant cardiovascular-initiated adverse reaction from this cause is no longer encountered, we believe, because radiologists avoid this problem in coronary arteriography when using the ionic high osmolality agents by using exclusively those with a suitable sodium content or by using low osmolar media.

Experimental work has shown that the new low osmolality agents, when used at approximately 300–350 mg I/ml produce an insignificant incidence of fibrillation.[49] In even more recent animals studies of digital subtraction technique in the coronary arteries, the absence of sodium produces a high incidence of fibrillation.[50] The low osmolality agent, ioxaglate, is the only one of the low osmolality agents in clinical use that already contains sodium.

Cholinesterase Inhibition

Many of the signs and symptoms produced by large doses of contrast media are similar to those found in experimental animals given cholinesterase poisons. However, the activity of contrast media as inhibitors of cholinesterase is only moderate,[51] and in our laboratory we did not find a synergism between an anticholinesterase, physostigmine, and contrast media.[52] Proof of this mechanism is not documented in clinical practice.

Complement Activation

Activation of complement by contrast media has been amply demonstrated in experimental animals and in humans, both in vitro and in vivo.[46,47,53–60] The pathway for activation of complement is controversial; it does not clearly follow the classic or alternative pathways, and the activation occurs nonsequentially, with falls in levels of CH_{50}, C3, C4, CLq, and factor B at the same time.[58,60–64] Controversy exists over whether the decrease in complement components is due to true activation, to nonsequential cleavage of components by the contrast medium, or by proteases involving the coagulation cascade.[62–65] The activation occurs in vitro in serum depleted of C4 and C2 and in agammaglobulinemic serum.[63] In normal sera activation is induced of CH_{50}, C2, 3, 4, and 5 in a dose-dependent manner.

Good correlation between contrast media reactions and complement activation is lacking.[46,47,53,55,56,65–67] Despite falls in serum complement in humans, reactions often did not occur, and the correlation between complement levels and contrast media plasma levels was not observed.[46]

With the suspicion that complement activation is not directly related to the pathogenesis of most reactions, other factors that might be related to complement activation have been sought.[20,63,68] Other homeostatic mechanisms of the individual may have to fail for the activation of complement by contrast media to lead to a clinical reaction, or an end-organ sensitivity may have to be postulated for some individuals. Anaphylactoid reactions to contrast media may be the result of complex interactions among several mediator systems, as yet unknown.[69] The ionic media varied in their activation of complement in vitro and in vivo,[46,47,62,63,68] the most active being iodipamide.[57,63] Lasser postulated that patients consuming their C1 esterase inhibitor prior to a contrast media injection and therefore having lower levels of inhibitor are prone to an adverse reaction.[58] In a rabbit model, pretreatment with corticosteroid caused marked increments in C1 esterase inhibitor levels and an increase in factor XII while protecting the rabbits from a lethal dose of iodipamide.[70] The C1 esterase inhibitor is known to be the major inhibitor of esterase activity and to play a role in the inhibition of the protease activities of plasma, factor XII, factor XI, and kallikrein.[6,20,59] Lasser also had evidence of an acceleration of conversion of pre-kallikrein.[6]

The liberation of histamine from complement activation has been considered, but it is not certain that histamine is released on this basis or some other one[71,72] because histamine release is not dependent on complement.[71] Complement activation, histamine release, and the incidence of reactions could not be correlated.[66] Reactions may be due to the responsiveness of tissues of individuals to the products of complement, or histamine, or some other activator. The role of prostaglandins may need more investigation.[40,73,74]

Drug Synergism

The idea that patients with heart disease are at increased risk for a contrast medium reaction is widely held. In the animal laboratory at Rochester we have found synergism between contrast media and cardiac glycosides.[45,52,75] In the presence of ouabain, sodium diatrizoate was found to show a dose-dependent inhibition of sodium-potassium-stimulated adenosine triphosphate (ATPase), an enzyme not particularly sensitive to direct inhibition by contrast media.[76] Because many patients with cardiac disease are receiving a cardiac glycoside, which may cause the toxic effects of arrhythmias and ventricular fibrillation, the idea that patients on large, near-toxic doses of digitalis drugs are more likely to have arrhythmias and cardiac arrest when contrast media are administered is a reasonable one. To our knowledge no clinical studies are available to support this idea of contrast media glycoside synergism, although one report did not detect a higher incidence of ECG change in digitalized patients undergoing IV urography.[77] Another synergism, that of IV chlorpromazine and nonionic contrast medium in the subarachnoid space, is not considered pertinent to adverse reactions of intravascular contrast media.

Hemodynamic and Heart Changes

The hemodynamic changes encountered with the IV injection of contrast medium are considerable and depend on the site and speed of injection, as well as the volume and concentration of the contrast media. There is an extensive literature on this topic;[78] this reference gives a review and list of appropriate references. After an IV injection, a systemic hypotension results, peripheral vascular resistance is lowered, pulmonary artery pressure increases, and tachycardia occurs. The hematocrit falls temporarily, and the blood volume expands. These changes and more occur when the injection is into the right heart or pulmonary artery. Cardiac output is decreased due to the accompanying decrease in cardiac contractility. Injection into the left heart and thoracic aorta produce similar changes, with additional right and left atrial pressures and left ventricular pressure increases.[78]

In two crucial regional circulations, the coronary and cerebral, pronounced circulatory changes are produced with contrast media injection into these circulations. An increase in coronary arterial blood flow, a decrease in cardiac contractility and cardiac output, and a lowering of systemic blood pressure occur with coronary artery injection. The ECG reveals changes ranging from minor to profound, the most serious by far being ventricular fibrillation. In cerebral angiography, slowing of the heart rate and rhythm and decrease in blood pressure are found. In peripheral arteriography, increases in regional blood flow are noted.

These hemodynamic alterations are thought to produce adverse reactions of warmth, faintness, and dizziness at one end of the spectrum to serious arrhythmias, shock, and cardiac arrest at the other extreme. Many serious reactions are believed to have a cardiovascular cause, particularly in the elderly whose regional circulations are already compromised by arteriosclerosis, or in the very young because of the more profound fluid shifts and changes in blood volume in this group. The high osmolality of the ionic contrast media are in the main responsible for the hemodynamic changes caused by the fluid shifts. Their high osmolar nature draws fluid from the extravascular space, the red blood cells, and endothelium to cause most of these alterations and to initiate reflexes that cause others. The nature of the cation and the anion also plays a role, but the main factor is believed to be the hyperosmolality.

A vagal effect on the heart characterized by bradycardia and hypotension is encountered rarely after contrast media administration, whereas the usual hemodynamic response is hypotension and tachycardia.[78] The vagal effect is similar to that experienced on occasion from painful stimuli, such as venipuncture or rectal distention or even psychic factors. A slow heart rate characterizes this condition, and the treatment of choice is atropine in an adequate dosage.[79]

Histamine Release

Histamine is released by contrast media in vitro from mast cells and basophils[69,80–84] and in vivo in animals and in humans.[42,65,66,71,72,80,85–87] However, correlation of histamine release and plasma histamine levels with patients' symptoms or any objective side effects has not been made, with histamine being released in both reactors and nonreactors.[42,65,66,72,81,85,88] Interrelationships between histamine release and complement activation have been shown both in vivo and in vitro.[71,72,81]

The interest in histamine release as a cause of contrast media reactions has been the basis for antihistamine premedication. There have been numerous publications attesting to the value of antihistamine drugs in preventing reactions, whereas other studies have found no value in this regime. Most of the previous experience has been with an H1 histamine receptor antagonist, and possibly the combination of H1 and H2 antagonists will be of more value but this has not been conclusively established.

Injection of Impurities

That noxious substances—allergic, vasoactive, or toxic in some way—might be incriminated in contrast media reactions is a possibility that has been raised but not conclusively proven.[89] Certainly, such substances have been shown to come from plastic devices.[90,91] Until more evidence is available, the radiologist should not leave contrast media in contact with plastic disposable syringes for more than a few minutes before injection, thus minimizing the interaction of contrast media with the plastic, the rubber on the plunger, or the lubricant used to prevent the sticking of the plunger in the barrel. Washing out syringes with several milliliters of saline before placing the contrast media in the syringe should minimize the injection of these possible impurities. It is also advisable to store bottles of contrast media in an upright position so as to prevent the contrast media from being in contact with the rubber stopper. The heating of contrast material likely increases the chance of having impurities come into contrast media from plastic and rubber.[92] The minute particles of glass and other solid materials present in the ampules of contrast media are not considered a cause of acute adverse reactions, although thromboembolic lesions have been shown in animal studies.[93]

Protein Binding and Enzyme Interface

Contrast media bind to serum albumin and inhibit activation of certain enzyme systems, such as B glucuronidase, alcohol dehydrogenase, glucose 6 phosphate dehydrogenase, adenosine triphosphatase, and carbonic acid anhydrase.[94–96] The binding of contrast media to serum protein is correlated to general toxicity and neurotoxicity as shown by animal tests.[95,97] The relation of this binding and the effects on enzyme systems to contrast media reactions in humans has not been elucidated.

Adverse reactions are not thought to be caused by the small amounts of free iodide ion present in the contrast media formulations or the free ion produced by the deiodination in the body.[98,99]

LOW OSMOLALITY CONTRAST MEDIA

After many years of attempting to develop better intravascular contrast agents, considerable progress was made with the development in 1969 of the first nonionic contrast medium by Almén and the Nyegaard Company of Norway. They developed a contrast medium that had approximately half or less of the high osmolality of the ionic monomers media, which had represented the state-of-the-art since intravenous urography was first introduced by Swick in 1927.

Although the total number of patients receiving the new low osmolality agents is but a small fraction of those having received the high osmolality agents, there is now enough evidence to show that the incidence of adverse reactions is lower with their use, particularly with the use of the nonionic low osmolality agents. The first study with significantly larger numbers of patients is that of Schrott (Table 9-2). A second study comes from Katayama in Japan (Table 9-3), and a third is still in progress in Australia (Table 9-4).

The following paragraphs assemble some evidence of why this is probably occurring, and why lower incidence of all adverse reactions, serious adverse reactions, and fatalities are to be anticipated.

Anxiety and Emotional Factors

The low osmolality contrast media, particularly the nonionic media, produce fewer and less severe feelings of arm pain, warmth and burning sensations, and

Table 9-2 Incidence of Adverse Reactions in a West German Study

	Total	*High-Risk Patients*	*Patients Requiring Hospitalization*
Ionic media*	5.0	–	0.05
Nonionic media†	2.1%	2.7	0.012%

*Percentages for ionic examinations from Shehadi *Radiology* 1988;137:229.
†Number of nonionic examinations: 50,660.

Source: "Iohexol in Excretory Urography" by KM Schrott et al in *Fortschritte der Medizin* (1986; 104:153–156), Copyright © 1986, Verlag Fortschritte der Medizin MB Schwappach und Company.

Table 9-3 Incidence of Adverse Reactions in a Japanese Study

	Total	*Severe*	*Very Severe*	*Prior Reactors*	*Patients with Allergic History*
Ionic media*	13.5%	0.45%	0.05%‡	45.1%	24.2%
Nonionic media†	4.2%	0.10%	0.01%‡	12.3%	8.75%

*Total ionic examinations: 77,040.
†Total nonionic examinations: 42,581.
‡One death in each of these two groups.

Source: "Clinical Survey on Adverse Reactions of Iodinated Contrast Media in Advance and Future Trends of Contrast Media" by H Katayama, Proceedings of the International Symposium on Contrast Media, Tokyo, November 6-7, 1987.

Table 9-4 Incidence of Adverse Reactions in an Australasian Study

		Low-Risk Patients			*High-Risk Patients*		
	Total	*Mild*	*Moderate*	*Severe*	*Mild*	*Moderate*	*Severe*
Ionic media*	3.65	3.22	0.24	0.09†	5.63	1.91	0.31
Nonionic media‡	1.53	1.26	0.11	0	1.54	0.12	0.03

*Total ionic examinations: 46,262, ionic low-risk examinations: 45,304, ionic high-risk examinations: 958.
†Two deaths in this group.
‡Total nonionic examinations: 14,738, nonionic low-risk examinations: 7,731, nonionic high-risk examinations: 7,007.

Source: Professor Frederick John Palmer, Department of Diagnostic Radiology, The Prince Henry Hospital, Anzac Parade, Little Bay NSW 2036.

bad taste in the mouth. If these are the unpleasant feelings that induce anxiety, which in turn plays a role in adverse reactions, such anxiety should be encountered less frequently with these newer contrast media. However, if the anxiety is caused by the patient's fearful anticipation of an unpleasant experience, then the newer media should have no effect on the incidence of reactions of any kind for the anticipation of the needle puncture and other aspects of the hospital environment and the examination will be the same.

Blood-Brain Barrier

The low osmolality media have less effect on the BBB than the higher osmolality media. Less toxic effect is therefore expected from cerebral angiography, but I doubt if the reduction of reactions from low osmolality media given intravenously is related to their lesser action on the CNS.

Blood Complications

Because the low osmolality media produce less red cell deformation and less red cell aggregation than the high osmolality media, the adverse reactions expected from this mechanism should be fewer and less severe.[26,27]

Bronchospasm

The nonionic media—iohexol and iopamidol—produced much less bronchospasm (about ⅑) than produced by the ionic high osmolar iothalamate.[100] Pulmonary artery pressure and associated hemodynamic effects are less elevated after the injection of contrast media when nonionic media are used.[37–39]

Calcium

A decrease in ionic calcium in vitro was not produced by the nonionic, iopamidol.[101] The low osmolality agents—the nonionic, iopamidol, and the dimeric, ioxaglic acid—did not produce lowering of the ionized calcium in the general circulation in clinical urography as the conventional ionic agents did.[46]

Cholinesterase

The nonionic low osmolality media—iohexol and iopamidol—were found to inhibit cholinesterase less than did the monomeric ionic medium, iothalamate, and the dimeric ionic low osmolality medium, ioxaglate.[102]

Complement Activation and Protein Binding

Avid protein binding ionic contrast media have induced complement activation more than the weak protein binders.[63,95,97] In another study lipid solubility of the contrast medium was correlated to some extent with activity of the complement system.[47] The new nonionics—iohexol and iopamidol—are more hydrophilic and less lipophilic and bind to protein more weakly. Metrizamide activated complement slightly less than ionic sodium iothalamate and less than the ionics, diatrizoate, acetrizoate, and iopamidol.[103] Iohexol activated human serum complement to a much lesser extent than metrizamide and ioxaglate.[96] The ionic medium most active in activating complement, the cholangiographic agent iodipamide, is seldom used any more. This contrast medium examination has largely been supplanted by ultrasonic and radioactive nuclide examinations.

Drug Synergism

The nonionic, metrizamide, showed synergism with cardiac glycosides, but this contrast medium is no longer used for angiography or urography now that better and less expensive low osmolality agents are available.[52,75]

Hemodynamics

Many reports of animal experiments and human studies find that the low osmolality contrast media produce lesser alterations in the hemodynamic status, both for the pulmonary and the general circulation, primarily because they cause less pronounced fluid shifts. (Three issues of *Investigative Radiology* document a large number of studies, and provide references to other work.[104–106]) With the milder fluid shifts there is less tendency to initiate cardiovascular reflexes. The studies of coronary arteriography indicate lower cardiac toxicity whether measured by contractility and force of ejection or electrical alterations. One exception to this is the work of Morris, who found that when the coronary arteries of the dog were perfused with diluted nonionic contrast media, as would be used in intra-arterial digital subtraction coronary arteriography in man, increased ventricular fibrillation was produced.[50] The lack of sodium ions in the nonionic media was evidently the cause of this finding, because the low osmolality medium, ioxaglate, which contains sodium, did not produce this incidence of fibrillation and the addition of sodium to the nonionic also reduces the fibrillation.

Histamine Release

Histamine release from nonionic contrast media has been reported to be lower than from the ionic monomers, diatrizoate and iothalamate.[107] The same study showed the dimeric ionic, ioxaglate, to be a more potent releaser of histamine on a molar basis, but not on a gram-for-gram of iodine basis than the nonionic agents.[107] Another study found histamine release by ioxaglate to be greater at all concentrations.[96]

REFERENCES

1. Dunn CR, Lasser EC, Sell S et al: Failure to induce hypersensitivity reactions to opaque contrast media analogs in guinea pigs. *Invest Radiol* 1975;10:317–322.

2. Lasser EC: Basic mechanism of contrast media reactions. *Radiology* 1968;91:63–65.

3. Kleinknecht D, Deloux J, Homberg JC: Acute renal failure after intravenous urography: detection of antibodies against contrast media. *Clin Nephrol* 1974;2:116–119.

4. Wakkers-Garritsen BG, Houwerziji J, Nater JP et al: IgE-mediated adverse reactivity to a radiographic contrast medium. *Ann Allergy* 1976;36:122–126.

5. Harboe M, Folling I, Haugen OA et al: Sudden death caused by interaction between a macroglobulin and a divalent drug. *Lancet* 1976;2:285–288.

6. Lasser EC: Adverse reactions to intravascular administration of contrast media. *Allergy* 1981;36:369–373.

7. Brasch RC, Caldwell JL, Fudenberg HH: Antibodies to radiographic contrast agents. Induction and characterization of rabbit antibody. *Invest Radiol* 1976;11:1–9.

8. Brasch RC, Caldwell JL: The allergic theory of radiocontrast agent toxicity: demonstration of antibody activity in serum of patients suffering major radiocontrast agent reactions. *Invest Radiol* 1976;11:347–356.

9. Sweeney MJ, Klotz SD: Frequency of IgE mediated radio contrast dye reactions. *J Allergy Clin Immunol* 1983;71:147.

10. Inman GKE: A comparison of urographic CM with particular reference to the aetiology and prevention of certain side effects. *Br J Radiol* 1952;25:625–631.

11. Lalli AF: The use of chlordiazepoxide hydrochloride (Librium) in preparation for urography. *J Assoc Can Radiol* 1974;25:44–46.

12. Lalli AF: Urographic contrast media reactions and anxiety. *Radiology* 1974;112:267–271.

13. Olsson O: Contrast media in diagnosis and the attendant risks. *Acta Radiol* 1954; 116(suppl):75–83.

14. Littner MR, Rosenfield AT, Ulreich S et al: Evaluation of bronchospasm during excretory urography. *Radiology* 1977;124:17–21.

15. Littner MR, Ulreich S, Putman CE et al: Bronchospasm during excretory urography: lack of specificity for the methylglucamine cation. *AJR* 1981;137:477–481.

16. Berg GR, Hutter AM, Pfister RC: Brief recording: electrocardiographic abnormalities associated with intravenous urography. *N Engl J Med* 1973;289:87–88.

17. Lawton G, Phillips T, Davies R: Alterations in heart rate and rhythm at urography with sodium diatrizoate. *Acta Radiol Diagn* 1982;23:107–110.

18. Mindell H: Personal communication, 1985.

19. Hesselink JR, Hayman LA, Chung KJ et al: Myocardial ischemia during intravenous DSA in patients with cardiac disease. *Radiology* 1984;153:577–582.

20. Broman T, Olsson O: The tolerance of cerebral blood vessels to a contrast medium of the Diodrast group: an experimental study of the effect on the blood brain barrier. *Acta Radiol* 1948;30:326–342.

21. Parvez Z, Moncado R, Fareed J et al: Antiplatelet action of intravascular contrast media. Implications in diagnostic procedures. *Invest Radiol* 1984;19:208–211.

22. Rao VM, Rao AK, Steiner RM et al: The effect of ionic and nonionic contrast media on the sickling phenomenon. *Radiology* 1982;144:291–293.

23. Gafter U, Creter D, Zevin D et al: Inhibition of platelet aggregation by contrast media. *Radiology* 1979;132:341–342.

24. Motomiya T, Yamazaki H: Inhibitory effect of Urografin 76 on platelet function and thrombus formation in vascular catheters. *Angiology* 1980;31:283–290.

25. Lasser EC, Lang JH, Lyon SG et al: Changes in complement and coagulation factors in a patient suffering a severe anaphylactoid reaction to injected contrast material: some considerations of pathogenesis. *Invest Radiol* 1980;15:S6–S12.

26. Aspelin P: Effects of ionic and nonionic contrast media on morphology of human erythrocytes. *Acta Radiol Diagn* 1978;19:675–687.

27. Aspelin P, Teitel P, Almén T: Effect of iohexol on red cell deformability in vitro. *Acta Radiol* 1980;362(suppl):127–130.

28. Bernstein EF: The respiratory factor in angiographic media toxicity. *Radiology* 1965; 84:670–677.

29. Read RC, Johnson JA, Vick JA et al: Vascular effects of hypertonic solutions. *Circ Res* 1960;8:538–548.

30. Boden WE: Anaphylactoid pulmonary edema ("shock lung") and hypotension after radiologic contrast media injection. *Chest* 1982;81:759–761.

31. Greganti MA, Flowers WM JR: Acute pulmonary edema after the intravenous administration of contrast media. *Radiology* 1979;132:583–585.

32. Wood BP, Smith W: Pulmonary edema in infants following injection of contrast media for urography. *Radiology* 1981;139:377–379.

33. Bernstein EF, Palmer JD, Aaberg TA et al: Studies of the toxicity of Hypaque-90 percent following rapid intravenous injection. *Radiology* 1961;76:88–95.

34. Jacobs L, Comroe JH Jr: Reflex apnea, bradycardia, hypotension produced by serotonin and phenyldiagnanide acting on the nodose ganglion of the cat. *Circ Res* 1971;29:145–155.

35. Read RC: Cause of death in angiography. *J Thor Cardiovasc Surg* 1959;38:685–695.

36. Almén T, Aspelin P, Nilsson P: Aortic and pulmonary arterial pressure after injection of contrast media into the right atrium of the rabbit. *Acta Radiol* 1980;362(suppl):37–41.

37. Almén T, Aspelin P: Cardiovascular effects of ionic monomeric, ionic dimeric, and non-ionic contrast media. Effects in animals on myocardial contractile force, pulmonary and aortic blood pressure and aortic endothelium. *Invest Radiol* 1975;10:557–563.

38. Di Donato M, Bongrani S, Cucchini F et al: Cardiovascular effects induced by the injection of a new nonionic contrast medium (iopamidol): experimental study in dogs. *Invest Radiol* 1979;14:309–315.

39. Senac JP, Prefant C, Adda M et al: Comparative study of the effects of two contrast media of different osmolarity on pulmonary hemodynamics and lung function, in Amiel M (ed): *Contrast Media in Radiology*. Berlin, Springer-Verlag, 1982, pp 288–290.

40. Paajanen H: The effect of ionic and nonionic contrast media on the metabolism of prostaglandin E_2 in rat lungs. *Invest Radiol* 1984;19:216–220.

41. Violante MR, Thomson KR, Fischer HW et al: Ventricular fibrillation from diatrizoate with and without chelating agents. *Radiology* 1978;128:497–498.

42. Thomson KR, Violante MR, Kenyon T et al: Reduction in ventricular fibrillation using calcium-enriched Renografin 76. *Invest Radiol* 1978;13:238–240.

43. Wolf GL, Hirschfeld JW: Changes in QT_c interval induced with Renografin 76 and Hypaque 76 during coronary arteriography. *J Am Coll Cardiol* 1983;1:1489–1492.

44. Wolpers HG, Baller D, Ensink FBM et al: Influence of arteriographic contrast media on the Na^{+}/CA^{++} –ratio in blood. *Cardiovasc Intervent Radiol* 1981;4:8–13.

45. Mallette LE, Gomez LS: Systemic hypocalcemia after clinical injections of radiographic contrast media: amelioration by omission of calcium chelating agents. *Radiology* 1983;147:677–679.

46. Freyria A-M, Pinet A, Belleville J et al: Effects of five different contrast agents on serum complement and calcium levels after excretory urography. *J Allergy Clin Immunol* 1982;69:397–403.

47. Higgins CB, Feld GK: Direct chronotropic and dromotropic actions of contrast media: ineffectiveness of atropine in the prevention of bradyarrhythmias and conduction disturbances. *Radiology* 1976;121:205–209.

48. Higgins CB, Gerber KH, Mattrey RF et al: Evaluation of the hemodynamic effects of intravenous administration of ionic and nonionic contrast materials. *Radiology* 1982;1142:681–686.

49. Almén T, Härtel M, Göran N et al: Effects of metrizamide on silver staining of aortic endothelium. *Acta Radiol* 1973;355(suppl):233–238.

50. Morris T, Hayakawa K, Sahler L, et al: Incidence of fibrillation with isotonic contrast media for intra-arterial coronary digital subtraction angiography. *Diagn Imaging Clin Med* 1986;55:109–113.

51. Lasser EC, Lang JH: Inhibition of acetylcholinesterase by some organic contrast media. A preliminary communication. *Invest Radiol* 1966;1:237–242.

52. Harnish PP, Morris TW, Fischer HW et al: Drugs providing protection from severe contrast media reactions. *Invest Radiol* 1980;15:248–259.

53. Arroyave CM, Bhat KN, Crown R: Activation of the alternative pathway of the complement system by radiographic contrast media. *J Immunol* 1976;117:1866–1869.

54. Arroyave CM, Tan EM: Mechanism of complement activation by radiographic contrast media. *Clin Exp Immunol* 1977;29:89–94.

55. Arroyave CM, Schatz M, Simon RA: Activation of the complement system by radiographic contrast media: studies in vivo and in vitro. *J Allergy Clin Immunol* 1979;63:276–280.

56. Heideman M, Jacobsson B, Lindholm N: Activation of the complement system by water soluble contrast media. A preliminary report. *Acta Radiol Diagn* 1976;17:733–736.

57. Lang JH, Lasser EC, Kolb WP: Activation of serum complement by contrast media. *Invest Radiol* 1976;11:303–308.

58. Lasser EC, Slivka J, Lang JH et al: Complement and coagulation: causative considerations in contrast catastrophes. *AJR* 1979;132:171–176.

59. Lieberman P, Siegle RL: Complement activation following intravenous contrast material administration. *J Allergy Clin Immunol* 1979;64:13–17.

60. Till G, Rother U, Gemsa D: Activation of complement by radiographic contrast media: generation of chemotactic and anaphylatoxin activities. *Int Arch Allergy Appl Immunol* 1978; 56:543–550.

61. Bhat KN, Arroyave CM, Crown R: Reaction to radiographic contrast agents: new developments in etiology. *Ann Allergy* 1976;37:169–173.

62. Hasselbacher P, Hahn J: In vitro effects of radiographic contrast media on the complement system. *J Allergy Clin Immunol* 1960;66:217–222.

63. Kolb WP, Lang JH, Lasser EC: Nonimmunologic complement activation in normal human serum induced by radiographic contrast media. *J Immunol* 1978;121:1232–1238.

64. Neoh SH, Sage MR, Willis RB et al: The in vitro activation of complement by radiologic contrast materials and its inhibition by epsilon aminocaproic acid. *Invest Radiol* 1981;16:152–158.

65. Siegle RL, Lieberman P: Measurement of histamine, complement component and immune complexes during patient reactions to iodinated contrast material. *Invest Radiol* 1976;11:98–101.

66. Simon RA, Schatz M, Stevenson DD et al: Radiographic contrast media infusions. Measurement of histamine, complement, and fibrin split products and correlation with clinical parameters. *J Allergy Clin Immunol* 1979;63:281–288.

67. Small P, Satin R, Palayew MJ et al: Prophylactic antihistamines in the management of radiographic contrast reactions. *Clin Allergy* 1982;12:289–294.

68. Schulze B: Serine proteases as mediators of radiographic contrast media toxicity. *Invest Radiol* 1980;15:S18–S20.

69. Ring J, Sovak M: Release of serotonin from human platelets in vitro by radiographic contrast media. *Invest Radiol* 1981;16:245–248.

70. Lasser EC, Lang JH, Lyon SG et al: Glucocorticoid induced elevations of C1-esterase inhibitor: a mechanism for protection against lethal dose range contrast challenge in rabbits. *Invest Radiol* 1981;16:20–23.

71. Ring J, Arroyave CM, Frizler MJ et al: In vitro histamine and serotonin release by radiographic contrast media (RCM). Complement-dependent and independent-release reaction and changes in ultrastructure of human blood cells. *Clin Exp Immunol* 1978;32:105–118.

72. Ring J, Endrich B, Intaglietta M: Histamine release, complement consumption, and microvascular changes after radiographic contrast media infusion in rabbits. *J Lab Clin Med* 1978; 92:584–594.

73. Paajanen H, Kormano M: Prostacyclin (PGI_2) and thromboxane (TXA_2) levels during intravenous contrast medium administration in CT. *Eur J Radiol* 1985;5:243–245.

74. Paajanen H, Kormano M, Uotila P: Modification of platelet aggregation and thromboxane synthesis by intravascular contrast media. *Invest Radiol* 1984;19:333–337.

75. Fischer HW, Morris TW, King AN et al: Deleterious synergism of a cardiac glycoside and sodium diatrizoate. *Invest Radiol* 1978;13:340–346.

76. Harnish PP, DiStefano V: Pharmacological action of radiographic contrast media reduced CSF production in the dog. *J Pharm Exp Thera* 1985;232:88–93.

77. Pfister RC, Hutter AM Jr, Newhouse JH et al: Contrast-medium-induced electrocardiographic abnormalities: comparison of bolus and infusion of methylglucamine iodamide and methylglucamine/sodium diatrizoate. *AJR* 1983;140:149–153.

78. Fischer HW: Hemodynamic reactions to contrast media. *Radiology* 1968;91:66–73.

79. Fischer HW, Colgan FJ: Causes of contrast media reactions. *Radiology* 1976;121:223.

80. Lasser EC, Walters AJ, Lang JH: An experimental basis for histamine release in contrast media reactions. *Radiology* 1974;110:49–59.

81. Ring J, Simon RA, Arroyave CM: Increased in vitro histamine release by radiographic contrast media in patients with history of incompatibility. *Clin Exp Immunol* 1978;34:302–309.

82. Ring J, Arroyave CM: Alteration of human blood cells and changes in plasma mediators produced by radiographic contrast media. *Z Immunitaetsforsch* 1979;155:200–211.

83. Robbins AH, Rosenfield AT, Pizzolato NF et al: Drip infusion urography with meglumine iodamide. *AJR* 1978;131:1043–1046.

84. Rockoff SD, Brasch R, Kuhn C et al: Contrast media as histamine liberators. I. Mast cell histamine release in vitro by sodium salts of contrast media. *Invest Radiol* 1970;5:503–509.

85. Brasch RC, Rockoff SD, Kuhn C et al: Contrast media as histamine liberators. II. Histamine release into venous plasma during intravenous urography in man. *Invest Radiol* 1970; 5:510–513.

86. Lasser EC, Walters A, Reuter SR et al: Histamine release by contrast media. *Radiology* 1971;100:683–686.

87. Rice M, Lieberman P, Siegle R et al: In vitro studies of reactions to radiocontrast agents (ICM) (abstracted). *J Allergy Clin Immunol* 1981;67:S71.

88. Seidel G, Groppe G, Meyer-Burgdorff C: Contrast media as histamine liberators in man. *Agents Actions* 1974;4:143–150.

89. Hamilton G: Contamination of contrast agents by rubber components of 50-ml disposable syringes. *Radiology* 1984;152:539–540.

90. Blais P: DEHP in blood bags and medical plastics. Their limitations. *Can Res* 1981;14:13–18, 22–25.

91. Petersen MC, Vine J, Ashley JJ et al: Leaching of 2-(2-hydroxy-ethylmercapto) benzothiazole into contents of disposable syringes. *J Pharm Sci* 1981;70:1139–1143.

92. Fischer HW: Comments on Hamilton's letter. *Radiology* 1984;152:540.

93. Winding O, Gronvall J, Faarup P et al: Sequelae of intrinsic foreign-body contamination during selective renal angiography in rabbits. *Radiology* 1980;134:321–326.

94. Lang JH, Lasser EC: Inhibition of adenosine triphosphatase and carbonic anhydrase by contrast media. *Invest Radiol* 1975;10:314–316.

95. Lasser EC, Lang JH: Contrast protein interactions. *Invest Radiol* 1970;5:446–451.

96. Mützel W, Siefert HM, Speck U: Biochemical-pharmacologic properties of iohexol. *Acta Radiol* 1980;362(suppl):111–115.

97. Rapaport S, Levitan H: Neurotoxicity of x-ray contrast media: relation to lipid solubility and blood-brain barrier permeability. *AJR* 1974;122:186–193.

98. Lang JH, Lasser EC, Talner LB: Inorganic iodide in contrast media. *Invest Radiol* 1974;9:51–55.

99. Talner LB, Lang JH, Brasch RC et al: Elevated salivary iodine and salivary gland enlargement due to iodinated contrast media. *AJR* 1971;112:380–382.

100. Dawson P, Pitfield J, Britton J: Contrast media and bronchospasm: a study with iopamidol. *Clin Radiol* 1983;34:227–230.

101. Morris TW, Ciaravino V, Sahler LG, Cioffi B: Factors modifying contrast media induced fibrillation. *Acta Radiol Diagn* 1985;26:1–5.

102. Dawson P: Contrast media and enzyme inhibition. I. Cholinesterase. *Br J Radiol* 1983; 56:653–656.

103. Siegle RL, Lieberman P: Leukocyte histamine release related to ionic and nonionic contrast material and similar molecules. *Invest Radiol* 1984;19:S105.

104. *Invest Radiol* 1984;19 (ioxaglate).

105. *Invest Radiol* 1984;19 (iopamidol).

106. *Invest Radiol* 1985;20 (iohexol).

107. Assem ESK, Bray K, Dawson P: The release of histamine from human basophils by radiological contrast agents. *Br J Radiol* 1983;56:647–652.

Chapter 10

Conventional Angiography

Peter Dawson

In spite of the advent in recent years of a number of new techniques for the study of blood vessels, angiography remains the gold standard. Its development depended entirely on the evolution of contrast agents safe enough for intravascular injection, and the realization of this fact came remarkably early. Roentgen reported his discovery of x-rays to the Wurzburg Physical Medical Society on December 28, 1895;[1] by January 1896, less than 1 month later, Haschek and Lindenthal in Vienna had not only reproduced his results and perceived the need for intravascular contrast media but had also devised the first such material and performed the first angiogram with it.[2] The primitive contrast agent was "Teichmann's paste," a mixture of cinnebar, petroleum jelly, and chalk, and the angiogram actually was performed on an amputated hand, rather than on a living patient. They had not, therefore, quite pioneered modern angiography, but their achievement and its impact were remarkable and important. Further cadaver studies were performed by Morton, who published a book on the subject as early as 1896.[3] Kassabian[4] also produced a large textbook in 1907 with a section devoted to blood vessels, which he had studied in cadavers using bismuth subnitrate, lead oxide, and metallic mercury as contrast agents. In 1910 Franck and Alwens[5] injected a suspension of bismuth and oil intravenously in dogs and observed its passage through the right heart and into the lungs.

Interest in developing clinical angiography was clearly running high, but progress in finding a clinically usable agent was slow. In 1923 Sicard and Forestier used Lipiodol for intravenous (IV) injections in both dogs and man.[6] Other than inducing coughing, this substance seemed to produce no ill effects, but, not surprisingly, carotid injections of Lipiodol in dogs proved to be fatal. In the same year Berberich and Hirsch[7] reported the successful use of strontium bromide for arteriography and venography in humans, and in 1924 Brooks[8] used sodium iodide as a contrast agent for lower limb arteriography. Sodium iodide was a toxic and painful material necessitating the use of general anesthesia. Yet, Brooks' experiments were an important landmark because they used iodine for the first time as the contrast-providing element in angiography, though it had already been used in the previous year for IV urography by Osborne and colleagues[9] and 6 years previously in 1918 by Cameron and colleagues for direct cystography and pyelography.[10]

Iodine, and, indeed, barium are excellent choices as contrast-providing elements even when judged on purely physical grounds. Their K-shell electron binding energies are 32 and 33 KeV respectively. This means that the cross-section for a photoelectric reaction with a photon of mean energy in the diagnostic range is very high. It was clear, however, that iodine could not reasonably be delivered in a simple inorganic form, but that some low-toxicity carricr molecule would be necessary. Such a compound was first supplied in 1925–1926 when Binz and Rath in Berlin[11] synthesized a large number of pyridine compounds, some of them iodinated, in a search for agents active against syphilis. They found that some of these compounds (Fig. 10-1) were at least partially selectively excreted in the urine (Uroselectans), leading to their initial use as IV urographic agents by Moses Swick in 1929.[12,13] They were soon used in angiography and proved to be a great advance on all that had preceded them. These monoiodinated compounds were rapidly superseded in the early 1930s by di-iodinated compounds (Fig. 10-2), which offered the improvement not only of a greater iodine content but also of higher solubility, allowing more concentrated solutions to be prepared.

Although there were many technical developments in angiography in the next two decades, there was surprisingly little further progress in contrast agent development. Moses Swick had suggested the use of iodinated hippuric acid as an IV urographic agent,[12,14] but it was found to offer no improvement in toxicity over the agents already in use. It was not until the early 1950s that the chemical problem of producing a low-toxicity iodinated benzene ring-based compound was solved by Wallingford[14] and Hoppe.[15] From this time forward angiographers had a range of outstandingly good intravascular contrast agents at their disposal (Fig. 10-3).

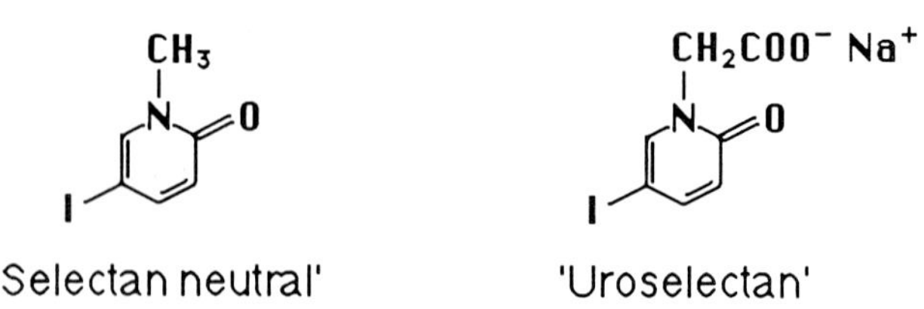

Figure 10-1 Monoiodinated pyridine compounds synthesized by Binz and Rath used for IV urography by Moses Swick.

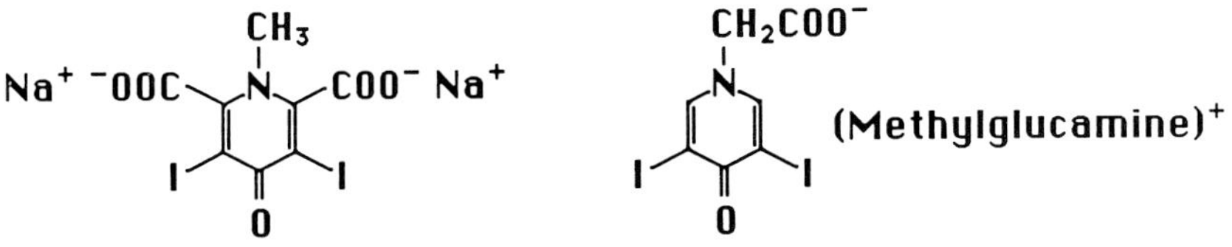

Iodomethamate 'Uroselectan B'

Iodopyracet 'Diodone'

(Binz & Räth / von Lichtenberg 1930)

Figure 10-2 Di-iodinated pyridine compounds. State of the art materials from the early 1930s to the early 1950s.

Although these agents were not entirely free of toxic effects and indeed were associated with occasional major adverse reactions and deaths,[16] they were among the safest products in the pharmacopoeia in terms of the large doses—whether expressed in milliliters, moles, or grams—that could be injected into patients, often rapidly, with, in the great majority of cases, no significant ill effect. They had the further considerable merit of low cost. Nevertheless, their shortcomings—principally, undesirable subjective side effects, including pain, experienced by patients—led to a search for even better agents. The principal thrust of this effort was

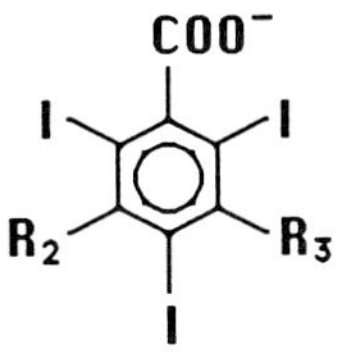

R_2	R_3	Proper name	Commercial name
H	$NHCOCH_3$	Acetrizoate	Urokon, Diaginol
CH_3CONH	$NHCOCH_3$	Diatrizoate	Urografin, Hypaque
CH_3CONH	$CONHCH_3$	Iothalamate	Conray
CH_3CONH	$N(CH_3)COCH_3$	Metrizoate	Isopaque, Triosil

Figure 10-3 The structures of conventional ionic contrast agents.

directed against their high osmolality,[17,18] and it led to the latest generation of low-osmolality compounds[18] listed in Table 10-1. Their chemistry and properties have been discussed in detail in previous chapters.

Such compounds are still somewhat hyperosmolar with respect to plasma[18] and possess some residual intrinsic chemotoxicity.[19–21] However, they are such excellent agents that there are grounds for doubting whether anything significantly better will be developed.

Table 10-2 shows how the toxicity, expressed in terms of lethal dose 50% (LD50), has been improved from the time of the pyridine compounds of the early 1930s to the nonionic low-osmolality compounds of the present day. The improvement is approximately of one order of magnitude. It might be argued that this is a relatively small improvement, given the scale of change in other areas in that half-century period. However, to suggest this would be not so much to criticize the efforts of recent workers but, rather, to pay tribute to the first workers in this field who gave us such excellent agents so early.

PROPERTIES OF THE IDEAL INTRAVASCULAR CONTRAST AGENT

The perfect angiographic agent should be an excellent absorber of x-rays within the diagnostic energy range, and iodine-containing compounds meet this requirement. The iodine should be carried on a stable molecule of low intrinsic toxicity that is not metabolized to any other toxic product and that is excreted entirely from the body. For purposes other than angiography, the route of excretion may of course be important in that the excreting system—renal tract or biliary tree, for example—may be visualized. Solutions of the agent should, ideally, be iso-osmolar with plasma.

Table 10-1 Low-Osmolality Contrast Agents

LOW-OSMOLALITY IONIC AGENT(S)
- Sodium meglumine ioxaglate (Hexabrix—Guerbet, Paris/Mallinckrodt).

FIRST-GENERATION NONIONIC AGENT(S)
- Metrizamide (Amipaque—Nycomed, Oslo).

SECOND-GENERATION NONIONIC AGENT(S)
- Iohexol (Omnipaque—Nycomed, Oslo/Winthrop).
- Iopamidol (Niopam/Isovue—Bracco, Milan/Squibb).
- Iopromide (Ultravist—Schering, Berlin/Berlex).

Table 10-2 Reduction of Contrast Agent Toxicity (expressed as LD50) over a Half Century

Agent	*Lethal Dose 50% (LD50), g iodine/Kg mouse*
Sodium iodomethamate (Uroselectan B)	2.0
Iodopyracet (Diodone)	3.2
Sodium acetrizoate (Urokon-Diaginol)	5.3
Sodium diatrizoate (Urografin/Hypaque)	8.4
Sodium meglumine ioxaglate (Hexabrix)	12.0
Metrizamide (Amipaque)	14.0
Iopamidol (Niopam)	21.0
Iohexol (Omnipaque)	24.0

Although the modern agents do not meet all these requirements in full, the nonionic agents meet most of them. True, they are formulated in solutions with osmolalities in the useful iodine range of approximately three times that of plasma, but this is a great improvement on the six or seven times found in the same iodine range with conventional agents. The modern agents have very low intrinsic chemotoxicity[19] and are stable, undergoing no metabolism (are not degraded before excretion) and being excreted rapidly by passive glomerular filtration. The low chemotoxicity of the second generation nonionic agents is largely the result of a uniform distribution of hydrophilic groups around the molecule, masking the large hydrophobic iodine atoms.[22]

CLINICAL APPLICATION

The importance of clinical angiography in the development of modern medicine would be difficult to overstate. Virtually no region of the body is now outside its scope, and the development of much surgery of the heart and central nervous system (CNS) was largely made possible by its development. Recent advances in imaging techniques, such as computed tomography, magnetic resonance imaging, ultrasound imaging, and duplex Doppler, make it unlikely that angiography will ever quite reclaim its former pre-eminence in radiodiagnosis, but it has been revitalized by three new factors. These are (1) modern contrast agents, discussed above and in other chapters; (2) the advent of digital angiography, discussed in Chapter 13 by Crummy; and (3) the growth of interventional angiographic techniques.

Choice of Contrast Agent

A bewildering variety of contrast agents—conventional ionic, low-osmolality ionic, and low-osmolality nonionic—are now available to the angiographer and the choice seems at first sight to be difficult. However, some simple principles may be followed.

1. Among the conventional contrast agents there is little to distinguish among the various anions in general, but the iothalamates seem to have a somewhat lower neurotoxicity than the diatrizoates and are to be preferred for neuroangiography.
2. Among the conventional agents, meglumine salts are less toxic than are sodium salts to vascular endothelium and to neural tissues and are therefore to be preferred for neuroangiography and phlebography.
3. Neither pure sodium nor pure meglumine salts are suitable for coronary angiography because of the increased risk of ventricular fibrillation. Mixed sodium and meglumine salts (or nonionic agents) should be used instead.
4. The new low-osmolality agents are superior to the conventional agents in terms of patient comfort, manifestations of systemic toxicity, and organ-specific toxicity as demonstrated in a large number of studies.[18,23–27] The only factor preventing their universal use in all patients is their substantially greater cost.
5. There are some differences between the low-osmolality ionic agent (ioxaglate) and the low-osmolality nonionic agents. The ionic agent seems to manifest greater systemic toxicity when used intravenously than the nonionics[28] but on arterial injection, ioxaglate seems to be of comparably low toxicity. For the angiographer, therefore, the two types of new agent may be considered to be approximately equivalent.

In addition to these general considerations, such specific factors as weight, patient age, underlying disease and other risk factors, the organ or system to be investigated, and the nature of the procedure need to be considered when choosing an agent. (Table 10-3.)

Weight

Most angiographers set some arbitrary upper limit to the total dose of contrast material per kilogram of patient that they are willing to administer in a short

Table 10-3 Summary of Indications for Use of Low-Osmolality Contrast Agents

Children
Elderly patients
Impaired cardiac, hepatic, and renal function
Previous reaction to contrast material
Atopy and asthma
Established allergy to other agents
Diabetes
Sickle cell disease
Complex, e.g., interventional procedures requiring high doses of contrast
Angioplasty
Painful angiography, such as digital (hand) angiography

period. This limit might be 1000 mg iodine/Kg for the conventional agents. Although caution must always be exercised, it seems reasonable to suggest that, because of their lower osmolality, their lower (or zero) sodium load, and their lower intrinsic toxicity, a limit of perhaps twice this can be taken when the low-osmolality agents are used.[29,30] This extra margin of safety offered by the new agents is very useful in small infants and in some complex interventional procedures in adults.[30]

Age

In infants and small children, contrast overload poses a real danger. Weight considerations alone might lead one to set an upper dose limit of a conventional contrast agent of only 10 ml, for example. Even this dose may present a threat to some of these patients because they may have poor cardiac status and they do have immature renal function. There is no doubt that the low-osmolality agents offer a greater margin of safety for the reasons discussed above.[29] Clearly, they may make possible procedures on small children that would have been life threatening with the conventional agents.

Similar considerations apply at the other extreme of life when patients are very likely to have multisystem impairment or disorders.

Disease

The above dose/weight considerations are of particular importance in patients of any weight and age with underlying cardiac, renal, or hepatic impairment. Such patients should ideally be offered the benefit of a new low-osmolality agent routinely.[29,31]

Patients with sickle cell disease, myeloma, and diabetes also have underlying disease states that make preferable the use of a new low-osmolality agent.[29,31,32]

Other Risk Factors

Patients who have previously experienced adverse reactions to iodinated contrast agents of any kind, patients with allergies to other materials, and patients with atopy or asthma seem to be at increased risk of adverse reactions to contrast media.[16] Also at increased risk, intriguingly, seem to be members of certain ethnic groups.[16] The status of the new agents in terms of the frequency with which they elicit significant adverse reactions has not yet been entirely clarified, but evidence is mounting that they are significantly less likely to be associated with such reactions[27,33] and are therefore to be preferred in these high-risk groups.

System/Organ Undergoing Examination

Conventional contrast agents in full concentration may cause severe pain on arterial injection. This pain may cause the patient to move, thus degrading the quality of the study, and vascular spasm may be induced. In such procedures as translumbar aortography (TLA) such movement during the procedure may actually be dangerous. It may indeed not be possible to perform some angiographic procedures using local anesthesia.

Because the pain induced by contrast agents seems to be primarily related to their osmolality,[34] it is not surprising therefore that the low-osmolality agents rarely produce such severe pain as to necessitate general anesthesia. Digital angiographic systems with their high-contrast sensitivity, however, allow more dilute concentrations of contrast agents to be used, and consequently, the need for low-osmolality agents is not usually necessary simply to eliminate pain.

Neuroangiography and Cardiac Angiography

Neuroangiography and cardiac angiography are discussed in detail in other chapters. Low-osmolality agents are associated with a lower incidence of all neurotoxic and cardiotoxic effects and, ideally, are to be preferred to the conventional agents in these applications.

Renal Angiography

Contrast agents seem to have a nephrotoxic potential.[35] Particularly when delivered directly into the renal artery in selective renal arteriography, they cause glomerular and tubular injury as indicated by the appearance in the urine of circulating plasma and tubular-specific proteins. The low-osmolality contrast agents, particularly the second-generation nonionic agents, produce significantly less dramatic effects and may be said on this basis to have lower nephrotoxicity. However, the relationship, if any, between these clearly nephrotoxic phenomena and the occasional episodes of clinical renal function impairment seen following contrast administration is by no means clear. The basis of contrast agent nephrotoxicity remains obscure, although several risk factors have been established. These include already impaired renal function, old age, diabetes, and dehydration. Patients in these high-risk categories should also be included in the list of patients who should be given the benefit of a modern low-osmolality contrast agent in any procedure.[35] On the basis of our present understanding, however, we suggest that if the radiologist could only take one precaution against the danger of contrast agent nephrotoxicity it should be to ensure that the patient is not dehydrated.[36]

Type of Procedure

Interventional Procedures

Some angiographic interventional procedures are complex and may involve the administration of large doses of contrast. For reasons discussed above, the new agents should be used when such a procedure is attempted. Many complex procedures may now be performed, which without these new contrast agents and without digital systems that allow smaller doses of dilute contrast agent to be used for each imaging sequence, would have been impossible in the past.[30]

Angioplasty

New agents have been recommended for angioplasty and are used routinely by many radiologists. Conventional agents injected into a blocked or stenosed arterial tree may cause particularly intense pain. Furthermore, they provide a new insult to an endothelium just damaged by the balloon dilation and may increase the risk of postprocedure thrombosis. However, the greater anticoagulant and antiplatelet properties of the conventional contrast agents may be advantageous in angioplasty.[21]

TOXIC EFFECTS AND ADVERSE REACTIONS

The function of contrast media is to provide contrast in blood vessels in angiography and in tissues or body cavities in other radiologic studies. Any effects that they may have on cells, tissues, organs, biochemistry, or physiology are not only irrelevant to their purpose but also undesirable; they must be viewed, no matter how apparently trivial, as aspects of toxicity. Unfortunately, even the best of modern agents possess some residual toxicity mediated by their still higher-than-plasma osmolality, their chemotoxicity, or a combination of the two. A great variety of effects on enzyme systems,[20,22,37] on cells,[19,38–40] on the complement system,[41,42] and on physiology[19,23] are well described. How precisely these effects relate to the various subjective experiences of patients, such as pain, flushing, and nausea, remain unclear. Likewise, the question of how impairment of the function of such organs as the heart, brain, and kidneys is mediated is obscure. Some of the toxic manifestations of contrast agents, such as flushing, are experienced to some degree routinely by most patients and although some authors describe these as mild "adverse reactions" this term is better reserved for more serious phenomena. It is important to understand, however, that there is a wide spectrum of adverse effects. It is uncertain whether the entire spectrum from mild to severe is mediated by the same events or whether the more serious and occasional adverse effects, sometimes described as "idiosyncratic reactions" (Chapter 9), are mediated by some fundamentally different phenomena. In any case, it is clear that there is a large element of arbitrariness in the clinical classification of adverse effects as minor, intermediate, and severe.

A further problem in understanding toxic effects springs from the difficulty of separating organ-specific effects from systemic effects. For example, the negative inotropic and chronotropic effects of contrast agents on the heart, a specific cardiotoxicity, may contribute greatly to systemic hypotension, but this may not be clinically obvious.

Common Side Effects

Of the common side effects the most important to the angiographer is pain, which may, on occasion, be severe enough to demand general anesthesia. Pain, although the detailed mechanisms have not been unraveled, is certainly a largely osmolality-mediated phenomenon. Not surprisingly, therefore, it is virtually eliminated by the use of low-osmolality contrast agents.[17,18,34]

Major Adverse Effects

These effects are discussed in Chapter 9 by Fischer. They are much less commonly associated with arterial than with venous injections. This may be because, following venous injection, the pulmonary bed, which is rich in histamine stores, is exposed early to high concentrations of contrast. Or, it might be due to the fact that endothelium is less damaged generally by injection into arteries than into veins because of more rapid blood flow, such damage being claimed to be the initiating event for some major reactions.[19,43] Such events are nevertheless a source of constant anxiety for angiographers. What evidence is available indicates that low-osmolality contrast agents are safer in this regard.[27,33]

Nonionic Agents and Thrombus Formation

One doubt about the safety of nonionic agents in angiography has been raised by Robertson,[44] who noted a greater likelihood of thrombus formation in syringes of contrast contaminated with blood when the contrast agent is nonionic. Such thrombi might inadvertently be injected into the patient with disastrous consequences. This occurrence is predictable on the basis of less effect on the coagulation cascade and platelet aggregation by the nonionic agents,[21] but there are many complex interacting factors in practice, such as the amount of contaminating blood, how well it is mixed with contrast, and the material of the syringe. One element of the low toxicity of nonionic agents is their small effect on platelets and coagulation.[19] It seems it is not possible to have a low-toxicity medium that has powerful anticoagulant and antiplatelet effects.[19,21] To the extent that there may be an occasional problem in this regard with the nonionic agents, the answer is to use scrupulous angiographic technique, rather than to hesitate to use the safer and better tolerated new agents when they are indicated.

REFERENCES

1. Roentgen WC: Über eine neue Art von Strahlen. Sitzgsber. *Physikmed Ges Würtsburg* 1895; 137:132–141.

2. Haschek E, Lindenthal OT: A contribution to the practical use of the photography of Roentgen. *Wien Klin Wochenschr* 1896;9:63–64.

3. Morton WG, Hammer EW: *The X-ray, or, Photography of the Invisible and Its Value in Surgery*. New York, American Technical Book Company, 1896.

4. Kassabian MK: *Roentgen Rays and Electrotherapeutics, with Chapters on Radium and Phototherapy*. Philadelphia, JB Lippincott, 1907.

5. Franck O, Alwens W: Kreislaufstudien am Roentgenschirm. *Münch Med Wochenschr* 1910; 51:950–955.

6. Sicard JA, Forestier G: Injections intravasculaires d'huile iodee sous controle radiologique. *CR Soc Biol* (Paris) 1923;88:1200.

7. Berberich J, Hirsch S: Die roentgenographische darstellung der arterien und venen am lebenden. *Münch Klin Wochenschr* 1923;49:2226–2229.

8. Brooks B: Intra-arterial injection of sodium iodide. *JAMA* 1924;82:1016–1020.

9. Osborne ED, Sutherland CG, Scholl AJ et al: Roentgenography of the urinary tract during excretion of sodium iodide. *JAMA* 1923;80:368–373.

10. Cameron DF: Aqueous solutions of potassium and sodium iodides as opaque mediums in roentgenology; preliminary report. *JAMA* 1918;70:754–755.

11. Binz A: The chemistry of Uroselectan. *J Urol* 1931;25:297–301.

12. Grainger RG: Intravascular contrast media—the past, the present and the future. *Br J Radiol* 1982;55:1–18.

13. Marshall VF: The controversial history of excretory urography, in Witten DM, Myers GH, Utz DC (eds): Emmett's *Clinical Urography*, vol I, ed 4. Philadelphia, WB Saunders, 1977, pp 2–5.

14. Wallingford VH: The development of organic iodide compounds as X-ray contrast media. *J Pharmacol Assoc* 1953;42:721–728.

15. Hoppe JO: Some pharmacological aspects of radio-opaque compounds. *Ann NY Acad Sci* 1959; 78:727–739.

16. Ansell G, Tweedie MCK, West CR et al: The current status of reactions to intravenous contrast media. *Invest Radiol* 1980;15:32–38.

17. Grainger RG: The osmolality of intravascular contrast media. *Br J Radiol* 1980;53:739–746.

18. Dawson P, Grainger RG, Pitfield J: The new low osmolar contrast media: a simple guide. *Clin Radiol* 1983;34:221–226.

19. Dawson P: Chemotoxicity of contrast media and clinical adverse effects: a review. *Invest Radiol* 1985;20:52–59.

20. Dawson P, Edgerton D: Contrast media and enzyme inhibition. I. Cholinesterase. *Br J Radiol* 1983;56:653–656.

21. Dawson P, Hewitt HP, Mackie IJ et al: Contrast, coagulation and fibrinolysis. *Invest Radiol* 1986;21:248–252.

22. Howell MJ, Dawson P: Contrast agents and enzyme inhibition. II. Mechanisms. *Br J Radiol* 1985;58:845–848.

23. Hayward R, Dawson P: Contrast agents in angiocardiography. *Br Heart J* 1985;52:361–368.

24. Iopamidol: intravascular and intrathecal applications. Worldwide symposium. *Invest Radiol* 1983;19(suppl):160–287.

25. Ioxaglic acid: a new low osmolality contrast medium. *Invest Radiol* 1984;19(suppl):289–392.

26. A worldwide clinical assessment of a new non-ionic contrast medium: Iohexol. *Invest Radiol* 1985;20(suppl):1–121.

27. McClennan BL: Low osmolality contrast media: premises and promises. *Radiology* 1987; 162:1–8.

28. Manhire R, Dawson P, Dennet R: Contrast agent induced emesis. *Clin Radiol* 1984; 35:369–370.

29. Dawson P, Pitfield J: Hexabrix and the sodium problem. *Br J Radiol* 1982;55:933–934.

30. Dawson P, Hemingway AP: Contrast agent doses in interventional radiology. *J Intervent Radiol*, 1987. In Press.

31. Grainger RG: The clinical and financial implications of the low osmolar radiological contrast media. *Clin Radiol* 1984;35:251–252.

32. Rao VM, Rao AK, Steiner RM et al: The comparative effect of ionic and non-ionic contrast media on the sickling phenomenon. *Radiology* 1982;144:291–293.

33. Schrott KM, Behrends B, Clauss W et al: Iohexol in excretory urography: results of a drug monitoring programme. *Fortschrift Medezin* 1986;4:153–156.

34. Mutzel W, Speck U: Effects of ionic and non-ionic contrast media after selective peripheral and cerebral arterial injections, in Taenzer V, Zeitler E (eds): *Contrast Media in Urography, Angiography and Computerised Tomography*. Stuttgart, Georg Thieme Verlag, 1983.

35. Dawson P: Contrast agent nephrotoxicity. An appraisal. *Br J Radiol* 1985;58:121–124.

36. Trewhella M, Forsling M, Rickards D et al: Dehydration, antidiuretic hormone and the intravenous urogram. *Br J Radiol* 1987;60:445–447.

37. Lasser EC, Lang JH: Acetylcholinesterase inhibition by some organic contrast agents. *Invest Radiol* 1966;2:237–242.

38. Dawson P, Harrison MJ, Weisblatt E: Effect of contrast media on red cell filtrability and morphology. *Br J Radiol* 1983;56:707–710.

39. Lasser EC, Walters A, Lang JH: An experimental basis for histamine release in contrast media reactions. *Radiology* 1974;110:49–59.

40. Assem ESK, Bray K, Dawson P: The release of histamine from human basophils by radiological contrast agents. *Br J Radiol* 1983;56:647–652.

41. Arroyave CM, Schatz M, Simon RA: Activation of the complement system by radiographic contrast media: studies in vivo and in vitro. *J All Clin Immunol* 1979;63:276–280.

42. Dawson P, Turner MW, Bradshaw A et al: Complement activation and generation of C3a anaphylatoxin by radiological contrast agents. *Br J Radiol* 1983;56:447–448.

43. Lasser EC, Lang JH, Hamblin AE et al: Activation systems in contrast idiosyncracy. *Invest Radiol* 1980;15:2–5.

44. Robertson HJF. Blood clot formation in angiographic syringes containing non-ionic contrast media. *Radiology* 1987;163:621–622.

Chapter 11

Neuroangiography

Michael R. Sage

HISTORICAL ASPECTS

The first intra-arterial injection of radiopaque material into the carotid artery and hence the intracranial circulation was performed by Egas Moniz in 1927.[1,2] Initially he attempted percutaneous injection of 7 ml of 70% strontium bromide but did not achieve vessel opacification. Subsequently, in his ninth patient, after direct cut-down on the carotid artery, good opacification was obtained by injecting 5 ml of 25% sodium iodide. Moniz described his first 90 cases in a book published in 1931.[1]

In searching for a suitable contrast material, Moniz examined solutions of iodine and bromine salts. He chose iodine because of its higher atomic number and therefore greater opacity.[2] Although sodium iodide proved to be the initial medium of choice, it irritated blood vessels and produced pain on injection.[1,2] Therefore, a colloidal preparation of thorium dioxide was introduced for cerebral angiography because it was found to be better tolerated. For a time, thorium dioxide (Thorotrast) was the contrast agent of choice for cerebral angiography.[2] It became apparent, however, that Thorotrast had many disadvantages, including the induction of neoplasms due to the natural alpha radiation of thorium.[3]

Subsequently, in 1931 Binz, while developing iodine-carrying molecules for antibiotic purposes, found that a series of pyridone compounds were secreted in substantial amounts by the kidneys and liver.[4] Swick[5,6] and Von Lichtenberg[7] explored the use of these molecules for urography.[8] Solubility was increased and toxicity reduced by replacing the methyl group with an acetyl group, and as a result Uroselectan became the first successful urographic agent. Uroselectan and other improved pyridine products, such as Diodrast (diodone) and Neo-iopax (iodoxyl, Uroselectan B), were later employed in general angiography and in cerebral angiography.[2]

IONIC CONTRAST MEDIA

It was not until the early 1950s that the suggestions and work of Swick,[5,6] Wallingford,[9] and Hoppe et al.[10] led to the first tri-iodinated contrast agent. Acetrizoic acid was developed, which resulted in great improvements in tolerability and opacification. Further derivatives of tri-iodobenzoic acid were then developed. These were made soluble by salification with sodium, meglumine, or both, with added calcium or magnesium. Such ionic contrast media dissociate in solution to form an anion and a cation; only the anion carries iodine atoms and is radiopaque. The cation has no function except to serve as a solubilizing agent. All conventional ionic contrast media are monomeric salts of tri-iodinated, substituted benzoic acid, and the ionic, intravascular contrast media formulations, now commercially available, were recently documented.[11]* Currently, the most extensively used ionic contrast media for neuroangiography are methylglucamine salts of iothalamate and diatrizoate.

In 1966, a dimer of iothalamic acid—iocarmate (Dimer X), a methylglucamine salt of iocarmic acid—was introduced.[12] This substance, however, did not gain widespread acceptance.

An ionic contrast medium with a low osmolality was developed by the synthesis of a mono-ionized dimer, Hexabrix or ioxaglate (Guerbet). This medium has six atoms of iodine per molecule. As each molecule dissociates into two ions, Hexabrix provides the same iodine:particle ratio of 3:1 as the nonionic preparations. This dimer has an intermediate position between conventional ionic and nonionic contrast media.[8]

NONIONIC MONOMERS AND DIMERS

A major advance in water-soluble contrast media was the synthesis of metrizamide (Amipaque) by Nyegaard in Oslo, following a suggestion by Almén.[13] By transforming the ionizing carboxyl group of conventional ionic contrast medium salts into a nondissociating group, such as an amide (-CONH-) group, the solute concentration could be reduced without the loss of iodine content. Such nonionic contrast media do not require a salifying agent and therefore have a lower osmolality than corresponding ionic contrast media.[14] Unfortunately, metrizamide is unstable in solution, necessitating lyophilization and preparation of a fresh solution before each use.[15] A second generation of nonionic monomers, including iopamidol, which was developed by Bracco in Milan, and iohexol, which was

---

*This catalog of intravascular contrast media available as of 1986 is reproduced in Appendix I.

developed by Nyegaard, achieved improved water solubility.[15] Both iopamidol and iohexol are now widely used in neuroangiography.

More recently, nonionic dimers have been proposed and synthesized, including iodecol, iotrol (by Sovak), and iodixanol (by Nyegaard). These dimers have a lower osmolality than the nonionic monomers of equivalent iodine concentration. They are not yet readily available for routine neuroangiography.

The development of the nonionic contrast media has been a major advance in reducing neurotoxicity. However, an ideal contrast medium for neuroangiography does not yet exist.[16] Such a contrast medium would have to be more hydrophylic than those currently available to be completely inert and would have an osmolality at diagnostically useful concentrations under the pain threshold; because satisfying these requirements would require high viscosity, contrast media based on benzene chemistry will remain a compromise for the foreseeable future.[16]

CONTRAST MEDIUM APPLICATIONS

Conventional Selective Angiography

A concentration of about 140 mgI/ml at the site of interest is necessary for adequate x-ray attenuation in conventional radiology.[17] To achieve such attenuation, the concentration of contrast medium to be injected depends on the degree of dilution, proximal to and at the site, that is caused by fluid dynamics in the system.[17] In general, by performing selective angiography a relatively small volume of contrast medium, whether ionic or nonionic, can be used. Too, as the bolus is administered close to the site of interest, a relatively low iodine concentration of between 280–300 mgI/ml is used. The use of a rapid bolus reduces the degree of dilution and ensures adequate intravascular x-ray attenuation.

Nonionic contrast media are preferred for conventional selective neuroangiography because of their reduced toxicity due to their lower osmolality at an equivalent iodine concentration. However, although the nonionic contrast media have a lower osmolality at an equivalent iodine concentration when compared to the ionic contrast media, their viscosity is similar or greater.[11] Therefore, to ensure rapid injection, it is necessary to warm contrast media, particularly nonionic formulations, before injection.

Digital Angiography

In recent years, there has been a rapid increase in the use of digital imaging processing, logarithmic signal amplification, and mask-mode subtraction.[18] It is likely that, in the future, digital subtraction angiography (DSA) will replace conventional angiography.

Intravenous DSA

The advantages and disadvantages of intravenous (IV) DSA have been extensively reviewed.[18] Intravenous studies can be performed with either peripheral or central injections of contrast medium. Although dilution occurs following an IV bolus, the digital technique permits visualization of vessels containing as little as 2% iodine.[18] Both ionic and nonionic contrast media are used. Because of the dilution factor, however, a relatively large volume of contrast medium is required. Therefore, the cost of nonionic contrast media is relatively prohibitive in intravenous DSA.[19]

Intra-arterial DSA

Even with the finest DSA equipment, intra-arterial DSA offers advantages over intravenous DSA and in the future will probably be preferred over conventional angiography.[18] Although an aortic arch injection may be performed, in practice selective catheterization is more appropriate in the majority of clinical situations. Ionic or nonionic contrast media at a similar iodine concentration to that used for conventional angiography are used. Because of the digital technique, however, either a reduced volume of contrast medium is injected at a slower rate, or the contrast medium is diluted with isotonic saline or water before injection.

PATHOPHYSIOLOGY

Since the introduction of nonionic contrast media, there have been several excellent reviews of the neurotoxicity of water-soluble contrast media.[16,17,20,21] In general, the overall toxicity of nonionic contrast media is less than equivalent iodine concentrations of ionic contrast media.

The systemic and local responses to contrast media are sometimes divided into chemotoxic and osmotic effects.[22] The chemotoxicity of ionic contrast media depends on the chemical properties of the basic iodinated substance (anion) and on the nature and mixture of the cation. The chemotoxicity of nonionic contrast media is determined by the overall structure of the molecule. The osmotic effect of both ionic and nonionic contrast media depends on the number of solute particles present in solution. The role of viscosity is yet to be clearly defined.

General Systemic Response

The rapid injection of a large bolus of contrast medium for arch angiography causes systemic and visceral vasodilation. The fall in systemic blood pressure may be profound and is particularly significant in patients with cerebrovascular insufficiency.[17] Such hypotension usually reaches a maximum level in about a minute and is rapidly reversed. Bradycardia is commonly observed, although the severity of both hypotension and bradycardia is less with nonionic contrast media. The injection of a large bolus into the aorta leads to an increase in the plasma volume; such expansion is more marked with the hypertonic ionic contrast media. Other physiologic and toxic reactions to such systemic angiography are covered in other chapters.

Pain and Heat

Sensations of pain and heat often follow the selective injection of ionic contrast media during cerebral angiography.[22–24] The sensation of pain is less severe with nonionic contrast media.[25–27] The increased osmolality of ionic contrast media is therefore implicated as a major causative factor of pain.[22,23] The critical threshold for vascular pain is an osmolality of 650–750 mosm.[24] Although the algogenic effect of osmotic substances has been known for a long time,[28] chemotoxicity also plays a role in the pain-producing potential of ionic contrast media; it has been shown that sodium salts cause greater pain reaction than meglumine salts.[24] However, because meglumine salts are used in cerebral angiography, the osmotic effect is presumably the major causative factor of pain. Arteriography of the external carotid artery is considerably more uncomfortable than arteriography of the common or internal carotid artery due to vasodilation of the vessels of the soft tissues of the scalp, face, and neck.[23] However, Hagen and Klink feel that vasodilation is not an important causative factor in the origin of pain but that biochemical mechanisms play a primary role.[22]

Multiple intracarotid injections of contrast media are sometimes made during cerebral angiography. Hangover effects of pain have been described for contrast media after repeat intra-arterial injections.[29] Clinical studies of peripheral angiography, which have been supported by animal studies, indicate that increased vascular pain occurs when ionic contrast media are injected repeatedly at intervals of less than 30 minutes. This phenomenon, which is independent of the osmolality of contrast media,[29] subsides within about 60 minutes. However, no synergism regarding pain has been demonstrated with nonionic contrast media, at least up to the second injection. In contrast, cross-synergism with double injections of ionic dimers and nonionic contrast media occurs but only when the nonionic contrast medium is given as a second injection.

Therefore, nonionic contrast media produce less discomfort on both initial and subsequent injections when compared with equivalent ionic contrast media, provided that the nonionic injection is not preceded by an injection of ionic contrast medium.

Vasodilation and Vasospasm

Whether or not it is responsible for the pain associated with selective angiography, vasodilation following intra-arterial contrast media has been well documented.[23,30–32] Intracarotid injection of contrast medium produces vasodilation, primarily of the external carotid circulation.[16,23] Vasodilation of the internal carotid artery and the intracranial circulation seems to be absent or less marked,[30,31] but has been documented experimentally with ionic contrast media.[33] In general, because the osmolality of the contrast media seems to be the major factor in producing vasodilation,[16] such changes are greater with ionic than nonionic contrast media.

Although vasospasm of the intracranial vessels has not been observed with ionic contrast media, it has been suggested at least experimentally that nonionic contrast media may on occasions lead to vasospasm of the intracranial vessels.[34,35] This effect may warrant further consideration in the future.

Blood Pressure and Pulse

Hypotension commonly occurs after selective cerebral angiography with ionic contrast media.[16,30,31] Once again, this response is less marked or absent with the nonionic contrast media.[36] Hilal[30,31] concluded that the injection of ionic contrast media into the carotid artery produces two phases of hypotension; the first is caused by severe bradycardia resulting from irritation of receptors in the extracerebral carotid vessels, and the second phase is due to the effect of the contrast media on brain centers. In reviewing the possible mechanisms causing the hypotension and bradycardia, Sovak[16] concluded that the hemodynamic effects of intracarotid contrast media result from a composite action on the cardiovascular brain centers and on the receptors in the carotid body and the external carotid bed.[37,38] Once again, the increased response elicited by ionic contrast media seems to be related to both an osmotic and chemotoxic effect.[16,39]

Although hypotension seems to be the common response to intracarotid contrast media, Velaj et al. showed, in rabbits, a paradoxical increase in blood pressure followed by a hypotensive phase with variable bradycardia and a final return of blood pressure levels to mildly above baseline.[40] There is no consistent correlation between systemic blood pressure alterations and blood-brain barrier disruption.[40,41]

Because of these cardiovascular effects, intracarotid ionic contrast media commonly cause an alteration in cardiac rhythm, particularly bradycardia.[16,37–39] Although bradycardia seems to be more common than tachycardia,[42–45] others have described a tachycardial effect.[46,47] At times, the incidence of tachycardia has been higher than bradycardia, particularly after vertebral angiography.[48]

As with hypotension, it is likely that the cardiac reflexes are mediated by receptors located in both the carotid circulation and the vasomotor brain centers. One study showed that the hypertonicity of ionic contrast media is likely to be at least partly responsible for these cardiovascular responses. It found that, with increased osmolality of test mannitol solutions, bradycardia and hypotension increase.[16] The changes are more pronounced with the sodium salts than with the meglumine salts of ionic contrast media.[45]

The effect of nonionic contrast media on blood pressure and pulse has been shown to be much less marked than of ionic contrast media at equivalent iodine concentrations.[36,43,49–51]

Interaction with Blood

Interaction of contrast media and erythrocytes has been well documented, with resultant increased erythrocyte crenation,[52] aggregation,[16] and rouleaux formation.[53] These changes are a combination of hypertonicity and chemotoxicity,[16] with the smallest changes occurring with the most hydrophylic contrast media with the lowest osmolality. At least theoretically, such aggregation and clumping of red blood cells could lead to local stasis and occlusion of arterial branches within the microcirculation.[21] The effect of various contrast media on coagulation, fibrinolysis, and platelet function has been extensively studied.[54,55] In vitro studies show that ionic contrast media have a much greater effect on these parameters than equivalent concentrations of nonionic contrast media.[54] The major component in the inhibitory effect on coagulation seems to reside in the inherent toxicity of the contrast medium, with only a minor component being attributable to the high osmolality of ionic contrast media compared to nonionic contrast media. Similarly, the strong inhibitory effect of ionic contrast media on platelet aggregation cannot be explained by osmolality and seems to be related to inherent toxicity of the contrast media itself.[54] All in vitro coagulation and platelet function studies clearly indicate that the new nonionic contrast media represent a major step forward in improving biocompatibility.[54]

Although in vitro studies have clearly demonstrated contrast media's effect on coagulation, fibrinolysis, and platelet function, one in vivo study during cerebral angiography showed no statistically significant effect of either ionic or nonionic contrast media on systemic coagulation or platelet function.[54]

In addition to affecting the components of blood, radiographic contrast media increase its viscosity.[52] At least theoretically, an increase in blood viscosity could lead to a decrease in blood flow during the passage of contrast medium and hence increase the time exposure of the microcirculation to such agents. The new nonionic contrast media have a similar or greater viscosity than equivalent concentrations of ionic contrast media, thereby suggesting that the role of viscosity in pathophysiology and hence toxicity is probably not great. However, the relative viscosity of contrast media used in neuroangiography warrants further consideration.

Cerebral Blood Flow

As previously discussed, intracarotid injections of contrast media, particularly hypertonic ionic contrast media, may lead to a variety of hemodynamic responses, including vasodilation or vasospasm; hypotension; bradycardia or tachycardia; increased blood viscosity; red cell crenation and aggregation; and alterations in blood coagulation, fibrinolysis, and platelet aggregation. Any or all of these responses may lead to some alteration in cerebral blood flow.

Unfortunately, the literature related to the effects of contrast media on cerebral blood flow has some discrepancies,[16] presumably reflecting the difficulty in fully controlling the interplay of various autoregulative mechanisms governing local and systemic hemodynamics in experimental models. For example, Hilal[30,31] suggests that the greater vasodilation of the external carotid circulation, compared to the internal carotid circulation, leads to a decrease of internal carotid flow in favor of the dilated external carotid bed if a common carotid injection is made. In contrast, a recent experimental study indicated that intracarotid infusions of hypertonic solutions, such as ionic contrast media, induce cerebral vasodilation and flow increases by impairing autoregulation.[56] In addition, Herrschaft et al. reported increased cerebral blood flow with carotid angiography.[57]

Although Gonsett[20] indicates that ionic contrast media do cause an increase in cerebral blood flow, most of the blood is shunted from the arterioles to the venules via bypass channels, and hence capillary circulation is slowed down. This finding is supported by the fact that blood-brain barrier lesions produced by contrast media are preferentially localized on arterioles and venules, rather than on capillaries.

Therefore, it seems that ionic contrast media do lead to some alteration in cerebral blood flow, presumably an increase. However, cerebral blood flow is remarkably stable even when potent adrenergic manipulation is attempted or profound functional shifts of regional blood flow occur.[16] It is likely that, with nonionic contrast media at least, the effect on cerebral blood flow is probably not great. A reduction in regional cerebral blood flow has been demonstrated in areas where increased blood-brain barrier permeability is produced experimentally with hypertonic mannitol[58] and hypertonic ionic contrast media,[40] but not with nonionic contrast media.[40]

BLOOD-BRAIN BARRIER

Intracarotid Injection

The blood-brain interface makes up the blood-brain barrier (BBB).[59] In nonneural tissues, the endothelium of the capillary wall allows free passage of ions and poorly fat-soluble nonelectrolytes, up to the molecular size of albumin, between the blood and interstitial fluid.[59] In contrast, the endothelium of the cerebral capillaries has a continuous basement membrane, with cells being connected by a continuous belt of tight junctions,[59,60] and vesicular transport (pinocytosis) is minimal.[61] Because of these morphologic characteristics, the endothelium of the cerebral capillaries has the permeability properties of an expanded plasma membrane.[61–64] It acts as a selective filter,[16] controlling the free passage of any substance between blood and brain, hence maintaining the homeostasis of the neuronal environment.

The BBB is the contact surface between the plasma and the luminal membrane of the endothelium. The likelihood of a particular substance entering this membrane and hence the extracellular fluid of the brain depends on its relative affinity for four molecules present at the interface; namely, plasma water, membrane lipid, plasma protein, and membrane protein.[65] Fat-soluble substances, such as xenon, pass freely across the barrier, whereas substances that bind to protein, such as bilirubin, do not. In general, water-soluble substances, including contrast media, also do not cross the normal BBB although certain water-soluble, metabolically important substances, such as glucose, pass across by active transport by membrane proteins. Therefore, in the presence of an intact BBB, minimal contrast medium passes into the brain after intracarotid injection. A concentration of less than 2% of that of plasma has been demonstrated in nephrectomized rats after an intracarotid injection.[66]

Increased permeability of the BBB after carotid injection of various ionic contrast media has been well documented in experimental animals.[41,67–75] More recently, using computed tomography (CT), such disruption has been demonstrated during human cerebral angiography with ionic contrast media.[76–78] Such damage facilitates the passage of contrast media into the brain, which probably accounts for its neurotoxicity.[21,30,76,79,80]

It is now generally agreed that the hyperosmolality of contrast media is the primary cause of opening of the BBB.[16,68,70,81] The mechanism of breakdown is uncertain, but it has been suggested that the hypertonic ionic contrast media lead to shrinkage of the endothelial cells and subsequent separation of the tight junctions.[20,82,83] The action is probably multifactorial, however, as increased pinocytosis (vesicular transport) has been observed.[20,83,84] Transient cerebral dehydration followed by cerebrovascular dilation may also contribute to BBB opening,[81] but there is no consistent correlation between systemic blood pressure alterations and BBB disruption.[40] Some authors have suggested that the red cell aggregation, which occurs with hypertonic contrast media, may play an important role in BBB disruption, with sludging leading to anoxic change and a subsequent increase in capillary cell permeability.[40,85,86]

The disruption of the BBB by hypertonic contrast media and other solutions has been shown to be reversible.[68,70,81,83] After the carotid infusion of hypertonic solutions, the BBB is reconstituted in 5 minutes to 2 hours, depending on the conditions of the infusion and the marker used for BBB testing.[21]

As hypertonic insults that produce little increase in permeability to large molecules, such as albumin-conjugated dyes, may produce a larger increase in permeability to smaller markers, such as sodium 22 and phosphorus 32,[21] the BBB opening does not appear to be an all-or-none phenomenon.

Rapoport, Thompson, and Bidinger[68] have investigated the threshold concentration for barrier opening by hypertonic contrast media, demonstrating increased permeability to Evans blue-albumin with an osmolality range from 0.8 to 1.5 osmol. Although the osmolality of contrast media is important in determining the BBB disruption, the infusion time of hypertonic solutions is also a significant factor.[41,81] Therefore, both the osmolality and duration of injection are important factors during cerebral angiography in determining alterations in the permeability of the BBB. A decrease in regional cerebral blood flow has been demonstrated in areas where BBB opening is demonstrated following hypertonic contrast media.[40] The intracarotid injection of ionic contrast media does seem to produce shunting of the blood from the arterioles to the venules via bypass channels, and capillary circulation is slowed down. This effect on the microcirculation probably explains why BBB lesions are preferentially localized on arterioles (60%) and venules (25%), rather than on capillaries (12%).[20]

Although ionic contrast media used for cerebral angiography lead to disruption or increase in permeability of the BBB in both experimental animals and humans, equivalent concentrations of nonionic contrast media produce little or no change in the integrity of the BBB in experimental animals.[40,73–75,87] This lack of effect is probably explained by the fact that the osmolality of nonionic contrast media falls below the threshold of 0.8 osmol/kg suggested by Rapoport et al.[68]

Although the difference in osmolality between ionic and nonionic contrast media is probably the major factor in determining their effect on the BBB, there also seems to be a chemotoxic factor. Hypertonic solutions of glucose, sodium chloride,[80] and mannitol[72,88,89] produce similar but less pronounced effects on the BBB than ionic contrast media with an equivalent osmolality. Accordingly, the effect on the BBB and the toxicity cannot be explained by osmotic action alone.[40,90,91]

The actual molecular structure of various ionic contrast media is important in determining their effect on the BBB.[20] Gonsett[20] using both radioisotope (P32) and fluorescence markers (Evans blue) observed BBB toxicity in decreasing order for acetrizoate, diatrizoate, metrizoate, iothalamate, and iodamide. Moreover, for each contrast medium, the sodium salt was more toxic than the meglumine salt. The addition of calcium and magnesium ions to the contrast media seemed to reduce their effect on the BBB.[20]

Therefore, if ionic contrast media are to be used for cerebral angiography, after considering relative osmolality, pure methylglucamine salts should be preferred, as even a mixed salt with a small amount of sodium has been shown to disrupt the BBB in humans.[76]

Intravenous Injection

During intravenous DSA of the cerebral vessels, a large IV bolus of contrast medium is given. After the bolus injection, the contrast medium is rapidly distributed throughout the vascular and extracellular spaces of the nonneural tissue.[92] Peak blood levels are reached almost immediately after injection; levels fall rapidly during the next 2 minutes as the contrast medium equilibrates between the plasma and the extracellular fluid.[93] After this period there is a more gradual fall in plasma level related to renal excretion. In normal, nonneural tissues practically no water-soluble contrast medium is bound to cell membranes or taken up by the cells, but instead it is rapidly and extensively distributed outside blood vessels into volumes approaching that of the extracellular fluid.[94–97]

In neural tissue, the BBB prevents rapid distribution of water-soluble contrast media into the brain extracellular fluid. After IV infusion, contrast material is delivered to the brain by the carotid arteries. Although alteration of the cerebral capillaries[84] and the permeability of the BBB[98] has been demonstrated in experimental animals after large doses of IV contrast medium, Neuwelt et al.[88] reported no significant neurologic deficit in patients given large doses of IV contrast media after deliberate breakdown of the BBB with mannitol.

NEUROLOGIC EFFECTS

Many of the neurologic complications of cerebral angiography are probably not related to the contrast media but to the catheterization procedure.[21] Other complications may result from the hemodynamic effects of contrast media discussed previously, such as hypotension, bradycardia, vasodilation or vasospasm, alterations in regional cerebral blood flow, and an increase in blood viscosity and aggregation clumping of red blood cells.

The actual role of contrast media in such complications is not clear. However, contrast media, particularly ionic contrast media, have been shown experimentally to alter neuronal function severely when introduced directly into the nervous system.[21] Direct application of contrast media to the cortex, even in isotonic solutions, can cause spontaneous cortical spiking.[21] The bathing of hippocampal neurones[99] and spinal preparations[100] indicates that contrast media can cause a combination of excitatory effects associated with their chemical molecular structure and inhibitory effects associated with their hypertonicity.[21] The neuronal mechanisms that mediate these effects are unknown, but some researchers favor a specific disinhibitory effect.[100,101] The demonstration of inhibition of specific enzymes, such as plasma cholinesterase, supports this hypothesis.[102,103] The marked sensitivity of nervous tissue to direct exposure to ionic contrast media in experimental animals was borne out by the use of ionic methylglucamine iothalamate for ventriculography and myelography, which led to seizures, clonic spasms, and occasionally more serious complications.[21,104,105]

Experimentally, nonionic contrast media have been shown to have a far less direct neurotoxic effect than equivalent concentrations of ionic contrast media.[106–109] This has been borne out by their current routine use in ventriculography and myelography.[110–113]

In neuroangiography, an intact BBB prevents the contrast media from coming into direct contact with the nervous tissue itself as the water-soluble contrast media are prevented from passing from the blood into the extracellular fluid of the brain or spine. Therefore, the integrity of the BBB during neuroangiography is probably the major factor in determining the neurotoxicity of the contrast media. When angiography is performed in the presence of a pathologic breakdown in the BBB, seizures are sometimes provoked.[114] Seizures have also been observed when the BBB is deliberately broken down with hypertonic mannitol.[115] This finding would suggest that, if the integrity of the BBB is altered either by pathologic conditions or by the hypertonicity of the contrast medium used, the contrast medium is allowed to cross into the extracellular fluid of the brain, resulting in a direct chemotoxic effect. Rapoport et al.[68] found that the threshold concentration for increased permeability to Evans blue with carotid injection of various contrast media ranged from 0.8–1.5 osmol/kg, whereas ionic contrast media commonly used for cerebral angiography have an osmolality of 1.4–1.5 osmol/kg. As the hypertonic ionic contrast media are more likely to lead to disruption of BBB, it is not surprising that they are more neurotoxic than corresponding iodine concentrations of nonionic contrast media.

After the carotid infusion of hypertonic solutions, the BBB is reconstituted in 5 minutes to 3 hours depending on the conditions of the infusion and the marker used for BBB testing.[116–118] Therefore, repeated injections of ionic contrast media within several minutes of each other may increase the risk of neurotoxic effects. This is possibly because barrier opening occurs with the first injection, allowing contrast medium increased entry into the extracellular fluid of the brain with subsequent injections.[2,21] The sodium salts of ionic contrast media have been shown to be more neurotoxic than equivalent concentrations of methylglucamine salts. Therefore, having crossed the BBB, neurotoxicity of ionic contrast media is due to a chemotoxic effect that is related to its molecular structure, rather than to an osmotic effect. This is supported by the production of seizures in experimental animals with large doses of hypertonic ionic CM but not with equivalent doses of hypertonic saline.

Most neurologic deficits associated with neuroangiography are thought to be ischemic complications of the catheter technique, rather than the result of the toxic effects of the contrast medium.[21] Seizures have been reported in 0.2% of cerebral angiograms and 0.4% of arch aortograms,[21,114] and are probably related to the contrast medium crossing the BBB because of pre-existing pathology or osmotic breakdown.

Transient cortical blindness sometimes occurs as a result of uncomplicated vertebral angiography and even with retrograde brachial or aortic arch injections.[119] This complication is thought to be a direct effect of the contrast medium on the occipital lobe, which is supported by CT documentation of contrast media in the occipital cortex in one case.[120]

Patients with ischemic cerebrovascular disease or subarachnoid hemorrhage have a greater incidence of neurologic complications from neuroangiography.[21,121,122] Although this incidence may be related to increased risk of arterial embolism or hemodynamic effects, such as spasm, there may be greater sensitivity to the contrast medium itself, perhaps due to increased BBB permeability.

Spinal cord injury is a well-known risk of both aortography and, more specifically, spinal angiography. Although such injury may be the result of catheter technique, the risk is greater for patients studied in the prone position, which allows gravitational layering of contrast media within the aorta.[123] The risk is also greater in the presence of distal aortic occlusion.[124] Most authors therefore attribute permanent spinal cord injury to the direct neurotoxic effect of the contrast media.[21]

REFERENCES

1. Bull JWD: The history of neuroradiology. *Proc Roy Soc Med* 1970;63:637–643.

2. Fischer HW: Contrast media, in Newton TH, Potts DG (eds): *Radiology of the Skull and Brain*, vol 2. *Angiography*, Book 1. *Technical Aspects*. St. Louis, CV Mosby, 1974, pp 893–907.

3. Thomas SF, Henry GW, Kaplan HS: Hepatolienography: past, present, and future. *Radiology* 1951;57:669–683.

4. Binz A: The chemistry of Uroselectan. *J Urol* 1931;25:297–301.

5. Swick M: Darstellung der Niere und Harnwege in Roentgenbild durch intravenose Einbringung eines neuen kontranstoffes: des Uroselectans. *Klin Urochenschr* 1929;8:2087–2089.

6. Swick M: Excretion urography by means of the intravenous and oral administration of sodium orth-iodohippurate with some physiological considerations. *Surg Gynecol Obstet* 1933;56:62–65.

7. Von Lichtenberg A, Swick M: Klinische prufung des uroselectans. *Klin Wochenschr* 1929; 8:2087–2089.

8. Grainger RG: Intravascular contrast media—the past, the present and the future. *Br J Radiol* 1982;55:1–18.

9. Wallingford VH: The development of organic iodide compounds as x-ray contrast media. *J Am Pharmacol Assoc* 1953;42:721–728.

10. Hoppe JO, Larsen HA, Coulston FJ: Observations on the toxicity of a new urographic contrast medium, sodium 3, 5-diacetamido-2, 4, 6, tri-iodobenzoate (Hypaque sodium) and related compounds. *J Pharmacol Exp Ther* 1956;116:394–403.

11. Fischer HW: Catalog of intravascular contrast media. *Radiology* 1986;159:561–563.

12. Irstam L, Sellden U: Adverse effects of lumbar myelography with Amipaque and Dimer X. *Acta Radiol (Diagn)* 1976;17:145–159.

13. Almén T: Contrast agent design. Some aspects on the synthesis of water-soluble contrast agents of low osmolality. *J Theor Biol* 1969;24:216–226.

14. Sage MR: Kinetics of water-soluble contrast media in the central nervous system. *AJNR* 1983; 4:897–906. *AJR* 1983;141:815–824.

15. Haavaldsen J: Iohexol. Introduction. *Acta Radiol* 1980;362(suppl):9–11.

16. Sovak M: Contrast media for imaging of the central nervous system, in Sovak M (ed): *Handbook of Experimental Pharmacology*. Berlin, Springer Verlag 1984, pp 295–340.

17. Kendall BK: Development in contrast media applied to neuroradiology. *Br Med Bull* 1980; 36:273–278.

18. Seeger JF, Carmody RF: Digital subtraction of the arteries of the head and neck. *Radiol Clin N Am* 1985;23:193–210.

19. White RI, Halden WJ: Liquid gold: low-osmolality contrast media. *Radiology* 1986;159: 559–560.

20. Gonsett RE: The neurotoxicity of water-soluble contrast media: actual concepts and future, in Amiel M (ed): *Contrast Media in Radiology*. Berlin, Springer Verlag 1982, pp 115–122.

21. Junck L, Marshall WH: Neurotoxicity of radiological contrast agents. *Ann Neurol* 1983; 13:469–484.

22. Hagen B, Klink G: Contrast media and pain: hypotheses on the genesis of pain occurring on intra-arterial administration of contrast media, in Taenzer V, Zeitler E (eds): *Contrast Media in Urography, Angiography and Computerized Tomography*. Stuttgart, George Thieme Verlag, 1983, pp 50–56.

23. Grainger RG: A clinical trial of a new low osmolality contrast medium. Sodium and meglumine ioxaglate (Hexabrix) compared with meglumine iothalamate (Conray) for carotid arteriography. *Br J Radiol* 1979;52:781–786.

24. Speck U, Siefert HM, Klink G: Contrast media and pain in peripheral angiography. *Invest Radiol* 1980;15:335–339.

25. Kido DK, Potts DG, Bryan RN, et al: Iohexol cerebral angiography. Multicentre clinical trial. *Invest Radiol* 1985;20(suppl):55–57.

26. Drayer BP, Velaj R, Bird R, et al: Comparative safety of intracarotid iopamidol, iothalamate, meglumine and diatrizoate meglumine for cerebral angiography. *Invest Radiol* 1984;19(suppl): 212–218.

27. Pinto R, Berenstein A: The use of iopamidol in cerebral angiography. *Invest Radiol* 1984; 19(suppl):222–224.

28. Armstrong D, Jepson JB, Keele CA, et al: Observations on chemical excitants of cutaneous pain in man. *J Physiol (Lond)* 1953;120:326–351.

29. Hagen B, Siefert HM, Mutzel W, et al: Increased pain reactions as hangover phenomena after intra-arterial injections of contrast media in rats, in Taenzer V, Zeitler E (eds): *Contrast Media in Urography, Angiography and Computerized Tomography*. Stuttgart, George Thieme Verlag, 1983, pp 57–61.

30. Hilal SK: Haemodynamic responses in the cerebral vessels to angiographic contrast media. *Acta Radiol* 1966;5:211–231.

31. Hilal SK: Haemodynamic changes associated with the intra-arterial injection of contrast media. *Radiology* 1966;86:615–633.

32. Lindgren P, Saltzman GF, Tornell G: Vascular reaction to water-soluble contrast media. *Acta Radiol* 1968;7:152–160.

33. Du Boulay GH, Kendall BK, Symon L, et al: The vasodilator action of angiography with Urografin 60% on the basal arteries of the brain of the baboon. *Neuroradiology* 1975;9:133.

34. Skalpe IO: The toxicity of non-ionic water-soluble monomeric and dimeric contrast media in selective vertebral angiography. *Neuroradiology* 1983;24:214–233.

35. Du Boulay GH, Wallis A: Cerebral arterial vasoconstriction due to contrast media. *Symposium Neuroradiologicum*, Stockholm, 1986.

36. Ingstrup HM, Laulund S: Clinical testing of omnipaque and amipaque in external carotid and vertebral angiography. *AJNR* 1983;4:1097–1099.

37. Higgins CB, Schmidt WS: Identification and evaluation of the contribution of the chemoreflex in the haemodynamic response to intracarotid administration of contrast materials in the conscious dog: comparison with the response to nicotine. *Invest Radiol* 1979;14:438–446.

38. Lynch PR, Harrington GJ, Michie C: Cardiovascular reflexes associated with cerebral angiography. *Invest Radiol* 1969;4:156–160.

39. Morris TW, Francis M, Fischer HW: A comparison of the cardiovascular responses to carotid injections of ionic and non-ionic contrast media. *Invest Radiol* 1979;14:217–223.

40. Velaj R, Drayer B, Albright R, et al: Comparative neurotoxicity of angiographic contrast media. *Neurology* 1985;35:1290–1298.

41. Jeppsson PG, Olin T: Neurotoxicity of roentgen contrast media. Study of the blood-brain barrier in the rabbit following selective injection of contrast media into the internal carotid artery. *Acta Radiol (Diagn)* 1970;10:17–34.

42. Lodin H, Ottander HG: Electrocardiograms in carotid angiography with Urografin. *Acta Radiol (Diagn)* 1967;6:519–523.

43. Skalpe IO, Lundervold A, Tjorstad K: Cerebral angiography with non-ionic (metrizamide) and ionic (meglumine metrizoate) water-soluble contrast media. A comparative study with double-blind technique. *Neuroradiology* 1977;14:15–19.

44. Greitz T, Tornell G: Bradycardial reactions during cerebral angiography. A comparison of Isopaque sodium, Isopaque B, Hypaque and Urografin. *Acta Radiol* 1967;270(suppl):75–86.

45. Hilal SK: Cerebral haemodynamics assessed by angiography, in Newton TH, Potts DG (eds): *Radiology of the Skull and Brain*. St. Louis, CV Mosby, 1974, pp 1049–1085.

46. Hayakawa K, Nishimura Y, Yoshida M, et al: ECG changes during cerebral angiography. *Neuroradiology* 1984;26:369–373.

47. Nakstad P, Sortland O, Aaerud O, et al: Cerebral angiography with the non-ionic water-soluble contrast medium Iohexol and Meglumine-Ca-metrizoate. A randomized double-blind parallel study in man. *Neuroradiology* 1982;23:199–202.

48. Lodin H: Electrocardiograms on vertebral angiography. *Acta Radiol (Diagn)* 1968;7: 117–123.

49. Amundsen P, Dugstad G, Stettebo M: Clinical testing of Amipaque for cerebral angiography. *Neuroradiology* 1978;15:89–93.

50. Thron A, Ratzka M, Voigt K, et al: Iohexol and ioxaglate in cerebral angiography, in Taenzer V, Zeitler E (eds): *Contrast Media in Urography, Angiography and Computerized Tomography*. Stuttgart, George Thieme Verlag, 1983, pp 115–119.

51. Valk J, Crezee F, Olislagers-De'slegte RGM: Comparison of iohexol 300 mgI/ml and Hexabrix 320 mgI/ml in cerebral angiography. *Neuroradiology* 1984;26:217–221.

52. Rand PW, Lacombe E: Effects of angiocardiographic injections on blood viscosity. *Radiology* 1965;85:1022–1032.

53. Bernstein EF, Evans RL, Saltzman GF: Physico-chemical properties of blood following exposure to methylglucamine iodipamide, and other contrast media. *Acta Radiol* 1964;2:401–419.

54. Stormorken H, Skalpe IO, Testart MC: Effect of various contrast media on coagulation, fibrinolysis and platelet function. An in vitro and in vivo study. *Invest Radiol* 1986;21:348–354.

55. Paajanen H, Kormano M, Votila P: Modification of platelet aggregation and thromboxane synthesis by intravascular contrast media. *Invest Radiol* 1984;19:333–337.

56. Hardebo JE, Nilsson B: Haemodynamic changes in brain caused by local infusion of hyperosmolar solutions, in particular relation to blood-brain barrier opening. *Brain Research* 1980;181:49–59.

57. Herrschaft H, Gleim F, Schmidt H: Effects of angiographic contrast media on regional cerebral blood flow and haemodynamics in man. *Neuroradiology* 1974;7:95–103.

58. Pappius HM, Savaki HE, Fieschi C, et al: Osmotic opening of the blood-brain barrier and local cerebral glucose utilisation. *Ann Neurol* 1979;5:211–219.

59. Bradbury M: *The Concept of a Blood-Brain Barrier*. New York, Wiley, 1979.

60. Reese TS, Karnovsky MJ: Fine structural localisation of a blood-brain barrier to exogenous peroxidase. *J Cell Biol* 1967;34:207–217.

61. Majno G: Ultrastructure of the vascular membrane, in Hamilton WF, Daw P (eds): *Handbook of Physiology, section 2: Circulation,* vol 3. Washington, DC, American Physiological Society, 1965, pp 2293–2375.

62. Karnovsky MJ: The ultrastructural basis of transcapillary exchanges. *J Gen Physiol* 1968; 52:64–95.

63. Fromter E, Diamond J: Route of passive ion permeation in epithelia. *Nature (New Biol)* 1972; 235:9–13.

64. Bradbury MWB: Why a blood-brain barrier? *Trends Neurosci* 1979;2:36–38.

65. Oldendorf WH: The blood-brain barrier and its relevance to modern nuclear medicine, in Magistretti PL (ed): *Functional Radionuclide Imaging of the Brain*, ed 5. New York, Raven Press, 1983, pp 1–10.

66. Fenstermacher JD, Bradbury MWB, Du Boulay GH, et al: The distribution of 125I metrizamide and 125I diatrizoate between blood, brain and cerebrospinal fluid in the rabbit. *Neuroradiology* 1980;19:171–180.

67. Harrington G, Michie C, Lynch PR, et al: Blood-brain barrier changes associated with unilateral cerebral angiography. *Invest Radiol* 1966;1:431–440.

68. Rapoport SI, Thompson HK, Bidinger JM: Equi-osmolal opening of the blood-brain barrier in the rabbit by different contrast media. *Acta Radiol (Diagn)* 1974;15:21–32.

69. Salvesen S, Nilsen PL, Holtermann H: Effects of calcium and magnesium ions on the systemic and local toxicities of the N-methylglucamine (meglumine) salts of metrizoic acid (Isopaque). *Acta Radiol* 1967;270(suppl):180–193.

70. Waldron RL, Bridenbaugh RB, Dempsey EW: Effect of angiographic contrast media at the cellular level in the brain: hypertonic vs chemical action. *AJR* 1974;122:469–476.

71. Waldron RL, Bryan RN: Effect of contrast agents on the blood-brain barrier. An electron microscopic study. *Radiology* 1975;116:195–198.

72. Sage MR, Wilcox J, Evill CA, et al: Comparison of the degree and variability of osmotic breakdown of the blood-brain barrier due to mannitol and methylglucamine iothalamate. *Invest Radiol* 1982;17:276–281.

73. Sage MR, Wilcox J, Evill CA, et al: Comparison of blood-brain barrier disruption following intracarotid metrizamide and meglumine iothalamate (Conray 280). *Aust Radiol* 1982;26:225–229.

74. Sage MR, Wilcox J: Comparison of blood-brain barrier disruption following intracarotid iohexol and meglumine iothalamate. *Aust Radiol* 1984;28:6–8.

75. Wycherley A, Wilcox J, Sage MR: Comparison of blood-brain barrier disruption in the rabbit following intracarotid iopamidol and methylglucamine iothalamate. *Aust Radiol* 1984;27:294–296.

76. Sage MR, Drayer BP, Dubois PJ, et al: Increased permeability of the blood-brain barrier following carotid Renografin 76. *AJNR* 1981;2:272–274.

77. Shibakiri I, Yamada R, Itami M: Study of blood-brain barrier damage by computed tomography. *Nippon Igaku Hoshasen Gakki Zasshi* 1980;40:497–499.

78. Numaguchi Y, Fleming MS, Hasou K, et al: Blood-brain barrier disruption due to cerebral arteriography: CT findings. *J Comput Asst Tomogr* 1984;8:936–939.

79. Doust BD, Fischer HW: Comparison of cerebral toxicity of monomeric and trimeric forms of sodium iothalamate. *Br J Radiol* 1971;144:764–766.

80. Bassett RC, Rogers JS, Cherry GR, et al: The effects of contrast media on the blood-brain barrier. *J Neurosurg* 1953;10:38–47.

81. Rapoport SI, Fredericks WR, Ohno K, et al: Quantitative aspects of reversible osmotic opening of the blood-brain barrier. *Am J Physiol* 1980;238:R421–R431.

82. Brinker RA: Neuroangiographic contrast agents, in Miller RE, Skucas J (eds): *Radiographic Contrast Agents*. Baltimore, University Park Press, 1977, pp 365–375.

83. Houthoff HJ, Go KG, Gerrits PO: The mechanisms of blood-brain barrier impairment by hyperosmolar perfusion. *Acta Neuropathol (Berl)* 1982;56:99–112.

84. Burns EM, Dobben GD, Kruckeberg MS, et al: Effects of ionic and nonionic contrast media on blood-brain barrier integrity, abstracted. *Invest Radiol* 1980;15:395.

85. Johnson JH, Knisely MH: Intravascular agglutination of the flowing blood following the injection of radiopaque contrast media. *Neurology (Minneap)* 1962;12:560–570.

86. Margolis G: Pathogenesis of contrast media injury: insights provided by neurotoxicity studies. *Invest Radiol* 1970;5:392–406.

87. Wilcox J, Evill CA, Sage MR: Effects of intracarotid ionic and non-ionic contrast material on the blood-brain barrier in a rabbit model. *Neuroradiology* 1986;3:86–88.

88. Neuwelt EA, Maravilla KR, Frenkel EP, et al: Use of enhanced computerised tomography to evaluate osmotic blood-brain barrier disruption. *Neurosurgery* 1980;6:49–56.

89. Neuwelt EA, Maravilla KR, Frenkel EP, et al: Osmotic blood-brain barrier disruption. Computerised tomographic monitoring of chemotherapeutic agent delivery. *J Clin Invest* 1979; 64:684–688.

90. Fischer HW, Cornell SH: The toxicity of the sodium and methylglucamine salts of diatrizoate, iothalamate and metrizoate. An experimental study of their circulatory effects following intracarotid injection. *Radiology* 1965;85:1013–1021.

91. Almén T: Toxicity of radiocontrast agents, in PK Knoefel (ed): *Radiocontrast Agents*. New York, Pergamon Press, 1971, pp 443–550.

92. Kormano MJ: Kinetics of contrast media after bolus injection and infusion, in Felix R, Kazner E, Wegener OH (eds): *Contrast Media in Computed Tomography*. Amsterdam, Excerpta Medica, 1974, pp 38–45.

93. Cattell WR, Fry EK, Spencer AG, et al: Excretion urography. 1. Factors determining the excretion of Hypaque. *Br J Radiol* 1967;40:561–571.

94. McChesney EW, Hoppe HO: Studies of the tissue distribution and excretion of sodium diatrizoate in laboratory animals. *AJR* 1957;78:137–144.

95. Kormano M, Dean PB: Extravascular contrast material: the major component of contrast enhancement. *Radiology* 1976;121:379–382.

96. Newhouse JH: Fluid compartment distribution of intravenous iothalamate in the dog. *Invest Radiol* 1977;12:364–367.

97. Dean PB, Kormano M: Intravenous bolus of 125I-labelled meglumine diatrizoate. Early extravascular distribution. *Acta Radiol (Diagn)* 1977;18:293–304.

98. Zamani AA, Morris JH, Kido DK, et al: Permeability of the blood-brain barrier to different doses of diatrizoate meglumine 60%, abstracted. *Invest Radiol* 1981;16:380.

99. Hershowitz N, Bryan RN: Neurotoxic effects of water-soluble contrast agents on rat hippocampus: extracellular recordings. *Invest Radiol* 1982;17:271–275.

100. Bryan RN, Dauth W, Gilman S, et al: Effects of radiographic contrast agents on spinal cord physiology. *Invest Radiol* 1981;16:234–239.

101. Bryan RN, Johnston D: Epileptogenic effects of radiographic contrast agents: experimental study. *AJNR* 1982;3:117–120.

102. Guidolet J, Barbe R, Borsson F, et al: Subcellular localization of uro-angiographic contrast by 125I-labelled media. *Invest Radiol* 1980;5:5215–5219.

103. Lasser EC, Lang JH: Physiologic significance of contrast-protein interactions. I. Study in vitro of some enzyme effects. *Invest Radiol* 1970;5:514–517.

104. Campbell RL, Campbell JA, Heimburger RF, et al: Ventriculography and myelography with absorbable radiopaque medium. *Radiology* 1964;82:286–289.

105. Praestholm J: Complications of myelography with Conray meglumine. *Acta Radiol* 1972; 13:860–864.

106. Oftedal S, Kayed K: Epileptogenic effect of water-soluble contrast media. An experimental investigation in cats. *Acta Radiol* 1973;355(suppl):45–55.

107. Sawhney BB, Oftedal S: Reactions to suboccipital injection of water-soluble contrast media in rabbits. *Acta Radiol* 1973;355(suppl):67–81.

108. Oftedal S: Toxicity of water-soluble contrast media injected suboccipitally in cats. *Acta Radiol* 1973;355(suppl):84–92.

109. Grepe A, Widen L: Neurotoxic effect of intracranial subarachnoid application of metrizamide and meglumine iocarmate. *Acta Radiol* 1973;355(suppl):102–118.

110. Turski PA, Sackett JF, Gentry LR, et al: Clinical comparison of metrizamide and iopamidol for myelography. *AJNR* 1983;4:309–311.

111. Sackett JF, Strother CM: *New Techniques in Myelography*. Hagerstown, MD, Harper & Row, 1979.

112. Lamb JT, Holland IM: Myelography with iopamidol. *AJNR* 1983;4:851–853.

113. Lamb JT: Iohexol vs iopamidol for myelography. *Invest Radiol* 1985;20:S37–S43.

114. Olivecrona H: Complications of cerebral angiography. *Neuroradiology* 1977;14:175–181.

115. Neuwelt EA, Frenkel EP, Diehl J, et al: Reversible osmotic blood-brain barrier disruption in humans: implications for chemotherapy of malignant brain tumors. *Neurosurgery* 1980;7:44–52.

116. Broman T, Olsson O: The tolerance of cerebral blood vessels to a contrast medium of the Diadrast group. *Acta Radiol* 1948;30(suppl):326–342.

117. Hardebo JE: A time study in rat on the opening and reclosure of the blood-brain barrier after hypertensive or hypertonic insult. *Exp Neurol* 1980;70:155–166.

118. Harris AB: Steroids and blood-brain barrier alterations in sodium acetrizoate injury. *Arch Neurol* 1967;17:282–295.

119. Horwitz NH, Wener L: Temporary cortical blindness following angiography. *J Neurosurg* 1974;40:583–586.

120. Studdard WE, Davis DO, Young SW: Cortical blindness after cerebral angiography: case report. *J Neurosurg* 1981;54:240–244.

121. Feild JR, Lee L, McBurney RF: Complications of 1000 brachial arteriograms. *J Neurosurg* 1972;36:324–337.

122. Mani RL, Eisenberg RL: Complications of catheter cerebral arteriography analysis of 5000 procedures. II. Relation of complication rates to clinical and arteriographic diagnosis. *AJR* 1978; 131:867–869.

123. Margolis G: Pathogenesis of contrast media injury: insights provided by neurotoxicity studies. *Invest Radiol* 1970;5:392–406.

124. Tornell G: Spinal cord tolerance to roentgen contrast media particularly during aortography with temporary occlusion of the aorta. *Acta Radiol* 1969;8:257–283.

Chapter 12

Coronary Angiography

Kenneth R. Thomson

For the most part, catheter placement in coronary angiography is easy. It is the injection of iodinated contrast media that is a major cause of problems related to the electrical and mechanical functions of the heart. The changes caused by ionic media have been reviewed recently[1] and are summarized here as they relate to nonionic contrast media. The effects of contrast media on the heart and circulation are listed in Table 12-1.

ELECTROCARDIOGRAM ALTERATIONS

In humans sinus bradycardia occurs after coronary arteriography with ionic contrast media injections. The effect is greatest with right coronary injections, and sinus arrest may occur.[2] In an experimental preparation, infusion of ionic contrast medium into the arteries supplying the sinoatrial and atrioventricular nodes caused

Table 12-1 Cardiac Effects of Iodinated Contrast Media

ELECTROPHYSIOLOGIC
Heart rate
ST segment, T wave, and QT interval alterations
Arrhythmias
METABOLIC
Lactate extraction ratio
Ionized calcium activity
HEMODYNAMIC
Myocardial contractility
Aortic and left ventricular pressure
Coronary blood flow

a dose-dependent decrease in heart rate. The nonionic contrast medium, metrizamide, caused significantly less slowing of the heart rate.[3] That the heart rate slowing is primarily due to osmolality of the injectate was shown by Popio et al.[4] In their experiments the electrophysiologic effects of ionic contrast media were reproduced by injection of hyperosmolar solutions of dextrose (equiosmolar with Renografin 76) but not by a saline solution containing 190 mEq/L of sodium.[4]

ST segment changes commonly occur after the injection of ionic contrast media. ST elevation or depression and T wave flattening and prolongation of the QT interval have been reported in older studies. The changes are most marked after direct coronary injections and are reduced when nonionic contrast is used. Ioxaglate, a monoacid dimer, is similar to nonionic contrast in this regard primarily because of its lower osmolality.

Such effects as transient bundle branch block have also been reported.[5] At times the depressive effect of ionic contrast may transiently improve a ventricular arrhythmia. More recent randomized double-blind studies in humans have shown that iopamidol, a nonionic contrast medium, caused significantly less change in heart rate, T wave amplitude, and QT interval than did sodium methylglucamine diatrizoate (Renografin 76).[6]

VENTRICULAR FIBRILLATION

There is no clear correlation between observed changes in the ECG and the onset of ventricular fibrillation, but ventricular fibrillation has been used as a primary indicator of contrast media toxicity. Following coronary artery injection of ionic contrast media in dogs,[7] temporary hyperpolarization of resting potentials and prolongation of action potentials occurred, which may be explained by a contrast-induced local deficiency of potassium and calcium ions and by a relative prevalence of sodium ions in coronary blood. Some ionic contrast media contain the chelating agents, sodium citrate and disodium edetate, both of which have the ability to bind ionized calcium. Coronary arteriography in dogs showed that compounds with added calcium or compounds without chelating agents were associated with less ventricular fibrillation than other ionic contrast media containing these chelating agents.[8,9] Ionic contrast media containing calcium and magnesium ions were associated with some lessening of the ECG and rhythm changes.[10] Formulations of contrast media containing metrizoate salts and calcium ions have been shown to be as well tolerated in laboratory and clinical studies as the preferred diatrizoate formulations.[11,12] When coronary sinus ionized calcium levels were measured in dogs during coronary angiography, a nonionic contrast medium (iopamidol) caused only a slight fall in calcium ion activity, whereas sodium methylglucamine diatrizoate caused significant lowering of calcium ion activity.[13] Higgins showed that significant changes in conduction occur when diatrizoate and calcium channel blockers (Verapamil) were given during coronary arteriography. The injections of iohexol and Verapamil produced only slight prolongation of the PR interval.[14]

The presence of sodium ions in a concentration of approximately 190 mEq/L is least toxic for monomeric ionic contrast media.[15] However, dimeric ionic contrast media, such as iocarmate, contain nearly twice as much sodium, and nonionic contrast media contain none at all. Because the threshold for ventricular fibrillation is higher with nonionic agents than with ionic agents[16] clearly the optimum sodium concentration depends on the type of contrast being used.

For diatrizoate formulations a sodium-to-methylglucamine ratio of 1:6.6 and a sodium content of 190 mmol/L has been found to be best tolerated as a coronary arteriographic agent.[17] To minimize calcium binding the contrast medium should be free of citrate.

MYOCARDIAL CONTRACTILITY

Ionic media cause a significant initial fall in cardiac contractility followed by a later augmentation of contractility. The initial fall may be profound and is related at least in part to the osmolality of the injected solution.[10] These changes were more marked in the presence of ischemia.[14] The nonionic contrast media produce a monophasic increase in contractility, which returns to normal within 1 minute, even in the presence of ischemia. In the presence of a critical coronary artery stenosis, an ionic contrast containing calcium ions (Isopaque) produced only slight increases in left ventricular contractile force similar to that produced by nonionic contrast media.[18] In the same series of experiments the monoacid dimer, ioxaglate, caused significant changes in left ventricular pressure, contractile force, and left ventricular dimension.

INTIMAL DAMAGE

Studies show that the plasma levels and urinary excretion of prostacyclin increase after both contrast medium injection and catheterization without contrast medium injection.[19] These findings suggest chemical and physical trauma to the intima. In other experiments iopamidol caused less inhibition of adenosine diphosphate-induced platelet aggregation than did ioxaglate or diatrizoate.[20] This effect was due to hypertonicity of the contrast media as similar effects were seen with hypertonic saline. In equiosmolar concentrations the monoacid dimer, ioxaglate, was the most potent inhibitor of platelet function of the materials tested.

Hyperosmolar ionic contrast media have been implicated in postphlebography thrombosis. The fibrolytic activity of saphenous vein intima is diminished by exposure to diatrizoate.[21] In a culture of human endothelial cells labeled with ^{51}Cr, the release of ^{51}Cr following contrast media challenge was up to six times higher with diatrizoate or metrizoate than with iopamidol, iohexol, or ioxaglate. The nonionic contrast media produced an effect equal to or lower than that of physiologic saline. The strongest osmolality-independent toxic effect was due to ioxaglate.[22]

MYOCARDIAL METABOLISM

Coronary angiography does not alter myocardial oxygen consumption[23] or the levels of SGOT, SGPT, LDH, and CPK.[11] Earlier workers did not find any significant difference in alterations of lactate and fatty acid metabolism caused by ionic versus nonionic contrast media. However, recently Gertz et al. have shown that a nonionic contrast, iopamidol, increased the myocardial lactate extraction ratio after ventricular injection.[24] Ventricular injection of sodium methylglucamine diatrizoate caused a decrease in myocardial lactate extraction ratio that persisted for 5 minutes. Both agents caused a fall in the arterial level of free fatty acids, and the normal physiologic response to this stimulus is an increase in the myocardial lactate extraction ratio. Gertz interpreted the fall in ratio caused by diatrizoate as indicating the presence of ischemia.

WATER AND ELECTROLYTE SHIFT: LUNG WATER—CORONARY BLOOD FLOW

When hyperosmolar contrast media are injected into the coronary circulation the coronary sinus hematocrit and serum protein are profoundly lowered,[25] and potassium levels and sodium ion activity significantly decrease.[13] These changes are short lived and are caused by water shifts directly related to the osmolality of the contrast. Similar but less marked changes follow the injection of hypo-osmolar ionic and nonionic contrast.

Ionic contrast media cause a persistent increase in lung water beginning 1 minute after intravenous injection and peaking at 2–3 minutes. This event has been implicated in the pathogenesis of pulmonary edema following angiography. In animal experiments nonionic contrast media caused no such increase in lung water and in fact produced a slight transient decrease.[26]

All contrast media produce an increase in coronary blood flow that is directly related to the osmolality of the contrast.[27] The increased flow is independent of changes in heart rate and ventricular and systemic pressure.

BLOOD PRESSURE EFFECTS

Hyperosmolar solutions injected into the pulmonary artery cause a transient rise in pulmonary arterial pressure. The rise in pressure is believed to be due to water shifts affecting the internal viscosity of the red cells and to aggregation of the red cells in response to the hyperosmolar stimulus. Each mechanism makes it more difficult for blood to pass through the lung capillaries and increases arterial resistance.[28] The rise in pressure is less with nonionic contrast media because of their lower osmolality. In patients with pulmonary arterial hypertension the use of large doses of a hyperosmolar contrast media may cause acute right heart failure. This is particularly true of cardiac digital subtraction angiography.

Comparison of diatrizoate and nonionic iohexol following ventricular and coronary artery injections showed that both agents caused a decrease in systemic arterial pressure, but the magnitude of the decrease and duration were greater with diatrizoate.[29,30] Neither agent caused a significant alteration in diastolic pressure. The fall in systolic pressure was accompanied by a significant fall in heart rate. Low osmolar agents, which include ioxaglate and other nonionic agents, are associated with a lower incidence of sensations of warmth than diatrizoate in full concentration. The sensation of warmth is directly related to the dose and osmolality.

CHOICE OF CONTRAST MEDIA

For diatrizoate formulations a sodium-to-methylglucamine ratio of 1:6.6 and a sodium content of 190 mmol/L has been found to be best tolerated as a coronary arteriographic agent.[17]

It has been established that a diatrizoate contrast medium with only methylglucamine ions is toxic to cardiac tissue[15] and that if the sodium content is reduced below 40 mmol/L an increased risk of ventricular fibrillation occurs. Formulations of contrast media containing metrizoate salts and calcium ions have been shown to be as well tolerated in laboratory and clinical studies and are the preferred diatrizoate formulations.[11,12] The diatrizoate or metrizoate compounds should not contain calcium-chelating agents.

If cost were unimportant then the nonionic contrast media would be the agents of first choice for all arteriography. In every parameter studied to date they have less (or similar) toxic effects in both animal and human experiments. In the coronary circulation the nonionic agents produce less effect on cardiac function than ionic agents, and those effects that nonionic agents do cause are of a shorter duration than those caused by ionic agents. In particular, disturbances of heart rate, blood pressure, and left ventricular function are significantly less when nonionic agents are used.

Because of their cost, it has been proposed that the nonionic contrast should be used in high risk patients only. From a hemodynamic point of view these patients are easy to identify, and in our private catheterization laboratory (Epworth Hospital), iopamidol has been used for the indications listed in Table 12-2 for the past 12 months. During that time three patients developed ventricular fibrillation after diatrizoate injections, and one patient with a history of previous severe reaction who was given iopamidol developed hypotension that persisted for 24 hours after the procedure. No deaths occurred in the 1,394 patients studied in 1986. As a result of the observed lower incidence of ECG and hemodynamic alterations with iopamidol, two cardiologists decided to use iopamidol exclusively in their patients.

At present a reluctance to the widespread use of nonionic contrast media exists on the basis of their significantly increased cost. The data in animals and humans indicate clearly that the nonionic contrast media are superior. Yet even in those who have a previous reaction to ionic contrast,[31] the standard ionic media are widely used and are associated with a relatively low incidence of serious or life-threatening side effects.[32] Unfortunately, the monoacid dimer ioxaglate, although it exhibits some lessened side effects due to its lower osmolality, has more adverse effects than some ionic contrast media, particularly with regard to histamine release, nausea, and vomiting. The legal liability of a physician whose patient has a serious reaction to ionic contrast media, when the reaction may have been avoided had a nonionic contrast been used instead, has not been determined.

In Australia, should nonionic contrast agents be used exclusively, the estimated additional cost to prevent the loss of one life is A$1 million (equivalent to approximately US$800,000). This is an amount that our health resources can ill afford.

For the optimum cost benefit the choice of contrast agent—ionic or nonionic—must be made by the physician and patient, given the individual circumstances of each occasion in which contrast medium is to be used. However, as for me and my family, we want a nonionic!

Table 12-2 Indications for Nonionic Contrast Media for Coronary Angiography

Previous reaction to contrast media
Significant allergic history
Left ventricular dysfunction
Known severe coronary disease
Suspected left main coronary artery disease
Coronary artery angioplasty
Coronary graft studies
Internal mammary arteriograms
Right heart failure
Renal impairment

REFERENCES

1. Fischer HF, Thomson KR: Contrast media in coronary arteriography: A review. *Invest Radiol* 1978;13:450–459.

2. Benchimol A, McNally EM: Hemodynamic and electrocardiographic effects of selective coronary arteriography in man. *N Engl J Med* 1966;274:1217–1224.

3. Higgins CB: Effects of contrast media on the conducting system of the heart. *Radiology* 1977;124:599–606.

4. Popio KA, Ross AM, Oravec JM, et al: Identification and description of separate mechanisms for two components of Renografin cardiotoxicity. *Circulation* 1978;58:520–528.

5. Gensini GG, DiGiorgi S: Myocardial toxicity of contrast agents used in angiography. *Radiology* 1964;82:24.

6. Ciuffo AA, Fuchs RM, Guzman PA, et al: Benefits of nonionic contrast in coronary arteriography: preliminary results of a randomized double-blind trial comparing Iopamidol with Renografin 76. *Invest Radiol* 1984;19(suppl):197–202.

7. Wolpers HG, Baller D, Hoeft A, et al: The effect of ion composition on cellular membrane potentials during selective coronary arteriography. *Invest Radiol* 1984;19:291–295.

8. Thomson KR, Violante MR, Kenyon T, et al: Reduction in ventricular fibrillation using calcium-enriched Renografin 76. *Invest Radiol* 1978;13:238–240.

9. Violante MR, Thomson KR, Fischer HW, et al: Ventricular fibrillation from diatrizoate with and without chelating agents. *Radiology* 1978;128:497–498.

10. Tragardh B, Almén T, Lynch PR: Addition of calcium or other cations and of oxygen to ionic and non-ionic contrast media: effects on cardiac function during coronary angiography. *Invest Radiol* 1975;10:231–238.

11. Baltaxe HA, Sos TA, McGrath MB: Effects of the intracoronary and intraventricular injections of a commonly available vs. a newly available contrast medium. *Invest Radiol* 1976;11:172–181.

12. Salveson S, Nilsen PL, Holterman H: Ameliorating effect of calcium and magnesium ions on the toxicity of Isopaque sodium. II. Studies on the isolated heart and auricles of the rabbit. *Acta Radiol* 1976;270(suppl):30–43.

13. Thomson KR, Evill CA, Fritzsche J, et al: Comparison of iopamidol, ioxaglate and diatrizoate during coronary arteriography in dogs. *Invest Radiol* 1980;15:234–241.

14. Higgins CB: Overview of cardiovascular effects of contrast media: comparison of ionic and nonionic media. *Invest Radiol* 1984;19(suppl): 187–190.

15. Snyder CF, Formanek A, Frech RS, et al: The role of sodium in promoting ventricular arrhythmia during selective coronary arteriography. *Am J Roentgenol* 1971;113:567–571.

16. Wolf GL, Mulry CS, Kilzer L, et al: New angiographic agents with less fibrillatory propensity. *Invest Radiol* 1980;16:320–323.

17. Hildner FJ, Scherlag B, Samet P: Evaluation of Renografin M76 as a contrast agent for angiocardiography. *Radiology* 1971;100:329–334.

18. Green CE, Higgins CB, Kelley MJ, et al: Effects of administration of contrast materials on left ventricular function in the presence of severe coronary artery stenosis: Advantages of newer contrast materials over standard materials. *Cardiovasc Intervent Radiol* 1981;4:110–116.

19. Roy L, Knapp HR, Robertson RM, et al: Endogenous biosynthesis of prostacyclin during cardiac catheterisation and angiography in man. *Circulation* 1985;71:435–440.

20. Paajanen H, Kormano M, Uotila P: Modification of platelet aggregation and thromboxane synthesis by intravascular contrast media. *Invest Radiol* 1984;19:333–337.

21. Whitehouse WM Jr, Queral LA, Flinn WR: The effect of sodium diatrizoate on the fibrinolytic activity of saphenous vein intima. *J Surg Res* 1981;30:391–397.

22. Laerum F: Injurious effects of contrast media on human vascular endothelium. *Invest Radiol* 1985;20(suppl):98–99.

23. Griggs DM Jr, Nakamura Y, Lennissen RLA, et al: Effects of radiopaque material on phasic coronary flow and myocardial oxygen consumption. *Clin Res* 1966;14:247.

24. Gertz EW, Wisneski JA, Neese R, et al: The effects of iopamidol on myocardial metabolism. *Invest Radiol* 1984;19(suppl):191–196.

25. Lehan PH, Harman A, Oldewurtel HA: Myocardial water shifts induced by coronary arteriography. *J Clin Invest* 1963;42:950.

26. Slutsky RA, Strich G: Extravascular lung water: effects of intravenous ionic and nonionic (iopamidol) contrast media during ischaemia. *Radiology* 1985;155:11–14.

27. Wolf GL, Gerlings ED, Wilson WJ: Depression of myocardial contractility induced by hypertonic coronary injections in the isolated perfused dog heart. *Radiology* 1973;101:655–658.

28. Almén T, Aspelin P: Cardiovascular effects of ionic monomeric, ionic dimeric and nonionic contrast media. *Invest Radiol* 1975;10:557–563.

29. Bettman MA, Higgins CB: Comparison of an ionic with a nonionic contrast agent for cardiac angiography. Results of a multicenter trial. *Invest Radiol* 1985;20(suppl):70–74.

30. Partidge JB, Robinson PJ, Turnbull CM, et al: Clinical cardiovascular experiences with iopamidol: a new nonionic contrast medium. *Clin Radiol* 1981;32:451–455.

31. Holtås S: Iohexol in patients with previous adverse reactions to contrast media. *Invest Radiol* 1984;19:563–565.

32. Davis K, Kennedy DK, Kemp HG Jr, et al: Complications of coronary arteriography from the collaborative study of coronary artery surgery. *Circulation* 1974;59:1105–1112.

Chapter 13

Digital Subtraction Angiography

Andrew B. Crummy

During the 1970s a new class of images, with substantially increased contrast detection and moderate spatial resolution, was developed. One such technique that has had great clinical impact is computed tomography. Digital subtraction angiography (DSA) is another such image. Because of the increased contrast resolution of DSA, the contrast agent requirements for the performance of digital arteriograms are considerably different from those of standard film arteriography.

During the developmental phase of DSA at the University of Wisconsin, the original experiments were carried out in a remotely located research laboratory. Safety considerations precluded arterial catheterization so only venous injections were performed. The quality of the initial intravenous (IV) studies was sufficient to encourage us and others to offer intravenous digital subtraction arteriography (IV-DSA) as a clinical service, and widespread interest in the examination developed.[1,2,3] When we moved to a new hospital in 1979 it was possible to install the DSA apparatus in a clinical area, so DSA in conjunction with intra-arterial injection became feasible. Investigation of intra-arterial digital subtraction arteriography (IA-DSA) was vigorously pursued, and it rapidly became a widely used clinical technique.[4]

It is important to recognize that DSA is a method for recording angiographic data, and as such it is independent of the route of injection. However, the site of injection is extremely important because the contrast agent requirements differ appreciably depending on whether an intravenous or intra-arterial route is employed. Fundamentally, the advantage of DSA is the decreased contrast agent requirement for a comparable study when compared with film recording.

INTRAVENOUS DIGITAL SUBTRACTION ARTERIOGRAPHY (IV-DSA)

Because in IV-DSA the contrast bolus is propelled by cardiac action, good cardiac output is essential for a satisfactory examination. Patients in frank or borderline congestive heart failure are not candidates for this type of examination.

There has been much discussion as to whether peripheral injection, usually into one of the veins of the antecubital fossa, or central injection into the superior vena cava or the right atrium is preferable.[5] Our initial studies used a 16- or 18-gauge angiocath that had been placed in the vein of the antecubital fossa. Two problems became apparent with this technique. On occasion, extravasation of a large volume of contrast agent into the soft tissues occurred. This generally resulted in considerable pain, and although none of our patients developed soft tissue necrosis, that complication nevertheless remained a potential hazard.

In addition, following a peripheral injection, some of the contrast would remain stagnated in the veins of the extremity. This was readily demonstrated by fluoroscopic observation. The amount that did not reach the central circulation was variable and could be decreased if the patient did a Valsalva maneuver in response to the sensation of heat that was experienced with the injection of a moderate volume of hyperosmotic contrast agent. In order to reduce the amount of peripheral stagnation we altered our injection technique. First, a volume of saline (25–40 cc) would be aspirated into the injector syringe, which was vertical with the injection port dependent. Then, the contrast agent would be layered beneath the saline. Thus when the injection was made there was a volume of saline behind the contrast agent to act as a propellant. This injection modification reduced the amount of peripheral sequestration of contrast agent, but the possibility of extravasation remained, so this approach was not entirely satisfactory.

In order to overcome these difficulties we switched to percutaneously inserted catheters, which at first were placed in the superior vena cava and then in the right atrium. It is currently our practice to do all our studies with a right atrial injection. If a satisfactory arm vein is not present we do not hesitate to utilize the transfemoral approach.

A right atrial (RA) injection has several advantages. First, the right atrium is a capacious reservoir, and the likelihood of extravasation if one employs a pigtail catheter is very low. In addition, because the injection can be made quite rapidly, the possibility of having a compact bolus is increased. Because none of the contrast agent remains stagnated in the slow flow peripheral venous system, the entire amount injected is part of the bolus.

In the average-sized adult, utilizing a second-generation DSA machine, we have found consistently good results with an RA injection of 35 cc of contrast agent that contains approximately 370 mg I/cc at the rate of 25 cc/sec. Unless the patient is extremely large it is not necessary to increase this volume. Of course in very small patients, particularly children, one can reduce the size of the bolus in proportion to the patient's size.

Although the interest in IV-DSA has decreased somewhat in recent years, we find that in selected cases it remains an excellent approach.[6] For example, we use IV-DSA to screen for renovascular hypertension and in some patients suspected of having extracranial cerebrovascular disease. It is also useful in studying the relationship of the renal arteries to an abdominal aortic aneurysm and the state of the iliac vessels in such patients. IV-DSA is a simple and very satisfactory method for follow-up of various vascular interventions, such as percutaneous transluminal angioplasty or bypass grafting.

Because three or four injections are usually performed, a relatively large volume of contrast agent may be required for IV digital subtraction arteriograms, particularly when compared to the IA-DSA. For this reason one must keep in mind the potential for nephrotoxicity in patients who are poorly hydrated or have compromised renal function. In addition, the high volume of hyperosmotic contrast agent may precipitate congestive heart failure. One must be acutely aware of the clinical status of the patient if an IV-DSA is to be undertaken.

INTRA-ARTERIAL DIGITAL SUBTRACTION ARTERIOGRAPHY (IA-DSA)

Because of the increased contrast level that can be achieved with intra-arterial injection, IA-DSA has a spatial resolution that approaches that of conventional film screen angiography.[4] Yet, the contrast requirement can be fulfilled with much less agent when compared with film studies. The amount of iodine needed to achieve the desired contrast level may be such that, if a dense agent (370 mg I/cc) is used, the volume required may be quite low. Under these circumstances the bolus passes rapidly so that opacification is present in only a few frames. In addition, mixing of the contrast agent with the blood may be incomplete so that streaming artifacts may be encountered. These problems may be obviated by using a larger volume of a less dense agent (140–280 mg I/cc).

Just as with film arteriography, the amount and volume of contrast agent required depend upon the size of the patient and the site of the injection. The size of the intensifier field is also a factor. Because contrast resolution increases as the size of the image intensifier field decreases, the format used is very important. Injections of 16–20 cc of agent with 280 mg I/cc at the aortic bifurcation are generally satisfactory to opacify the distal vessels of the lower extremities if one is using a 14-inch field. Nevertheless, with poor "run off," satisfactory imaging may require reduction of field size because a reduction in field size will increase contrast detection. This of course will necessitate an increase in the number of injections to cover the extremity. Therefore, the contrast volume required for an examination is quite variable. In general, if the largest format available for a peripheral vascular DSA examination is 9 inches, more contrast agent is required than if a standard examination is done.[7] When a 14-inch image intensifier format is used, the DSA examination uses less contrast agent.[8]

Injection of 6 cc of (140–280 mgI/cc) contrast agent over 2 seconds is satisfactory for selective study of the renal artery. Likewise, evaluation of the portal venous system after a superior mesenteric artery injection requires a larger volume than if one wishes to see the arterial side alone. Because of the infinite variety of sites for injection and the variation in size of patients, image intensifier fields, and clinical problems, no attempt is made to have an encyclopedic recitation of the various injection volumes and concentrations. If one is working actively in the field an appreciation of the various contrast agent requirements is rapidly obtained.

NONIONIC VERSUS IONIC CONTRAST AGENTS FOR DSA

Any advantage that a nonionic agent has over an ionic agent is operative whether one uses film screen or digital subtraction recording for arteriography. Because DSA requires less contrast agent for any particular examination, any shortcoming that is volume related will be decreased with DSA. It is apparent that the new agents cause distinctly less discomfort. Under such circumstances the patient is much less likely to move, a factor of major importance in reducing motion artifacts in DSA. For example, in peripheral arteriography, motion of the feet, because of discomfort experienced following an intra-arterial injection in patients with severe ischemia, may be a major problem. With the use of nonionic agents this problem is virtually eliminated. However, reduced patient movement with nonionic agents has not been proved with a controlled study. Also, the intense feeling of heat associated with a large-volume IV injection is reduced.

The smaller osmotic load of a nonionic agent lessens the likelihood of its precipitating congestive heart failure. A similar risk reduction is achieved in IV- or IA-DSA with an ionic agent, because of the decreased contrast agent requirement. This risk can be further reduced with the use of a nonionic agent. It is less apparent whether there is a similar reduction in renal toxicity.

DSA can have considerable impact on the cost of using nonionic contrast agents. Because of the reduced volume of contrast agent required for the DSA examination the cost of the agent per examination is substantially reduced. When one considers that the cost of a nonionic agent is approximately ten times that of the ionic agents, DSA clearly has an economic advantage. At the present time we are using nonionic agents almost exclusively in our DSA examinations.

REFERENCES

1. Crummy AB, Strother CM, Sackett JF, et al: Computerized fluoroscopy: digital subtraction for intravenous angiocardiography and arteriography. *AJR* 1980;135:1131–1140.

2. Brennecke R, Brown TK, Bursch J, et al: Computerized video image processing with application to cardioangiographic Roentgen image series, in Nagel HH (ed): *Digital Image Processing*. New York, Springer-Verlag, 1977, p 244.

3. Ovitt T, Capp MP, Christenson P, et al: Development of digital video subtraction system for intravenous angiography. *Soc Photo-optical Instrum Engineers* 1979;206:73.

4. Crummy AB, Stieghorst MF, Turski PA, et al: Digital subtraction angiography: current status and use of intra-arterial injection. *Radiology* 1982;145:303–307.

5. Modic MT, Weinstein MA, Pavlicek W, et al: Intravenous digital subtraction angiography: peripheral versus central injection of contrast material. *Radiology* 1983;147:711–715.

6. Carmody RF, Yang PJ, Seeger JF et al: Digital subtraction angiography: update 1986. *Invest Radiol* 1986;21:899–905.

7. Garvey CJ, Wilkins RA, Lewis JD: Peripheral vascular disease: prospective study of intra-arterial digital subtraction angiography using a 9-inch intensifier. *Radiology* 1986;159:423–427.

8. Jensen SR, Crummy AB, Voegeli D et al: Unpublished data.

Chapter 14

Phlebography

Michael A. Bettmann

The initial use of contrast agents was to visualize arteries and veins,[1] although with time the uses clearly multiplied. Contrast phlebography was employed as early as the 1920s, utilizing contrast agents that unfortunately had high degrees of toxicity. In the following decades, the interaction of contrast agents with the veins has provided insight into the utility of various agents as an informative sidelight to the information obtained from the examination. This pattern has continued to the present day, with some objective information about the newer contrast agents derived from the effects observed during phlebography or from effects observed experimentally which are relevant to the interaction of contrast agents with the vessel wall.

The types of side effects encountered with contrast infusion into the veins of the lower extremity range from discomfort to cardiovascular alterations to venous thrombosis. The effects on the cardiovascular system are dealt with in detail in Chapters 8 and 12. The direct cardiovascular effects of contrast infusion into the veins of the lower extremity may be clinically noteworthy but are in general indirect. Volumes of infusion tend to be large, creating a significant osmotic load. This in itself suggests that direct cardiac effects, as well as reflex cardiovascular effects, are to be expected. Similarly, effects on the kidney might be expected because of the considerable contrast volume that often must be infused. Despite these and other less well-defined systemic effects, however, the major concern with contrast infusion for phlebography centers on the interaction of the contrast agents with the vessel wall. The etiology of postphlebographic thrombosis is complex, but is at least in part related to the interaction of contrast material with the vessel wall and is a major focus of this chapter.

ROLE OF CONTRAST PHLEBOGRAPHY

By far the most common use of contrast phlebography is when there is clinical suspicion of deep vein thrombosis of the leg (DVT) (Fig. 14-1). DVT is an

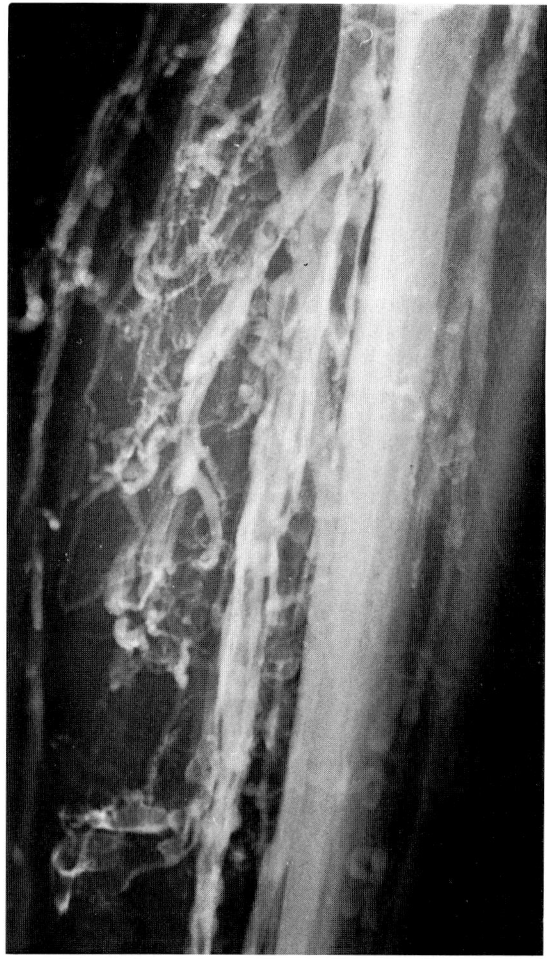

Figure 14-1 A 60-year-old woman with extensive deep and muscular venous thrombosis. Note the thrombi completely occluding the popliteal veins and filling peroneal and muscular veins. Venous drainage is mainly through superficial veins.

important disease that is both common and difficult to diagnose. That is, this disease may be lethal, may cause significant sequelae in the form of the post-phlebitic syndrome, and may also go completely undiagnosed on clinical grounds.[2] Further, symptoms strongly suggestive of this disease may be entirely misleading.[3] Other diagnostic modalities are available for DVT, but have various drawbacks ranging from poor sensitivity to poor specificity.[4–8] Contrast phlebography is also on occasion used to define the presence or absence of incompetent perforators leading to symptomatic varicosities (Fig. 14-2). Occasionally, retro-

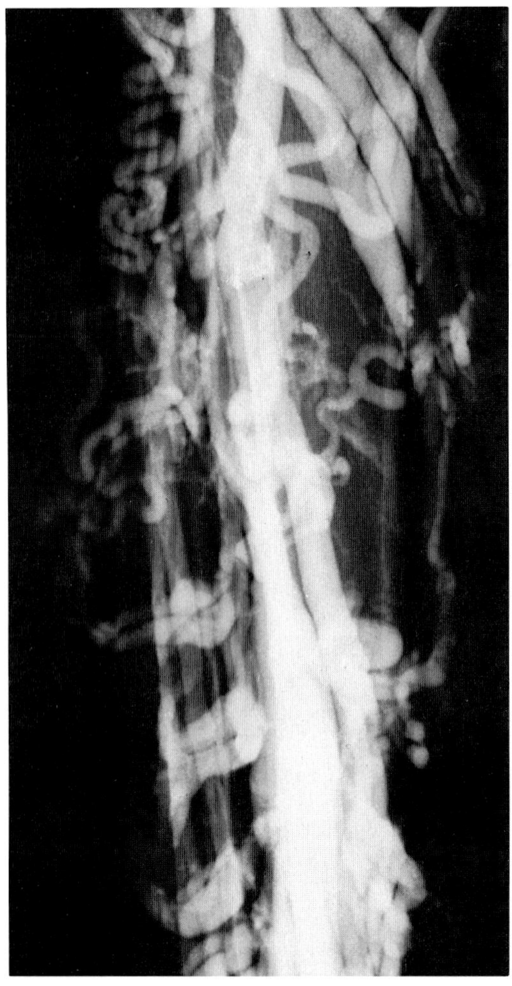

Figure 14-2 A 45-year-old woman with painful calf and varicose veins. A film of the high calf shows no evidence of venous thrombosis, but several serpiginous varicosities, draining from the deep system to the superficial, are present.

grade phlebography is used to define the adequacy of venous valve function,[9] and in some centers antegrade phlebography is used preoperatively for definition of the greater saphenous vein before in situ arterial bypass graft surgery. Various techniques have been used for phlebography, and the complication rate is at least in part dependent on these techniques.[10–13]

TECHNIQUE

As with other applications of contrast, there is a minimum iodine concentration necessary for adequate visualization of veins. The actual iodine concentration that is necessary for appropriate vessel definition depends on factors such as venous volume, rate of blood flow, rate of contrast infusion, soft tissue size of the individual patient, film-screen combination utilized, and radiographic technique. In general, until the late 1970s, a contrast agent containing 280 mg I/ml was routinely utilized. It was shown, however, that the incidence of discomfort could be markedly decreased by utilizing a contrast agent with an iodine concentration in the range of 200–225 mg I/ml.[14] Use of this lower iodine concentration did not lead to a loss of accuracy for the procedure. Although this is somewhat difficult to prove, it was empirically established by noting that the incidence of DVT was identical with contrast media at both concentrations. This is not to say that variations in technique, injection rate, and patient body habitus may not necessitate higher iodine concentrations in some patients, whereas in other patients an even lower concentration may be satisfactory.

On arterial infusion, blood is essentially replaced by contrast for brief periods of time. With contrast phlebography, however, primarily because of the very large volume of the veins of the lower extremity, ranging from 300–800 ml,[15] the aim is not to replace blood flow, but rather to allow mixing of contrast with blood and thereby to achieve complete visualization. Again, this mixing can be achieved with low concentration of contrast in certain patients. For example, in thin young women with no varicosities, contrast might routinely be utilized at an iodine concentration of 150 mg/ml. In patients with rapid flow, such as those with cellulitis and suspected superimposed deep vein thrombosis, flow tends to be very rapid and therefore a higher iodine concentration must routinely be used. Similarly, in particularly large patients or in patients with extensive varicosities, an iodine concentration of at least 200 mg/ml and perhaps greater must be used to afford satisfactory visualization of all major veins.

The technique that is currently most often used in the United States is that first described by Rabinov and Paulin.[11] With this method, a single leg is studied, the leg is nonweight bearing, and a tourniquet is not routinely in place during contrast infusion. Contrast is slowly infused to allow mixing with blood, and all major deep and superficial veins, including muscular veins that are important components of the deep system, are visualized and evaluated. Because of the multiplicity of veins, a moderate amount of vessel overlap occurs. If, alternatively, a tourniquet is used and the superficial veins are therefore excluded from evaluation, less overlap is likely to occur, and evaluation is likely to be somewhat simplified. A lower iodine concentration might then be employed routinely. The major drawback to this latter technique, however, is that phlebography is the only method that allows evaluation of all the major veins of the lower extremity, and most thrombi that embolize are thought to begin in the muscular sinusoids of the lower extremity. Since these veins drain the superficial muscles, they fill on phlebography from the superficial veins. Limiting the visualization of these muscular veins limits the unique contributions of contrast phlebography.

COMPLICATIONS

Pain

The most common complication of phlebography is pain and discomfort during infusion.[14] At its worst, this sensation is experienced as an intensely warm cramping feeling during the contrast infusion. The degree of discomfort depends on the type and concentration of contrast agent used.[14,16] It may be related to contact of the contrast with the vessel wall, as well as to distention of the veins, and consequently, unlike peripheral arterial injection, this discomfort is generally slow in onset. As with arteriography, the discomfort generally abates rapidly with the termination of the in-flow of contrast. The primary determinant of this discomfort is the osmolality of the contrast agent. Lowering the osmolality, therefore, clearly lowers both the incidence and the severity of discomfort.[14]

Postphlebography Syndrome

Another complication that is clearly related to the concentration of the contrast is the postphlebography syndrome.[14] On the basis of symptoms, this has sometimes been thought to represent postphlebographic phlebitis. It is characterized by pain and swelling in the low calf, generally beginning 6–12 hours after the contrast infusion. This complication, as some others, is noted only if looked for, and it must be distinguished from local inflammation relating to contrast extravasation. The postphlebography syndrome clinically resembles superficial phlebitis and is probably in fact a chemical phlebitis which may relate to the relatively high concentration of contrast that comes into contact with the superficial veins in the low calf. This syndrome is generally self-limited and responds to symptomatic treatment. Repeat contrast phlebography rarely demonstrates fresh thrombus. As with pain and discomfort, the incidence of this complication is markedly lessened merely through dilution of the contrast agent, as is illustrated in Table 14-1.

Contrast Extravasation

Contrast extravasation into the foot in the area of the infusion site can be a troublesome complication. This extravasation is generally painful, attracting the

Table 14-1 Complications of Phlebography with Four Contrast Formulations

	<%>			
Complication	*Meglumine Diatrizoate 60%, <282 mg I/ml>*	*Meglumine Diatrizoate 45%, <212 mg I/ml>*	*Meglumine Iothalamate 43%, <202 mg I/ml>*	*Iopamidol, <200 mg I/ml>*
Discomfort				
Mild/none	39	75	94	100
Moderate	39	20	4	0
Severe	22	5	2	0
Postphlebography syndrome	27	7.5	4	2
Positive 125-I-fibrinogen	39	9	10	10
Postphlebographic thrombosis	26	9	<8	<4

Source: Adapted with permission from "Contrast Venography of the Leg: Diagnostic Efficacy, Tolerance, and Complication Rates with Ionic and Nonionic Contrast Media" by MA Bettmann et al in *Radiology* (1987;165:113–116), Copyright © 1987, Radiological Society of North America Inc.

attention of the patient and therefore the physician. In patients with decreased sensitivity or marked discomfort before the infusion, however, large amounts of contrast material may extravasate. This can produce a significant inflammatory response in the soft tissues, similar to a chemical burn. In general, this complication resolves with symptomatic treatment. In patients with poor in-flow (e.g., severe peripheral vascular disease) or poor healing (e.g., poorly controlled diabetes) resolution of this complication may be delayed, and skin grafting may on rare occasions be necessary. Rare cases have also been reported that required amputation,[17] but this almost certainly was due in large part to pre-existent poor tissue perfusion. Decreasing the iodine concentration and therefore the osmolality decreases the reaction to extravasated contrast, and the same is true for low osmolar agents. A theoretical problem, however, and perhaps a practical one, is that agents with lower osmolality produce less discomfort on extravasation. The volume that enters the soft tissues may therefore be increased, and the tissue reaction commensurately marked, with diminished realization of extravasation.

Postphlebographic Thrombosis

The most significant complication of contrast phlebography and one that has provoked the greatest concern and interest is postphlebographic thrombosis.[18] Although inflammatory effects of contrast interaction with veins have long been recognized,[19] this specific complication was not noted until the 1970s.[20] It is impossible to diagnose clinically with any accuracy, and perhaps for this reason the true incidence is somewhat difficult to determine. Depending on the contrast agent used and the method of monitoring and diagnosing this complication, the incidence has ranged from 0–60%.[21] Studies have shown that the incidence depends on both the type of contrast agent used and the concentration.

This complication is a particularly interesting one because it allows an objective assessment and comparison of different contrast agents. One problem that arises in comparing results from different studies, however, is that different parameters are used for diagnosis. For example, the 125-I fibrinogen uptake test has been used as a means of monitoring for the occurrence of postphlebographic thrombosis. This test is extremely sensitive, but is known to have limitations in its specificity.[22] The presence of a positive test, then, indicates fibrin deposition, but not necessarily deposition in thrombus in the venous system. It may in fact indicate thrombi that are too small to be visualized by phlebography, but the importance of such thrombi clinically must be questioned. Alternatively, it may be positive because of an unrelated soft tissue hematoma. Further, various methods of monitoring have been used. Disregarding clinical diagnosis, the different techniques used have varying sensitivities; fairly marked differences in the incidence of postphlebographic thrombosis are therefore not terribly surprising.

COMPARISON OF CONTRAST AGENTS

Contrast agents have long been known to interact with the vessel wall,[18,23] and such interactions were thought to lead to sclerosis, which was initially felt to be of therapeutic benefit.[23] As the understanding of DVT increased, the detrimental nature of this effect of the contrast agents was recognized. Following the realization that contrast agents could cause venous thrombosis, a difficult realization as this complication is rarely clinically evident, it was noted that there were differences among contrast agents. For example, lowering the osmolality of standard contrast agents led to a decrease in the incidence of postphlebographic thrombosis.[14,16] Subsequently, it was shown that the nonionic contrast agent, metrizamide, had a far lower incidence of both positive fibrinogen uptake scans and of proven venous thrombosis than did a conventional ionic formulation.[24] As noted, several authors have subsequently investigated this process clinically with various contrast agents. Again using various methods of diagnosis, the almost invariable conclusion has been that lower osmolar agents lessen the incidence of this complication, although the specific factors leading to this alteration, such as lower osmolality or the specific contrast formulation, have not been clarified. The increased comfort with the low osmolar agents and a lower incidence of positive scans have been felt to be sufficient justification for their use.

One study examined the differences between a dilute conventional contrast agent and a nonionic agent in more detail.[21] This study indicated that the incidence of positive scans and objective postphlebographic thrombosis was equal with the dilute conventional agent and the particular nonionic formulation (Table 14-1), implying, perhaps, that there is a threshold to postphlebographic thrombosis in terms of osmolality. Below a certain level of osmolality, this complication may be significantly decreased, but does not seem to be further decreased with still lower osmolality, as was achieved with the nonionic formulation used at a low iodine concentration. Alternatively, it is unlikely but possible that a specific nonionic formulation may actually have a detrimental effect that counteracts the effect of the lowered osmolality. An additional interesting finding of this study, compared with prior studies, was that there seems to be a difference between conventional formulations at equivalent osmolality. The patients studied with an iothalamate preparation had a slightly lower incidence of clinically significant side effects than did those studied with a diatrizoate formulation. This is a somewhat surprising finding, as these two conventional formulations are so similar chemically.

SUMMARY

Postphlebographic thrombosis, then, is clearly the most significant complication of contrast phlebography. It is related both to the osmolality and to the formulation of the contrast agents. The thrombi that occur are generally in the superficial veins of the calf, although deep vein thrombosis, usually confined to the calf, occurs in about one-third of the patients.[15] This complication is especially disturbing as it is exactly the disease that the test is designed to diagnose, not precipitate or exacerbate.

The etiology of both the postphlebographic syndrome and postphlebographic thrombosis is complex. Clinically, lowering osmolality decreases both the incidence of discomfort and the incidence of these two objective complications. These adverse effects correlate with experimental studies of interactions between contrast agents and the vascular endothelium. Many studies dating from at least the 1950s have investigated this interaction.[25–27] These early studics concluded that contrast agents were capable of causing marked endothelial damage.[27] On more careful evaluation, this damage seems to occur either with infusions of contrast of very long duration or when contrast comes in contact with endothelium in the absence of blood flow. Some electron microscopic studies have suggested that contrast does damage the endothelium, even following relatively brief exposures.[28,29] Perhaps more importantly, contrast agents have been shown to alter endothelial function in various ways.[29–33] The endothelium is known to be a very active structure metabolically. It possesses potent lytic properties, which have been shown to be suppressed following the infusion of contrast agents.[34] Further, various functional parameters of the endothelium have been shown in our studies to be altered with relatively short, low concentration infusions of contrast agents. Many of these vasoactive properties play important roles in both causing vasodilation and in suppressing platelet aggregation and consequent clot formation on the endothelium. These studies suggest that it is possible to alter endothelial function without causing gross morphologic damage. The differences among contrast media are primarily due to osmolality, but alterations among different formulations were also observed. Perhaps most interestingly, the conventional diatrizoate formulation with sodium as the sole cation caused significantly greater alteration in most measured parameters than did the same contrast agent with sodium and meglumine or meglumine alone as cations. Additionally, inherent differences between certain low osmolality (ratio 3) formulations were observed. Additives to the contrast agent, such as citrate, seem to have little effect on endothelial function at the concentrations at which they are normally present.

Studies demonstrate that the current contrast agents for phlebography are fairly well understood and are safe, but that room for improvement in both understanding and clinical safety still exists.

REFERENCES

1. Abrams HL: Introduction and historical notes, in Abrams HL (ed): *Abrams Angiography*. Boston, Little, Brown and Co, 1983, pp 3–12.

2. Hull RD, Hirsh J, Carter CJ, et al: Diagnostic efficacy of impedance plethysmography for clinically suspected deep-vein thrombosis. A randomized trial. *Ann Intern Med* 1985;102:21–28.

3. O'Donnell TF Jr, Abbott WM, Athanasoulis CA et al: Diagnosis of deep venous thrombosis in the outpatient by venography. *Surg Gynecol Obstet* 1980;150:69–74.

4. Singer I, Royal HD, Uren RF et al: Radionuclide plethysmography and Tc-99m red blood cell venography in venous thrombosis: comparison with contrast venography. *Radiology* 1984;150:213–217.

5. Erdman WA, Weinreb JC, Cohen JM et al: Venous thrombosis: clinical and experimental MR imaging. *Radiology* 1986;161:233–238.

6. Vogel P, Laing FC, Jeffrey RB Jr et al: Deep venous thrombosis of the lower extremity: US evaluation. *Radiology* 1987;163:747–751.

7. Hull RD, Hirsh J, Carter CJ et al: Diagnostic efficacy of impedance plethysmography for clinically suspected deep-vein thrombosis. *Ann Intern Med* 1985;102:21–28.

8. Rosebrough SF, Grossman ZD, McAffee JG, et al: Aged venous thrombi: radioimmunoimaging with fibrin-specific monoclonal antibody. *Radiology* 1987;162:575–577.

9. Kistner RL, Ferris EB, Randhawa G et al: A method of performing descending venography. *J Vasc Surg* 1986;4:464–468.

10. Lea TM, McAllister V, Tonge K: Simplified phlebography in deep venous thrombosis. *Clin Radiol* 1971;22:490–494.

11. Rabinov K, Paulin S: Roentgen diagnosis of venous thrombosis in the leg. *Arch Surg* 1972; 104:134–144.

12. Coel MN: Adequacy of lower limb opacification: comparison of supine and upright phlebography. *Am J Roentgenol* 1980;134:163–165.

13. Le Veen RF, Dobry CA, Wolf GL: Pressure-infusion venography of the leg with remote-control fluoroscopy. *Radiology* 1981;138:730–731.

14. Bettmann MA, Paulin S: Leg phlebography: the incidence, nature, and modification of undesirable side effects. *Radiology* 1972;122:101–104.

15. Asmussen E, Christensen EH, Nielson M: The regulation of circulation in different postures. *Surgery* 1940;8:604.

16. Bettmann MA, Salzman EW, Rosenthal D, et al: Reduction of venous thrombosis complicating phlebography. *Am J Roentgenol* 1980;134:1169–1172.

17. Lea TM: Gangrene following peripheral phlebography of the leg. *Br J Radiol* 1970;43:528–530.

18. Albrechtsson U, Olsson CG: Thrombotic side effects of lower limb phlebography. *Lancet* 1976;1:723–724.

19. Dougherty J, Homans J: Venography, a clinical study. *Surg Gynecol Obstet* 1940;71:697–702.

20. Athanasoulis CA: Phlebography for the diagnosis of deep leg vein thrombosis, in Fratontoni J, Wessler S (eds): *Prophylactic Therapy of Deep Vein Thrombosis and Pulmonary Embolism*. US Dept of Health, Education, and Welfare publication No. (NIH)76-866, 1975.

21. Bettmann MA, Robbins A, Braun S et al: Contrast venography of the leg: diagnostic efficacy, tolerance, and complication rates with ionic and nonionic contrast media. *Radiology* 1987;165:113–116.

22. Harris WH, Salzman EW, Athanasoulis CA, et al: Comparison of 125-I-fibrinogen count scanning with phlebography for detection of venous thrombi after elective hip surgery. *N Engl J Med* 1975;292:665–667.

23. dos Santos JC: La phlebographie direct. *J Int de Chir* 1938;3:625–635.

24. Laerum F, Holm HA: Post phlebographic thrombosis: a double-blind study with methylglucamine metrizoate and metrizamide. *Radiology* 1981;140:651–654.

25. Zinner G, Gottlob R: Morphologic changes in vessel endothelia caused by contrast media. *Angiology* 1959;10:207–213.

26. Mersereau WA, Robertson HR: Observations on venous endothelial injury following the injection of various contrast media in the rat. *J Neurosurg* 1961;18:289–294.

27. Ritchie WGM, Stewart GJ, Lynch PR: The effect of contrast media on normal and inflamed canine veins. *Invest Radiol* 1974;9:444–455.

28. Parvez Z, Khan T, Moncada R: Ultrastructural changes in rat aortic endothelium during contrast media infusion. *Invest Radiol* 1985;20:407–412.

29. Gospos C, Freudenberg N, Straubesand J et al: The effect of contrast media on the aortic endothelium of rats. *Radiology* 1983;147:685–688.

30. Laerum F: Acute damage to human endothelial cells by brief exposure to contrast media *in vitro*. *Radiology* 1983;147:681–684.

31. Nordby A, Halgunset J, Haugen OA: Effects of radiographic contrast media on monolayer cell cultures. *Invest Radiol* 1986;21:234–239.

32. Laerum F, Borsum T, Reisvang A: Human endothelial cell culture as an evaluation system for the toxicity of intravascular contrast media. *Invest Radiol* 1986;18:199–206.

33. Bettmann MA, Gordon J: Effects of contrast agents on endothelial cell function (abs.). *Radiology* 1985;157:211.

34. Whitehouse WM Jr, Queral LA, Flinn WR et al: The effect of sodium diatrizoate on the fibrinolytic activity of saphenous vein intima. *J Surg Res* 1981;30:391–397.

Chapter 15

Agents in Computed Tomography

W. Dennis Foley

X-ray computed tomography (CT) is a form of sophisticated cross-sectional tomographic radiography. The basic advantage of CT over conventional radiography is that it enables a cross-sectional image display and superior contrast resolution. The cross-sectional image display avoids the problem of superimposition inherent in projection radiography. Lesions that would otherwise not be identified on projection radiography can be detected and accurately located by CT. The CT number or Hounsfield unit (HU) scale characterizes tissue attenuation on a numeric scale with reference to water. CT number values for identical tissues imaged on different type of CT scanners vary, a reflection of different calibration standards and variances in the projection reconstruction algorithm used. However, the CT number scale, because of the low scatter content incident on the detectors and the sensitivity and dynamic range of the detection system, produces images with far greater contrast sensitivity and latitude than conventional film radiography. If adequate x-ray technique is used, CT can detect contrast differences of 0.5% (5 HU), whereas the best contrast differential achieved by projection film radiography is 4–5%.

Early expectations were that CT, because of its intrinsically high contrast sensitivity, would provide accurate lesion detection and localization without the need for contrast enhancement. However, it became obvious from early clinical experience that contrast enhancement was both useful and necessary for many applications.[1] In addition, contrast enhancement techniques had to be modified to reflect the sensitivity of CT, the time course of an examination, pharmacokinetics of contrast distribution, and patient tolerance. These factors are examined in greater detail in later sections of this chapter.

BASIC PRINCIPLES OF ATTENUATION

X-ray attenuation in CT is primarily a reflection of two effects—photoelectric absorption and Compton scatter.[2] In soft tissue at the beam energies employed in

CT (100–140 kVp), almost all photon attenuation is due to the Compton reaction. In the Compton reaction, the incident x-ray photon collides with a peripheral electron, resulting in the release of a secondary scattered photon—itself contrast degrading—and of an ejected energized electron. The energy of the incident photon is shared by the scattered photon and electron. Photon attenuation due to the Compton effect is directly dependent on electron density, which in biologic tissues reflects physical tissue density. Photoelectric absorption does occur, however, particularly in the thyroid gland, bone, and hematomas. This is due to photon absorption in the inner electron shells of the higher Z materials; iodine, calcium, and iron, respectively. The likelihood of a photoelectric interreaction is dependent on beam energy and atomic number and will occur with lower beam energies (e.g., kVps producing effective energy in the 32 to 40 keV range) and with higher atomic number materials. However, low beam energies significantly increase x-ray tube loading and absorbed dose and are not routinely employed except in dual energy applications, such as quantification of bone mineral density or liver iron.

Intravascular contrast agents for CT are iodinated triiodo benzoic acid derivatives. These may be in the ionic or nonionic form. Ionic contrast agents are either disassociated monomers that have high osmolality (1500 mosmol/kg H_2O) with reference to plasma (300 mosmol/kg H_2O) or monoacid dimers, such as Hexabrix, that are low osmolality agents (580 mosmol/kg H_2O). Nonionic dimeric agents, such as iopamidol or iohexol, have an osmolality similar to Hexabrix.[3] Contrast agents enhance vessels and tissues by means of the photoelectric effect, with the degree of enhancement dependent on local iodine concentration (Fig. 15-1). Despite differences in osmolality, the iodine content of these agents is equivalent. Thus, the radiologic effects are identical, although the physiologic effects are dissimilar. When employed in CT, intravascular contrast agents are distributed in both the vascular and, in the nonneural tissues, in the extravascular spaces. Factors affecting tissue attenuation are the rate and amount of contrast material administered, blood iodine level, the rate of redistribution of contrast material from the vascular to the extravascular space (in the nonneural tissues), and timing of the CT scan.

LESION DETECTABILITY

Four major factors—image noise, contrast, partial volume effect, and slice registration—affect lesion detectability in CT. Noise is dependent on incident photon flux and detector efficiency. Contrast is determined by regional differences in tissue attenuation, which, on unenhanced scans of soft tissue structures (except the thyroid gland), reflect regional differences in tissue electron density. When intravenous (IV) injection of iodinated contrast material is employed for CT, the recorded contrast reflects both regional differences in electron density and the

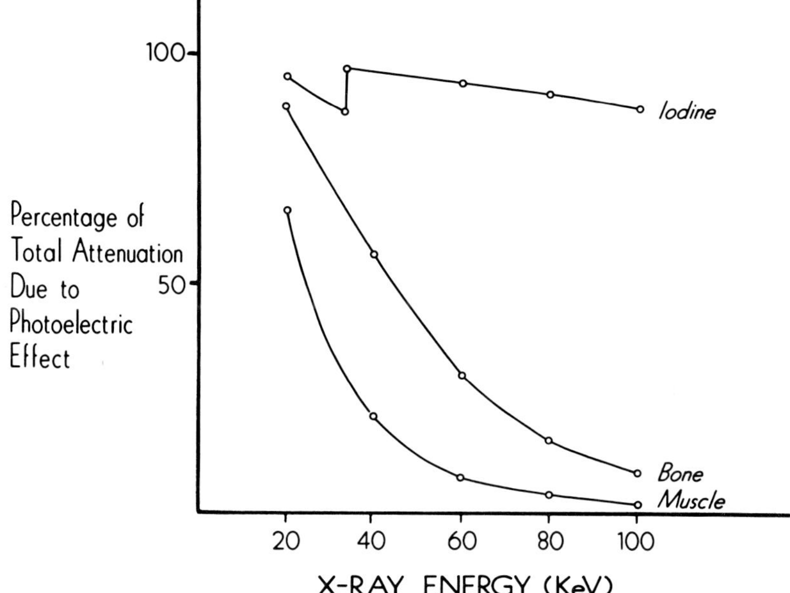

Figure 15-1 Graph representing the percentage of total x-ray photon attenuation caused by the photoelectric effect as a function of x-ray energy. Notice that the majority of photon attenuation due to iodine in the diagnostic energy range (60–80 keV) is due to the photoelectric effect. The Compton reaction accounts for the majority of photon attenuation in bone and soft tissue.

---

additive (or subtractive) effect of regional differences in vascular and extravascular iodine. Differential iodine concentration between focal lesions and surround may accentuate pre-existent differences in tissue density. Thus, a focal, hypodense hepatic lesion seen on a noncontrast study may be more apparent after IV contrast enhancement. In addition, a focal lesion that is either barely perceptible or imperceptible on a noncontrast study may become perceptible after contrast enhancement. However, differential iodine concentration may result in the reverse effect. If there is greater iodine concentration in a lesion of lower electron density with reference to the surround, "contrast enhancement" may result in an isodense lesion.

The ability of a CT system to record differential attenuation between a focal lesion and the surround depends on such technical factors as beam energy, photon output, detector sensitivity and conversion efficiency, electronic noise, the reconstruction algorithm employed, matrix size, and characteristics of the display. However, partial volume effect—the relation between lesion diameter and slice thickness—may be the most important determinant in detectibility of small lesions.[4] Lesions with high intrinsic contrast (differential attenuation) may be undetected if the photon count incident on the detectors reflects absorption by both the focal lesion and normal surround (Fig. 15-2). Partial volume averaging can

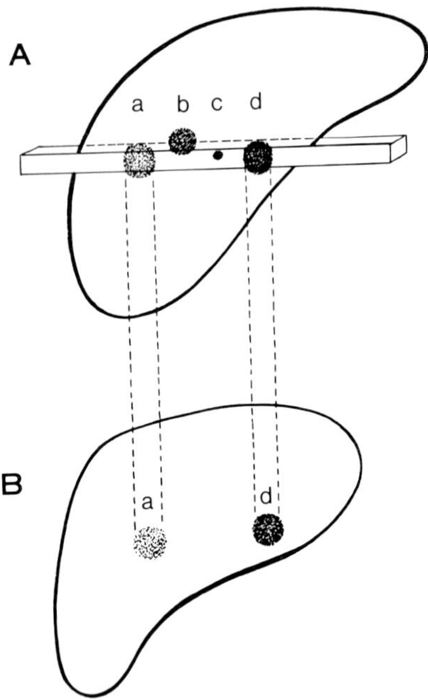

Figure 15-2 Schematic demonstrating the effect of contrast and partial volume effect on lesion detectability. (*A*) Four lesions varying in size and contrast are shown in relation to a projected CT slice through the liver. (*B*) The resultant CT appearance is demonstrated.

Lesions *a* and *d* are of equal size, and each occupies the full slice thickness. The differing contrast is accurately portrayed in the resultant image. Lesion *b* is slightly smaller than either *a* or *d* and intermediate in contrast. However, lesion *b* is only slightly within the slice profile, and the resultant attenuation difference is too small to be recorded. Lesion *c* is of equal contrast with reference to background as lesion *d*, but is considerably smaller. Even though lesion *c* is fully within the slice profile, partial volume averaging of its attenuation with surrounding background results in failure to detect it in the CT image. *Source:* Reprinted with permission from *Seminars in Liver Disease* (1982; 2[1]:14–28), Copyright © 1982, Thieme-Stratton Inc.

---

only be counteracted by either increasing local contrast with exogenous agents or by using narrower slice collimation.

Slice registration refers to two factors: (1) the accuracy of table repositioning and (2) the ability of a patient to employ equivalent respiration between individual scans so that thorax and abdomen studies that are programmed to occur at contiguous levels, do in fact, result in anatomically contiguous slices. Misregistration occurs when unequal respiratory excursions are employed in the interscan interval. This will result in anatomic overlap and missing segments in scan slices that are sequentially, but not anatomically, contiguous. Misregistration may result in the failure to detect small pulmonary nodules or small hepatic lesions.

CONTRAST MEDIA PHARMACOKINETICS

Arterial iodine concentration is a direct reflection of the rate and amount of contrast material administered, cardiac output, plasma expansion induced by hyperosmolar contrast agents, and extravascular redistribution and renal filtration of the injected contrast. For single-level dynamic CT obtained during the first circulation of the injected bolus, bolus delivery and cardiac output are the critical factors affecting arterial iodine concentration.[5] Plasma expansion and concomitant extravascular redistribution and renal filtration of contrast material are important factors affecting vascular and tissue iodine concentration during sustained bolus injections employed in incremental (contiguous-level) dynamic scans.

The major clinical indication for performing a single-level dynamic scan for vascular diagnosis is evaluation for suspected thoracic aortic dissection.[6] Other possible indications include the demonstration of renal vein tumor thrombus and arteriovenous fistulas.[7] Timing should encompass the arrival, time to peak, and clearance phases of the circulating bolus in the region of interest. In general, a 100-HU elevation of vascular CT number is desirable. In suspected aortic dissection, this high level of aortic enhancement in combination with a rapid dynamic scan allows accurate detection of an intimal flap, distinction of the intimal flap from artifact, and recording of differential flow rates in the true and false lumen (Fig. 15-3). In the nonanemic patient, the baseline vascular CT number is 40 to 45 HU. Thus, for the postcontrast single-level dynamic scan, elevation of the aortic CT number to 140 HU is desirable. In patients with normal cardiac output, this level of vascular enhancement can be achieved by a rapid IV injection of 50 cc of 60% contrast material into the antecubital vein. Injection rates of 8 cc/sec with a total injection volume of 50 cc are easily achieved by mechanical volume flow rate injectors when utilizing 18- to 19-gauge plastic venous catheters. The palm of the patient's hand should be positioned against the side of the gantry so that the arm is approximately at 90° to the chest wall. In this position, venous flow through the axillary and subclavian veins is not constricted at the thoracic outlet, as can occur when the arm is elevated above the shoulder. This is a simple but important technical consideration. Following injection, the patient's arm may then be placed comfortably above the shoulder. The technologist should be able to operate the injector and the CT console function keys simultaneously, allowing the physician to monitor the contrast injection directly to ensure that there is no contrast extravasation. Both contrast injection and scanning should be terminated immediately if contrast extravasation occurs.

Single-level dynamic scans for vascular diagnosis rely on contrast changes induced during the first circulation of the injected bolus. The first scan of a single-level dynamic sequence should be obtained immediately before the arrival of the contrast bolus, with subsequent scans obtained in rapid succession. For thoracic aortic studies, the delay between the beginning of the injection and the beginning

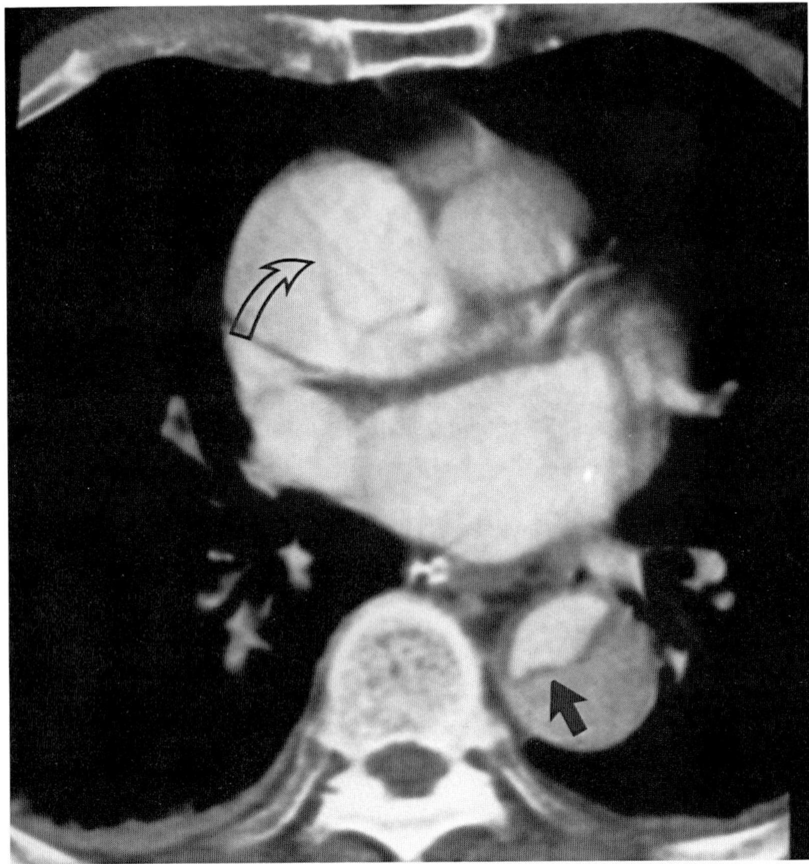

Figure 15-3 Image from a single-level dynamic scan of a patient with a Type A aortic dissection. The scan level is at the upper left atrium immediately cephalad to the aortic root. There is an irregular intimal flap in the ascending aorta (*open arrow*) and a clearly defined intimal flap in the descending aorta (*closed arrows*). Differential contrast flow in the true and false lumen is clearly demonstrated in the descending aorta.

of the scan is usually 5 seconds. Short scan times of adequate technical quality with short interscan delays are required. In patients with normal cardiac output, the contrast bolus is usually cleared by 10 to 15 seconds. However, in patients with poor cardiac output and large central blood volume, the injected contrast bolus may not clear for 25 to 30 seconds.

Renal vein enhancement is usually delayed 5 to 10 seconds after arrival of the contrast bolus in the renal artery. A single-level dynamic scan through the left renal vein may be used in the diagnosis of renal vein tumor thrombosis if it is not evident on preliminary noncontrast or incremental dynamic scans through the kidneys. In addition, single-level dynamic scans can document the patency of distal end-to-side splenorenal shunts when simultaneous enhancement of the splenic and renal veins at the distal anastomotic site is demonstrated.[8]

The infrarenal inferior vena cava receives venous blood flow from the lower extremities, pelvis, and lumbar region. Contrast transit time through these peripheral tissues is markedly delayed relative to the kidney; the lower abdominal inferior vena cava does not enhance for a period of up to 45 seconds after the beginning of the upper extremity venous injection. This is important because an infrarenal aortocaval fistula can be diagnosed by dynamic CT when anomalous early enhancement of the distal inferior vena cava is recognized (Fig. 15-4).[9] In normal patients, rapid circulation of contrast through the kidneys results in caval contrast enhancement at the ostea of both renal veins within 15 to 20 seconds after injection, which is contrasted to the central, nonenhanced, nonadmixed blood flowing from the extremities, pelvis, and lumbar region (Fig. 15-5). This is a normal finding that should not be mistaken for caval thrombus.

The most common indication for a single-level dynamic scan for tissue diagnosis is suspected hepatic hemangioma. In this case, dynamic scans are performed both during the first circulation and the recirculation of the injected contrast bolus. Both the pattern of vascular enhancement during the first circulation and the pattern of vascular and tissue enhancement during recirculation are analyzed. Characteristic findings of a hemangioma (hypodense lesion on a noncontrast scan, peripheral hyperdense rim on early scans with delayed central fill-in) have been observed in only 55% of hemangiomas.[10] Other findings, including early central enhancement and delayed inhomogeneous enhancement without central fill-in, have been documented in proven hemangiomas. These "noncharacteristic findings" presumably reflect biologic variability, such as scar and intralesional thrombus and hemorrhage. The delayed central fill-in that occurs during contrast recirculation may take from 5 to 20 minutes or longer and is directly related to lesion size. This examination is best performed as a dedicated study as a bolus of 75 cc of contrast material is injected. This large recirculating bolus makes delayed central fill-in easier to perceive on CT. The relatively low sensitivity of CT in demonstrating the characteristic findings of a relatively common entity, which may be an isolated finding in patients with extrahepatic malignancy, has put the clinical role of CT scanning in some doubt. Early experience with a nuclear medicine technique using labeled red blood cells and single photon emission computed tomography suggests that this technique is very sensitive and specific and could supplant the use of both enhanced CT and MRI in this clinical situation.

When dynamic scanning is used, it is most commonly incremental; that is, a rapid sequence of scans is obtained at contiguous levels after bolus contrast administration. Incremental scanning combines the advantages of vascular enhancement and the rapid survey of an organ or particular region during early peak visceral enhancement. However, the bolus delivery technique is modified in that

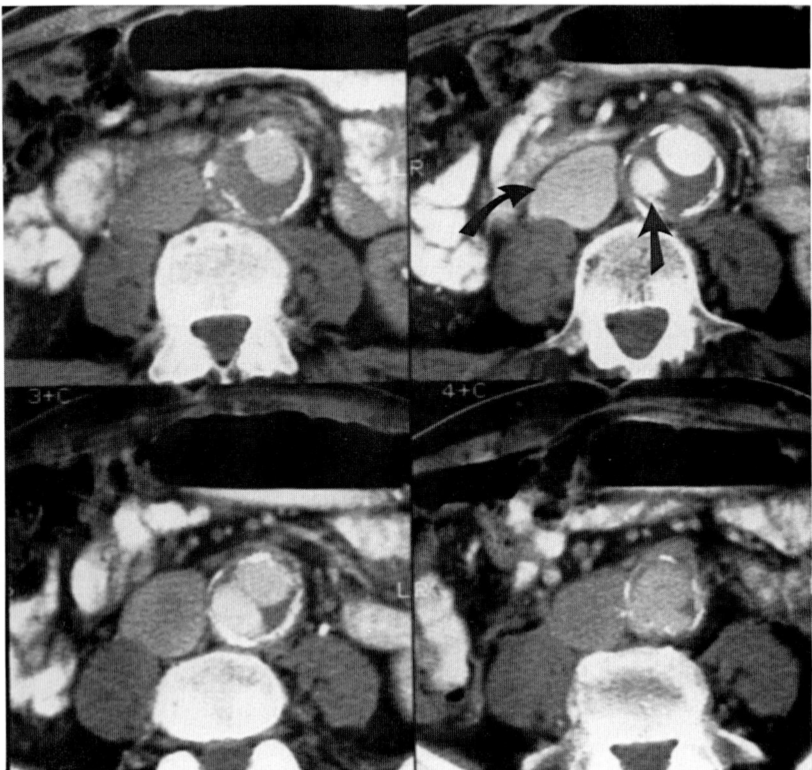

Figure 15-4 Aortic graft with pseudoaneurysm (*open arrow*) and aortocaval fistula recognized by early enhancement of the inferior vena cava (*curved arrow*). This patient had a prior aortic aneurysmectomy with end-to-end interposition graft covered by the wall of the native calcified aortic wall. There is postoperative hematoma, a normal finding, between the graft and the native aortic wall. However, a second discrete false lumen is opacified simultaneously with early abnormal enhancement of the inferior vena cava. Findings indicate a pseudoaneurysm arising from the distal anastomosis and an aortocaval fistula. The distal anastomosis was above the aortic bifurcation. *Source:* Reprinted with permission from "CT Detection of Aortocaval Fistula" by WD Middleton, DF Smith, and WD Foley in *Journal of Computed Assisted Tomography* (1987;11:344–347), Copyright © 1987, Raven Press.

the initial rapid bolus injection is followed by a continuous injection with the aim of delivering the total contrast load within a short time period. Volume flow rate injectors with biphasic injector controls are ideal for this purpose. A technique has been developed at our institution in which 180 cc of 60% contrast material (50 g of iodine) is injected over 2 minutes and is employed for survey dynamic scans of the thorax, upper abdomen, or pelvis. With this technique, the arterial iodine concentration is a reflection of both the first circulation bolus and continuous replenishment.[11]

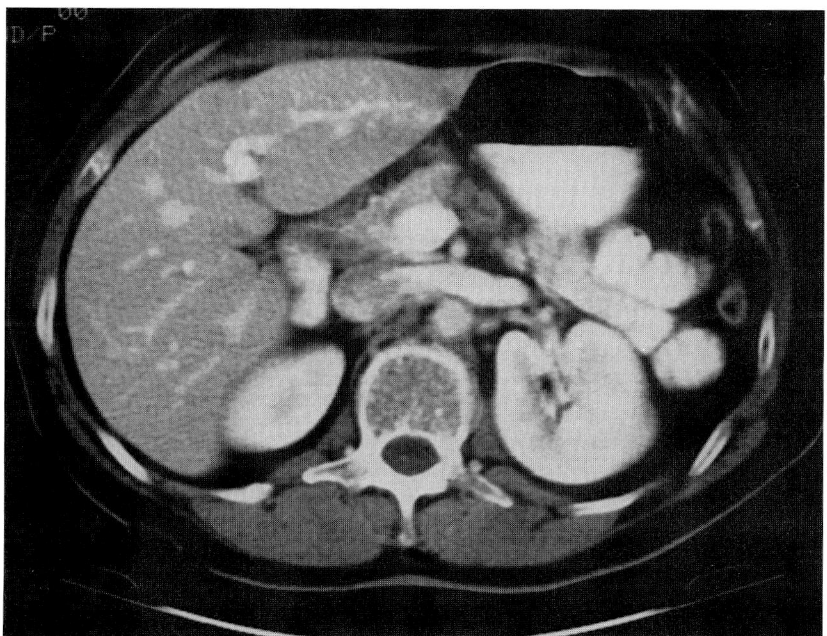

Figure 15-5 Apparent thrombus in the central inferior vena cava reflects relatively rapid flow through the renal circulation in comparison to slower circulation in the lumbar region, pelvis, and lower extremities. Nonenhanced blood in the central inferior vena cava is contrasted to enhanced blood flowing into the inferior vena cava from the renal vein.

When applied to thoracic CT scanning, this technique (50 g iodine load injected over 2 minutes) results in sustained vascular enhancement throughout the survey dynamic scan. As a result, major brachiocephalic arteries, the thoracic aorta and central pulmonary vessels, and the cardiac chambers are all adequately enhanced up to a value of 100 HU. In order to use this injected bolus adequately, the CT scanner should be capable of surveying the region of interest with rapid scans of acceptable technical quality that will allow the contrast differential between vascular and nonvascular tissue to be defined clearly. This is of critical importance in the diagnosis of hilar lymphadenopathy for staging of lung carcinoma (Fig. 15-6), as well as in the identification of largely thrombus-filled aortic aneurysms in which only the neck of the aneurysm may be opacified. These small aneurysm lumens may not be identified when the contrast differential between the vascular space and surrounding clot is relatively less as would occur during an infusion examination. Essentially, contrast would be lost by partial volume averaging.

In abdominal CT scanning, evaluation of the liver for focal hepatic lesions is often a major objective. Many authors have evaluated methods of lesion detection

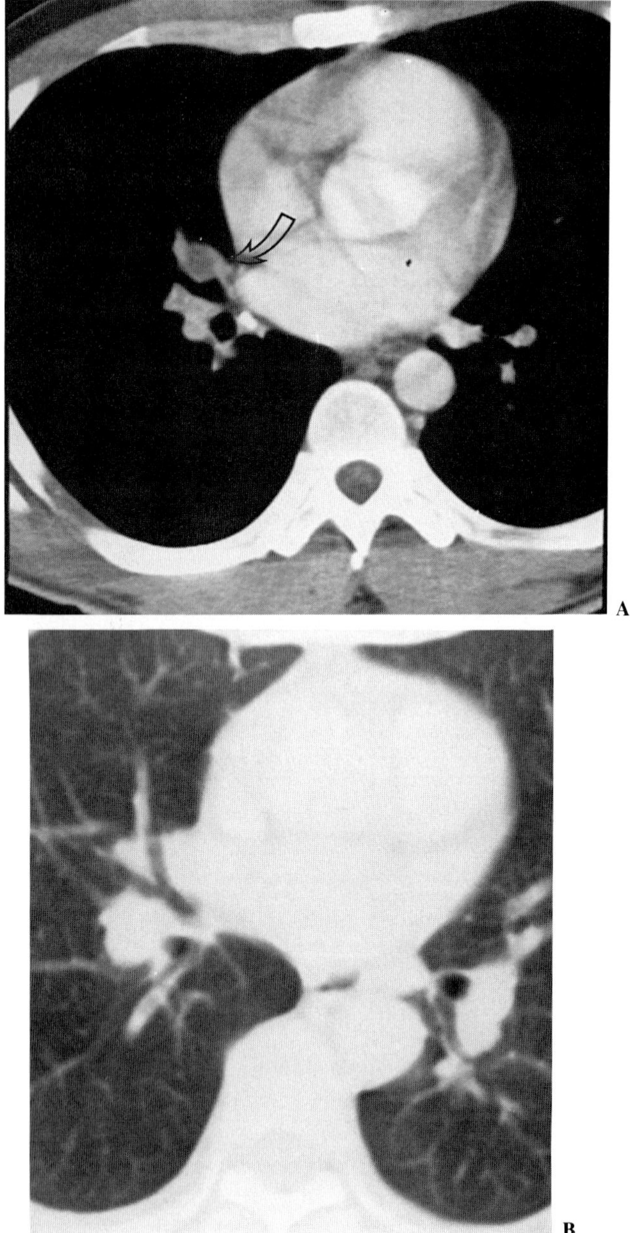

Figure 15-6 (*A*) Inferior right hilar lymphadenopathy demonstrated by a dynamic scan. There is an enlarged lymph node adjacent to the right middle lobe bronchus (*open arrow*). Enhanced lower lobe arteries are seen posterior and lateral to the lower lobe bronchus. (*B*) No bronchial distortion or mass is evident on the lung window image. A small calcified lymph node is medial to the lower lobe bronchus.

and appropriate contrast enhancement. There is an emerging consensus that the incremental dynamic scan during bolus contrast administration maximizes lesion detectibility for most metastases, with the exception of those such as pancreatic islet cell tumor or carcinoid tumor that are hypervascular. The delayed iodine scan[12] in which residual extravascular contrast material, including that which is within hepatocytes, as well as in the interstitial spaces of the liver, provides contrast enhancement of normal tissues has also been demonstrated to have a sensitivity equivalent to the early dynamic scan with bolus contrast administration.[13] However, the delayed iodine scan is critically dependent upon the amount of iodine delivered and requires at least 60 g of iodine for adequate lesion detection. A recent study in which the average amount of iodine administered was 75 g (a combination of abdominal angiography and CT arterial portography using superior mesenteric contrast injection) reported that lesion detectability was superior with the delayed iodine scan in comparison to an early dynamic scan following bolus contrast delivery. However, a 75-g iodine load is not routinely administered to patients and although it may be physiologically "safe" it is not at this time recommended for routine use. Delayed hepatic CT scanning may be performed for patients who are operative candidates for suspected isolated hepatic metastases or for those patients who have equivocal findings on survey dynamic CT scanning.

Delivery of a 50-g iodine load over 2 minutes produces a peak aortic enhancement at the end of contrast injection. It is well known that, during the arterial phase of the first circulation of an injected bolus, splenic enhancement is greater than hepatic enhancement. However, during the portal venous phase, hepatic enhancement equals and subsequently exceeds splenic enhancement. With continued IV contrast replenishment, splenic enhancement remains greater than hepatic enhancement, even though hepatic enhancement now reflects contributions from both the hepatic arterial and the portal venous supply. On a noncontrast scan, hepatic attenuation is 5 to 10 HU greater than splenic attenuation. With an incremental dynamic scan, this is reversed, and hepatic attenuation is 5–10 HU less than splenic attenuation. The criterion for fatty liver on a noncontrast scan is that hepatic attenuation is equal to or less than splenic attenuation. However, due to the contrast reversal achieved by continuous bolus contrast delivery and continuous replenishment, hepatic attenuation should be at least 25 HU less than splenic attenuation in order to diagnose confidently fatty liver.[14]

Hepatic enhancement reflects both sinusoidal opacification and extravascular redistribution of the injected contrast.[15] The rate of extravascular contrast redistribution in the liver is relatively rapid, with half of an injected bolus present in the extravascular space within 1 minute after injection. With the continuous bolus injection technique, hepatic enhancement reaches a relative plateau 1 minute after the beginning of contrast injection. Hepatic enhancement is then maintained at this plateau level for 2–2½ minutes (Fig. 15-7). It is important to note that hepatic enhancement is maintained after the IV contrast injection ends, even

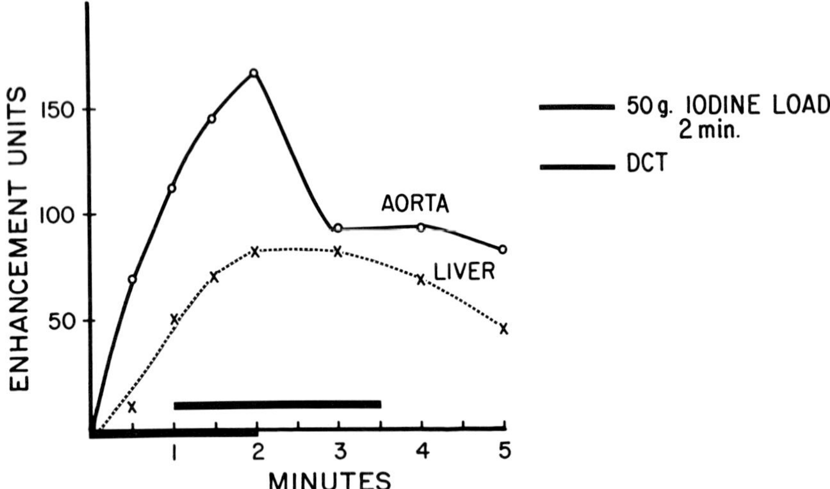

Figure 15-7 Graft of vascular and hepatic enhancement achieved by bolus injection of 50-g iodine load over 2 minutes. Sixty percent contrast material is initially injected at 5 cc/sec for 10 seconds. Subsequently, contrast material is injected at 1 cc/sec. Peak vascular enhancement occurs at the end of injection with a subsequent biphasic decrease. The initial rapid drop reflects intravascular-to-extravascular redistribution of contrast material. When intravascular-to-extravascular equilibrium has been achieved, the subsequent decline in aortic enhancement reflects renal filtration. Hepatic enhancement is delayed in comparison to vascular enhancement. However, a relative plateau of hepatic enhancement is maintained after the end of contrast injection, presumably due to the continuous extravascular redistribution of contrast material before equilibrium is achieved.

The optimal dynamic scan is performed between 1 and 3½ minutes after the beginning of injection. Note that vascular enhancement is always greater than parenchyma enhancement, enabling focal lesions to be identified accurately within specific lobes and segments. *Source:* Reprinted with permission from "Contrast Enhancement Technique for Dynamic Hepatic CT Scanning" by WD Foley et al in *Radiology* (1983;147:797–803), Copyright © 1983, Radiological Society of North America Inc.

---

though vascular enhancement shows a biphasic decrease. There is a relatively rapid drop to levels isodense with hepatic parenchyma and then a slower rate of decrease that parallels the decrease in hepatic enhancement and is due to renal filtration. The rapid drop in vascular enhancement presumably reflects intravascular to extravascular redistribution before equilibration of contrast material concentration between the two spaces. At 3 to 3½ minutes after the contrast injection begins, both vascular and hepatic enhancement begin a gradual decline caused by renal filtration of the contrast material. At this time, there is a dynamic interreaction in which contrast material leaves the extravascular space to re-enter the intravascular space, replacing contrast material that has been excreted by the kidney.

The rate of extravascular redistribution of contrast material in the liver and spleen is relatively rapid; it is faster than has been observed in muscle, fat, and lymphoid tissue. This rapid rate is of considerable benefit in the diagnosis of enlarged mediastinal, abdominal, and pelvic lymph nodes in that high levels of contrast differentiation between major adjacent vessels and lymph nodes are achieved by the dynamic scan approach. This contrast differentiation maximizes sensitivity in detecting enlarged pulmonary hilar lymph nodes and in distinguishing retroperitoneal varices or hypogastric vessels from lymph nodes.

Most hepatic metastases arise from relatively hypovascular primary lesions. These include tumors of the colon, lung, breast, pancreas, and endometrium. It would be expected that hypovascular metastases would be hypodense in relation to normal hepatic parenchyma during an incremental dynamic scan with bolus contrast administration.[16] Less contrast material is delivered to hypovascular metastases than to normal hepatic parenchyma. In fact, contrast differentiation between tumor and liver is maximized, and parenchymal enhancement is greatest in the 1- to 3-minute plateau time period before the equilibrium phase between the intravascular and extravascular spaces (Fig. 15-8).[17] During the equilibrium phase, contrast material diffuses into the extravascular tumor compartment. The rate of contrast diffusion probably reflects both neovascularity and the relative proportion of viable cellular tumor and necrotic material. The greater the degree of neovascularity and the proportion of cellular as compared to necrotic tumor, the greater the extent of contrast diffusion into the interstitium of the tumor. When the concentration of iodine in the vascular and extravascular spaces of both tumor and normal liver is equal, the tumor will be isodense. Some tumors are sufficiently necrotic that contrast diffusion is limited and an isodense lesion does not result. In addition, necrotic tissue has less intrinsic photon absorption than normal hepatic parenchyma, so that even if contrast equilibration occurs, there is sufficient intrinsic contrast differentiation for the lesion to be detected.

Hepatic metastases, particularly from colon carcinoma, usually have a relatively vascular rim that represents the most viable cellular portion of the tumor. With bolus contrast administration and the incremental dynamic scan technique, the tumor rim may become isodense with hepatic parenchyma during the plateau phase. Nevertheless, contrast differentiation between the central hypodense and presumably necrotic portion of the tumor and normal hepatic parenchyma is accentuated. Rim enhancement of tumors can be appreciated by both IV injection techniques, but more particularly by selective hepatic artery injection.[18] In fact, contrast enhancement of the tumor rim may be the only positive finding of a hepatic metastasis in patients with diffuse fatty liver and multifocal metastases in whom attenuation of the fatty liver and central necrotic tumor is equal (Fig. 15-9).

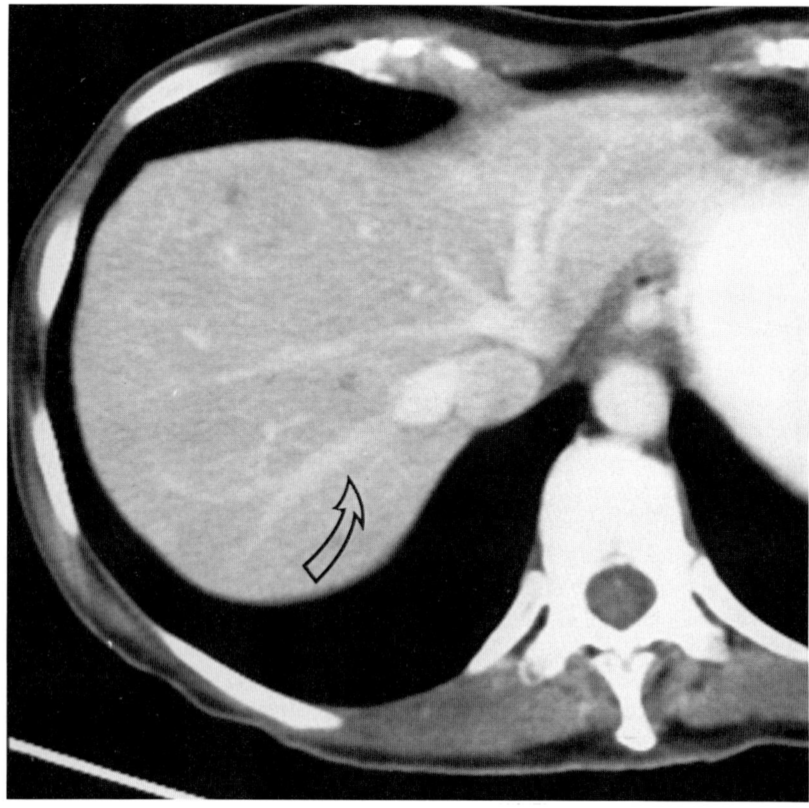

Figure 15-8 Small hepatic metastases from primary breast carcinoma detected by the dynamic incremental scan technique. Metastases are approximately 5 mm in diameter and are detected with a slice thickness of 10 mm. Note the anomalous accessory right hepatic vein (*curved arrow*). This case illustrates both good contrast differentiation and anatomic localization of metastases.

Comparison of lesion detectability during early incremental dynamic scans and during scans obtained 5 to 10 minutes after injection demonstrates a significant percentage of isodense lesions with the latter technique. This technique of early delayed scanning should be distinguished from the 4- to 6-hour delayed scanning that may be used to detect hepatic metastases. The 5- to 10-minute delayed abdominal scan is often used in the staging of lung carcinoma in which the dynamic scan during bolus contrast administration is performed for the thorax, with the early delayed scans being obtained through the upper abdomen to identify hepatic or adrenal metastases. Because of the time delay between injection and scanning, this technique becomes comparable to infusion studies in which a 42- to 50-g iodine load is administered and nondynamic scans at a rate of three to four per

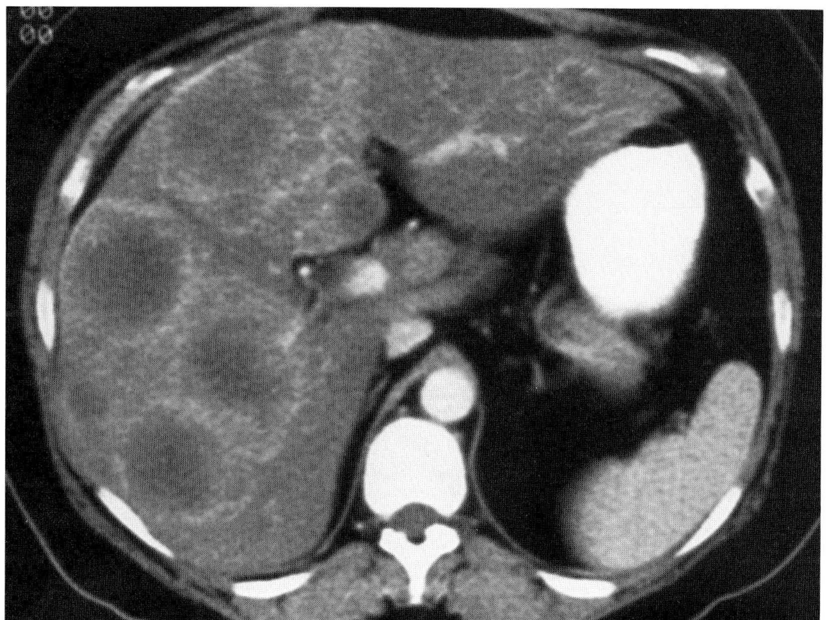

Figure 15-9 A patient with diffuse fatty liver and multifocal metastases. The margins of the metastases are defined by rim enhancement. The fatty liver, which is clearly shown by the marked contrast difference between hepatic and splenic parenchyma, significantly decreases contrast between the central necrotic tumor and hepatic parenchyma.

---

minute are performed during and subsequent to the infusion (Fig. 15-10). When this early delayed scan technique was compared to dynamic upper abdominal scanning performed on a separate day, a significant percentage of isodense lesions was produced by the early delayed technique. In essence, bolus and infusion techniques deliver the same total load of contrast material. The bolus method is preferred as it improves lesion detectability and can be obtained easily with most modern CT scanners with very acceptable image quality.

A goal in implementing dynamic hepatic CT scanning is to enhance positively both central and peripheral intrahepatic vessels so that detected lesions can be accurately assigned to specific hepatic lobes and segments. This is important for patients who are potential operative candidates for focal hepatic resection for suspected isolated metastases. In addition, positive enhancement of intrahepatic vessels avoids confusion in distinguishing small peripheral focal lesions from peripheral intrahepatic vessels, a situation that can occur with infusion techniques, as well as noncontrast scans.

Hepatic resection for suspected isolated colon metastases has been demonstrated to improve survival in selected patients. The most sensitive technique for

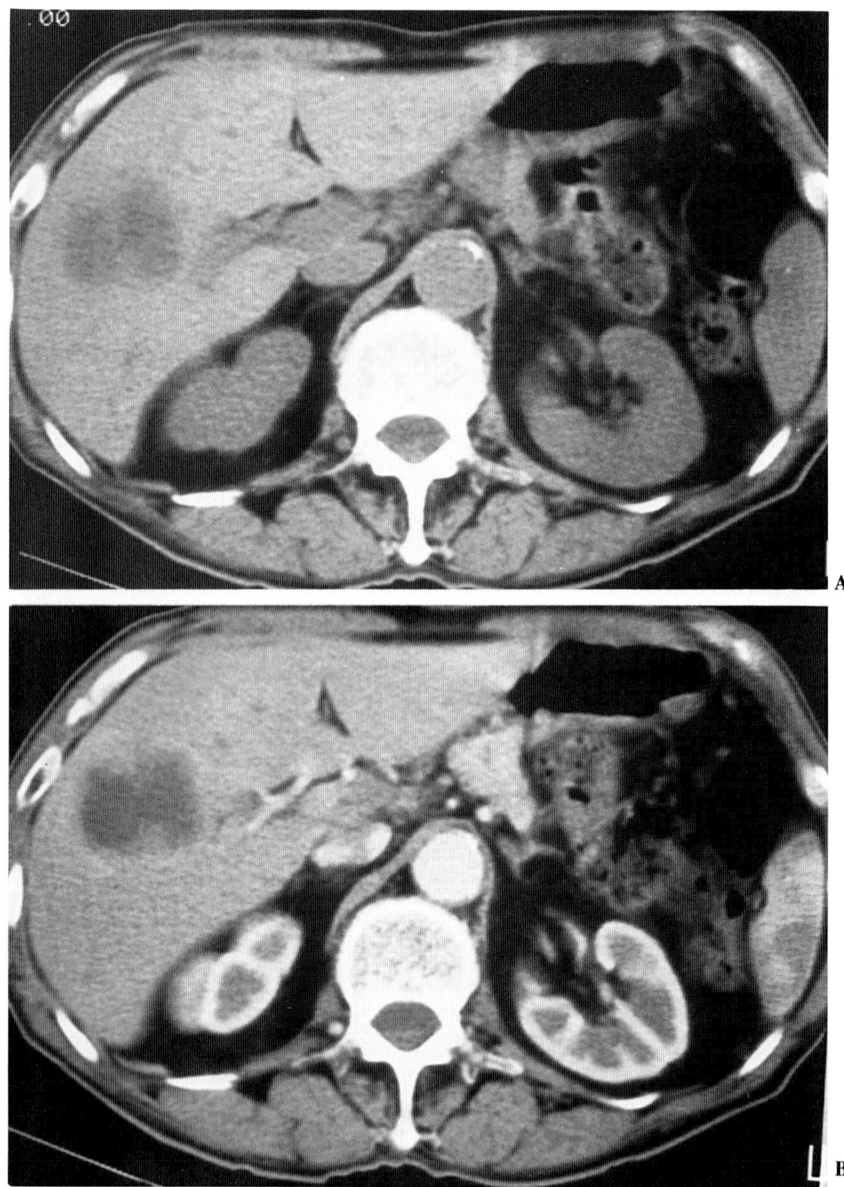

Figure 15-10 Comparison of (*A*) noncontrast, (*B*) dynamic postcontrast, and (*C*) "5 minute early delayed'' postcontrast CT scan in patient with hepatic metastasis. Note the improvement in contrast with (*B*) dynamic scan image. The focal lesion is almost isodense on (*C*) early delayed scan. This study also highlights the cortical angionephrogram in the dynamic scan and the uniform nephrogram in the ''5 minute early delayed'' scan.

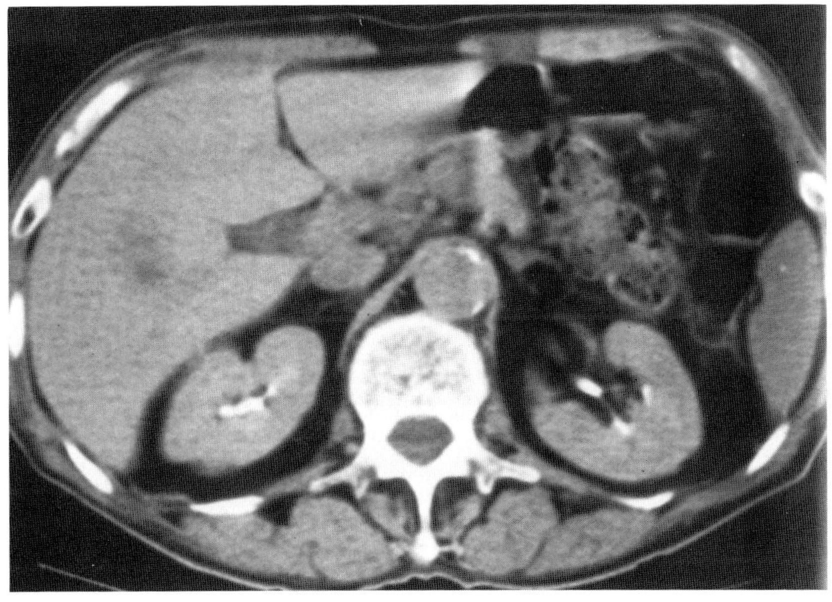

C

documenting the number and location of hepatic metastases is arterial injection CT. By selective injections, this method enhances hepatic vessels, parenchyma, or tumor significantly more than IV bolus enhancement. Arterial injection CT is usually combined with preoperative angiography, which is performed as a vascular mapping procedure. For colon metastases that, apart from the vascular rim, are usually hypovascular relative to hepatic parenchyma, arterial portography CT using superior mesenteric artery contrast injection is the preferred approach.[19] Hepatic and portal veins and normal parenchyma are selectively enhanced to a greater degree than with IV injection, providing better contrast to hypodense lesions (Fig. 15-11). The catheter tip should be positioned distal to any anomalous right hepatic arterial branches. An injection of 120 cc of 60% contrast material at a rate of 1 cc/sec with scanning beginning 30 seconds after the beginning of injection is the technique that has been adopted at our institution. Additional metastatic lesions not documented by prior IV contrast-enhanced CT can be demonstrated with arterial portography CT. If equivocal findings occur on arterial portography CT, delayed hepatic CT scanning 4–6 hours after the procedure is suggested. In a recent study, delayed hepatic CT scanning was more sensitive than arterial portography CT.[13] Arterial portography CT can be degraded by focal perfusion defects caused by the incomplete admixture of opacified superior mesenteric venous blood and nonopacified splenic blood, resulting in hypoperfu-

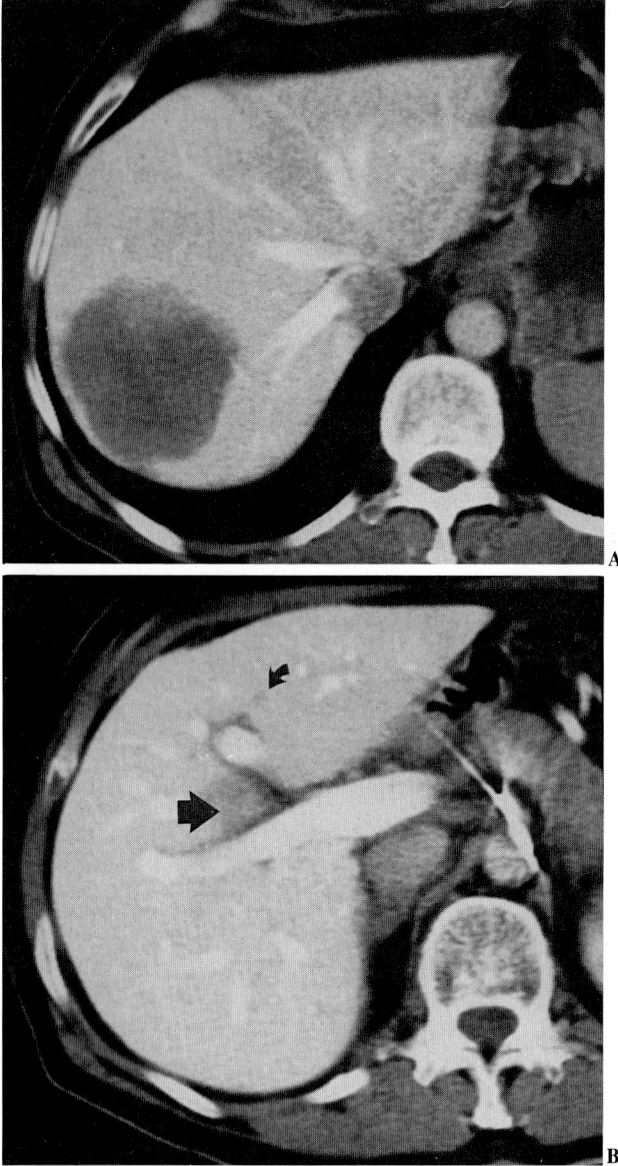

Figure 15-11 Example of arterial portography CT in patient with metastatic colon carcinoma. Notice the denser opacification of the central hepatic veins as compared to the inferior vena cava and the aorta, a characteristic of selective mesenteric artery contrast injection. (*A*) The large posterior superior right lobe lesion had been documented on prior IV contrast-enhanced CT. (*B*) Additional lesions in the lateral segment of the left lobe (*curved arrow*) and in the quadrate lobe (*straight arrow*) were only identified on arterial portography CT.

sion of the left hepatic lobe. In addition, central metastases, by compression of portal vein branches, can result in perfusion defects that simulate or mask additional metastases.

Lesions that are hypervascular may become isodense during IV contrast-enhanced dynamic hepatic CT scanning.[20] This has been documented for pancreatic islet cell tumors and carcinoid tumors and could be expected for other vascular neoplasms, including renal cell carcinoma, thyroid carcinoma, and possibly hepatoma (Fig. 15-12). In patients in whom these diagnoses are suspected, a precontrast and postcontrast scan technique is advised. Approximately 20% of islet cell tumors and carcinoids are isodense on postcontrast dynamic scan. However, in the remaining 80%, there is sufficient differential vascularity and heterogeneity to enable these tumors to be detected. Contrast enhancement should be performed in these patients even after a positive noncontrast study, particularly if such patients as those with carcinoid and hepatoma are potential candidates for hepatic resection. Additional tumors are likely to be identified by a contrast-enhanced study. The best preoperative contrast-enhanced CT technique is arterial injection CT via a catheter positioned in the proper hepatic artery. However, arterial injection CT may give incomplete information if there are anomalous right or left hepatic artery branches. Incomplete admixture of injected contrast can result in perfusion abnormalities that simulate or mask focal lesions.[21]

Contrast enhancement is important not only in detection of focal hepatic lesions but is also critical in detecting small pancreatic carcinomas and in identifying a dilated pancreatic duct (Fig. 15-13).[22] Thin slice collimation and contrast enhancement are used in pancreatic CT, reflecting the small size of the organ being examined and the dimensions of the pancreatic duct.

Contrast enhancement is used for renal CT unless noncontrast scans are employed to detect renal calculi or intrarenal hematomas. During continuous bolus injection of contrast material, renal parenchyma enhancement is greater than vascular enhancement. This effect is opposite to that obtained with other abdominal visceral organs and reflects renal filtration and tubular concentration that occur in addition to vascular enhancement during the injection period. An enhanced cortical angionephrogram, which is caused by preferential renal cortical flow, disappears several minutes after the end of injection. The major value of the dynamic scan in renal CT is detection of renal vein tumor thrombus. Demarcating the extent of a central relatively hypodense renal carcinoma may be difficult during the cortical angionephrogram phase as the medullary pyramids are also relatively hypodense. Delineation is best achieved when the kidney parenchyma is uniformly enhanced.

The normal uterine myometrium shows significant contrast enhancement both during the vascular and equilibrium phases. Myometrial enhancement is significantly greater than enhancement of other smooth muscle. This fact has been employed in the evaluation of endometrial cancer infiltration as hypodense tumor

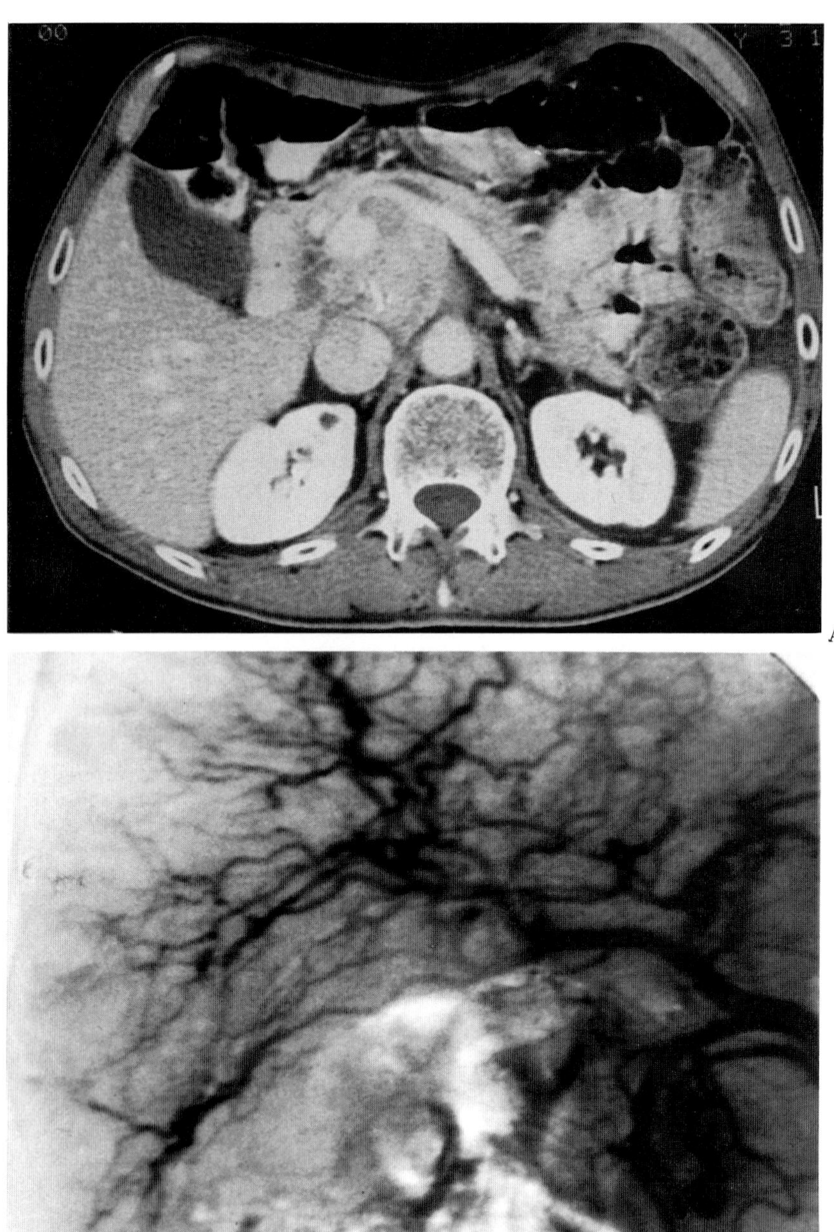

Figure 15-12 A patient with focally calcified pancreatic islet cell tumor in the pancreatic head and uncinate process with multiple hepatic metastases. (*A*) No focal hepatic lesions are seen on the dynamic scan. Selective hepatic artery digital subtraction arteriography [arterial phase (*B*) and capillary phase (*C*)] clearly defines multiple hypervascular metastases not seen on the dynamic CT scan.

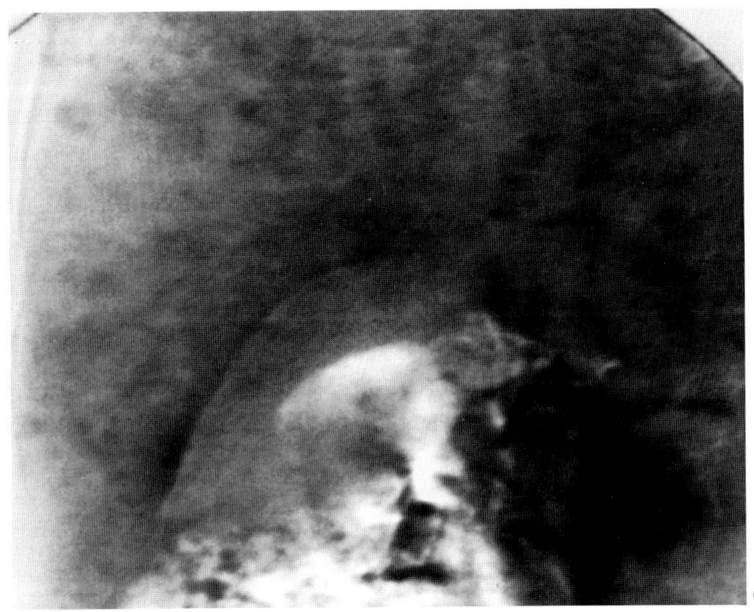

C

is differentiated from enhanced myometrium, thus placing the patient into a stage II category.[23]

Survey dynamic scan with bolus contrast administration should be used not only for suspected neoplasm but also in all cooperative patients in whom a survey study of the abdomen, pelvis, or thorax is planned. This technique can also be used in studies of patients with suspected traumatic or inflammatory conditions. In essence, the same total load of contrast material is given as for an infusion study, and although image quality is comparable to nondynamic studies, lesion detectibility in the vast majority of instances should be improved. Improved CT scanner performance now allows for survey dynamic scans of the abdomen/thorax, abdomen/pelvis, and pelvis/abdomen. In patients undergoing staging CT studies for lung carcinoma, survey dynamic scans beginning at the lower margin of the liver and incremented through the liver and encompassing the thorax are now performed. This scan sequence allows a dedicated hepatic CT study to be performed while the residual contrast enhancement of mediastinal vessels achieved at the 3- to 4-minute time interval after the beginning of injection is still adequate to diagnose hilar lymphadenopathy. The abdomen/pelvis survey dynamic scan can be performed in patients with colon cancer; it combines the advantages of dynamic hepatic CT with the ability to differentiate dilated vessels clearly from enlarged

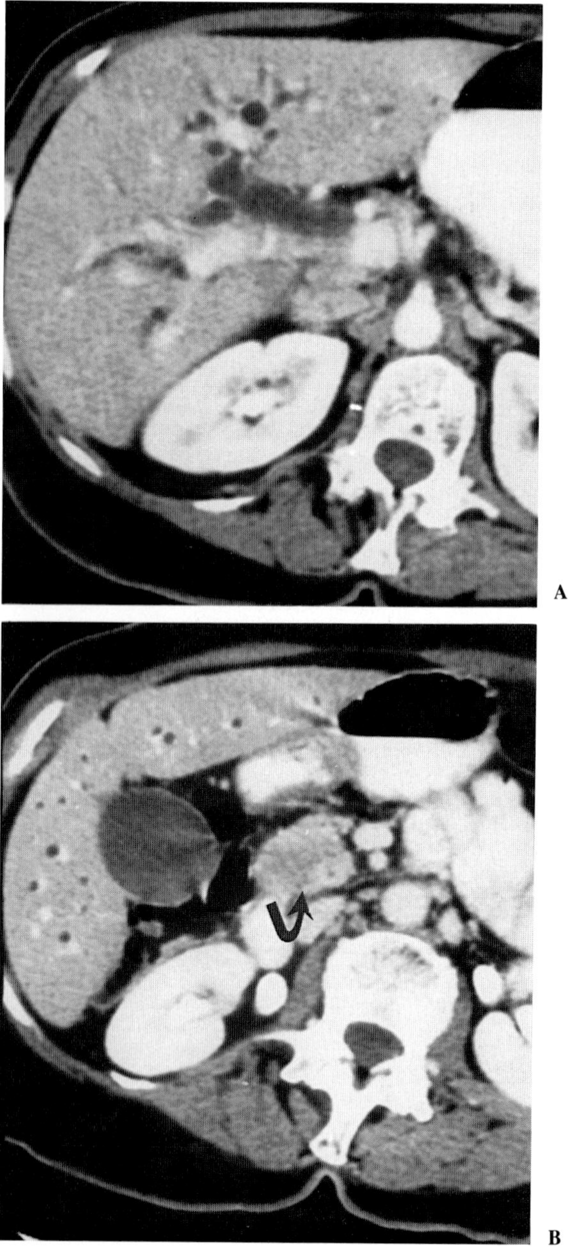

Figure 15-13 Pancreatic head carcinoma with malignant biliary obstruction. (*A*) Dilated bile ducts are identified adjacent to enhanced portal vessels. No focal metastases are seen. (*B*) Note the hypodense tumor mass (*curved arrow*) in a normal-sized pancreatic head and slightly deformed uncinate process.

lymph nodes in the abdominal retroperitoneum. The pelvis/abdomen dynamic technique can be used in the staging of pelvic malignancy, allowing differentiation of enlarged lymph nodes from tortuous enlarged hypogastric vessels in the early dynamic phase. However, the ability to detect focal hepatic metastases is limited as the liver is scanned in the "early delayed phase" of this dynamic sequence. Nevertheless, prostate, bladder, cervix, or endometrial cancer do not often produce hepatic metastases without evidence of regional extension or lymphadenopathy.

NEUROLOGIC COMPUTED TOMOGRAPHY

As the normal blood-brain barrier is intact, no extravascular enhancement of neural structures, apart from the pituitary gland, occurs (Fig. 15-14). Focal lesions in which pathologic vessels leak contrast material into the interstitium

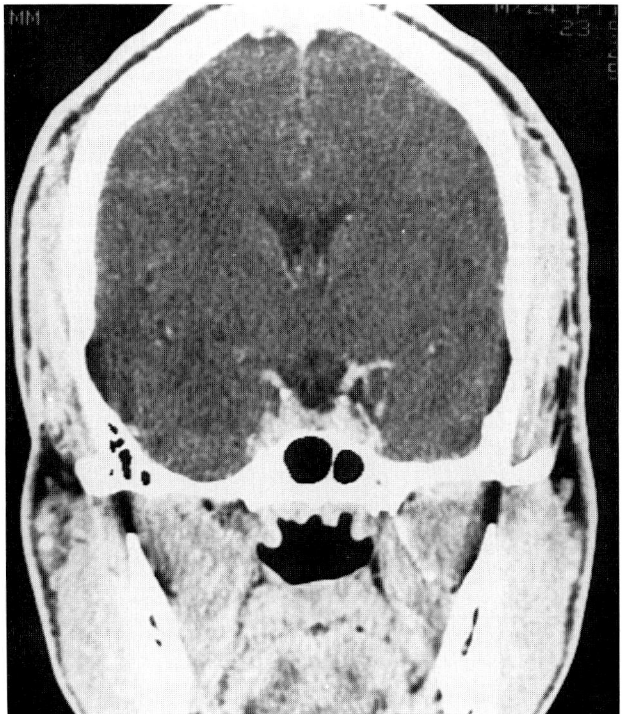

Figure 15-14 Contrast-enhanced coronal CT scan. Note the enhanced normal pituitary gland situated between the cavernous sinuses and carotid arteries. There is relatively minimal enhancement of brain parenchyma caused by intravascular iodine. This is clearly contrasted to enhanced muscle tissue at the base of skull and tongue reflecting both intravascular and extravascular iodine.

enhance at a rate determined by the degree of blood-brain barrier disruption, the total amount of contrast material administered, and the timing of the CT scan.[24] Any focal lesion that has adequate blood supply and pathologic vascularity, particularly focal tumor and abscess, will demonstrate abnormal enhancement compared to the background brain parenchyma. The dynamic scan, as used in thoracic and abdominal CT imaging, is not necessary. Contrast material administered by standard infusion (300 cc of 30% contrast material, 42 g of iodine) is generally sufficient to enhance pathologic neural lesions significantly when scans are obtained between 5 and 30 minutes after the beginning of an infusion. Some authorities advocate the use of a high dose contrast infusion (600 cc of 30% contrast material, 80 g iodine) in conjunction with delayed CT scanning obtained 1 to 2 hours after contrast administration. In these circumstances, contrast enhancement of small focal lesions relative to normal neural tissue seems to be maximized, probably due to continued slow leakage of contrast material across the abnormal blood-brain barrier.

PATIENT TOLERANCE

With IV administration of contrast material and the requirement for detection of lesions as small as 5 mm in diameter, relatively large volume loads of ionic contrast material are administered for enhanced CT scans. In the past, a volume load of 3 cc per kilogram per hour was accepted as a safe and tolerable amount for patients with normal renal function undergoing angiography. This norm was established on the basis of considerable clinical experience.[25] Adequate hydration of the patient receiving these loads of contrast material was stressed. During angiographic procedures, contrast material is administered as a series of test doses during fluoroscopy or as one or a number of nonselective or selective transcatheter arterial contrast injections. The delays inherent in angiography relating to catheter positioning under fluoroscopy, patient preparation for definitive angiographic sequences, and image recording on film provide a convenient time parameter (that is, hours) with which to encapsulate a presumed safe dose of contrast material.

In contrast, CT studies are performed as rapid survey procedures. Contrast material is administered as a single bolus injection or infusion, and a total load of contrast material equivalent to that which would have been administered during angiography is often given within a matter of minutes. With ionic contrast media that are hyperosmolar plasma expanders, the patient's tolerance to an acute intravascular volume expansion is a more critical factor than during angiography. Patients with normal cardiac function can tolerate an acute intravascular volume expansion of 1 L with no abnormal effects on cardiac function. The usual load of ionic contrast material administered by bolus contrast injection for incremental dynamic scan (180 cc of 60% contrast material, 50 g of iodine) is equivalent to 750 ml of normal saline. This is well within the expected tolerance of patients with normal cardiac function. In patients with poor cardiac function, use of nonionic contrast material with osmolality only slightly greater than plasma significantly decreases the amount of acute intravascular volume expansion resulting from the usual bolus contrast delivery.

Considerable controversy still exists regarding the potential renal toxicity of radiographic contrast agents. Several investigators have cited as evidence of direct renal toxicity the measurement of elevated enzyme levels in the urine in an animal model preparation following selective contrast material injection into the renal artery. These elevated levels are thought to reflect direct damage to the tubular cells. Clinical experience, on the other hand, indicates that patients with normal renal function who are adequately hydrated can clearly tolerate a 50-g iodine load of contrast material. Furthermore, volume loads of 5 to 6 cc per kilogram (80 g of iodine) have been administered to a large patient population (275 patients) undergoing delayed neurologic CT for suspected cerebral metastases. Of these patients, 232 had normal baseline renal function, and in only 2 patients was there significant elevation in serum creatinine or reduction of urine output at 24 hours, 48 hours, 72 hours, and 7 days postprocedure. In a subset of 39 of these patients who were nondiabetic but had serum creatinine levels varying between 1.5 and 4.5 mg%, only 1 demonstrated a transient elevation of creatinine, which subsequently returned to baseline values. In the other 38 patients, serum creatinine level and urine output were unaffected.[26] This data and the considerable prior experience with angiography suggest that a 50-g iodine load administered over 2 minutes to patients with normal cardiac and renal function should not produce any adverse effect. This has been our own clinical experience over the past 5 years with approximately 5,000 patients who have had dynamic abdominal CT scans.

Explicit guidelines on the use of IV contrast bolus injections or infusions in patients with abnormal renal function who require abdominal or thoracic CT scanning have not been developed. In our own practice, noncontrast scans are obtained as baseline studies. If the noncontrast study of the liver is negative, as in an ultrasonographic study, and exclusion of hepatic metastases is an important clinical issue (as would be the case in potential resection for primary pancreatic carcinoma), a magnetic resonance imaging (MRI) study would be performed. A technically adequate hepatic MRI that was negative would still not exclude the possibility of a focal lesion that could be detected by dynamic bolus-enhanced CT. In these circumstances, the contrast-enhanced dynamic CT scan would be obtained with due regard for patient hydration. In patients with focal lesions identified on noncontrast scans who are scheduled for chemotherapy with sequential CT to monitor response, noncontrast scans are favored for both baseline and serial studies.

EXPERIMENTAL CONTRAST AGENTS

The water-soluble urographic contrast material utilized for body CT contrast enhancement is not organ specific, and the pattern of contrast enhancement of focal lesions and normal adjacent parenchyma is time dependent and often unpredictable. It is only in neurologic CT that lesion detectability is not dependent on the time course of contrast enhancement.

The perfect body CT contrast agent would be organ specific, would produce a predictable enhancement of normal parenchyma and/or focal lesion that would not be time dependent, and would be nontoxic to human subjects. Such a contrast agent is not available for routine clinical use at this time. However, investigators have pursued different approaches to producing an organ-specific CT contrast agent, and these are summarized in this section. The greatest need for an organ-specific CT contrast agent is for hepatic imaging, which is the focus of discussion. The experimental agents produced so far have a reticuloendothelial cell distribution, enhancing both the liver and spleen. Focal lesions are hypodense relative to normal parenchyma.

Perfluorocytlbromide (PFOB)

Perfluorocytlbromides are inert dense liquids that have been investigated as potential blood substitutes. For use as CT contrast agents, they have been emulsified with yolk phospholipids. The average particle size is 0.5 microns. The agent is radiopaque due to its bromine content. Clearance is via reticuloendothelial cell metabolism; the free agent is cleared by lung respiration.[27]

In animal experiments, the agent has no acute hemodynamic effects. It is macrophage avid. At a dose of 5 mg PFOB/kg body weight given intravenously, hepatic attenuation increases by 49 HU and splenic attenuation by 250 HU. Its half-life is 5 days. Doses required for blood flow enhancement are significantly greater than for organ enhancement. However, because of the slow clearance of the injected agent, blood flow enhancement is obtained over a time course of hours following initial administration.[28]

Initial clinical trials of PFOB as an hepatosplenic CT contrast are being performed in Europe. It is expected that clinical trials in the United States will commence within the near future.

Contrast-Carrying Liposomes

Liposomes are vesicles made from biodegradable lipids that can be produced in highly precise sizes using a microfluidizer.[29] Iodinated contrast agents can be incorporated within the vesicles. For the microemulsified liposomes (MELS), the degree of incorporation is dependent on the individual iodinated agent. In general, a higher iodine-to-lipid ratio is achieved with nonionic agents than with ionic agents. A relatively prolonged retention of iodinated contrast-agent-carrying liposomes can be achieved. Contrast-carrying liposomes are biodegradable and are cleared by reticuloendothelial cell metabolism and renal filtration of the released iodinated contrast agent. Their theoretical advantage over perfluorocytlbromides is their greater radiopacity per volume of suspension. However, the agent cannot be produced in large bulk amounts with a predictable iodine-to-lipid ratio.

EOE_{13}

This agent is an aqueous emulsion of the iodinated ester of poppy seed oil. Currently, this agent is used as an investigational drug at the National Institutes of Health.[30] It has a reticuloendothelial cell distribution resulting in predictable increase in attenuation of liver and spleen. Clearance is via reticuloendothelial cell metabolism.

The agent is usually administered under a corticosteroid cover, and approximately 10% of patients develop a pyrogenic response. In human trials, the sensitivity of EOE_{13}-enhanced hepatic CT is equivalent to that of the 4- to 6-hour delayed hepatic iodine CT scanning after large-dose (80-g) iodine load administration.[13] As EOE_{13} is specifically distributed to the reticuloendothelial system, only a relatively small amount of iodine (approximately 4 g) is required to produce a 40 HU elevation in hepatic CT attenuation. A similar increase requires a 40-g iodine load of urographic water-soluble contrast material.

EOE_{13} is useful in detecting not only hepatic metastases, particularly small hepatic metastases, but also in detecting focal lymphomatous deposits in both the liver and spleen.

Iodipamide Ethyl Ester

This agent, which was developed at the University of Rochester, is an esterified benzoic acid derivative in particulate suspension. Particle size varies from 0.5–1.5 microns. The agent has a reticuloendothelial cell distribution. In dog studies, IV injection at a dose of 75 mg of iodine/kg of body weight results in a 60-HU elevation in hepatic CT number. This level of enhancement is maintained for 4–6 hours, with continuous slow clearance occurring over the subsequent week.[31]

High Anatomic Number Particulate Material

Various metallic oxides, including cerium oxide, gadolinium oxide, and dysprosium oxide, that have a reticuloendothelial cell distribution have been evaluated in preliminary animal experiments.[32] There is a predictable increase in hepatic attenuation. The potential long-term toxicity of these agents for human use is a serious concern and has not yet been systematically evaluated.

Iodinated Starch

Iodinated starch granules use a triiodobenzoyl ester of potato starch in particulate suspension with a particle size of 5 microns or less.[33] The suspension has an iodine content of 62% by weight. The agent has a reticuloendothelial cell distribution. However, there is prolonged retention of the agent. It is unlikely that clinical efficacy studies will ever be undertaken due to the potential problems of long-term toxicity.

CHOLANGIOGRAPHIC AGENTS

IV cholangiography is no longer a useful radiographic procedure as there are now better and alternative noninvasive (ultrasonography, isotope cholangiography) and invasive (percutaneous transhepatic and endoscopic retrograde cholangiography) techniques for examining the extrahepatic biliary ductal system. Nevertheless, an agent developed for IV cholangiography, such as iosefamate meglumine, is a potentially useful hepatic-specific contrast agent as it is selectively excreted by hepatocytes. Unfortunately, hepatic enhancement is relatively limited. In addition, the relatively greater frequency of side effects using cholangiographic agents in comparison to urographic agents has precluded their use in human clinical trials.

SUMMARY

In many cases, focal lesions do not provide sufficient intrinsic contrast to be detected by nonenhanced CT studies. Contrast-enhancement techniques have been developed that exploit differences in vascularity and interstitial diffusion of contrast between normal parenchyma and focal lesions in both body and neurologic CT. To achieve vascular and organ enhancement, contrast loads equivalent to those administered in angiographic procedures have been adopted in CT scanning. Bolus contrast injection using volume flow rate injectors can be integrated into protocols using CT scanners programmed to operate in the dynamic mode. In general, this achieves the best combination of vascular and organ enhancement to detect focal lesions, to locate these lesions accurately, and, in certain instances, to characterize them. In neurologic CT, disruption of the blood-brain barrier results in lesion enhancement.

The recommended volume load of water-soluble contrast material (50 g of iodine, 180 cc of 60% diatrizoate or iothalamate) is tolerated by patients with normal cardiorenal function. Selective use of contrast enhancement in patients with biochemical renal failure is advised.

Various experimental hepatic-specific contrast agents have been developed and tested in animal models. Such agents do not have time-dependent enhancement of focal lesions and normal parenchyma and, if available clinically, could avoid the need for dynamic scan techniques. However, such agents are still experimental and their efficacy and potential toxicity have not been systematically evaluated in clinical trials.

REFERENCES

1. Alfidi R, MacIntyre WJ, Meaney TF, et al: Experimental studies to determine application of CAT scanning to the human body. *AJR* 1975;124:199–207.

2. Zatz LM: Basic principles of computed tomography, in Newton TH, Potts DG (eds): *Radiology of the Skull and Brain: Technical Aspects of Computed Tomography*. St. Louis, CV Mosby, 1981, pp 3853–3876.

3. Spataro RF: Newer contrast agents for urography. *Radiol Clin N Am* 1984;22:365–380.

4. Plewes DB, Dean PB: Detectability of spherical objects by computed tomography. *Radiology* 1979;133:785–786.

5. Claussen CD, Banzer D, Pfretzchner C, et al: Bolus geometry and dynamics after intravenous contrast medium injection. *Radiology* 1984;153:365.

6. Thorsen MK, Sandretto MA, Lawson TL, et al: Evaluation of dissecting aortic aneurysm: accuracy of CT diagnosis. *Radiology* 1983;148:773–777.

7. Takayasu K, Moriyama N, Ishikawa T, et al: Large spontaneous intrahepatic arterioportal fistula demonstrated by rapid sequential computed tomographic scan: reported two cases. *Radiology* 1985;156:863.

8. Foley WD, Gleysteen JJ, Lawson TL, et al: Dynamic CT scanning and pulsed Doppler ultrasonography in the evaluation of splenorenal shunt patency. *J Comput Assist Tomogr* 1983;7:106–112.

9. Middleton WD, Smith DF, Foley WD: CT detection of aortocaval fistula. *J Comput Assist Tomogr* 1987;11:344–347.

10. Freeny PC, Marks WM: Hepatic hemangioma: dynamic bolus CT. *AJR* 1986;147:711–719.

11. Foley WD, Berland LL, Lawson TL, et al: Contrast enhancement technique for dynamic hepatic CT scanning. *Radiology* 1983;147:797–803.

12. Bernardino ME, Irwin BC, Steinberg HV, et al: Delayed hepatic CT scanning: increased confidence and improved detection of hepatic metastasis. *Radiology* 1986;159:71–74.

13. Miller DL, Simmons JT, Chang R, et al: Hepatic metastasis detection: comparison of three CT contrast enhancement methods. *Radiology* 1987;165:785–790.

14. Alpern MB, Lawson TL, Foley WD, et al: Efficiency of contrast enhanced incremental dynamic computed tomography in the detection of focal hepatic masses and fatty infiltration. *Radiology* 1986;158:45–50.

15. Kormano M, Dean PB: Extravascular contrast material: the major component of contrast enhancement. *Radiology* 1976;121:379–382.

16. Marchal GY, Baert AL, Wilms GE: CT of noncystic liver lesions: bolus enhancement. *AJR* 1980;135:57–65.

17. Burgener FA, Hamlin DJ: Contrast enhancement of hepatic tumors in CT: comparison between bolus and infusion techniques. *AJR* 1983;140:291–295.

18. Moss AA, Dean PB, Axel L, et al: Dynamic CT of hepatic masses with intravenous and intra-arterial contrast material. *AJR* 1982;138:847–852.

19. Matsui O, Kadoya M, Suzuki M, et al: Dynamic sequential computed tomography during arterial portography in the detection of hepatic neoplasms. *Radiology* 1983;146:721–727.

20. Bressler EL, Alpern MB, Glazer GM, et al: Hypervascular hepatic metastases: CT evaluation. *Radiology* 1987;162:49–51.

21. Freeny PC, Marks WM: Hepatic perfusion abnormalities during CT angiography: detection and interpretation. *Radiology* 1986;159:685–691.

22. Marchal GJF, Baert AL, Wilms GE: Intravenous pancreaticography in computed tomography. *J Comput Assist Tomogr* 1979;3:727–732.

23. Hamlin DJ, Burgener FA, Beecham JB: CT of intramural endometrial carcinoma: contrast enhancement is essential. *AJR* 1981;137:551.

24. Gado MH, Phelps ME, Coleman RE: An extravascular component of contrast enhancement in cranial computed tomography. Part II. Contrast-enhancement and the blood-tissue barrier. *Radiology* 1975;117:595–597.

25. Doust BD, Redman HC: The myth of one ML/KG in angiography: a study to determine the relationship of contrast medium dosage to complications. *Radiology* 1972;142:557–560.

26. Hayman LA, Evans RA, Fahr LM, et al: Renal consequences of rapid high dose contrast CT. *AJR* 1980;134:553–555.

27. Mattrey RF, Long DM, Multer F, et al: Perfluorocytlbromide: a reticuloendothelial-specific and tumor-imaging agent for computed tomography. *Radiology* 1982;145:755–758.

28. Mattrey RF, Long DM, Peck WW: Perfluorocytlbromide as a blood pool contrast agent for liver, spleen and vascular imaging in computed tomography. *J Comput Assist Tomogr* 1984; 8:739–744.

29. Cheng KT, Seltzer SE, Adams DF, et al: The production and evaluation of contrast-carrying liposomes made with an automatic high pressure system. *Invest Radiol* 1987;22:47–55.

30. Miller DL, Vermess M, Doppmann JL, et al: CT of the liver and spleen with EOE_{13}. Review of 225 examinations. *AJR* 1984;143:235–243.

31. Sands MS, Violante MR, Gadeholt G: Computed tomographic enhancement of liver and spleen in the dog with iodipamide ethyl ester particulate suspension. *Invest Radiol* 1987;22:408–416.

32. Seltzer SE, Adams DF, Davis MA, et al: Hepatic contrast agents for computed tomography: high atomic number particulate material. *J Comput Assist Tomogr* 1981;5:370–374.

33. Cohen Z, Seltzer SE, Davis MA: Iodinated starch particles: new contrast material for computed tomography of the liver. *J Comput Assist Tomogr* 1981;5:843–846.

Part III

Genitourinary Agents

Chapter 16

Urography

Robert F. Spataro

The number of intravenous excretory urograms (IVUs) has decreased in recent years because of a combination of three factors: cost containment in medicine, development of competitive imaging modalities, and apprehension about possible adverse effects of contrast media.[1] The number of IVUs has decreased between 30–50% from its peak utilization. Yet, although the death notice for the IVU as a useful diagnostic imaging procedure has been posted for many years, it appears to have been a premature epitaph.[1] The IVU has retained a significant role in the evaluation of urinary tract diseases and remains the best overall screening examination of the entire urinary tract. Hundreds of thousands of IVUs continue to be performed every year because of the diagnostic usefulness of this examination. When performed for appropriate clinical indications the IVU provides essential and appropriate diagnostic information and has been shown not to be overutilized as a diagnostic imaging procedure.[2]

Although cost containment and the development of new imaging modalities continue to decrease the utilization of IVU, the development of new low osmolality and nonionic contrast media (CM) has significantly decreased the adverse effects of contrast media and has improved the safety and diagnostic efficacy of the IVU. Use of these new safer contrast media removes much of the apprehension in the performance of contrast-enhanced studies, such as the IVU. The knowledge of the pharmacodynamics and physiology of contrast media in excretory urography is an essential part of every radiologist's knowledge and is the subject of this chapter.

In the chapter on urographic contrast agents in the first edition of this book, Peter Dure-Smith stated that the tri-iodinated contrast agents developed in the early 1950s "proved to be highly successful and considerably less toxic compared to the earlier compounds, although their use in urography is still dogged by the occasional unexplained serious or fatal reaction."[3] In the years between the first edition of this book and the current edition, tremendous strides have been made in the development of new nonionic and low osmolality contrast media. These new low osmolality CM, particularly nonionic CM, have significantly less systemic toxicity and have been shown to decrease the number of minor, moderate, and severe contrast media reactions. They are expected to decrease significantly the number of fatal reactions caused by contrast media injection for excretory urography. Moreover, new low osmolality and nonionic contrast media provide not only increased safety but significant improvement in the quality of excretory urography.

AVAILABLE CONTRAST MEDIA

Three major types of water-soluble contrast media are available for excretory urography. The generalized structural formulas of these three basic types of contrast media are illustrated in Figure 16-1 (also listed is a nonionic dimer—iotrol). The ionic monomeric contrast media were developed in the 1950s and are referred to as conventional ionic contrast media. They are derivatives of triiodinated benzoic acid and are ionized in solution. The three types of conventional ionic contrast media only differ in the side chain at position three. The cation associated with the conventional ionic CM is either sodium or methyl glucamine (meglumine) or a combination of the two. In solution, these conventional ionic CM dissociate into the benzoic acid anion and the sodium or methyl glucamine cation, giving a ratio of iodine atoms to particles in solution of 3:2, or 1.5 (Fig. 16-1). These CM are referred to as ratio 1.5 contrast media.

Nonionic contrast media, originally conceived and developed by Torsten Almén, are also triiodinated benzoic acid derivatives.[4,5] However, these new nonionic monomeric contrast media contain no ionizing group so that in solution no anions or cations are formed. In solution, therefore, there are three iodine atoms to one particle in solution, a ratio of 3 (Fig. 16-1). Because of the decreased number of particles in solution these new contrast agents have significantly lower osmolality for the same iodine concentration, which offers a significant advantage in decreasing physiologic changes and adverse side effects and improving urinary iodine concentrations. Table 16-1 compares the physical properties of ionic and nonionic CM.

A second new low osmolality ratio 3 CM, Hexabrix (ioxaglate), uses a different approach to achieve low osmolality. Hexabrix is a monoacidic dimer in which two benzoic acid rings are attached. It contains a total of six iodine atoms and only one ionizing carboxyl group in the molecule. In solution there are six iodine atoms for two particles in solution, again achieving a ratio of 3. However, in this case the contrast media are still ionic, producing a negatively charged anion and a positively charged cation. Hexabrix has lower osmolality but still has an ionic structure. The osmolality of Hexabrix is even lower than for the nonionic contrast media (Table 16-1). However, it has some disadvantages. Because of its ionicity,

		Number of Iodine Atoms	Number of Particles In Solution	Ratio of Iodine to Particles In Solution
Ionic Monoacidic Monomer (COO^- (Na^+ / Meglumine$^+$); CH_3-CO-HN; R; I)	Hypaque, Renografin (diatrizoate) Conray (iothalamate) Metrizoate	3	2	1.5
Ionic Monoacidic Dimer (COO^- (Na^+ / Meglumine$^+$); R; I)	Hexabrix (ioxaglate)	6	2	3
Nonionic Monomer (R_1, R_2, R_3; I)	Isovue (iopamidol) Omnipaque (iohexol) Amipaque (Metrizamide) Ioversol Iopromide	3	1	3
Nonionic Dimer (R; I)	Iotrol	6	1	6

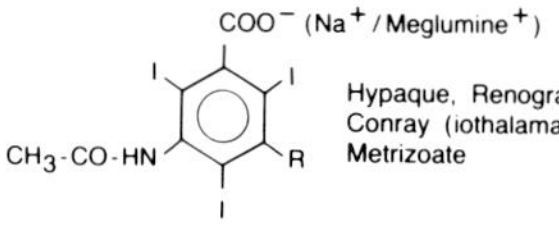

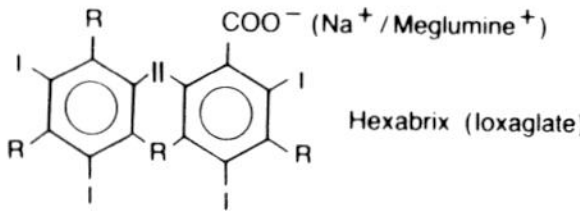

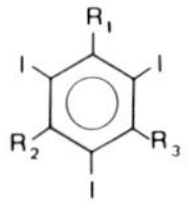

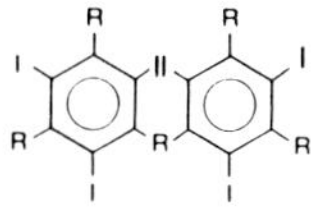

Figure 16-1 Generalized contrast media structures. *Source*: Copyright 1988. Urban & Schwarzenberg, Baltimore-Munich. Reproduced with permission from *Urologic Imaging and Interventional Techniques* edited by William Bush. All rights reserved.

it causes such side effects as nausea and vomiting and more CNS and systemic toxicity than nonionic ratio 3 contrast media.

Thus at the current time there are three major types of contrast media available for excretory urography—the conventional ratio 1.5 high osmolality ionic monomeric contrast media, such as Hypaque, Renografin, Conray, and others; the new low osmolality ratio 3 ionic dimeric contrast medium, Hexabrix; and the new nonionic monomeric ratio 3 low osmolality contrast media, such as Isovue, Omnipaque, ioversol, and iopromide. Isovue and Omnipaque are already FDA approved for intravascular use.

ADVERSE EFFECTS

Mortality and Morbidity

Table 16-2 lists the incidence of adverse reactions with the ionic contrast media. When all adverse reactions are considered, approximately 5% of patients experi-

Table 16-1 Physical Properties of Contrast Media

Contrast Media	*Iodine Content, <mgI/ml>*	*Osmolality, <mOsm/Kg H>*	*Molecular Weight*	*Viscosity, <37° (mPa.s)>*
Hypaque 50% (Sodium diatrizoate)	300	1500	614	2.5
Conray (Methylglucamine iothalamate)	282	1217	614	4.0
Renografin 60 (Sodium/ Methylglucamine diatrizoate)	288	1511	614	3.9
Isopaque 280 (Methylglucamine calcium metrizoate)	280	1500	628	4.0
Conray 400 (Sodium Iothalamate)	400	1965	614	4.5
Amipaque (Metrizamide)	280	450	789	5.0
Isovue (Iopamidol)	300	616	777	4.5
Iopromide	300	607	791	4.8
Omnipaque (Iohexol)	300	620	821	4.8
Ioversol (MP-328)	320	716	807	6.0
Hexabrix (Sodium/ methylglucamine ioxaglate)	320	580	1269	7.5
Iotrol	300	300	1626	9.1
Human blood plasma	—	300	—	—

Source: Copyright 1988. Urban & Schwarzenberg, Baltimore-Munich. Reproduced with permission from *Urologic Imaging and Interventional Techniques* edited by William Bush. All rights reserved.

ence some adverse side effects.[6,7] Intermediate reactions that may require treatment but not hospitalization occur in approximately 1% of patients, whereas severe reactions and cardiac arrests are even more infrequent.[6–13] Serious reactions requiring hospitalization and considered to be life threatening occur in approximately 0.05% or 1 in 2,000 patients. Cardiac arrest occurs in approximately 1 in 6,000 patients, a frequency of .017%.[8]

Table 16-3 shows the accumulated literature on the mortality rates associated with excretory urography with ionic contrast media. The reported mortality associated with excretory urography has varied between 1 per 117,000 patients to as low as 1 in 14,000 patients.[6,8,11,12,14–17] When the recent studies using higher doses of contrast media between 1970 and 1983 are summarized the death rate is approximately 1 in 40,000 patients or .0025%. Two recent large studies have shown death rates of 1 in 75,000 and 1 in 93,000, respectively.[8,16] Although the risk of death is relatively low with the conventional ionic contrast media, because these drugs are diagnostic and not therapeutic, even this low risk of mortality has led to apprehension in their use.

Table 16-2 Incidence of Adverse Reactions with Ionic Contrast Media

Reactions	*Number (Percentage)*
All adverse reactions to contrast media—include minor reactions that require no treatment, such as nausea, vomiting, limited urticaria, mild rash, lightheadedness, and mild dyspnea	1/20 (5%)
Intermediate reactions—include extensive urticaria, facial edema, bronchospasm, laryngeal edema, dyspnea, rigors, mild chest pain, and hypotension, which often require treatment but not hospitalization; these reactions are not considered life-threatening	1/100 (1%)
Severe reactions—include laryngeal edema, pulmonary edema, refractory hypotension, circulatory collapse, severe angina, myocardial infarction, cardiac arrhythmias, convulsions, coma, and respiratory and cardiac arrest; these reactions require hospitalization and are considered life-threatening	1/2000 (0.05%)
Cardiac arrest	1/6000 (0.017%)
Mortality	1/40,000 (0.0025%)

Source: Copyright 1988. Urban & Schwarzenberg, Baltimore-Munich. Reproduced with permission from *Urologic Imaging and Interventional Techniques* edited by William Bush. All rights reserved.

The new low osmolality and nonionic contrast media have been shown experimentally and clinically to have much less acute systemic toxicity, morbidity, and mortality than the conventional ionic contrast media. Table 16-4 shows experimental data on the acute systemic toxicity in mice. Conventional ionic media have acute IV lethal doses (LD50) in the range of 7 to 10 g I/kg body weight. Hexabrix

Table 16-3 Mortality Rates Reported in the Literature Associated with Excretory Urography with Conventional Ionic Contrast Media

Series	*Year*	*Number of Deaths*	*Number of Examinations*	*Mortality Rate*
Pendergrass et al.[14]	1958	99	11,546,000	1:117,000
Wolfromm et al.[15]	1966	15	912,300	1:61,000
Ansell[6]	1970	8	318,500	1:40,000
Witten et al.[17]	1973	1	33,000	1:33,000
Shehadi[11]	1975	6	81,278	1:14,000
Shehadi and Toniolo[12]	1980	11	214,033	1:19,458
Hobbs[8]	1981	5	466,600	1:93,320
Hartman et al.[16]	1982	4	300,000	1:75,000
TOTALS	all	149	13,871,711	1:93,099
	1970-1983	35	1,413,411	1:40,383

Table 16-4 Acute Intravenous Toxicity (LD50) of Contrast Media In Mice

Iothalamate	(Conray)	LD50 ≈ 9-10 g I/kg*
Diatrizoate	(Hypaque)	
	(Renografin)	LD50 ≈ 7-10 g I/kg†‡
Metrizoate	—	
Metrizamide	(Amipaque)	LD50 ≈ 17.5-18.7 g I/kg*
Iohexol	(Omnipaque)	LD50 ≈ 23-25 g I/kg$
Iopromide	—	LD50 ≈ 16.5 g I/kg‖
Iopamidol	(Isovue)	LD50 ≈ 20/23.6 g I/kg‡
Ioversol	(MP-328)	LD50 ≈ 20 g I/kg†
Ioxaglate	(Hexabrix)	LD50 > 12.8 g I/kg† LD50 ≈ 15 g I/kg$

*Salveson S: Acute toxicity of metrizamide. *Acta Radiol* 1973;335(suppl):5–13.
†Mallinckrodt Inc., St. Louis, MO 63134
‡E.R. Squibb and Sons, Inc., Princeton, NJ
$Salveson S: Acute intravenous toxicity of iohexol in the mouse and in the rat. *Acta Radiol* 1980;362(suppl):73–75.
‖Schering AG, West Germany
$Sovak M, et al: Combined methods for assessment of neurotoxicity. Testing of new nonionic radiographic media. *Invest Radiol* 1980;15(suppl):S248–253.

Source: Copyright 1988. Urban & Schwarzenberg, Baltimore-Munich. Reproduced with permission from *Urologic Imaging and Interventional Techniques* edited by William Bush. All rights reserved.

(ioxaglate), a low osmolality ratio 3 ionic dimer, has an LD50 of approximately 12–15 g I/kg. The nonionic ratio 3 contrast media Isovue (iopamidol), Omnipaque (iohexol), ioversol (MP-328), and iopromide have acute LD50 doses in the range of 20–25 mg I/kg body weight. Thus, the nonionic contrast media have relative acute systemic toxicity levels between two and three times lower than conventional ionic agents.

Although a large number of urograms have been performed with the new contrast media, the number is still much less than with conventional agents. Initial data on morbidity and mortality are still beginning to be obtained. In a German clinical trial the overall reaction rate with nonionic contrast media was 2.1%, the percentage of moderate reactions 0.9%, and the number of severe reactions .01%.[18] The incidence of severe reactions with nonionic agents in this study was five times less than with conventional contrast media.

In an Australian study of 97,000 patients performed by the Royal Australian College of Radiology comparing nonionic and ionic contrast media, the rate of reactions with nonionic contrast media was 1.3%; intermediate reactions were 0.116%; and no severe reactions were encountered (personal communication, Geoffrey Benness, June 1988). In high-risk patients the number of mild reactions was 5.6% with conventional agents versus 1.5% with nonionic media, intermediate reactions were 1.98% versus 0.12% and severe reactions were 0.3% versus 0.03% when conventional ionic CM were compared to nonionic CM. In this study the nonionic contrast media were five to ten times less likely to cause a particularly serious reaction than conventional ionic agents. Furthermore, using nonionic CM in high-risk patients resulted in fewer overall and severe reactions than using conventional ionic CM in normal patients. In another large study involving over 100,000 patients in Japan the severity of all adverse reactions, including very severe reactions, was reduced considerably with nonionic contrast media. Very severe reactions occurred in 0.05% of patients with ionic CM versus 0.01% with nonionic CM (personal communication, H. Katayama).

These studies show that the number of overall reactions with the new nonionic contrast media are at least one-third to one-fifth that of conventional contrast media and the number of severe reactions may be at least three to five times or even less common than with conventional agents.

What is the ultimate mortality rate associated with the new nonionic low osmolality contrast media? Initial reported mortality rates with these new media have shown mortality rates in the range of approximately one death in 250,000 patients or .0004%.[19] This would give a relative safety of about five times that of conventional ionic contrast media. Although these results are preliminary, it is quite apparent from the experimental and clinical data that the new low osmolality and particularly nonionic ratio 3 contrast media will be significantly safer in both morbidity and mortality. This should lead to a renewed utilization of contrast media for all contrast studies, including IVU, as the fear of adverse contrast reactions and patient intolerance is lowered.

High-Risk Patients

Certain patients—those with a previous history of allergies, significant cardiovascular disease, diabetes, and previous reactions to contrast media—are at risk for adverse reactions to contrast media.[6,7] Several studies examined the use of nonionic contrast media in patients with previous reactions to conventional ionic contrast media.[20,21] These studies have shown that these high-risk patients with previous adverse reactions to contrast media have little in the way of adverse reactions to nonionic contrast media. Similar findings have been noted by H.W. Fischer and the author in a study in progress. Although the nonionic contrast media have been shown to be safer in patients with previous adverse reactions, new pretreatment regimens have also been shown to provide increased safety with conventional ionic contrast media. Pretreatment with corticosteroids may decrease the incidence of adverse reactions of all types.[22] Prophylaxis regimes with prednisone, Benadryl, and ephedrine decrease the incidence of adverse reactions in patients with previous contrast media reactions.[23,24] Larger clinical trials are needed to determine whether a two-dose corticosteroid pretreatment regimen is a reasonable alternative to nonionic contrast media.[22]

RENAL HANDLING AND EXCRETION

Conventional ionic contrast agents are excreted via the kidney by glomerular filtration. There is no significant reabsorption. There is also no significant tubular secretion. Similarly, both nonionic contrast agents and the monoacidic dimer are handled by the kidney, by glomerular filtration without significant reabsorption or secretion. The amount of contrast agent filtered is determined by the glomerular filtration rate (GFR), and the plasma concentration of the contrast agent is determined by this formula:

$$\text{Filtered load} = \text{GFR} \times P_d \text{ where } P_d = \text{the plasma contrast concentration.}$$

The filtered load of contrast media and the amount of contrast media in the urine are dependent on the plasma contrast concentration, which in turn is highly dependent on the rate and dose of contrast media injected. Increasing the dose of contrast injected increases the plasma concentration.[25] Given an equivalent dose of contrast, the rapidity of injection is directly related to the peak plasma concentration and thus the peak in contrast media filtered load. A rapid IV bolus gives a significantly higher peak plasma contrast concentration than a slow injection or infusion of a similar dose (Fig. 16-2).[26] The larger the dose and the faster the injection, the higher the plasma level and the higher the initial filtered load of contrast media into the renal tubules.

Once the contrast medium is filtered, the only modification during tubular transit is that it is concentrated by tubular reabsorption of water. Under normal conditions, salt and water reabsorption in the proximal nephron causes 80–90% of filtered water to be reabsorbed, producing an increasing concentration of contrast media in the proximal tubule of between five and ten times that of the plasma concentration.[27] In the distal nephron, additional salt and water reabsorption occurs under the influence of the antidiuretic hormone (ADH), which may concentrate the contrast in the urine up to 40 times that of the initial peak plasma and filtered concentration.

Under conditions of osmotic diuresis, tubular transit of fluid is increased, decreasing significantly the reabsorption of salt and water in the proximal nephron. Conventional ionic contrast agents cause significant osmotic diuresis, leading to less reabsorption of salt and water than would occur under conditions of normal diuresis. At doses of 500 mg I/kg, the contrast media concentration at the proximal nephron is only about twice that of plasma. It is in the tubular modifica-

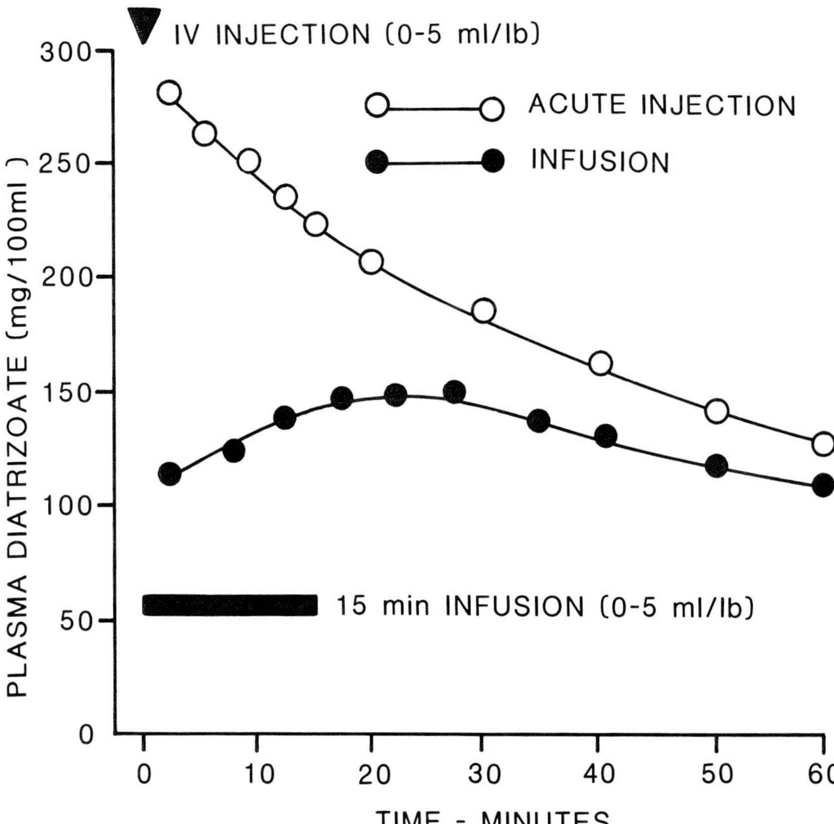

Figure 16-2 Comparison of the plasma concentrations of diatrizoate after equivalent doses given by rapid IV injection and drip infusion. *Source*: Reprinted with permission from *Investigative Radiology* (1970;5:479), Copyright © 1970, JB Lippincott Company.

tion of the filtered contrast that the new low osmolality nonionic contrast media provide a significant advantage in excretory urography. Because of the lower osmolality and correspondingly decreased osmotic diuresis the tubular transit is decreased and there is more salt and water reabsorption in the proximal nephron with ratio 3 contrast media than with ratio 1.5 CM. As a result, there are significantly higher concentrations of contrast media in the urine with the new low osmolality nonionic contrast media.[28]

The urinary contrast media (iodine) concentration does not increase continuously with increasing plasma contrast concentrations, but reaches a peak due to the osmotic diuresis and decreased tubular reabsorption at high peak plasma concentrations. With ionic contrast media the peak urinary iodine concentration may be reached at an injected dose level in the range of 300 mg I/kg.[29] However, the total contrast excretion rate, which is the urine flow rate times the urinary contrast concentration, may not peak until as high a dose as 600 mg I/kg body weight (Fig. 16-3).[28,29] At these higher concentrations the actual concentration of contrast media in the urine may actually decrease, but the urine flow rate increases, causing the total urinary iodine amount to increase dramatically. The density of the nephrogram and pyelogram is related not only to the concentration of contrast but also to the total amount of contrast media (iodine) in the kidney and urinary collecting systems. Ratio 3 CM, because of their lower osmolality and diuresis, do not reach peak urinary iodine concentrations in the urine until an injected dose of 500-600 mg I/kg (Fig. 16-4), a significant advantage over conventional ionic CM.[28] Total contrast excretion rate may not peak until much higher.

URINARY OPACIFICATION

Contrast enhancement of the kidney and collecting structures during excretory urography is a complicated multifactorial phenomenon. The density or opacifica-

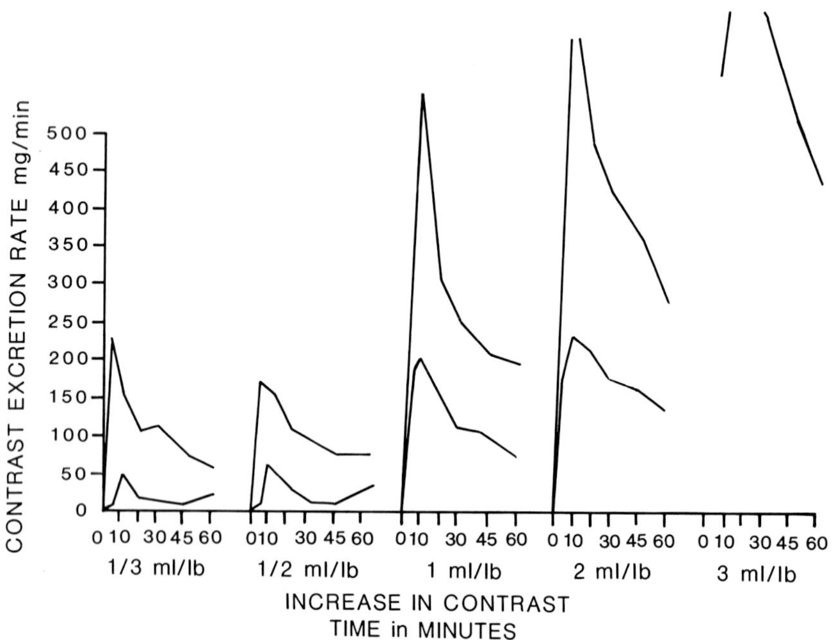

Figure 16-3 Minimal and maximum contrast material excretion rates in 34 experiments following a rapid injection of ⅓, ½, 1, 2, and 3 ml of CM/lb body weight in dogs. *Source*: Reprinted with permission from "The Bolus Effect in Excretory Urography" by P Dure-Smith et al in *Radiology* (1971;101:29–34), Copyright © 1971, Radiological Society of North America Inc.

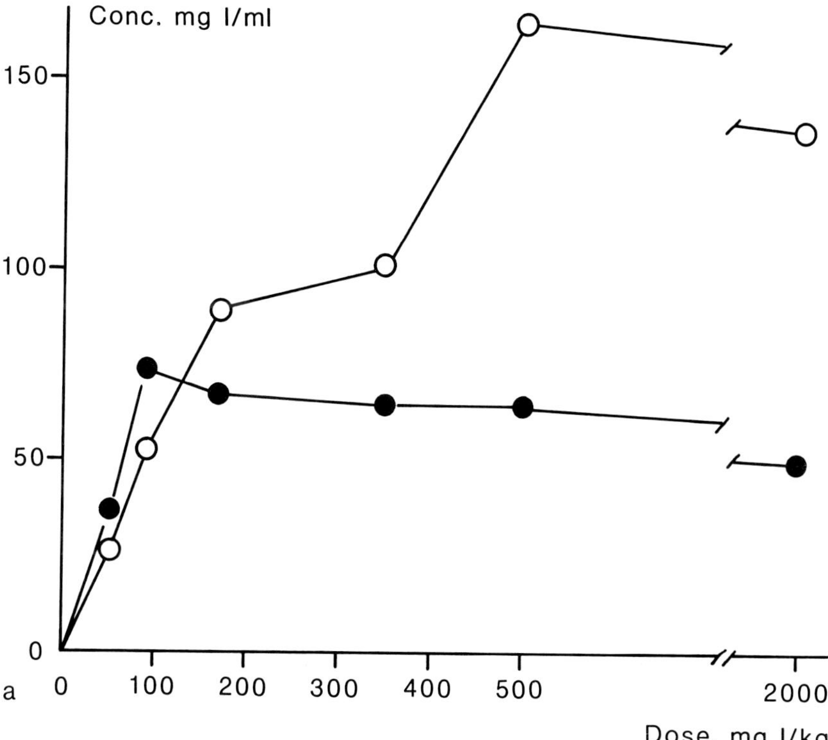

Figure 16-4 (*A*) Urinary iodine concentration. (*B*) Urinary flow as a function of dose during the first 30 minutes after injection of contrast media into rabbits. *Source*: Reprinted from *Radiocontrast Agents* (p 144) by M Sovak with permission of Springer-Verlag, © 1984.

tion of the nephrogram and collecting systems is not determined simply by the concentration of contrast media in the urine but also by the number of atoms of iodine in the path of the x-ray beam. The kidney and collecting structures are not rigid tubes and the proximal and distal nephrons and collecting tubules, and the collecting structures of the urinary tract are distensible and can increase in size. When the volume of a tube is similar and the concentration of iodine is higher in one of the tubes, the attenuation of the x-ray beam will be greater in the tube with the higher concentration and it will appear denser on the radiograph. In general, within the diagnostic range of iodine concentrations in clinical excretory urography, the denser the opacification of the structure, the better the diagnostic information.[28]

New ratio 3 low osmolar contrast media produce a significantly higher iodine concentration in the urinary tract than conventional ionic ratio 1.5 contrast media, although the ratio 1.5 contrast media may result in larger volumes and thereby

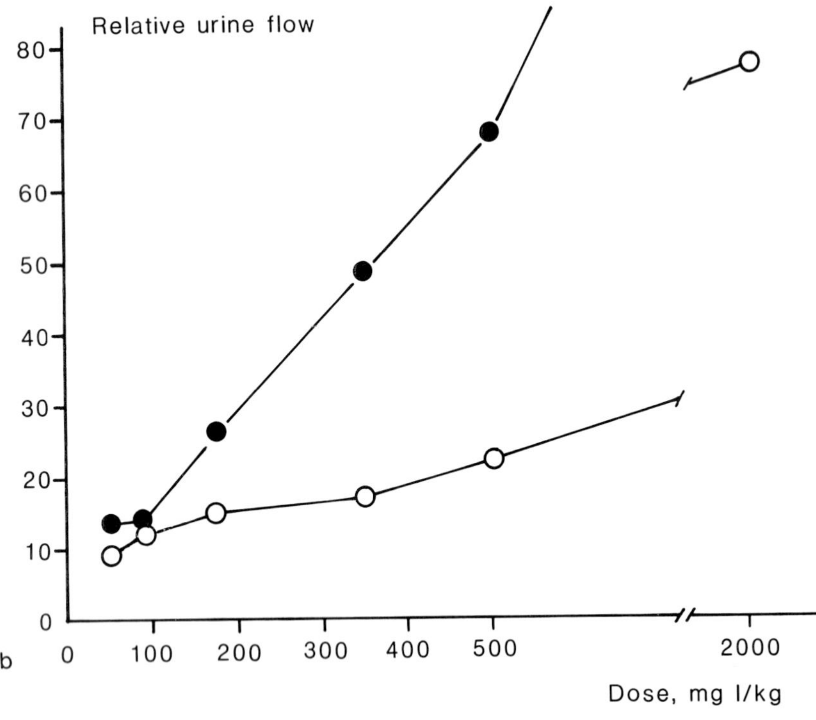

increased contrast in the urinary tract. Both theoretically and experimentally it has been shown that the new ratio 3 low osmolar contrast media produce significantly denser opacification of the urinary tract. The higher urinary concentrations of contrast media produced by the low osmolar CM outweigh the increased volume of the conventional agents produced by diuresis.[28,30,31]

Dose

Animal experiments and human studies have shown that the maximum iodine concentration in urine may be obtained with ionic CM with a dose of about 300 mg I/kg (1 ml CM/kg).[32] Because the total urinary iodine (contrast) excretion may continue to rise until doses nearer 500 mg I/kg, the optimal urographic dose may be higher than the dose at which the maximum concentration is obtained. A dose of 450 mg I/kg (1.5 mg CM/kg) provides a near-optimal urinary iodine excretion rate. In clinical practice, a dose of 50 ml of CM (.75 ml/kg; 200 mg I/kg) is adequate for visualization. A dose of 300 mg I/kg (1 ml CM/kg) or approximately 75 ml in an average patient provides near-optimal iodine concentration with conventional ratio 1.5 CM, but not yet peak urine flow and total urinary iodine excretion. A dose of 100 ml of CM (1.5 ml/kg; 450 mg I/kg) gives near-optimal urography in the majority of patients with conventional ratio 1.5 agents because the total urinary CM excretion rate is greater than with lower doses. Similarly, a dose of 100 ml of low osmolality ratio 3 CM provides a superior urogram than a similar dose of conventional CM, whereas a 50 ml dose of low osmolality CM provides an adequate but not optimal IVU.

Experimental Urography

Detailed animal studies have provided data on the comparative urinary excretion of low osmolality and conventional contrast media.[30,31,33–43] These studies have shown that low osmolality CM provide significantly higher urinary iodine concentrations, cause less urinary diuresis, and provide denser opacification of the urinary collecting system than conventional ionic contrast media. Detailed urinary excretion studies in rabbits that showed acute changes in urinary iodine concentration, urinary flow rates, urinary iodine excretion rates, blood iodine concentrations, and cumulative iodine excreted illustrate the pharmacodynamics of contrast media of conventional and new contrast agents.[30]

Figure 16-5 illustrates the urinary iodine concentration over time of Hypaque 50% (diatrizoate), Isovue (iopamidol), and Hexabrix (ioxaglate). Three minutes after a bolus injection of contrast media, Hexabrix and Isovue (iopamidol) give twice the iodine concentration in the urine as diatrizoate. The higher urinary iodine concentrations produced by the low osmolality CM continue to be significantly higher for 3 hours after contrast media injection. Between 10 minutes and 2 hours after injection the monoacidic dimer, Hexabrix (ioxaglate), gives higher urinary iodine concentrations than the nonionic CM. During this time period the urinary iodine concentration produced by the monoacidic dimer continues to be more than twice as high as with the conventional agents. The urinary iodine concentration with the nonionic contrast media continues to be higher, but is only approximately 70% higher than the conventional agents. Although the low osmolality contrast media give higher urinary iodine concentrations than conventional agents, Hexabrix, which has the lowest osmolality and contains 30% sodium, gives even higher concentrations than the nonionic contrast media because of the ability of the kidney to resorb sodium and water. In clinical urography this same relative difference in iodine concentrations is also seen. Nonionic contrast media provide approximately 70% higher urinary iodine concentrations than ionic CM, whereas Hexabrix provides approximately 100% higher urinary iodine concentration.[31]

Figure 16-6 demonstrates the urine flow rate after bolus injection of contrast media. There is a marked increase in the flow of urine because of the diuretic effect of contrast media. The urine flow rate in the first minute after contrast injection

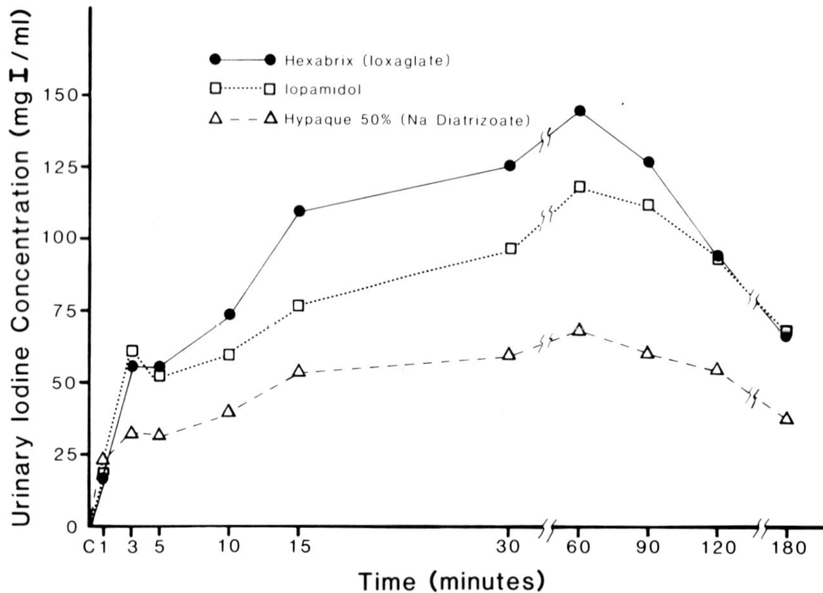

Figure 16-5 Urinary iodine concentration (mg I/ml) versus time comparing conventional ratio 1.5 CM and newer ratio 3 CM after bolus injection. *Source*: Reprinted with permission from *Investigative Radiology* (1982;17:494–500), Copyright © 1982, JB Lippincott Company.

increases nearly 20 times with conventional media, such as Hypaque 50 (diatrizoate), is about 10 times greater 5 minutes after injection, and steadily decreases for the next 3 hours. The low osmolality contrast media show a significantly lower urine flow rate than the conventional agents because of the lower osmolality and lower urinary diuresis. The urine flow rate with the ratio 3 agents is approximately one-half that of conventional agents.

Figure 16-7 illustrates the total urinary iodine excretion rate, which is the product of the urinary iodine concentration and the urine flow rate. The urinary iodine excretion rate is similar for all three contrast agents, indicating that the renal handling of the three agents is similar; that is, all three contrast agents are excreted by glomerular filtration with no significant reabsorption of the contrast molecules. Thus, although at equal doses the urinary iodine excretion rate is similar for the three contrast agents, the urinary iodine concentration is significantly higher and the urine flow rate significantly lower with the low osmolality contrast agents.

The blood iodine concentration (Fig. 16-8) after bolus injection increases rapidly during the injection phase. There is then a rapid decrease in blood iodine concentration as equilibrium of the contrast agent is reached rapidly, with water shifting into the vascular space from the interstitial space and CM moving from the intravascular space to the interstitial space. Most of the initial fall in blood iodine

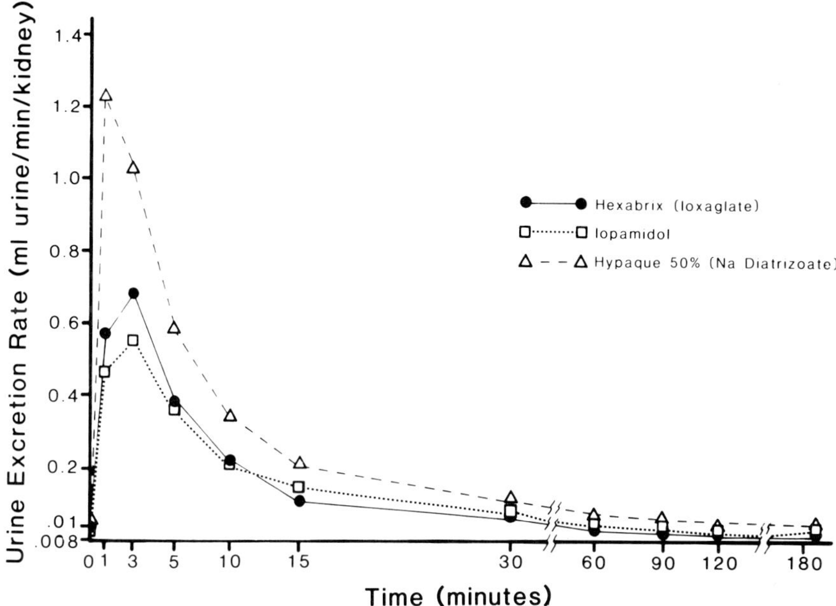

Figure 16-6 Urine excretion rate (ml urine/min/kidney) versus time comparing conventional ratio 1.5 and newer ratio 3 CM after bolus injection. *Source*: Reprinted with permission from *Investigative Radiology* (1982;17:494–500), Copyright © 1982, JB Lippincott Company.

concentration is not due to renal excretion but to equilibration of contrast media with the interstitial space.[44] After equilibration of CM in the intravascular and interstitial space, the slow decrease in blood iodine concentration is then mostly the result of renal excretion of contrast media.

Figure 16-9 demonstrates the cumulative percentage of contrast media (iodine) excreted in urine. During the first 15 minutes after a bolus injection of contrast, only approximately 30% of the contrast media is cleared in the urine, and 70% remains in the body, mostly in the interstitial space. The peak iodine concentration provides the highest peak urinary concentration, although only a small percentage of the total iodine dose injected is cleared by the kidneys during the first 15 minutes. Thus, the peak blood concentration provides the most optimal enhancement and excretion.

As shown in Figures 16-8 and 16-9, although there is a drop in blood iodine concentration of more than 50% in the first 5 minutes after bolus injection, only about 15% is excreted in the urine. Therefore, a bolus CM concentration passing through the kidney is necessary to provide an optimal urogram. The total dose injected is not the most important factor because a higher dose infused over a long period of time will not provide the peak iodine excretion rate that is seen with a

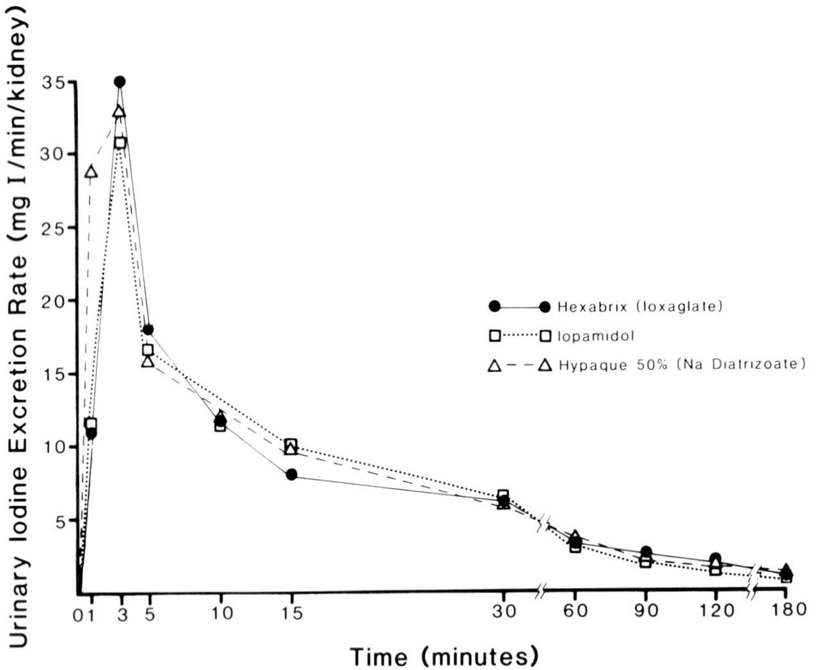

Figure 16-7 Urinary iodine excretion rate (mg I/min/kidney) versus time comparing conventional ratio 1.5 CM and ratio 3 CM after bolus injection. *Source*: Reprinted with permission from *Investigative Radiology* (1982;17:494–500), Copyright © 1982, JB Lippincott Company.

bolus injection. That is why a bolus injection provides an optimal urogram at equivalent doses compared to an infusion.

In summary, detailed animal experiments of renal excretion of conventional ionic and new nonionic contrast media have demonstrated that the low osmolar nonionic CM provide significantly higher urinary iodine concentrations. The urinary iodine concentration is more than double that produced by conventional CM in the first few minutes and approximately 70% higher in the first 30 minutes. The urinary iodine concentration is more than 100% higher for Hexabrix than with conventional CM and approximately 30% higher with Hexabrix compared to a nonionic CM after the first 10 minutes. Thus, these agents have the potential to provide denser kidney enhancement, providing a better nephrogram and denser opacification of the collecting system with a potential to providing better resolution of the urinary system.

COMPARATIVE CLINICAL UROGRAPHY

A number of clinical trials have been performed comparing conventional contrast agents and the new low osmolality agents both in adult and pediatric pa-

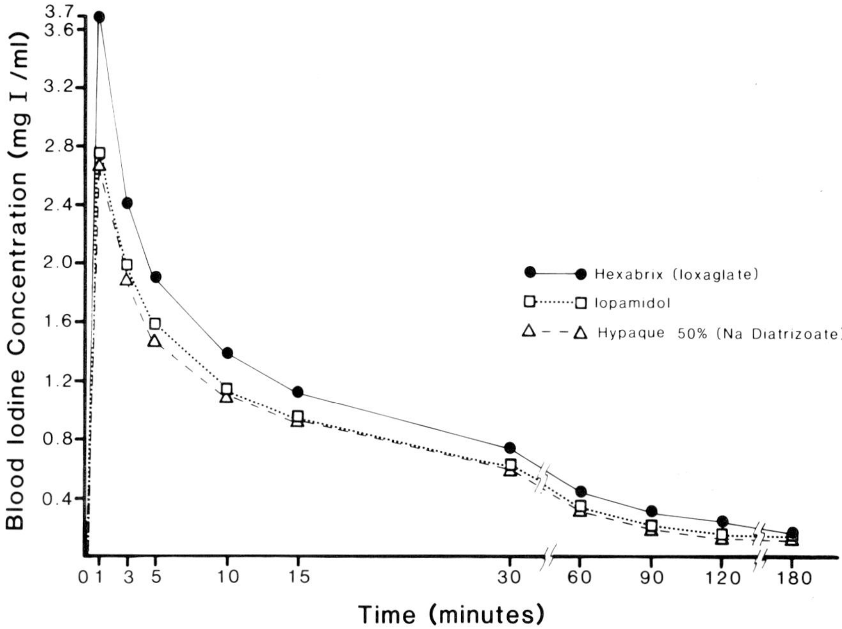

Figure 16-8 Blood iodine concentration (mg I/ml) versus time after bolus injection of 450 mg I/kg. *Source*: Reprinted with permission from *Investigative Radiology* (1982;17:494–500), Copyright © 1982, JB Lippincott Company.

tients.[45–60] All of these clinical studies have shown that the use of low osmolality contrast media results in better patient tolerance with fewer adverse side effects and reactions. Some of the studies at low doses have shown equivalent urographic quality. The preponderance of clinical experience and our own work with large numbers of patients in clinical urography conclude, however, that low osmolality contrast media provide better quality urography by producing higher urinary iodine concentrations and denser opacification of the collecting system. Ratio 3 CM provide better resolution of the urinary collecting systems with equivalent or slightly better nephrographic enhancement of the parenchyma. The only major disadvantage to ratio 3 contrast media is their significant increased cost. Currently in the United States they cost approximately 15 times as much as conventional ionic contrast media. This has led to great debate about the cost-benefit ratio and the medical and economic considerations in using these new contrast media in urography and in other contrast examinations.[19,61–63]

Much of the following discussion of contrast media in urography is based on our own data in controlled trials that used all of the available contrast media for urography in the United States.[31,53,54] In our trials we studied the following parameters—urinary iodine concentration, radiographic quality, blood pressure

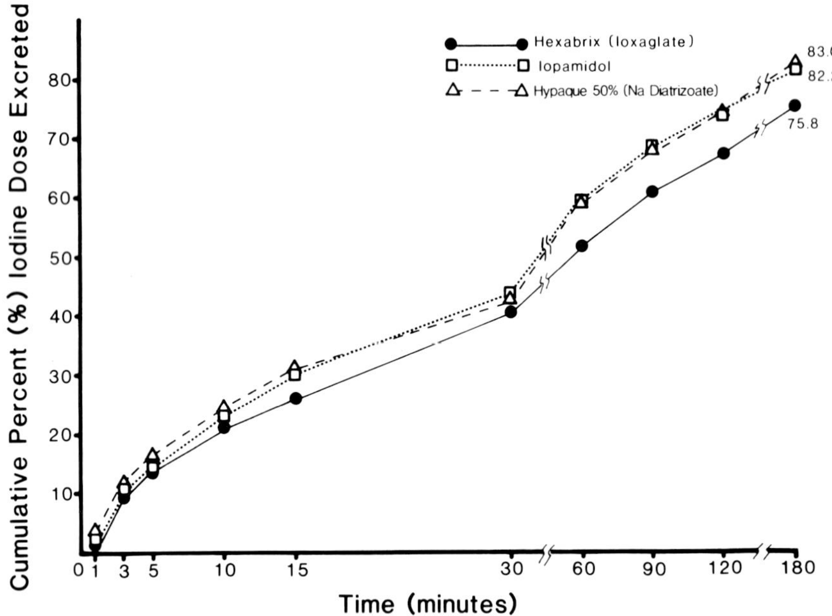

Figure 16-9 Cumulative percent of iodine dose excreted versus time. *Source*: Reprinted with permission from *Investigative Radiology* (1982;17:494–500), Copyright © 1982, JB Lippincott Company.

and pulse changes, adverse side effects, and patient tolerance—with different contrast media during urography.

Adverse Effects

The low osmolality CM provide better patient safety, tolerance, and acceptability than conventional ionic CM. We have found significantly fewer side effects with the new CM, a finding confirmed by others.[51–60] Tachycardia is a common effect of CM. Both Hexabrix and the nonionic CM, ioversol, cause significantly less tachycardia than diatrizoate.[54] Systemic hypotension, which is commonly seen after CM injection was significantly reduced with both Hexabrix and ioversol. In our studies nausea was seen in approximately 35% of patients with conventional CM. Hexabrix produced nausea in approximately 20% of patients, whereas the nonionic CM—Isovue, Omnipaque, and ioversol—caused nausea in only approximately 5% of patients.[53,54] Vomiting was seen in 6–10% of patients with conventional CM and a similar number with Hexabrix, but was rarely seen with nonionic CM. Severe body heat and other unpleasant sensations resulting from injection of conventional CM were seen in approximately 85% of patients, but only 30-45% of patients reported any body heat with low osmolality CM. Uncomfortable body heat was rated more than twice as severe with conventional CM. Lightheadedness was much less common with low osmolality CM. The number of patients experiencing a significantly unpleasant or uncomfortable experience during urography was significantly higher with ionic CM. Although both types of ratio 3 low osmolality CM were better tolerated by patients, nonionic ratio 3 CM seemed to be better tolerated than the ionic dimer Hexabrix. Thus, ratio 3 low osmolality CM have distinct advantages of improved patient tolerance and fewer hemodynamic changes during IVU than conventional ionic CM.

Urinary Iodine Concentration

In 66 normal patients studied with Renografin 60 (meglumine/sodium diatrizoate) we found an average iodine concentration of 33 mg I/ml in voided urine specimens approximately 20 minutes after a bolus injection of a dose of 1.5 ml CM/kg body weight (450 mg I/kg). A similar dose of 1.5 ml CM/kg body weight of Hexabrix produced a measured iodine concentration of 74 mg I/ml urine, whereas ioversol (MP-328) produced an iodine concentration of 50.2 mg I/ml urine.

In a study of 63 normal patients at a fixed dose of 100 ml of contrast media injected in a bolus, Reno-M-60 (meglumine diatrizoate) produced an iodine concentration of 31 mg I/ml of urine. A normal population of 60 patients injected with Omnipaque (iohexol) produced an iodine concentration of 52 mg I/ml of urine.

Another clinical trial studied 87 high-risk patients—severely debilitated and elderly patients and patients with significant cardiovascular disease, asthma and highly allergic histories, previous adverse contrast media reactions, and impaired renal function—in whom the new low osmolality contrast media were being routinely used. It showed that in patients injected with Isovue (iopamidol) at a dose of 100 ml of contrast a urinary iodine concentration of 43.2 mg I/ml was obtained. With Omnipaque (iohexol) in a similar high-risk group an iodine concentration of 43.8 mg I/ml was obtained. Thus, in high-risk patients, iodine concentrations with nonionic contrast media were 40% higher than in normal patients with conventional agents. In this group where urography generally is less than satisfactory, low osmolality contrast media produced higher iodine concentrations than the average CM concentration seen in normal patients with conventional contrast media and provided good urograms in patients normally expected to have poor studies.

Both nonionic contrast agents and Hexabrix produce significantly higher urinary iodine concentrations than conventional agents because of their lower osmolality. There are slight differences in the nonionics, probably due to differences in osmolality. Hexabrix differs from the nonionic agents both in its lower osmolality and its sodium content, which allows for further reabsorption of water.

Double-blind reading of urograms comparing ratio 3 contrast media with conventional 1.5 contrast media have shown that urographic quality is significantly better with the lower osmolar contrast media. In our studies comparing the nonionic contrast media, ioversol and Hexabrix, to the conventional contrast agent, Reno-M-60, the quality of the urograms produced by both ioversol and Hexabrix were rated significantly better.

Clinical Urography

Conventional ionic contrast media provide good quality excretory urograms. Nephrographic enhancement in patients with normal renal function is generally good. Opacification of the collecting system is usually good and provides good anatomic detail in the calyces, renal pelves, ureters, and the bladder. Because of the diuresis that occurs with the doses necessary for good opacification of the urinary system, there is good urine flow rate with good distention and opacification of the urinary collecting system and bladder.

The use of low osmolality contrast media results in a different appearance to the IVU. The nephrogram density is at least as good and in some studies has been better than with conventional agents. In normal patients the difference is small. In debilitated patients, however, the nephrogram quality provided by the high urinary concentrations with the low osmolality ratio 3 contrast media yields a significant advantage in diagnosis. The most dramatic change seen in excretory urography with low osmolality contrast media is in the opacification of the urinary collecting system. As contrast concentrates in the nephrons and collecting tubules, significantly higher urinary iodine concentrations are produced. In the collecting tubules this concentration is seen as a dense papillary blush, which is more often seen with the low osmolality contrast media.[54] Because of the higher urinary iodine concentration there is denser opacification of the calyces, renal pelves, and ureters. Because of the decreased diuresis there is slight delay in opacification of these structures when compared to conventional agents. Some authors have thought that, because of the decreased diuresis, urography with low osmolality contrast media would provide less distention of the renal pelvis and ureters and poor visualization of these structures. This has not proven to be the case, and in fact denser and better opacification of the renal calyces, pelves, and ureters has been observed by most authors without use of compression. The ability to (1) provide detailed anatomic information for such diseases as medullary sponge kidney, papillary necrosis, renal tuberculosis, early transitional cell carcinoma, and other abnormalities of the renal pelvis and ureters and (2) to determine the site and etiology of urinary obstruction is better with ratio 3 CM because of the denser opacification of the urinary collecting system. Bladder opacification is also denser. In particular, in postvoid radiographs, mucosal detail of the bladder is excellent with ratio 3 contrast media. Postexcretory urographic voiding cystourethrograms can be obtained more easily and are of better quality than those with conventional agents because of the higher urinary iodine concentration. In fact, IV injection of approximately 30 g of iodine can provide urinary iodine concentrations of 75 mg I/ml or higher (over 100 mg I/ml), which are in the range of those that provide excellent diagnostic quality for voiding cystourethrography and retrograde pyelography. Excellent quality urograms can be obtained in patients without any fluid restriction and even in patients who are being hydrated because of underlying renal or medical diseases.

The clinical adverse effects of conventional high osmolality ionic contrast media, including hemodynamic changes, systemic responses, cardiovascular effects, electrocardiographic changes, bronchospasm, effects on coagulation, enzyme inhibition, induction of thrombophlebitis, and nephrotoxic potential, are well described in the literature.[42,64–85] Some of these adverse effects, such as tachycardia, hypotension, vasodilation, changes in serum osmolality, and rapid fluid and electrolyte shifts, and such patient symptoms as heat and flushing are due to osmolality. Other effects, such as neurotoxicity, general systemic toxicity, nephrotoxicity, enzyme inhibition, and effects on coagulation and clotting, are due to chemotoxicity of contrast agents or a combination of chemotoxicity and osmolality.[64,65] It is clear from the accumulated literature that low osmolality contrast media and particularly nonionic contrast media produce much less adverse clinical effects in almost every area, with the possible exception of nephrotoxicity, which is at this time unclear.[73,78,80]

Unfortunately, the marked increased cost of nonionic agents makes a choice of a contrast medium for urography a difficult decision for the physician.[19,61,62] Because of the massive financial burden that would be created by use of these new ratio 3 CM agents in all patients, it has been necessary to develop guidelines as to those patients who would benefit most from the use of those agents. In general, throughout the United States those patients in whom it is preferable to use low osmolality CM include the following: (1) patients less than 1 year of age; (2) patients with previous adverse contrast media reactions; (3) patients with significant cardiovascular or cerebrovascular disease; (4) patients with asthma or severe allergic histories; (5) patients with multiple myeloma, sickle cell disease, and other hemoglobinopathies; (6) severely debilitated and elderly patients with fragile medical conditions; and (7) patients with impaired renal function. Other indications at the discretion of the attending radiologist of course are also valid when because of increased safety or efficacy the new ratio 3 CM are believed to be of advantage to the patient.

Excretory urography will continue to be a commonly used diagnostic procedure because of its proven diagnostic efficacy. Although conventional CM will continue to be utilized and will continue to provide good diagnostic studies and patient safety, the use of low osmolality CM will reduce the morbidity and mortality associated with CM injection, provide better quality IVUs, and decrease the apprehension that may exist when a contrast-enhanced examination is considered.

REFERENCES

1. Pollack HW, Banner MP: Current status of excretory urography. A premature epitaph? *Urol Clin N Am* 1985;12:585–601.

2. Doubilet P, McNeil BJ, Van Houten FX, et al: Excretory urography in current practice: evidence against overutilization. *Radiology* 1985;154:607–611.

3. Dure-Smith P: Urographic agents, in Skucas J, Miller R (eds): *Radiographic Contrast Agents*. Baltimore, University Park Press, 1977, pp 273–306.

4. Almén T: Experience from 10 years of development of water soluble nonionic contrast media. *Invest Radiol* 1980;15(suppl):S283–S288.

5. Almen T: Contrast agent design: some aspects of the synthesis of water soluble contrast agents of low osmolality. *J Theor Biol* 1969;24:215.

6. Ansell G: Adverse reactions to contrast agents. Scope of problem. *Invest Radiol* 1970;5: 374–384.

7. Ansell G, Tweedie MCK, West CR, et al: The current status of reactions to intravenous contrast media. *Invest Radiol* 1980;15(suppl):S32–S39.

8. Hobbs BB: Adverse reactions to intravenous contrast agents in Ontario, 1975-1979. *J L'Assoc Can Radiol* 1981;31:8–10.

9. Lalli AF: Contrast media reactions: data analysis and hypothesis. *Radiology* 1980;134:1–12.

10. Witten DM: Reactions to urographic contrast media. *JAMA* 1975;231:974–977.

11. Shehadi WH: Adverse reactions to intravascularly administered contrast media. *AJR* 1975; 124:145–152.

12. Shehadi WH, Toniolo G: Adverse reactions to contrast media. *Radiology* 1980;137:299–302.

13. Shehadi WH: Contrast media adverse reactions: occurrence, recurrence, and distribution patterns. *Radiology* 1982;143:11–17.

14. Pendergrass HP, Tondreau RL, Pendergrass EP, et al: Reactions associated with intravenous urography: historical and statistical review. *Radiology* 1958;71:1–12.

15. Wolfromm R, Dehouve A, Degand F, et al: Les accidents graves par injection intraveineuse de substances iodees pour urographie. *J Radiol Electrol* 1966;47:346–357.

16. Hartman GW, Hattern RR, Witten DM, et al: Mortality during excretory urography: Mayo Clinic experience. *AJR* 1982;139:919–922.

17. Witten DM, Hirsch FD, Hartman GW: Acute reactions to urographic contrast medium: incidence, clinical characteristics and relationship to history of hypersensitivity states. *AJR* 1973; 119:832–840.

18. Schrott KM, Behrends B, Clauss W, et al: Iohexol in excretory urography. *Fortschr Med* 1986;104:153–156.

19. McClennan BL: Low-osmolality contrast media: premises and promises. *Radiology* 1987;162:1–8.

20. Holtas S: Iohexol in patients with previous adverse reactions to contrast media. *Invest Radiol* 1984;19:563–565.

21. Rapoport S, Bookstein JJ, Higgins CB, et al: Experience with metrizamide in patients with previous severe anaphylactoid reactions to ionic contrast agents. *Radiology* 1982;143:321–325.

22. Lasser EC, Berry CC, Talner LB, et al: Pretreatment with corticosteroids to alleviate reactions to intravenous contrast material. *N Engl J Med* 1987;317:845–849.

23. Greenberger PA: Contrast media reactions. *J Allergy Clin Immunol* 1984;74:600–605.

24. Greenberger PA, Patterson R, Topev CM: Prophylaxis against repeated radiocontrast media reactions in 857 cases. *Arch Intern Med* 1985;145:2197–2200.

25. Dure-Smith P: The dose of contrast medium in intravenous urography. A physiologic assessment. *AJR* 1970;108:691–697.

26. Cattell WR: Excretory pathways for contrast media. *Invest Radiol* 1970;5:473–486.

27. Cattell WR, Fry IK, Spencer AG, et al: Excretion urography: I-Factors determining the excretion of Hypaque. *Br J Radiol* 1967;40:561–571.

28. Golman K, Almén T: Urographic contrast media and methods of investigative uroradiology, in Sovak M (ed): *Radiocontrast Agents*. New York, Springer-Verlag, 1984, pp 127–191.

29. Dure-Smith P, Simenhoff M, Zimskind PD, et al: The bolus effect in excretory urography. *Radiology* 1971;101:29–34.

30. Spataro RF, Fischer HW, Boylan L: Urography with low-osmolality contrast media. Comparative urinary excretion of Iopamidol, Hexabrix, and diatrizoate. *Invest Radiol* 1982;17:494–500.

31. Spataro RF: New and old contrast agents: pharmacology, tissue opacification and excretory urography. *Urol Radiol* 1988;10:2–5.

32. Fischer HW, Rothfield JH, Carr JD: Optimum dose in excretory urography. *AJR* 1971; 113:423–426.

33. Almén T, Härtel M, Golman K: Metrizamide in experimental urography. III. Effects of temporary ureteric stasis on urinary iodine concentration after intravenous injection of ionic and nonionic contrast media. *Acta Radiol* 1973;335(suppl):330–338.

34. Evill CA, Benness GT: Urographic excretion studies with metrizamide and "dimer." A high dose comparison in dogs. *Invest Radiol* 1977;12:169–174.

35. Evill CA, Benness GT: Urographic excretion studies: preliminary results with six-iodine, singly-ionizing sodium salt P-286. *Invest Radiol* 1978;13:325–327.

36. Golman K, Almén T: Metrizamide in experimental urography. *Acta Radiol* 1973; 335(suppl):312–322.

37. Golman K: Metrizamide in experimental urography. V. Renal excretion mechanism of a nonionic contrast medium in rabbit and cat. *Invest Radiol* 1976;11:187–194.

38. Golman K, Almén T, Denneberg T, et al: Metrizamide in urography. II. A comparison of ^{51}Cr-EDTA clearance and metrizamide clearance in man. *Invest Radiol* 1977;12:353–356.

39. Owman T: *Excretion of Urographic Contrast Media: Experimental Studies in the Rabbit. II. Comparisons of Various Contrast Media at Urography with Simulated Compression*, thesis, University Hospital, Lund, Sweden, 1978.

40. Sjoberg S, Almén T, Golman K: Excretion of contrast media for urography. Urine volume and iodine concentration during free urine flow in the rabbit. *Acta Radiol* 1980;362(suppl):93–98.

41. Sjoberg S, Almén T, Golman K: Excretion of contrast media for urography: iohexol and sodium diatrizoate during ureteric stenosis in rabbits. *Acta Radiol* 1980;362(suppl):99–104.

42. Spataro RF: Newer contrast agents for urography. *Radiol Clin N Am* 1984;22:365–380.

43. Wilcox J, Evill CA, Sage MR, et al: Urographic excretion studies with nonionic contrast agents. *Invest Radiol* 1983;18:207–210.

44. Kormano M, Dean PB: Extravascular contrast material: the major component of contrast enhancement. *Radiology* 1976;121:379–382.

45. Bjork L, Zachrisson BF: Low dose urography with a low osmolar contrast medium. *Acta Radiol Diagn* 1986;27:111–113.

46. Palma LD, Rossi M, Stacul F, et al: Iopamidol in urography. *Urol Radiol* 1982;4:1–3.

47. Dray RJ, Winfield AC, Muhletaler CA, et al: Advantages of nonionic contrast agents in adult urography. *Urology* 1984;24:297–299.

48. Levorstad K, Kolbenstvedt A, Loyning EW: Iohexol compared with metrizoate in urography. *Acta Radiol Diagn* 1983;24:337–341.

49. Loughran CF: Clinical intravenous urography; comparative trial of ioxaglate and iopamidol. *Radiology* 1986;161:455–458.

50. McClennan BL, Ling D, Rholl KS, et al: Urography with a low osmolality contrast agent. Comparison of Hexabrix with Conray 325. *Invest Radiol* 1986;21:144–150.

51. Rankin RN, Eng FWHT: Iohexol vs. Diatrizoate. A comparative study in intravenous urography. *Invest Radiol* 1985;20:S112–S114.

52. Robey G, Reilly BJ, Carusi PA, et al: Pediatric urography: comparison of metrizamide and methylglucamine diatrizoate. *Radiology* 1984;150:61–63.

53. Spataro RF, Katzberg RW, Fischer HW, et al: A comparison of contrast media in high dose clinical urography. Presented at the Scientific Meeting of the Society of Uroradiology, Orlando, FL, Jan 30, 1988.

54. Spataro RF, Katzberg RW, Fischer HW, et al: High-dose clinical urography with the low-osmolality contrast agent Hexabrix: Comparison with a conventional contrast agent. *Radiology* 1987;162:9–14.

55. Taenzer V, Heep H, Clauss W: Urography with non-ionic contrast media: I. Diagnostic quality and tolerance of iohexol in comparison with meglumine amidotrizoate, in Taenzer V, Zeitler E (eds): *Contrast Media in Urography, Angiography and Computerized Tomography*. Stuttgart, Thieme Verlag, 1983, pp 148–152.

56. Taenzer V, Meisel P, Hartwig P: Urography with non-ionic contrast media: II. Diagnostic quality and tolerance of iopromide in comparison with ioxaglate, in Taenzer V, Zeitler E (eds): *Contrast Media in Urography, Angiography and Computerized Tomography*. Stuttgart, Thieme Verlag, 1983, pp 153–155.

57. Thompson WM, Foster WL Jr, Halvorsen RA, et al: Iopamidol: New, nonionic contrast agent for excretory urography. *AJR* 1984;142:329–332.

58. Winfield AC, Dray RJ, Kirchner FK Jr, et al: Iohexol for excretory urography: a comparative study. *AJR* 1983;141:571–573.

59. Wolf K-J, Steidle B, Skutta T, et al: Iopromide. Clinical experience with a new non-ionic contrast medium. *Acta Radiol Diagn* 1983;24:55–62.

60. Dawson P, Heron C, Marshall J: Intravenous urography with low-osmolality contrast agents: theoretical considerations and clinical findings. *Clin Radiol* 1984;35:173–175.

61. Fischer HW, Spataro RF, Rosenberg PM: Medical and economic considerations in using a new contrast medium. *Arch Intern Med* 1986;146:1717–1721.

62. Grainger RG: The clinical and financial implications of the low osmolar radiological contrast media. *Clin Radiol* 1984;35:251–252.

63. Wolf GL: Safer, more expensive iodinated contrast agents: how do we decide? *Radiology* 1986;159:557–558.

64. Dawson P: New contrast agents chemistry and pharmacology. *Invest Radiol* 1984;19: S293–S300.

65. Dawson P: Chemotoxicity of contrast media and clinical adverse effects. *Invest Radiol* 1985;20:S84–S91.

66. Fischer HW: Hemodynamic reactions to angiographic media. *Radiology* 1968;91:66–73.

67. Fischer HW, Katzberg RW, Morris TW, et al: Systemic response to excretory urography. *Radiology* 1984;151:31–33.

68. Mancini GBJ, Ostrander DR, Slutsky RA, et al: Intravenous vs. left ventricular injection of ionic contrast material: hemodynamic implications for digital subtraction angiography. *AJR* 1983; 140:425–430.

69. Heron CW, Underwood SR, Dawson P: Electrocardiographic changes during intravenous urography: a study with sodium iothalamate and iohexol. *Clin Radiol* 1984;35:137–141.

70. Wolf GL, Mulry CS, Kilzer K, et al: New angiographic agents with less fibrillatory propensity. *Invest Radiol* 1981;16:320–323.

71. Harkonen S, Kjellstrand C: Contrast nephropathy. *Am J Nephrol* 1981;1:69–77.

72. Pfister RC, Hutter AM Jr, Newhouse JH, et al: Contrast-medium-induced electrocardiographic abnormalities: comparison of bolus and infusion of methylglucamine iodamide and methylglucamine/sodium diatrizoate. *AJR* 1983;140:149–153.

73. Mudge GH: Nephrotoxicity of urographic radiocontrast drugs. *Kidney Int* 1980;18:540–552.

74. Gale ME, Robbins AH, Hamburger RJ, et al: Renal toxicity of contrast agents: iopamidol, iothalamate, and diatrizoate. *AJR* 1984;142:333–335.

75. Cochran ST, Wong WS, Roe DJ: Predicting angiography-induced acute renal function impairment: clinical risk model. *AJR* 1983;141:1027–1033.

76. Smith H-J, Levorstad K, Berg KJ, et al: High dose urography in patients with renal failure. *Acta Radiol Diagn* 1985;26:213–220.

77. Stormorken H, Skalpe IO, Testart MC: Effect of various contrast media on coagulation, fibrinolysis, and platelet function. An in vitro and in vivo study. *Invest Radiol* 1986;21:348–354.

78. Kumar S, Hull JD, Lathi S, et al: Low incidence of renal failure after angiography. *Arch Intern Med* 1981;141:1268–1270.

79. Albrechtsson U, Olsson CG: Thrombosis following phlebography with ionic and non-ionic contrast media. *Acta Radiol Diagn* 1979;20:46–52.

80. Golman K, Almén T: Contrast media-induced nephrotoxicity. Survey and present state. *Invest Radiol* 1985;20:S92–S97.

81. Drayer BP, Velaj R, Bird R, et al: Comparative safety of intracarotid iopamidol, iothalamate meglumine and diatrizoate meglumine for cerebral angiography. *Invest Radiol* 1984;19:S212–S218.

82. Laerum F, Holm HA: Postphlebographic thrombosis. *Radiology* 1981;140:651–654.

83. Rao VM, Rao AK, Steiner RM, et al: The effect of ionic and nonionic contrast media on the sickling phenomenon. *Radiology* 1982;144:291–293.

84. Rao AK, Rao VM, Willis J, et al: Inhibition of platelet function by contrast media: iopamidol and ioxaglate versus iothalamate. *Radiology* 1985;156:311–313.

85. Longstaff AJ, Henson JHL: Bronchospasm following intravenous injection of ionic and nonionic low osmolality contrast media. *Clin Radiol* 1985;36:651–653.

Chapter 17

Cystography, Cystourethrography, and Retrograde Urethrography

William H. McAlister

CYSTOGRAPHY AND VOIDING CYSTOURETHROGRAPHY

For many years cystography and voiding cystourethrography (VCU) have provided useful diagnostic information. Some indications for VCU include urinary tract infections; voiding abnormalities, such as poor or interrupted stream; upper urinary tract disease demonstrated by other imaging studies; hematuria; prolapsing urethral mass; megalourethra; large bladder; prune belly syndrome; pelvic trauma; and severe lower tract congenital abnormalities, such as hypospadias or epispadias.

Wulff first described cystography in 1905,[1] and voiding cystourethrography was mentioned in 1921 and 1924.[2,3] Although many contrast agents were employed in these early studies, most are of historical interest only. These include silver compounds, bismuth compounds, barium sulfate and thorium, solutions of sodium and potassium iodide, diiodinated pyridone compounds, and iodized oils, e.g., Lipiodol.[3–6] Currently the agents used for cystourethrography are the same as those for intravenous excretory urography (triiodobenzoic acid compounds) but in lower concentrations.[7–10] Low osmolality ionic and nonionic iodine containing contrast media are too costly for frequent use. Air, nitrous oxide, or carbon dioxide in combination with either iodine-containing contrast agents or barium sulfate can be used for double-contrast bladder studies to evaluate bladder and prostate tumors and bladder diverticula.[11–15] Occasionally, thickening agents are added to the media in air contrast studies. Barium sulfate is no longer utilized in cystourethrography because intrarenal reflux of barium sulfate has been shown to evoke a serious inflammatory reaction in experimental animals.[16] Nuclear scintigraphy is often used for follow-up evaluation of vesicoureteral reflux because it delivers a lower radiation dose than conventional radiography.

Techniques

The bladder and urethra can be studied by the intravenous (IV) injection of contrast medium or by direct instillation of contrast medium. In both methods the bladder is distended with the contrast medium, and voiding films are obtained. The disadvantages of excretory urography are that minor to moderate degrees of reflux can be missed even on postvoid roentgenograms, and the method is not applicable to infants, young children, or uncooperative patients. In conventional or retrograde cystography the bladder is filled with contrast material via a feeding tube or straight or balloon catheter. Occasionally 1% Xylocaine or distention of the urethra with saline may facilitate passage of the catheter.[17] If urethral entry is not possible, percutaneous suprapubic catheterization can be used.[18] The contrast medium can be hand injected with a syringe and three-way stopcock or, preferably, administered by a gravity drip with the container not more than 70 cm above the patient. Neither method influences the incidence of reflux. However, gravity instillation allows more contrast medium to enter the bladder. Filling of the bladder can be monitored by serial overhead radiographs or preferably by intermittent fluoroscopic observations. Voiding is recorded by overhead radiographs, videotape, camera, or spot radiographs. Transient obstruction of the end of the penis by a Foley catheter in the fossa navicularis or compression of the penis with incontinence clamps during micturition may help demonstrate urethral strictures. Videorecordings are preferred in infants and young children and those with voiding dysfunction. A technique for visualizing the bladder wall with contrast medium diffused from a submucosal injection in the anterior wall of the rectum has also been described.[19]

Histologic Effects on the Bladder

Investigators have documented adverse effects of iodine-containing contrast media on the bladder of humans and experimental animals. Four patients with bladder injury secondary to contrast medium administration were presented by Talarico et al.[20] One patient had permanent damage to her bladder from a 20% sodium iodide solution; she subsequently underwent a cystectomy with an ileal conduit. A second patient developed reflux and a contracted bladder from a 5% sodium iodide solution. The third patient developed similar complications from sodium methiodal 30%. A fourth patient developed mucosal edema of the bladder that was documented cystoscopically after instillation of 40% Cysto-conray (iothalamate meglumine). Although not specifically dealing with the bladder, McClennan et al.[21] reported that uroepithelial cells obtained from the pelvicalyceal systems of ten patients undergoing retrograde pyelography with a 60%

solution of diatrizoate meglumine showed cellular shrinkage, nuclear pyknosis, fragmentation, and occasionally cytoplasmic vacuolization. Uroepithelial cells exposed in vitro to 50%, 25%, and 12.5% diatrizoate sodium and 60%, 30% and 15% diatrizoate meglumine exhibited the same cytologic changes as cells exposed in vivo. Fewer histologic effects were observed as the contrast agents were diluted. The cellular changes are due to an acute cellular reaction, rather than to prolonged retention of iodine in mucosal cells.[22] Iodine is organically bound in newer agents. In older agents there was more free or inorganic iodide.

Shopfner reported that Cystokon (acetrizoate sodium) produced bladder irritability, pain, small bladder capacity, mucosal irregularity, vesicoureteral reflux, and urethral spasm.[23] McAlister et al.[24] demonstrated that 30% acetrizoate sodium (Cystokon), diatrizoate sodium 25% (Hypaque sodium), and diatrizoate meglumine 30% (Hypaque M-30) in rat bladders evoked an inflammatory response that reached a peak at 48 hours. Because the osmolality of normal urine ranges from 100 to over 1000 mosm/l the histologic effects cannot be explained on the basis of osmolality alone. Siu et al.[10] studied the effects of five iodinated contrast media and physiologic saline in rabbits. These authors found that the bladder tolerated a greater volume of sodium chloride than iothalamate meglumine 43% (Cysto-Conray), iothalamate meglumine 30%, diatrizoate meglumine 30%, diatrizoate sodium 30%, or meglumine iocarmate (Dimer-X). The contrast media produced dehydrative, hydropic, and desquamative changes in rabbit bladder mucosa, with some of the effects lasting up to 3 months. McAlister et al.[9] studied less concentrated ionic and nonionic agents—iothalamate meglumine 17.2% (Cysto-Conray II, 81 mg I/ml), diluted iothalamate meglumine 10% (47 mg I/ml), and diatrizoate meglumine 10%, (Hypaque-Cysto, 47 mg I/ml)—and an isotonic nonionic agent, metrizamide (Amipaque, 190 mg I/ml), and found no adverse histologic effects in rat bladders. Interestingly, when ionic iodine-containing contrast media are placed into the peritoneal cavity, little inflammatory reaction is produced,[25] but when introduced into the soft tissues (muscle and subcutaneous tissue), they can evoke substantial inflammatory effects.[26]

These data indicate that lower concentrations of ionic iodine-containing contrast media are safe cystographic agents. Substantially higher concentrations are associated with some, usually reversible damage to the bladder mucosa.

Iodine Concentration

High iodine concentrations of the ionic agents ranging from 30–50% were initially needed in cystourethrography because of the poor resolution of early cinefluorographic systems. With improved image resolution the need for these high concentrations has diminished. The clinical and histologic studies that have documented less mucosal damage and satisfactory diagnostic imaging from lower concentrations have led to recommendations that concentrations of under 20% of the ionic iodine-containing agents be used in routine examinations.

The visualization of water-filled Foley balloons was studied by McAlister and Griffith in the bladders of dogs after instillation at different iodine concentrations of iothalamate meglumine ranging from 2.5–30%.[9] Concentrations greater than 20% (94 mg I/ml) obscured the Foley balloon in the bladder. Bladder masses were best demonstrated with hypotonic solutions with concentrations around 5% (24 mg I/ml). Additionally, concentrations of iothalamate meglumine ranging from 5–30% were placed in 3.1-mm diameter tubes simulating reflexing ureters. The tubes were placed in a 12-cm diameter water phantom simulating an AP abdomen of a 3- to 4-year-old child. Adequate visualization of the ureter-simulating tubes occurred at concentrations as low as 5%.

Dure-Smith et al.[27] obtained optical density measurements with constant radiographic exposure factors from wells filled with contrast medium in concentrations from 1.7–336 mg I/ml placed in a phantom. They found very little change in the density of ionic iodine concentrations above 100 mg/ml (approximately a 20% solution).

The existing data suggest that the lowest possible concentration of contrast medium not only evokes less inflammatory response but also allows easier identification of filling defects within the bladder while still maintaining adequate levels for the demonstration of ureteral anatomy, ureteral reflux, and mass lesions in the bladder. Commercially available packaged units include Cysto-Conray II 17.2% (iothalamate meglumine, 81 mg I/ml, 404 mosm/l) and Hypaque-Cysto (diatrizoate meglumine), which can be diluted to a 10% solution (47 mg I/ml, 234 mosm/l). Current packaging enables Cystografin (diatrizoate meglumine) (Squibb) to be diluted to an 18% solution (85 mg I/ml).

Absorption of Contrast Media from the Bladder

Iodine-containing contrast material is absorbed from the bladder during routine voiding cystourethrography. The lowest blood iodine levels were found in children who voided promptly.[28] Although patients who retain contrast in the bladder for periods of time up to 45 minutes show the highest levels of circulating iodine, there is no direct relationship between the length of time that a contrast agent remained in the bladder and the blood iodine levels. Blood iodine levels greater than 6,000 mg/dl and total blood iodine in excess of 90 mg produce pyelograms on abdominal roentgenograms.

Complications

Complications of cystography are infrequent but include urinary tract infection through introduction of organisms into the bladder via the catheter or exacerbation of a previous infection.[29] Glynn[30] reported a 6% incidence of fresh infections or exacerbation of a previous infection closely related to cystography. Organisms in the bladder can be forced from the bladder into the kidneys with reflux. Urethral organisms have been shown to pass up into the kidney in refluxing animals.[31] Sepsis, death, and appreciable renal shrinkage can occur from infections associated with cystography.[32] Increased pressures in the renal pelvis and intrarenal tubules associated with vesicoureteral reflux and intrarenal reflux decrease renal blood flow and contribute to ischemia and atrophy, as well as the establishment of renal infection.[33]

Catheter trauma is another complication. Urethral and bladder rupture or erosions can occur, resulting in dysuria, hematuria, and frequency. When the catheter is removed urinary retention may result from the conversion of a partial urethral obstruction into a complete one by edema related to catheterization. Transient cystic and solid masses in the urinary bladder have been identified by sonography after voiding cystourethrography.[34]

Bladder perforation or intramural extravasation may follow overdistention or excessive intravesical pressure. Bladder rupture, mucosal tears, and extravasation are seen especially with small unused or defunctionalized bladders in patients on dialysis or with upper urinary tract diversion.[35]

Allergic reactions, either generalized or localized, may occur secondary to the absorption of contrast medium, local anesthetics, or antibiotic additives. McAlister et al. reported urticaria developing in association with cystography with 5% Cystokon.[29]

Catheter retention problems with the Foley catheter include spontaneous deflation, an inability to deflate the balloons, or rupture of the catheter balloon with fragments remaining in the bladder. Percutaneous puncture under sonographic guidance or introduction of ether into the balloon may perforate the Foley balloon that cannot be deflated.[36] Too, feeding tubes may coil or knot in the bladder. An angiographic guide wire can be threaded into the catheter and the knot uncoiled under fluoroscopic control.[37]

Inflammatory response in the bladder from the contrast media is another complication.[26] Contrast medium may enter the peritoneal cavity after voiding, with filling of the vagina and retrograde flow through the fallopian tubes. Transurethral reflux may occur from cleansing procedures. In a study using neonates, the perineum was cleansed with soap mixed with contrast medium, and the contrast medium was subsequently shown in the bladder in abdominal radiographs. Other complications include accidental catheterization of the vagina or dilated ectopic ureteral orifice, kidney pain from reflux, radiation effects, and anuria following cystography, presumably related to edema of the bladder mucosa, especially the ureteral orifices. Finally, autonomic dysreflexia[38,39] is an uncontrolled sympathetic response to various stimuli, including retrograde cystography, below the level of a high spinal cord lesion. It can precipitate severe hypertension and in time myocardial failure or intracranial hemorrhage.

RETROGRADE URETHROGRAPHY

Examination of the urethra in retrograde fashion is useful in suspected urethral trauma, diagnosis and delineation of strictures, assessment of the postoperative appearance of the urethra in a variety of congenital and acquired diseases, demonstration of urethral fistulas, assessment of the urethra in the presence of prostatic enlargement or mass, neurogenic dysfunction, intersex problems, and urethral duplications. Retrograde urethrographic shortcomings include inadequate demonstration of partial obstructions, sphincter spasm limiting the evaluation of the posterior urethra, infection, false passage, pain, and radiation exposure to the operator's hands.

Contrast Media

The majority of the agents that have been employed are primarily of historical interest. Cunningham used an Argyrol (colloidal silver) solution to study the urethra in 1910.[40] Other agents used for retrograde urethrography include Methiodal, Diodrast, barium sulfate, iodopyracet, sodium iodide in peanut oil, Iodochloral (iodine with chlorine organically bound to peanut oil), semisolid medium of Flocks (iodized oil, alcohol, and tragacanth), semisolid medium of Lowsley and Kirwin (Hippuran, glycerine, and tragacanth), and Lipiodol.[6,41] In 1947 Crabtree reported 27 cases of urethral extravasation with 4 deaths from pulmonary oil embolization following Lipiodol extravasation into the venous system of the penis.[42] Interestingly enough, Lipiodol and other oils continued to be used, and additional deaths were reported as late as 1960.[43] Barium extravasation into the urethral blood vessels was the cause of a fatal pulmonary embolus described by Gaudin in 1949.[44] The retrograde injection of air also caused a fatal pulmonary embolus after the air entered the venous system. Agents utilized in Europe but rarely in the United States were Umbradil-viscos, which consisted of iodopyracet (Diodrast 35%) with sodium carboxymethylcellulose, lidocaine 0.25%, and distilled water, and 70% Endografin.[15] A number of additives have been placed in the contrast media to thicken them because it was felt that increasing the viscosity of urethrographic contrast media could produce better urethral distention. In 1967 Thixokon—a mixture of urokon and Amioca flour—was introduced, but was eventually withdrawn because of sterility questions.[45] Such agents as Lubafax (Burroughs-Wellcome) have been added and are still being added in some cases. Antibiotics not intended for systemic use have been added to cystographic agents for antibacterial coverage but present a hazard because these agents can be absorbed into the circulation. Renal failure and death have been reported after mixture of neomycin with contrast material for a retrograde study.[46] Contrast medium was seen to extravasate from the urethra into the venous system in 6 of 113 urethrograms, leading McClennan to conclude that anything that cannot be utilized safely intravenously should not be used in retrograde urethral studies because these materials can enter the bloodstream.[6]

Techniques

Retrograde urethrography can be accomplished in a variety of ways. Contrast material can be administered into the distal urethra using a vacuum suction cannula, various clamps (e.g., Brodny) around the distal penis, or either a Foley catheter inflated or a catheter-tip syringe inserted into the fossa navicularis. The Foley catheter method enjoys the greatest use. Contrast agents are administered via hand injections or infusion pumps, and imaging is accomplished by either overhead films or fluoroscopic observation and recording by videotape and camera spot films. If the catheter is already in place, a feeding tube can be inserted alongside and contrast medium injected to rule out extravasation or the presence of a fistula.

When examinations are performed using fluoroscopic control, the need for thickening agents markedly diminishes. Special problems present with retrograde studies in small infants; small end-hole catheters, such as 3.5 or 5 French feeding tubes, are used with these patients. Collodion on the tip of the penis helps control leakage of contrast material. Retrograde flush techniques occasionally are needed. Double-balloon techniques (one at the bladder neck and another at the external meatus) for demonstrating urethral diverticula in the lumen are rarely employed.

Histologic Effects

The histologic effects seen in the bladder have been previously discussed. In addition, Sorensen[47] in 1972 pointed out that the more viscous the contrast media for urethrography, the greater the incidence of local inflammatory response. When anesthetic solutions or contrast media used in urethrography are applied to the cheek pouches of hamsters, they both produce tissue injury and changes in the vessels consisting of vasospasm and thrombosis.[47]

Inflammatory changes in the urethra and the bladder have been reported with retrograde urethrography in experimental animals. Rapid distention of the bladder via a Foley catheter in the penile urethra of dogs with either a lactated Ringer's solution or contrast media resulted in a severe fibrinopurulent urethrocystitis.[48] In another study with dogs, Johnston demonstrated a local hemorrhagic cystitis and a diffuse transmural fibrinonecrotic cystitis in 11 of 14 dogs after retrograde urethrography.[49] Although the etiology is unclear, it is speculated that the inflammatory change resulted from overdistention of the bladder, catheter trauma, and the contrast medium. Jakse investigated the reaction of the periurethral tissue in rabbits to retrograde urography with 30–60% iodine-containing contrast media (sodium meglumine amidotrizoate). The study group included normal rabbits and rabbits with urethral infections and penetrating urethral injuries.[50] Jakse concluded that retrograde urethrography immediately after acute urethral trauma has limited adverse effects on the normal healing process if performed under aseptic conditions using a small volume of contrast material. However, a combination of urethrography and iatrogenic infection resulted in a significant augmentation of the inflammatory response. The dissolution of intracellular membranes and rise in endothelial permeability were shown to increase directly with contrast media concentrations and contact time in studies of the vascular endothelium of bat wings and cheek pouches of hamsters.[50]

The same agents used for intravenous urography and cystourethrography are recommended for retrograde urethrography. Many investigators use concentrations of iodine greater than that used for voiding cystourethrography—namely, 25–30%—although lower concentrations are preferable. The ideal contrast medium for retrograde urethrography fulfills seven criteria listed by Kaufman and Russell in 1956: adequate radiopacity, adequate viscosity to distend both sides of the stricture, nonirritating to the urinary mucosa, miscible in urine and water, harmless if introduced into the circulation, sterile, and inexpensive.[4] Except for increased viscosity, which is probably not necessary, current contrast media used for retrograde urethrography meet these criteria. The addition of thickening agents should be discouraged because of their possible entry into the venous system.

OTHER MEANS OF EXAMINING THE BLADDER AND URETHRA

Other means of examining the bladder and urethra include chain cystourethrography in women for stress incontinence.[51] Radiographs are taken with contrast media in the bladder and a chain resting on the base of the bladder and lining the urethra so that the urethrovesical angles can be measured. Alternatively, filling of a Foley catheter with barium allows visualization of the bladder base plate and ureterovesical angle during straining and standing.

Anatomy of the bladder and urethra during voiding can be delineated by suprapubic or transrectal sonography with urine or saline as the contrast agent.[39] A variety of conditions including posterior urethral valves have been diagnosed by this method.[52]

Vesicoureteral reflux and/or residual urine can be documented by the instillation of 5 ml of Lipiodol placed into the bladder (ascendant Lipiodol techniques). The patient is allowed to void, and 24 hours later an abdominal roentgenogram is taken. If Lipiodol is still present, this indicates residual urine. Reflux also can be documented by the presence of Lipiodol in the kidneys.

REFERENCES

1. Wulff P: Verwendbarkeit der X-Strahlen fur die diagnose der Blasendifformitaten. *Fortschr Roentgenstr* 1905;8:193.

2. Beclere H: Retrecissements de l'uretre. Injection a la gelobarine. *Bull Soc Radiol Med Paris* 1924;12:110–113.

3. Burrows EH: *Urethral Lesions in Infancy and Childhood*. Springfield, IL, Charles C Thomas, 1965.

4. Kaufman JJ, Russell M: Cystourethrography: clinical experience with the newer contrast media. *AJR* 1956;75:884–892.

5. Kjellberg SR, Ericsson NO, Rudke U: *The Lower Urinary Tract in Childhood*. Chicago, Year Book Publishers, 1957.

6. McClennan BL, Becker JA, Robinson T: Venous extravasation at retrograde urethrography: precautions. *J Urol* 1971;106:412–413.

7. Emmett JL,Witten DM: *Clinical Urography: An Atlas and Textbook of Roentgenologic Diagnosis*. Philadelphia, WB Saunders, 1971.

8. Grossman H, Merten DF, Effmann EL, et al: Isotonic water soluble contrast material for cystourethrography. *J Urol* 1982;128:1006–1008.

9. McAlister WH, Griffith RC: Cystographic contrast media: clinical and experimental studies. *AJR* 1983;141:997–1001.

10. Siu CM, Dunbar JS, Wright VJ, et al: Contrast media used in cystourethrography: experimental evaluation. *Invest Urol* 1975;12:434–441.

11. Doyle FH: Cystography in bladder tumours: a technique using "steripaque" and carbon dioxide. *Br J Radiol* 1961;34:205–215.

12. Doyle FH: Bladder cancer, double contrast cystography and a bladder analogue. *Br J Radiol* 1963;36:306–318.

13. Lang EK: Double contrast gas barium cystography in the assessment of diverticula of the bladder. *AJR* 1969;107:769–775.

14. Shawdon HH, Doyle FH, Shackman R: Double contrast cystography applied to the diagnosis of tumours in bladder diverticula. *Br J Urol* 1965;37:536–544.

15. Wong W, Saito T, Ogawa H: Radiologic detection of prostatic carcinoma by double contrast retrograde urethrocystography. *J Urol* 1975;114:746–751.

16. Brodeur AE, Goyer RA, Melick W: A potential hazard of barium cystography. *Radiology* 1965;85:1080–1084.

17. Lorenzo RL, Bradford BF: A helpful technique for difficult urethral catheterizations in boys for voiding cystourethrography. *Radiology* 1980;135:515.

18. Simon G, Berdon WE: Suprapubic bladder puncture for voiding cystourethrography. *J Pediat* 1972;81:555–558.

19. Shafik A: Anal cystography: new technique of cystography. *Urology* 1984;23:313–316.

20. Talarico RD, Rajendraprasad CP, Lavengood RW, et al: Cystourethrography. Reversible and irreversible damage to bladder. *NY State J Med* 1979;20:2080–2082.

21. McClennan BL, Oertel YC, Malmgren RA, et al: The effect of water soluble contrast material on urine cytology. *Acta Cytol* 1978;22:230–233.

22. Fischer S, Nielsen MH, Werdelin O, et al: Iodine-containing radiographic contrast media in rat bladder mucosa during excretory urography. A possible cause of exfoliation of abnormal epithelial cells. *Acta Cytol* 1982;20:537–541.

23. Shopfner CE: Clinical evaluation of cystourethrographic contrast media. *Radiology* 1967;88:491–497.

24. McAlister WH, Shackelford GD, Kissane J: The histologic effects of 30% Cystokon, Hypaque 25%, and Renografin-30 in the bladder. *Radiology* 1972;104:563–565.

25. McAlister WH, Shackelford GD, Kissane J: The histologic effects of some iodine-containing contrast media on the rat peritoneal cavity. *Radiology* 1972;105:581–582.

26. McAlister WH, Palmer K: The histologic effects of four commonly used media for excretory urography and an attempt to modify the responses. *Radiology* 1971;99:511–516.

27. Dure-Smith P, Rosen R, Stern A, et al: Physiology of the excretory urogram. A densitometric and subjective assessment of changes in contrast medium concentration. *Invest Radiol* 1974;9:104–108.

28. Currarino G, Weinberg A, Putnam R: Resorption of contrast material from the bladder during cystourethrography causing an excretory urogram. *Radiology* 1977;123:149–150.

29. McAlister WH, Siegel MJ: Complications of diagnostic radiology, in Kassner EG (ed): *Iatrogenic Disorders of the Fetus, Infant, and Child.* New York, Springer-Verlag, 1985, pp 1–38.

30. Glynn B, Gordon IRS: The risk of infection of the urinary tract as a result of micturating cystourethrography in children. *Ann Radiol* 1970;13:283–287.

31. Heptinstall RH: Experimental pyelonephritis: ascending infection of the rat kidney by organisms residing in the urethra. *Brit J Exp Path* 1964;45:436–441.

32. McAlister WH, Cacciarelli A, Shackelford GD: Complications associated with cystography in children. *Radiology* 1974;111:167–172.

33. Thomsen HS, Talner LB, Higgins CB: Intrarenal backflow during retrograde pyelography with graded intrapelvic pressure: a radiologic study. *Invest Radiol* 1982;17:593–603.

34. Markle BM, Catena L: Bladder pseudomass following cystography-related bladder trauma. *Radiology* 1986;159:265.

35. Matsumoto AH, Clark RL, Cuttino JT, Jr: Bladder mucosal tears during voiding cystourethrography in chronic renal failure. *Urol Radiol* 1986;8:81–84.

36. Higgins WL, Mace AH: Puncture of a nondeflatable Foley balloon using ultrasound guidance. *Radiology* 1984;151:801.

37. Harris VJ, Ramilo J: Guide wire manipulation of a knot in a catheter used for cystourethrography. *J Urol* 1976;116:529.

38. Fleischman S, Shah P: Autonomic dysreflexia: an unusual radiologic complication. *Radiology* 1977;124:695–697.

39. Perkash I, Friedland GW: Catheter-induced hyperreflexia in spinal cord injury patients: diagnosis by sonographic voiding cystourethrography. *Radiology* 1986;159:453–455.

40. Cunningham JH, Jr: The diagnosis of urethral structure by the roentgen rays. *Trans Am Assoc Genitourinary Surgeons* 1910;5:369–371.

41. McCallum RW, Colapinto V: *Urological Radiology of the Adult Male Lower Urinary Tract.* Springfield, IL, Charles C Thomas, 1976.

42. Crabtree EG: Venous invasion due to urethrograms made with Lipiodol. *J Urol* 1947; 57:380–389.

43. Ulm AH, Wagshul EC: Pulmonary embolization following urethrography with an oily medium. *Med Intell* 1960;263:137–139.

44. Gaudin H: Fatal embolism following urography. *J Urol* 1949;62:375–377.

45. Denman J: Thixokon: an improved urethrographic medium. *J Urol* 1957;78:93–96.

46. Schmidlapp CJ: Respiratory cardiac arrest after retrograde pyelography with Neomycin-containing medium. *J Urol* 1961;85:993–994.

47. Sorensen SE: Local toxic effects of anaesthetics and contrast media in urethrography. *Acta Radiol Diagn* 1972;12:225–240.

48. Barsanti JA, Crowell W, Losonsky J, et al: Complications of bladder distention during retrograde urethrography. *Am J Vet Res* 1981;42:819–821.

49. Johnston GR, Stevens JB, Jessen CR, et al: Complications of retrograde contrast urethrography in dogs and cats. *Am J Vet Res* 1983;44:1248–1256.

50. Jakse G, Marberger M, Simonis HJ, et al: Urethrography in urethral trauma: tissue reaction to extravasation of contrast dye and iatrogenic infection. *Eur Urol* 1981;7:178–183.

51. Rubesin SE, Pollack HM, Banner MP: Simplified chain cystourethrography. *Radiology* 1982;145:199–200.

52. McAlister WH: Demonstration of the dilated prostatic urethra in posterior urethral valve patients. *J Ultrasound Med* 1984;3:189–190.

Chapter 18

Hysterosalpingography

Heun Y. Yune

HISTORY AND TECHNIQUE

Silver salts and noniodine halogen compounds were the earliest contrast materials used in the developing stage of hysterosalpingography.

In 1910, 15 years after the discovery of x-rays by Roentgen, Rindfleisch injected an emulsion of bismuth into the uterine cavity in a suspected case of extrauterine pregnancy.[1] In 1914, Rubin[2] and Cary[3] separately demonstrated the feasibility of radiographic investigation of tubal patency and uterine neoplasms using Collargol as the contrast agent. In the same year Stiassny reported on the use of silver ointment for the same purpose.[4] In the following years, other agents containing thorium, bromine, and barium were tried and were considered to be of a quality inferior to Collargol.

Lipiodol was initially prepared by Lafay[5] in 1902 as intramuscularly injected depot for prolonged iodine treatment in certain diseases. The incidental discovery that Lipiodol was a satisfactory contrast agent opened a whole new area of its use in contrast radiography. In 1921 Sicard and Forestier reported their experience in the use of Lipiodol as a contrast agent in 5,000 cases.[6] They described the use of this contrast agent in hysterosalpingography 5 years later.[7] Heuser,[8] a year earlier in 1925, was the first investigator to utilize Lipiodol for hysterosalpingography.

The possible harmful effect of oily foreign material retained in body cavities was a concern to many investigators. This led to the trial use of aqueous iodinated contrast agents in the late 1920s and early 1930s in hysterosalpingography. However, they were not accepted as satisfactory because of their rapid disappearance from the uterus and fallopian tubes.

The hysterosalpingographic technique was influenced by the physical and biologic characteristics of available contrast media and influenced the search for new contrast agents that would be compatible with new and improved radiographic equipment.[9–12]

Initially, the primitive fluoroscopic equipment and the poor image quality of spot films available necessitated the use of overhead radiographs and nonabsorbable contrast media. For the evaluation of fallopian tube patency, 24-hour delayed filming was routinely required to observe spillage of injected contrast material into the peritoneal cavity. Therefore, it was difficult to determine the proper contrast dosage at the time of injection. Standard equipment included a long nozzle syringe or an intrauterine cannula; a sealing device, such as an acorn rubber tip to obliterate the cervical os around the cannula; and a cervical tenaculum to provide countertraction to maintain this seal during the injection of contrast agents. The introduction of the Malmström vacuum cannula[13] enabled visualization of the cervical canal, which was concealed by the intrauterine cannula previously.

The original technique of hysterosalpingography remained basically unchanged for nearly 30 years, during which time Lipiodol remained the contrast material of choice. Introduction of Ethiodol facilitated the injection procedure because of its lower viscosity. The development of the fluoroscopic image intensifier in the late 1950s greatly modified the technical aspects of hysterosalpingography.[12,14] The use of image intensification-television fluoroscopic monitoring enables the contrast agent injection to be terminated when a satisfactory volume is injected. Using fluoroscopy of a tightly coned-down area to monitor the flow of the contrast medium also eliminates unnecessary radiation exposure to other tissues and organs, especially when the uterus is in an abnormal position. The superior image quality of modern-day fluoroscopic spot films obviates the need for standard overhead radiographs. An aqueous contrast material is entirely compatible with current technique, and the entire examination takes only 20 to 30 minutes to perform.

The author's technique has been in use in the past 20 years with little modification and has been found to be least traumatic and essentially pain free. Readers are invited to review the details of this technique described in the author's previous report[15] and in the first edition of this book.[16] The minimal modification in this technique is the liberal use of glucagon (0.5 mg, IV), unless there is a contraindication, immediately before applying the vacuum cervical cup. Pelvic cramps and tubal spasms are rare observations since this modification was implemented.

CONTRAST AGENTS

The ideal contrast agent for hysterosalpingography would (1) have an adequate radiopacity, (2) have proper viscosity, (3) be readily absorbable from mucosal or serosal surfaces, and (4) be nontoxic and harmless.[17–23]

Oily Contrast Agents

Lipiodol provided excellent radiopacity. However, a small endometrial lesion could be easily overshadowed by the high degree of radiopacity of this contrast agent when the uterus was fully distended. Monitoring progression of the contrast material's flow required frequent radiographs obtained at various stages of contrast injection. Therefore, the contrast material had to be viscous. High viscosity required high injection pressure. Absorption of an oily contrast agent through normal serosa takes 3 to 6 months or longer. When the serosa is diseased or when a portion of the contrast material is retained in an obstructed fallopian tube, the medium can remain encapsulated for as long as 25 years. Although Lipiodol is remarkably nontoxic to various tissues and organs when large amounts have spilled into the peritoneal cavity, widespread oil granulomas, retention cysts, and adhesions were encountered.[24–26] Particularly, if free iodine were liberated from the organic oily compound, it could act as a chemical irritant to the tissues, which would compound the foreign body reaction. Yet, hypersensitivity or allergic reaction was rarely reported with the use of Lipiodol. The majority of reported fatalities in hysterosalpingography were related to the pulmonary oil embolism associated with intravasation of an oily contrast material into the uterine venous plexus during injection.[27–36]

Ethiodol is closely related to Lipiodol. Iodine is bonded to the unsaturated fatty acids of the vegetable poppy seed oil. The glyceryl ester of these fatty acids, which is the original Lipiodol, has a high viscosity. Ethiodol is produced when the glycerol group is replaced by an ethyl group. Its viscosity is lowered to approximately 1/40 to 1/20 that of Lipiodol (at 15°C). Injection of this contrast agent is achieved with a much lower pressure than with Lipiodol and, therefore, is more compatible with the modern fluoroscopic technique of hysterosalpingography. However, all potential risks associated with the use of oily contrast media are still present with its use.

Aqueous Contrast Agents

Compared with the oily contrast agents, the aqueous agents have distinct advantages: (1) their radiopacity can be modified according to their volume and concentration; (2) their low viscosity makes it possible to use very little pressure for injection; (3) they are readily absorbed through serosa or mucosa, usually within 1 hour; and (4) foreign body reaction and embolism associated with the use of oily contrast media are eliminated. In the early days, aqueous contrast agents were considered unsatisfactory because of their fast diffusion and absorption and resultant lack of adequate radiopacity to outline anatomic details. In general, they were also thought to be more prone to elicit abdominal pain, presumably from peritoneal irritation.

A strong interest in aqueous contrast agents was prompted by the introduction of the image-intensifier fluoroscopy technique. High quality spot radiographs can be obtained at a viscosity level of 1/50 or less of Lipiodol. In addition, fluoroscopic monitoring enables the examiner to observe the physiologic (and pathophysiologic) dynamic changes that were not apparent with conventional radiographic techniques. In fact, high radiopacity, which was previously preferred, actually interferes with the demonstration of more detailed anatomic structures, such as the mucosal plicae in the ampullary segments of the fallopian tubes. This evolution in hysterosalpingography technique has a close parallel to that of myelography. Today, most investigators use aqueous contrast agents exclusively because of their many advantages and reduced danger and condemn the use of oily contrast agents. Currently, there are several aqueous contrast agents available for hysterosalpingography.

All aqueous contrast agents have a tendency to crystallize and freeze the plunger of the syringe. They are best prewarmed in a water bath to body temperature before opening the vial cap. It is the author's impression that the combination of the contrast agent warmed to body temperature and a slow but continuous injection under low pressure is associated with a lower incidence of visible tubal and uterine contractions. Glucagon premedication usually eliminates these contractions temporarily.

The average dosage required for a complete examination is somewhere around 5 to 6 cc. In a small, nulliparous uterus, no more than 3 cc is required. However, a large, bulky uterus, large hydrosalpinx, or a blind space around the fallopian tube due to peritubal adhesions can accommodate more than 10 cc of contrast agent in itself and may require a total of 15 cc or more.

In the past the injection pressure was thought to be of great significance both for diagnostic purposes and for the prevention of complications.[37] An injection pressure in excess of 200 mm Hg was considered unnecessary and dangerous. This was based on the observation that an occluded fallopian tube ruptured when carbon dioxide was injected at this pressure. Today, using aqueous contrast agents, such a high pressure is entirely unnecessary. Even in the presence of complete obstruction of the fallopian tubes, an injection pressure far below 100 mm Hg is quite sufficient to demonstrate fully the exact location and nature of such an obstruction.

Sinografin

Sinografin was introduced in the late 1950s after the recognition that earlier viscous aqueous contrast agents used for hysterosalpingography were still associated with excessive abdominal pain.[23] Sinografin is an aqueous solution combining meglumine diatrizoate equivalent to 40% diatrizoic acid and meglumine iodipamide equivalent to 20% iodipamide. The iodine concentration is approximately 38 g/100 ml. Its viscosity is approximately 29 centipoises at 25°C. If the vial is not opened, Sinografin is quite stable. Its chemical composition can be altered, however, by prolonged exposure to air and also by direct exposure to strong light.

Other Aqueous Contrast Material

During the last several years we have used other standard, non-viscosity-enhanced aqueous contrast material, such as Renografin, Conray, or Hypaque in 60% preparation. It was our observation that the pain and cramps associated with hysterosalpingography may have a dual etiology; the first phase is elicited by stimulation of the cervix upon application of the vacuum cup cervical adaptor and the second phase upon injection of contrast material, which may trigger tubal spasms and irritation upon spillage into the peritoneal cavity. This was the main reason why we have elected to use glucagon premedication liberally. We arbitrarily chose the glucagon dosage to be half of the amount used for double-contrast barium enema. This has effectively eliminated severe functional spasms and greatly alleviated cramps. Another significant motivating factor in changing to ordinary aqueous contrast material was the economy of the procedure. These agents are significantly less expensive than those developed specifically for hysterosalpingography.

We have used these standard contrast media: meglumine salt of diatrizoic acid alone (Hypaque 60) or combined with sodium salt of diatrizoate (Renografin 60) or meglumine salt of iothalamate (Conray), which makes up to 60% weight/volume of the agent. They are hypertonic and thus elicit a shift of fluid through the mucosal or serosal surfaces. Their viscosity is low (approximately 6 centipoises at 25°C), which can be lowered further by warming to body temperature (4 cps at 37°C). Excellent mucosal details of the endocervical canal, uterine cavity, and the fallopian tubes are demonstrated on spot films obtained with a state-of-the-art fluoroscopic unit using these contrast agents.

The nonionic or low osmolality aqueous contrast agents recently introduced into the market have been tried in hysterosalpingography.[38] The author has compared the peritoneal reactive changes associated with these new aqueous contrast agents to those associated with the conventional high osmolality, ionic contrast agents (data to be published). No significant difference was found in histopathologic changes in these two groups. The approximately 20:1 price ratio difference between these two groups without any significant medical advantage does not justify routine use of these new agents in hysterosalpingography.

THERAPEUTIC EFFECTS

Although the exact mechanism is not known, the therapeutic effect of hysterosalpingography in the infertile patient group is the subject of considerable clinical interest.[39–41] Several mechanisms, either singularly or in combination, are considered to be potential causes of reversion to fertility. They include (1) bactericidal effect of the injected contrast media; (2) lubricating effect, especially of the oily contrast media; (3) chemical stimulation of the tubal mucosa; (4) mechanical stimulus to encourage tubal peristalsis; and (5) dilation of the tube where it is incompletely obstructed or removal of inspissated material, such as a mucous plug or blood clot.[42] Some of these mechanisms are experimentally and clinically proved and others are hypothetical. Whatever the real mechanism, the therapeutic effect is primarily limited to lesions of the fallopian tubes.[43] In the days of Lipiodol, the therapeutic effect was frequently attributed to the chemical and physical nature of the oil in the contrast agent,[44,45] but the same therapeutic effect is observed by those investigators who use aqueous contrast agents exclusively and, therefore, it is felt that the oil is not the only or the major reason. The therapeutic effect of hysterosalpingography cannot be claimed too liberally, as spontaneous reversion to fertility without any therapeutic efforts is estimated to be approximately 20%.[46–48] If, however, some degree of bilateral tubal obstruction was proved to be present before or at the time of hysterosalpingography and if the patient achieves conception within a month or two after hysterosalpingography, it is strong supportive evidence of this therapeutic effect.[49–51]

COMPLICATIONS

Complications associated with hysterosalpingography can be grouped into two major categories: (1) complications associated with the use of contrast agents and (2) complications associated with instrument manipulation. Although the first category is the main concern of this chapter, the presence of anatomic injury secondary to instrument manipulation is a significant contributing factor. Hence, the following review of this subject covers both categories.

Pain

Pain is a subjective response to an unpleasant stimulus and has a very wide range of severity and character. There are no scientific, objective scales to measure the degree of pain. Even if the examination was a painless one from the physical sense, if the patient is anxious and defensive, the examination can be quite traumatic and painful. In our experience, there are only a handful of patients who find hysterosalpingography to be totally pain free. Although earlier investigators' experience has suggested that the water-soluble contrast agents irritate the peritoneum when they spill into the pelvic cavity and result in abdominal or pelvic pain and sometimes radiating pain to the shoulders (irritation to the diaphragms),[19–21,52] such was not our experience with the use of current aqueous contrast agents. If, however, one uses a high degree of pressure in injecting the contrast agent, or if there is tubal obstruction and the intrauterine and intratubal pressure is increased, most patients will complain of discomfort, even if the injection is performed slowly. The combination of glucagon, gentle manipulation of the instruments, and a slow rate of injection eliminates or reduces the pain. In this author's experience during the past 20 years, use of an analgesic or tranquilizer was required in less than 3% of the patients examined.

Mechanical Injury

Perforation of the uterine wall or the cervical canal, although usually not life threatening, was not an infrequent complication of hysterosalpingography in the past.[52,53] It was caused by the use of uterine probes and intrauterine cannula in the old examination technique. Forceful dilation of a tight cervical canal or a preexisting lesion on the wall of the cervical canal or uterus was the usual cause of this complication. The current vacuum cervical adaptor technique eliminates the need for probing and dilating the cervical canal.

Bleeding in various amounts was always associated with the old technique because of the use of such instruments as the tenaculum, the cervical probe, and the intrauterine cannula, all of which are associated with physical injury. With the vacuum cervical adaptor technique, as long as the cervical mucosa is not already diseased and the vacuum cervical adaptor is in the proper position, bleeding is not a usual complication. A number of patients, however, show evidence of cervicitis with superficial mucosal erosions that will result in a light ''spotting'' when the vacuum cervical adaptor is released. If the bleeding through the cervical canal is more than ''spotting'' with the use of the vacuum cervical adaptor, it usually represents a bleeding lesion above the cervical canal, rather than a complication of the technique.

Chemical and Physical Reaction

The aqueous contrast agents currently used are extremely safe. Nevertheless, a few hypersensitivity reactions have been observed. In the author's personal experience, there were only three occasions (0.4%) in which an urticarial rash developed with or without itching upon injection of the contrast agent. If a patient is extremely hypersensitive to the contrast agent used, an anaphylactoid reaction may develop. It is advisable to have an emergency cart available in the immediate vicinity of the examination room. For a milder hypersensitive reaction, such as an urticarial rash, antihistamines are the treatment of choice. The complaint of severe pelvic cramps may be followed by vasovagal reflex. It is treated successfully with a moderately heavy dose of analgesics (i.e., intravenous Demerol).

Acute and chronic foreign body reactions are reported with all contrast agents. The aqueous contrast material is not exempt, but oily contrast materials are more prone to be associated with foreign body granuloma formation.[24,54,55] A granuloma is more apt to develop if the contrast material remains trapped in a blind space, such as an obstructed fallopian tube, or is encysted in the peritoneal cavity.[25,56] Pre-existing inflammatory conditions of the mucosa or the serosa are suggested as the major factors in developing the granulomatous reactions.[57] The viscosity-enhancing agent added to the old aqueous contrast agents may have played a rolc,[58] but it is also possible that other foreign materials, such as starch or talcum powder on gloves or small lint filaments introduced together with the contrast agent, may be the real source. The incidence of foreign body granuloma formation is extremely low in the hands of those investigators who use the aqueous contrast agents and observe strict aseptic technique.

Because of the use of iodinated contrast agents, if the patient is to undergo thyroid function tests, hysterosalpingography should be performed after these tests.[59]

Contrast Intravasation

Although the incidence is very low (less than 1%), intravasation of contrast material into either the uterine venous plexus or the lymphatics may occur.[28,30,32–34,36,60–62] Contributing factors are injured uterine mucosa before or at the time of hysterosalpingography or the use of excessive pressure for the contrast injection, especially when the uterine mucosa is thin or fragile, such as during the menstrual period, or when there are such lesions of the mucosa or submucosa as polyps, fibroids, endometriosis, or endometritis.[29,63,64] Even when intravasation occurs, readily absorbable aqueous contrast agents will cause no untoward reaction or serious damage.[65] If, however, an oily contrast agent is used and the intravasated quantity is significant, pulmonary oil embolism with cough, dyspnea, chest pain, cyanosis, and shock can develop.[30,31] Most of the reported deaths as a complication of hysterosalpingography were caused by pulmonary embolism by oily contrast media.[66,67] A patient with oil embolism to the retinal artery after hysterosalpingography, associated with a temporary loss of vision, has also been reported.[68]

Infection

Bacteria may be inadvertently introduced into the peritoneal cavity with the contrast agent injection. Hence, performing hysterosalpingography without the proper aseptic precautions, including perineal and vaginal preparation, is malpractice. Strict adherence to the aseptic technique is mandatory. Bacterial peritonitis is a recognized cause of mortality associated with hysterosalpingography.[27,69] If there is evidence of an active infectious process in the external genital region or the vagina, the examination should be postponed.

REFERENCES

1. Rindfleisch W: Darstellung des Cavum uteri. *Berlin Klin Wschr* 1910;47:780–781.

2. Rubin IC: Roentgendiagnostik der Uterustumoren mit Hilfe von intrauterinen Collargol Injektionen. *Zbl Gynaekol* 1914;38:658–660.

3. Cary WH: Note on determination of patency of fallopian tubes by the use of Collargol and the x-ray shadow. *Am J Obstet Dis Wom* 1914;69:462–464.

4. Stiassny S: Über Röntgendiagnostik der Uterustumoren. *Zbl Gynaekol* 1914;38:800.

5. Lafay L: Sur les huiles iodées et bromées. *Bull Gen Ther* 1902;142:631–635.

6. Sicard SA, Forestier J: Méthode radiographique de l'exploration de la cavité epidurale par le Lipiodol. *Rev Neurol* 1921;28:1264–1266.

7. Forestier J: Iodized oil (Lipiodol) in roentgenology. *Am J Roentgenol* 1926;15:352–354.

8. Heuser C: Lipiodol in the diagnosis of pregnancy. *Lancet* 1925;2:1111–1112.

9. Carelli H, Gandulfo R, Ocampo A: L'exploración Radiológica en Ginecologia. *Semana Méd* 1925;32:85–88.

10. Marshak RH, Poole CS, Goldberger MA: Hysterography and hysterosalpingography: an analysis of 2,500 cases with special emphasis on technique and safety of the procedure. *Surg Gynecol Obstet* 1950;91:182–192.

11. Parekh MC, Murthy YS, Arronet GH: Vacuum cervical adaptor. *Obstet Gynecol* 1970; 36:940–943.

12. Rozin S, Schwartz A: Television fluoroscopy in gynecological diagnosis. *Obstet Gynecol* 1965; 26:524–530.

13. Malmström T: A vacuum uterine cannula. *Obstet Gynecol* 1961;18:773–776.

14. Aaro LA, Stewart JR: Hysterosalpingography with image-intensified fluoroscopy. *Am J Obstet Gynecol* 1969;105:1124–1128.

15. Yune HY, Klatte EC, Cleary RE, et al: Hysterosalpingography in infertility. *Am J Roentgenol* 1974;121:642–651.

16. Yune HY: Hysterosalpingography, in Miller RE, Skucas J (eds): *Radiographic Contrast Agents*. Baltimore, University Park Press, pp 307–321.

17. Brown WE, Jennings AE, Bradbury JT: The absorption of radiopaque substances used in hysterosalpingography. *Am J Obstet Gynecol* 1949;58:1041–1053.

18. Freedman HL, Taffen CH, Friedman H, et al: Hypaque as a contrast medium for hysterosalpingography. *Fertil Steril* 1959;10:403–408.

19. Griffiths JL II: A clinical and radiologic evaluation comparing the use of two contrast media in hysterosalpingography. *Br J Radiol* 1969;42:835–837.

20. Rolland M, Carpenter F, Rich J: A new water-soluble opaque medium in the study of hysterograms and hysterosalpingograms. *Am J Obstet Gynecol* 1953;65:81–87.

21. Rubin IC, Myller E, Hartman CG: Salpix: a new approach to the ideal radiopaque medium for hysterosalpingography. *Fertil Steril* 1953;4:357–370.

22. Thomas HH, Dunn D: Salpix, as a medium in hysterosalpingography. *Fertil Steril* 1956;7:155–165.

23. Whitelaw MJ, Miller EB: New water-soluble medium (Sinografin) for hysterosalpingography. *Fertil Steril* 1959;10:227–239.

24. Lash AF: Lipiodol pelvic cysts. *Surg Gynecol Obstet* 1930;51:55–60.

25. Malter IJ, Fox RM: Prolonged oviduct retention of iodized contrast medium. *Obstet Gynecol* 1972;40:221–224.

26. Weisman A: Incidence of residual intraperitoneal Iodochloral after hysterosalpingography. *Fertil Steril* 1952;3:290–296.

27. Bang J: Complications of hysterosalpingography. *Acta Obstet Gynecol Scand* 1950;29:383–399.

28. Bloomfield A: Six cases of venous intravasation following intrauterine Lipiodol injection. *J Obstet Gynecol Br Emp* 1946;53:345–346.

29. Holt BB, Armstrong JT: Dangers and contraindications to hysterosalpingography. *Texas Med* 1970;66:44–45.

30. Ingersol F, Robbins LI: Oil embolism following hysterosalpingography. *Am J Obstet Gynecol* 1947;53:307–311.

31. Levinson JM: Pulmonary oil embolism following hysterosalpingography. *Fertil Steril* 1963;14:21–27.

32. Measday B: An analysis of the complications of hysterosalpingography. *J Obstet Gynecol Br Emp* 1960;67:663–667.

33. Shapiro JH: Pulmonary oil embolism: a complication of hysterosalpingography. *Am J Roentgenol* 1957;77:1055–1058.

34. Titus P: Oil embolism from hysterosalpingography. *Am J Obstet Gynecol* 1947;53:1067–1068.

35. Weigen JF, Thomas SF: *Complications of Diagnostic Radiology*. Springfield, IL, Charles C Thomas, 1973.

36. Zacharial F: Venous and lymphatic intravasation in hysterosalpingography. *Acta Obstet Gynecol Scand* 1955;34:131–149.

37. Brantley WM, Del Valle RA, Aaby GU, et al: Rupture of a silent pyosalpinx following hysterosalpingogram. *Obstet Gynecol* 1960;16:483–485.

38. Winfield AC, Maxson WS, Harding DR, et al.: Hexabrix as a contrast agent for hysterosalpingography. *Invest Radiol* 1984;19:S389–390.

39. Feldman HJ: Hysterosalpingogram for therapy with infertility. *Int J Fertil* 1960;5:289.

40. Rutherford RN: The therapeutic value of repetitive Lipiodol tubal insufflations. *West J Surg* 1948;56:145–154.

41. Serjeant B: Hysterosalpingography as a diagnostic and therapeutic procedure. *Med J Austr* 1957;1:103–105.

42. Mackey RA, Glass RH, Olson LE, et al: Pregnancy following hysterosalpingography with oil and water soluble dye. *Fertil Steril* 1971;22:504–507.

43. Vogt CJ: Hysterosalpingography: a safe diagnostic therapeutic gynecological procedure. *Am J Obstet Gynecol* 1954;67:1298–1306.

44. Gillespie HW: Therapeutic aspect of hysterosalpingography. *Br J Radiol* 1965;38:301–302.

45. Palmer A: Ethiodol hysterosalpingography for the treatment of infertility. *Fertil Steril* 1960;11:311–315.

46. Buxton L, Southam A: Critical survey of present methods of diagnosis and therapy in human infertility. *Am J Obstet Gynecol* 1955;70:741–752.

47. Mendizabal AF, Usubiaga I: Spontaneous pregnancies in sterile patients. *Int J Fertil* 1963; 8:513–516.

48. Weir WC, Weir DR: Natural history of infertility. *Fertil Steril* 1961;12:443–451.

49. Matters RF: Transuterine injection of Lipiodol in the treatment of sterility. *Med J Austr* 1948; 2:740–742.

50. Wahby O, Sobrero AJ, Epstein JA: Hysterosalpingography in relation to pregnancy and its outcome in infertile women. *Fertil Steril* 1966;17:520–530.

51. Weir WC, Weir DR: Therapeutic value of salpingograms in infertility. *Fertil Steril* 1951; 2:514–522.

52. Siegler AL: *Hysterosalpingography*. New York, Harper and Row, 1967.

53. Ottow B: Uber violente Cervix perforation bei der Hysterosalpingographie. *Zbl Gynaekol* 1936; 60:1154–1162.

54. Aaron JB, Levine W: Endometrial oil granuloma following hysterosalpingography. *Am J Obstet Gynecol* 1954;68:1594–1597.

55. Campbell JS, Nigram S, Hurtig A, et al: Mineral oil granulomas of the uterus and parametrium and granulomatous salpingitis with Schaumann bodies and oxalate deposits. *Fertil Steril* 1964;15:278–289.

56. Rubin IC: Lipoid granuloma in fallopian tubes localized by intrauterine Diodrast injection with special reference to the value of follow-up x-ray films. *Radiology* 1939;33:350–353.

57. Jones GES, Woodruff JD: Effect of a radiopaque water-soluble medium on the histopathology of the endometrium. *Am J Obstet Gynecol* 1960;80:337–340.

58. Bergman F, Gorton G, Norman O, et al: Foreign body granulomas following hysterosalpingography with a contrast medium containing carboxymethylcellulose. *Acta Radiol* 1955; 43:17–29.

59. Slater S, Paz-Carranza JB, Solomons E, et al: Effect of hysterosalpingography on assay of thyroid function. *Fertil Steril* 1959;10:144–149.

60. Hipona FA, Ditchek T: Uterine lymphogram following hysterosalpingography. *Am J Roentgenol* 1966;98:236–238.

61. Pujol y Brull A, Vanrell J, Carulla-Riera V: L'injection Accidentelle du systeme veineux uteroovarien au cours de l'hysterographie sur le vivant. *J Radiol Electr* 1929;13:38–44.

62. Rozin S: *Uterosalpingography in Gynecology*. Springfield, IL, Charles C Thomas, 1965.

63. Drukman A, Rozin S: Uterovenous and uterolymphatic intravasation in hysterosalpingography. *J Obstet Gynecol Br Emp* 1951;58:73–78.

64. Sampson JA: The escape of foreign material from the uterine cavity into the uterine veins. *Trans Am Gynecol Soc* 1918;43:16–23.

65. Avnet NL, Elkin M: Hysterosalpingography. *Radiol Clin N Am* 1967;5:105–120.

66. Farris AM, McMurry A: Uterosalpingography: report of a fatality. *Texas State Med J* 1947; 42:592–597.

67. Gajzago E: Ein im Anschluss an Hysterographie durch Oelembolie verusachter Todesfall. *Zbl Gynaekol* 1931;55:543–544.

68. Charawanamuttu AM, Hughes-Nurse J, Hamlett JD: Retinal embolism after hysterosalpingography. *Br J Ophthalmol* 1973;57:166–169.

69. Nielsen PH: Injuries caused by hysterosalpingography. *Acta Obstet Gynecol Scand* 1946; 26:565–597.

Chapter 19

Seminal Vesiculography

Heun Y. Yune

TECHNIQUE

The technique of vasoseminal vesiculography evolved from urologists' therapeutic attempts to correct inflammatory conditions in the seminal vesicle by injecting an antiseptic agent through the exposed vas deferens. The same technique was used for contrast injections in radiography.[1,2] Before the turn of the century, the method of transurethral catheterization of the ejaculatory duct was introduced, and in the decade that followed the first published reports of transvasoseminal vesiculography, retrograde transurethral vesiculography was also popularized. Initially colloidal silver was the contrast material used, followed by thorium, and iodinated oil was introduced as late as the 1950s.[3,4] Today most investigators favor the use of aqueous contrast materials.

Vasoseminal Route

Under local anesthesia, a skin incision is made and the spermatic cord is exposed for about 2 cm of its length in the scrotum approximately 4 cm distal to the external inguinal ring. The vas is identified on palpation as a hard strand of tissue approximately 2 mm in diameter. If the patient is to have a prophylactic vasectomy as, for example, in carcinoma of the prostate or tuberculous epididymitis, a thin polyethylene catheter may be introduced through an incision on the vas. However, because the majority of patients undergo seminal vesiculography today for evaluation of male infertility with azoospermia or oligospermia, vasopuncture rather than vasotomy is preferred. A short, beveled, 20-gauge or smaller caliber needle is introduced into the lumen of the vas, with the needle pointing toward the seminal vesicle. Before injection of the contrast agent, three to four injections of approximately 2–3 cc of warm saline solution are made to irrigate the tract and free inspissated viscous material in the ampulla and the seminal vesicle. After irrigation, 5 cc of 50 to 60% solution of a triiodinated aqueous contrast material is injected slowly through the same needle. The injection is carried out simultaneously on both sides. When approximately 3 to 4 cc is injected per side, a radiograph of the pelvis (centered on the top of the symphysis pubis) is obtained. If the radiograph obtained is satisfactory, the needle is withdrawn and the skin incision closed.

Through the same route, epididymography can also be performed.[5–8] A slightly smaller gauge needle is preferred, and the needle is directed toward the testis. A smaller quantity of contrast material is needed (1 to 1.5 cc), but of a slightly higher concentration (60 to 70% solution). Because the main objective of seminal vesiculography in the work-up of male infertility is to demonstrate the possible presence of an obstruction in the vas deferens, epididymography is performed when vasoseminal vesiculography demonstrates no such obstruction.[9] If there is an obstruction in the vas deferens, the contrast agent injection will encounter resistance, and leakage around the needle will be observed. In such an instance, a radiograph should be obtained before the full dose of contrast material is injected.

Observations for presence or absence of obstruction, contrast material leakage, and proper needle tip position are facilitated by fluoroscopy and spot film recording. The contrast material injection can be directly controlled by continuous fluoroscopic monitoring by utilizing a specially designed injection assembly that combines a long, flexible tubing and a small-gauge needle, such as a lymphangiogram set.

Transurethral Route

This is not a popular technique for various reasons. The identification of the ejaculatory duct through the urethra requires considerable experience and the use of a special instrument (McCarthy panendoscope with foroblique telescope).[10] The ejaculatory duct is very fragile and can be easily traumatized by the catheter (a 3 to 4 French urethral catheter is used). Because the retrograde injection through the ejaculatory duct fills the seminal vesicle preferentially, the opacification of the vas deferens in its entirety is not as easily accomplished as through the vasoseminal route.[11,12] Often the openings of the ejaculatory ducts cannot be seen clearly through the urethroscope, especially when the prostate is hypertrophied and the catheterization is performed blindly.[10,13–15]

CONTRAST AGENTS

Any water-soluble urographic contrast agent may be used. The most commonly used are meglumine and/or sodium diatrizoate or iothalamate solutions, such as

Hypaque, Renografin, Renovist, Urografin, Conray, etc., in 50 to 60% concentrations. If, during the injection, no resistance is encountered, it is recommended that the film exposure be made while the injection is progressing. Both the contrast injection and the filming can be accomplished simultaneously in one stage using the specially designed injection assembly and fluoroscopic monitoring as described above. With the availability of new, nonionic, low osmolality aqueous contrast materials that could conceivably reduce the incidence of contrast-material-induced local tissue irritation (see the section on complications below) it may be a prudent practice to use one of these new agents in cases of male infertility work-up. However, as yet, there is no scientific basis to support their use. In an animal study that the author has undertaken recently, the inflammatory reactions caused by the nonionic, low osmolality contrast materials when deposited into the peritoneal cavity were not significantly different from those caused by conventional high osmolality ionic agents (data to be published). Whether the findings will be the same when the contrast agents are administered in the vas to the seminal vesicle, to the epididymis, or in the interstices surrounding the vas is not yet known and requires investigation.

COMPLICATIONS

In addition to possible hypersensitivity reactions to the contrast agent used, a transitory, local tissue irritation may be encountered. This usually does not result in permanent tissue damage unless the contrast injection was made into the wall, rather than into the lumen of the vas deferens. The injection of the contrast agent should be performed with very little pressure, and if any resistance is encountered during the injection of the first 0.5 to 1 cc, either a fluoroscopic investigation or a film exposure of the injection site should be obtained. Direct fluoroscopic monitoring of the contrast material injection reduces the incidence of high volume contrast material injection into the wall. At the completion of injection, another 2 to 3 cc of warm saline irrigation is advisable to eliminate local tissue reaction.

If a vasotomy is performed, there is a risk of stricture development when the incision heals. Hence, injection through the vasoseminal route should not be made through a vasotomy in an infertility work-up.

Some authors advocate adding a 1% concentration of an aqueous solution of neomycin to the contrast agent to prevent infection. However, strict adherence to aseptic surgical technique obviates the need for this practice.

REFERENCES

1. Belfield WT: Skiagraphy of the seminal ducts. *JAMA* 1913;60:800–801.
2. Belfield WT: Vasotomy-radiography of the seminal duct. *JAMA* 1913;61:1867–1869.

3. Pereira A: Roentgen interpretation of vesiculograms. *Am J Roentgenol* 1953;69:361–379.

4. Young HH, Waters CA: X-ray studies of the seminal vesicles and vasa deferentia after urethroscopic injection of ejaculatory ducts with thorium: a new diagnostic method. *Am J Roentgenol* 1920;7:16–22.

5. Abeshouse BS, Heller E, Salik JO: Vasoepididymography and vasoseminal vesiculography. *J Urol* 1954;72:983–991.

6. Boreau J: L'epididymographie. *J Urol Nephrol (Paris)* 1953;59:416–423.

7. Boreau J, Elbim A, Hermann P, et al: L'epididymographie. *Presse Med* 1951;59:1406–1407.

8. Golji H: Clinical value of epididymo-vesiculography. *J Urol* 1957;78:445–455.

9. Emmett JL, Witten DM: *Clinical Urography*. Philadelphia, WB Saunders, 1971.

10. McCarthy JF, Ritter JS, Klemperer P: Anatomical and histological study of the verumontanum with special reference to the ejaculatory ducts. *J Urol* 1927;17:1–16.

11. McMahon S: An anatomical study by injection technique of the ejaculatory ducts and their relations. *J Urol* 1938;39:442–443.

12. Peterson AP: Retrograde catheterization in diagnosis and treatment of seminal vesiculitis. *J Urol* 1938;39:662–677.

13. Herbst RH, Merricks JW: Visualization and treatment of seminal vesiculitis by catheterization and dilatation of ejaculatory ducts. *J Urol* 1939;41:733–750.

14. Herbst RH, Merricks JW: Transurethral approach to the diagnosis and treatment of the seminal vesicles. *Ill Med J* 1940;78:393–396.

15. Merricks JW: The modern conception of the diagnosis and treatment of infections of the seminal vesicles: with roentgenographic visualization of these organs by catheterization of the ejaculatory ducts. *New Int Clin* 1940;2:193.

Part IV

Hepatobiliary Agents

Chapter 20

Pharmacokinetics

Francis A. Burgener

In contrast to direct cholangiography, where a contrast agent is injected directly into the biliary system (e.g., intraoperative and percutaneous transhepatic cholangiography or endoscopic retrograde cholangiopancreatography), indirect visualization of the biliary system is achieved by radiopaque compounds excreted into the bile. All agents currently used for that purpose are derivatives of triiodobenzoic acid (Fig. 20-1). The iodine in these compounds provides the opacification, and the remaining carbon, hydrogen, oxygen, and nitrogen atoms serve as a framework or carrier for the iodine atoms.

The structural arrangements of the atoms is important in providing stability, nontoxicity, and concentration in various organs. A high degree of stability of the iodine atoms in the molecule is achieved by attaching them to the aromatic nucleus, thereby preventing the release of relative toxic iodide by chemical breakdown or metabolic attack. With conventional radiography, minimum iodine bile concentrations of ⅓% and 1% are required for diagnostic visualization of the gallbladder and bile ducts, respectively.[1–3]

In intravenous (IV) cholangiography the contrast agent must be taken up from the blood into the hepatocyte, transported through the hepatocyte from the

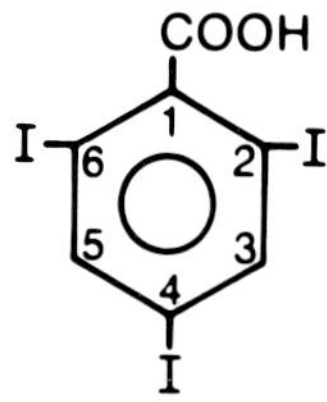

TRIIODOBENZOIC ACID

Figure 20-1 Triiodobenzoic acid. Parent compound of all biliary contrast agents.

sinusoidal to the canalicular membrane, and excreted into the bile in concentrations affording radiographic opacification. Oral cholecystographic agents, in addition, must be absorbed from the intestinal tract into the blood and then concentrated in the gallbladder. They are also largely metabolized in the liver.[3] For this reason the physicochemical properties of oral and IV biliary contrast agents are different as reflected by their different molecular structures in Figs. 20-2 and 20-3.

Biliary contrast agents do not have a prosthetic group at the number 5 position on the benzene ring as do urographic contrast agents. This difference seems to be an important factor in directing these compounds into the bile, rather than urine; at the same time it increases the oil/water distribution coefficient and their overall

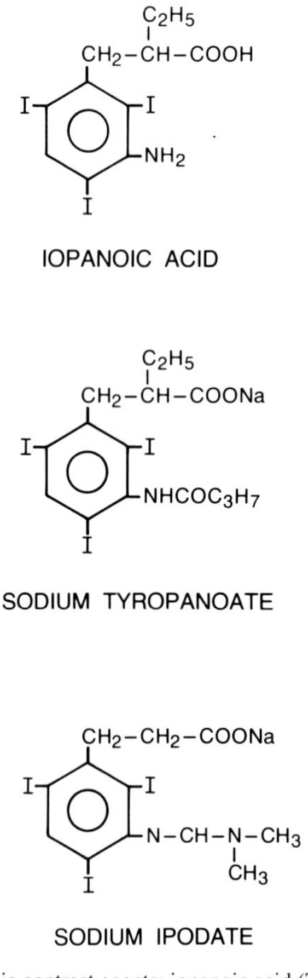

C_2H_5
$CH_2-CH-COOH$
I I
NH_2
I

IOPANOIC ACID

C_2H_5
$CH_2-CH-COONa$
I I
$NHCOC_3H_7$
I

SODIUM TYROPANOATE

$CH_2-CH_2-COONa$
I I
$N-CH-N-CH_3$
CH_3
I

SODIUM IPODATE

Figure 20-2 Oral cholecystographic contrast agents: iopanoic acid (Telepaque), sodium tyropanoate (Bilopaque), and sodium ipodate (Oragrafin).

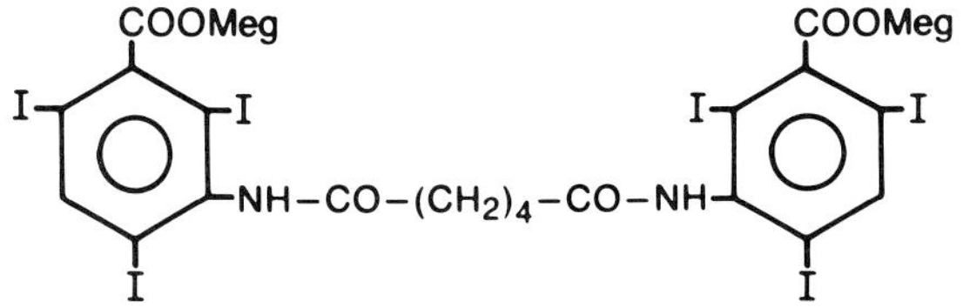

Figure 20-3 Intravenous cholangiographic contrast agents: iodipamide (Cholografin, Biligrafin) and iodoxamate (Cholovue).

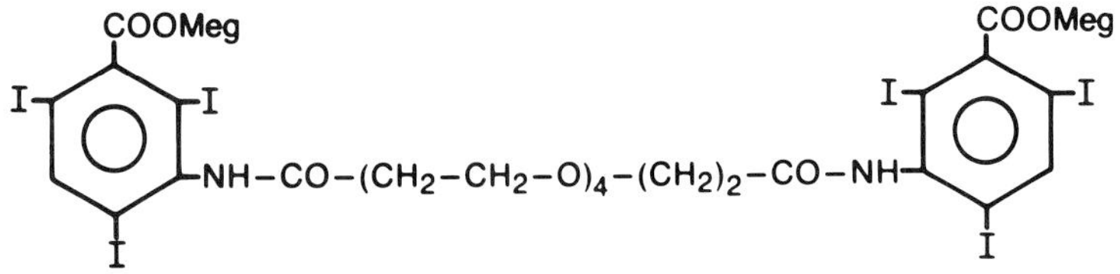

---

toxicity. Toxicity is reduced in urographic agents by the introduction of a suitable solubilizing group in this position on the benzene ring.[4]

Oral cholecystographic contrast agents differ from intravenously administered agents by the introduction of a lipophilic chain at the number 1 position on the benzene ring. This enables orally administered agents to pass through the intestinal wall and enter the portal circulation. In IV cholangiographic agents, two triiodinated benzene rings are linked together by a polymethylene chain of various length. Compared to their monomeric counterparts the hypertonic effect is reduced in these dimers by 25% (six iodine atoms are provided at the cost of three ions [one anion and two cations], versus a cost of four ions [two anions and cations each] in a monomer). The reduced osmolality of dimeric compounds is, however, achieved at the expense of a substantially higher viscosity.

INTESTINAL ABSORPTION

Absorption is dependent upon drug solubility. The solubility is affected by the physicochemical state of the compound, the pH of gastrointestinal (GI) contents, and the presence of bile salts. For those agents given in solid form, the rate of dissolution is usually the limiting factor in their absorption.

Oral cholecystographic agents are weak acids administered in solid form. As with other drugs, they must first dissolve in the aqueous milieu of the intestinal tract before they can be absorbed into the blood. The water solubility of iopanoic acid is poor and is increased for both tyropanoate and ipodate.

The effect of the physicochemical state of iopanoic acid on the rate of its solution and subsequent absorption is evident from clinical investigations showing that the sodium salt of this agent is more rapidly absorbed than the acid itself.[5,6] Sodium iopanoate precipitates in the low pH of the stomach as iopanoic acid. It has been postulated that iopanoic acid produced in this fashion is in an amorphous state with a greater surface area-to-weight ratio than in the crystalline form of the commercially available iopanoic acid.[5,6] Compared to the latter form, the solution rate of freshly precipitated iopanoic acid is significantly higher in the first 6 hours after precipitation.[7] After that time, however, the solution rate of these two preparations is the same.[7] The particles of freshly precipitated iopanoic acid are smaller than in commercial tablets and electron-microscopic examination of the iopanoic acid crystals confirms a higher surface area-to-weight ratio in the freshly precipitated form.[8]

Oral cholecystographic contrast agents are much more soluble in water in the ionized than the nonionized form.[8,9] With a rising pH both ionization and water solubility of these weak acids increase. Therefore, they are considerably better dissolved in the higher pH of the intestine than in the low pH of the stomach where they are largely insoluble and consequently not absorbed. Iopanoic acid, the least water-soluble oral cholecystographic contrast agent, is, in the absence of bile salts even at the relatively high pH of the small intestine, so poorly dissolved that sufficient absorption of the agent required for diagnostic opacification of the gallbladder may not occur.[8] If, however, iopanoic acid is completely dissolved in an alkaline solution and instilled in this form directly into the duodenum of dogs, then rapid absorption of this agent occurs, with 50% of the dose being excreted within 2 hours.[10]

The most important factor affecting solubility of oral contrast agents in the intestinal tract is the presence of bile salts. They are excreted by the liver into the bile by an extremely powerful active transport system and are stored in the gallbladder in the fasting state.[11] When the gallbladder is stimulated to contract after ingestion of a fatty meal, bile salts empty into the duodenum. Subsequently they are absorbed from the terminal ileum, taken up by the liver from the portal blood, and excreted into bile. With synthesis of approximately 600 mg of bile acids per day, the liver is able to maintain a bile salt pool of 2 to 3 g and to secrete into the bile 20 to 30 g/day.[11] Bile salts are capable of solubilizing water-insoluble compounds by their detergent-like ability to form micelles.[12] Micelles are polyanionic macromolecular aggregates formed by special orientation of the involved molecules. Bile salts increase the aqueous solubility of all oral cholecystographic agents. The presence of bile salts is most important for the dissolution and subsequent absorption of iopanoic acid, the least water-soluble oral contrast agent.[7]

As with most drugs, the oral cholecystographic agents are absorbed by passive diffusion across the GI mucosa.[8] The epithelial lining of the GI tract behaves as a simple lipid barrier that is permeable to the lipid-soluble nonionized form of the cholecystographic agent. The driving force for its absorption is the concentration gradient across this lipid barrier separating the intestinal lumen from the blood.[13] An increase in blood flow at this location enhances the absorption rate of drugs, including cholecystographic agents.[13]

With iopanoic acid the nonionized lipid-soluble moiety is approximately 100 times less soluble in the aqueous milieu of the intestinal tract than the ionized moiety.[8,9] The ratio of the nonionized to ionized form of iopanoic acid decreases with an increasing pH. Hence, this contrast agent is almost completely nonionized at the low pH of the stomach. Nevertheless, iopanoic acid is virtually not absorbed from the stomach despite a favorable ratio of the nonionized (lipid-soluble) to ionized (water-soluble) form of 1,000:1 in the gastric juice with a pH of 1.4 (calculated with the Henderson-Hasselbach equation). This apparent paradox is, however, easily explained by the extremely poor solubility of iopanoic acid in the low pH of the stomach. Absorption of the iopanoic acid occurs at the higher pH of the small intestine, where the agent is considerably better dissolved and the solute is largely present in the ionized form that is generally thought to be incapable of penetrating the lipid mucosal barrier. However, more recent experience with other drugs suggests that the ionized moiety may contribute also considerably to the absorption of weak acids, such as oral cholecystographic agents.[13] Estimates of weak electrolyte permeabilities in the small intestine indicate that the permeability of the ionized form is usually three to four orders of magnitude smaller than the nonionized form.[13] Because the concentration of the ionized moiety of a weak acid is several orders of magnitude larger in the small intestine than the corresponding nonionized moiety, the high concentration of the poorly permeant ion may more than offset the better permeability of the nonionized compound. The absorption of weak acids in ionized form seems to bypass the lipid membranes surrounding the epithelial cells, suggesting that the principal route for passive ionic permeation in epithelia is an extracellular channel, probably represented by the tight junction and the lateral intercellular spaces.[13]

DISTRIBUTION VOLUME

Both oral and IV biliary contrast agents have an affinity to a variety of proteins. In the plasma these contrast agents are largely and reversibly bound to albumin, thereby establishing an equilibrium between the bound and unbound fraction.[14] Because only the small unbound fraction of the agent is subject to renal excretion by glomerular filtration, this elimination pathway is delayed when compared to urographic contrast agents that are not bound to albumin.

The distribution volume of urographic and biliary contrast agents is similar and approximates the combined intravascular and interstitial (extracellular) fluid space.[3] This suggests that accumulation of biliary contrast agents in hepatic cells accounts only for a minor fraction of the distribution volume. Furthermore, there is no evidence that biliary contrast agents are concentrated in the hepatocytes above the level of their plasma concentrations.[15] This is also supported by the apparent failure of these agents to improve the contrast enhancement of the liver in computed tomography when compared to urographic agents.[16]

The distribution of contrast agents between the intravascular and interstitial space occurs by diffusion along concentration gradients. However, only the small fraction of the biliary agent that is not bound to albumin can diffuse freely from the intravascular to the interstitial compartment. Under these conditions, distribution volume corresponding approximately to the entire extracellular space is only attainable when the ratio between the unbound and bound agent is similar in both the intravascular and interstitial compartment. This indicates that a large portion of the biliary contrast agent must be bound also to proteins or other substances outside the blood vessels.[3]

HEPATIC UPTAKE

Biliary excretion of a solute requires its uptake from the blood through the sinusoidal membrane into the hepatocyte, its transport from the sinusoidal membrane through the hepatocyte to the canalicular membrane, and its excretion from the hepatocyte through the canalicular membrane into the bile canaliculus.[11] A paracellular excretion pathway from the sinusoid through the intercellular space into the bile canaliculus has also been suggested as an alternative route for some solutes excreted by passive osmotic filtration.[17] This hypothesis of a paracellular excretion pathway, however, can be discarded for the excretion of all biliary contrast agents because they seem to be excreted into the bile by an active and saturable transport mechanism.[18]

The mechanism for the uptake of biliary contrast agents from the blood into the hepatocytes is unknown. Similarities in the biliary excretion of biliary contrast agents and other organic anions, such as sulfobromophthalein (BSP) and bilirubin, exist that have been investigated in greater detail.[11] Because these substances seem to compete for the same biliary transport system, one can speculate that mechanisms involved in their hepatic uptake might also be similar. The first step involved in the biliary excretion of these organic anions is their dissociation from the plasma albumin that impedes the hepatic uptake. On the isolated perfused rabbit liver preparation, the presence of albumin in the perfusate retards the transfer of iodipamide from plasma to bile.[19]

The avidity of biliary contrast agents in plasma for albumin suggests that protein binding of these agents in the cell membranes and cytoplasm of hepatocytes may play an important role in their hepatic uptake. Saturable binding sites for organic anions have indeed been found in liver cell plasma membranes, but it is not clear whether these proteins are actually located on the sinusoidal membrane (and available for uptake) or on the canalicular membrane.[20,21] The uptake of organic anions by the hepatocytes meets the criteria of a carrier-mediated process that seems to be saturable.[11] It seems to exceed considerably the maximal capacity for biliary excretion and is therefore unlikely the rate-limiting step.

Organic anion binding proteins are found in the cytoplasm of hepatocytes and were originally designated as Y and Z proteins.[22] Y is most abundant in the liver, comprising about 5% of the total protein in the cytosol fraction, whereas Z is present in organs other than the liver.[23] The affinity of bilirubin and BSP is greater both in vivo and in vitro for Y (also known as ligundin and glutathione transferase) than Z.[24] The binding affinity of bilirubin to Y, however, is an order of magnitude less than to albumin.[23] Iopanoic acid and iodipamide show similar binding to Y and Z in vitro as bilirubin and BSP.[25]

A general correlation exists between the amount of Y in the liver and the rate of elimination of organic anions from plasma. Administration of phenobarbital to rats results in elevated levels of Y and accelerated BSP uptake by the liver.[23] Administration of phenobarbital also stimulates hepatic blood flow, bile secretion, glucoronyl transferase activity (responsible for conjugation of a variety of organic anions), and the synthesis of both bile acids and a variety of enzymes.[26] It is therefore not clear whether the increased biliary excretion rate of iopanoic acid found after pretreatment with phenobarbital is related to the increased amount of Y in the liver.[27] Nevertheless, Y and Z proteins seem to play an important role as hepatic acceptors in the poorly understood process of hepatic uptake and subsequent biliary excretion of contrast agents. Little is known about the intracellular transport of organic anions, including biliary contrast agents, from the sinusoidal to the canalicular pole of the hepatocyte.

BIOTRANSFORMATION IN THE LIVER

As with bilirubin, the poorly water-soluble oral cholecystographic agents are converted into water-soluble glucuronide conjugates in the hepatocytes before their excretion into the bile.[28] Conjugation occurs by the action of the microsomal enzyme, glucuronyl transferase, which catalyzes the transfer of glucuronic acid from the nucleotide uridine diphosphate glucuronic acid to the contrast agent.[28] Formation of the glucuronide reduces the amount of unconjugated oral cholecystographic agent in the hepatic cytoplasm, which may facilitate the transport of the agent from the blood to the hepatocyte by establishing a concentration gradient across the cell membrane. Such a concentration gradient, however, seems not to be crucial for the biliary excretion of oral cholecystographic agents, because the biliary clearance of IV cholangiographic contrast agents that are not conjugated in the liver is at least as efficient as that of oral cholecystographic agents.

The glucuronidation of oral cholangiographic agents, however, affects their biliary clearance rate. The IV administration of conjugated iopanoate produces good radiographic opacification of the gallbladder in 1 hour as compared to an 8-hour delay in visualization with an equivalent dose of unconjugated iopanoate.[29,30] Conjugation increases the water solubility of oral cholecystographic agents. Conjugates do not diffuse back into the liver from the bile canaliculus as readily as the unconjugated lipid-soluble compounds. Similarly conjugated oral cholecystographic agents are also poorly reabsorbed from the intestinal tract, thereby preventing extensive enterohepatic recirculation.[28]

HEPATIC EXCRETION

Bile production may be divided into two processes: canalicular and ductular bile formation.[11] Bile canaliculi lie between the walls of the hepatocytes that form their borders. Excretion of fluid by the hepatocytes into the canaliculi is accordingly called canalicular bile formation. Canaliculi are drained by the smallest intralobar bile ducts, the bile ductules. The process of fluid production and resorption distal to the canaliculi in the bile ductules and ducts is termed ductular bile formation. It is limited, besides water, to the secretion and reabsorption of inorganic electrolytes. In humans the daily bile production approximates 600 ml, of which 75% is of canalicular and 25% of ductular origin.

Secretion of bile salts into bile takes place through a carrier-mediated process. It is saturable with a maximum capacity (transport maximum or Tm) of 8.5 μmol/min/kg in the dog.[31] Bile salts are a major force in the canalicular bile formation because their secretion provides an osmotic drive for the filtration of water and electrolytes.[11] As does the biliary Tm, the choleretic effect of bile salts varies from species to species, being approximately 8 ml water per mmol bile salt in the dog.[32,33] Because bile acids form micelles (polyanionic macroaggregates) in the bile, most of the osmotic activity must be accounted for by their accompanying cations.

Biliary excretion of both oral and IV contrast agents is also a carrier-mediated process that is limited by a transport maximum.[18] Their biliary excretion rate increases hyperbolically with rising plasma concentrations until a plateau is approached. At this point a further increase in the plasma concentration produces no additional increase in biliary excretion.[8,34] At transport maximum the bile concentration of the contrast agent is approximately 100 times higher than the plasma concentration.[35–39] The biliary clearance of contrast agents, however, is even more effective at lower plasma concentrations that do not permit biliary excretion at maximum capacity. Under those conditions the bile:plasma ratio of biliary contrast agents can approach three orders of magnitude. These findings imply that biliary contrast agents are excreted by an active process that becomes saturated above a certain plasma concentration.

Interference exists in the biliary excretion of many organic anions, including bilirubin, BSP, and both oral and IV biliary contrast agents.[3,40] Stated differently, if two of these organic anions are administered simultaneously in dosages permitting their excretion at maximum capacity, the Tm of both these compounds decreases. Competition for conjugation could account for the decreased biliary excretion rate of bilirubin and an oral cholecystographic agent when administered together, because both these compounds are glucuronidized in the hepatocytes before their excretion. Conjugation cannot be responsible for the interference in biliary clearance between an IV cholangiographic agent and all other aforementioned organic anions, however, because IV cholangiographic agents are not metabolized in the liver and are excreted unchanged. It has been suggested that with the exception of bile acids, organic anions excreted into the bile share the same active secretion mechanism.[3] However, the assumption of a single organic anion transport system cannot reconcile completely more recent findings suggesting that (1) the biliary Tm of bilirubin is significantly lower than for biliary contrast agents[35,41] and (2) bilirubin depresses the biliary excretion of iodipamide considerably more than vice versa.[42]

The biliary excretion of bile acids differs from other organic anions in several ways. The biliary Tm of bile acids is approximately ten times higher than that of biliary contrast agents,[31,35] and bile acids administered in a physiologic range do not compete for biliary excretion with other organic anions. On the contrary, the administration of bile salts in physiologic dosages increases the Tm of biliary contrast agents, BSP, and bilirubin.[41–45] In mutant Corriedale sheep, which have an inherited defect in the biliary excretion of organic anions that is closely related to the Dubin-Johnson syndrome in humans (idiopathic conjugated hyperbilirubinemia caused by a reduced ability to transport organic anions from the liver cell into the bile), the maximum biliary excretion rate of BSP is low, whereas that of bile acid is normal.[46]

When bile acids are administered in nonphysiologic high dosages, the simultaneous excretion of biliary contrast agents, bilirubin, and BSP is depressed.[3,44,47] These observations suggest that bile acids are excreted primarily by a very potent and independent transport system. When, however, saturation of this hepatic excretion mechanism for bile acids is approached by unphysiologically high plasma levels, competition between bile acids and other organic anions for the nonspecific organic anion transport system will take place.[44]

Bile acid administration is associated with an increase in biliary Tm of other organic anions, such as biliary contrast agents, BSP, and bilirubin.[41–45] This effect, however, varies considerably from compound to compound: it is considerable for iopanoic acid[37,43] and bilirubin,[41] both of which are noncholeretic compounds, and is small for IV cholangiographic agents, which are all highly choleretic.[44] In dogs with complete bile diversion, an increase of the taurocholate infusion rate from 0.5 μmol/min/kg to 2.0 μmol/min/kg increases the biliary excretion rate of iopanoic acid by 100%,[37] whereas an increase of the taurocholate infusion rate from 0.3 μmol/min/kg to 0.6 μmol/min/kg increases the iodipamide excretion rate by only 20%.[44] With a further stepwise increase in taurocholate infusion rate to 2.4 μmol/min/kg, the biliary iodipamide excretion rate no longer changes and then falls significantly with unphysiologically high taurocholate infusion rates of 4.8 and 9.6 μmol/min/kg, respectively.[44]

The enhancing effect on the biliary Tm of organic anions by bile acids in a physiologic range can be explained theoretically by (1) induction of bile flow, (2) sequestration of the agent in micelles formed by bile acids, (3) direct effect on the organic anion transport system, and (4) recruitment of transporting hepatocytes.[11] There is strong evidence that the increase in Tm is not related to the increase in bile flow; canalicular choleresis induced by substances other than bile acids, such as theophylline and the bicyclic organic anion SC 2644, has no influence on the Tm of BSP or biliary contrast agents.[48,49] No correlation has been found either between the Tm of organic anions and their incorporation into mixed micelles formed by different bile acids.[50] Dehydrocholic acid, a bile acid that stimulates bile flow across the canaliculus but forms micelles poorly, increases the Tm of BSP and biliary contrast agents to the same extent as micelle-forming bile acids.[51] In contrast, glucodihydrofusidate, a steroid compound that forms mixed micelles in vitro and is excreted into the bile, does not increase the BSP Tm.[52] Therefore, it seems more likely that bile acids increase the biliary Tm of other organic anions by direct action on the transport system of these agents. The theoretical possibilities include the following. (1) The active transport of bile acids provides the driving force for organic anions through a co-transport system; (2) bile acids modify, possibly in an allosteric way, the carrier of the organic anions;[53] and (3) bile acids increase, by a recruitment process, the number of hepatocytes available for transport. At present, available kinetic data are insufficient to distinguish between these possibilities.

The biliary Tm of organic anions varies considerably from species to species and is dependent on bile salts.[54] In the dog, the biliary Tm of all oral and IV biliary contrast agents is similar in the presence of an optimal bile acid plasma and liver concentration, respectively. Under those conditions the canine Tm of biliary contrast agents amounts to approximately 1 μmol/min/kg, although both higher and lower values have been reported in the literature depending on the experimental design.[35–39] In hepatic dysfunction caused by either liver parenchymal disease or biliary obstruction, the biliary excretion rate of contrast agents diminishes progressively with the severity of the disease.[55,56] When under those conditions the basal canine bile flow falls below 2 μL/min/kg, then the biliary excretion of contrast agents becomes too small to afford radiographic opacification.[57]

Radiographic visualization of the biliary system, however, is not directly related to the excretion but to the concentration of the contrast agent in the bile. The concentration of contrast materials in the bile is determined, in addition to their rate of excretion, by the choleresis associated with their excretion and the basal bile flow. Intravenous cholangiographic contrast agents, such as iodipamide and iodoxamate, are very potent choleretics: their excretion into the bile is associated with a marked increase in bile flow.[35,39] The choleretic effect of these agents amounts to 22 ml/mmol.[58] Stated differently, each millimole of iodipamide or iodoxamate excreted into the bile drags along 22 ml of water. If there were no basal bile flow, the concentration of these agents would be constant and equal for any excretory rate, e.g., 1 mmol/22 ml or .045 mmol/ml. However, the concentration in bile is lower than this calculated value because the basal bile flow dilutes the contrast agents. The greater the rate of basal bile flow, the greater is the contrast dilution. The basal bile flow is proportionally more significant—causing greater dilution—when small amounts of the contrast agents are excreted into the bile. At high rates of contrast media excretion, the basal bile flow is less significant, causing less dilution of the contrast agents.

On the other end of the spectrum is the oral cholecystographic agent, iopanoic acid, which is not a choleretic.[37] The concentration of iopanoic acid is therefore only determined by the dilutional effect of the basal bile flow and its excretion rate. However, the biliary excretion rate of iopanoic acid is markedly dependent on the rate of excretion of bile salts that are potent choleretics. Experiments in dogs suggest that the concentration of iopanoic acid is reduced when its biliary excretion rate is increased by additional bile salt excretion because the bile salts seem to be more efficient in stimulating water flow than in augmenting iopanoic acid excretion.[37,49]

The choleretic effect of tyropanoate and ipodate, two other oral cholecystographic contrast agents, is between iodipamide and iopanoic acid. Both tyropanoate and ipodate induce a bile flow of approximately 8–11 ml for each millimole of contrast material excreted in bile.[38]

Ductular bile formation is limited to the excretion of water and inorganic electrolytes and is not involved in the excretion of biliary contrast agents.[11] Secretin, gastrin, histamine, cholinergic drugs, and vagal stimulation increase ductular bile fluid production, whereas anticholinergic drugs reduce ductular bile flow. The administration of atropine in the dog increases the bile iodipamide concentration by 12 to 16%, presumably by reducing ductular bile flow and its dilutory effect.[59]

CONCENTRATION IN THE GALLBLADDER

The gallbladder concentrates and stores bile secreted from the liver and discharges its content in response to a meal. Only approximately half of the secreted hepatic bile enters the gallbladder; the rest is discharged directly into the duodenum. Ninety percent of the water entering the gallbladder is reabsorbed. Concentration of bile in the gallbladder is an energy-consuming process.[60] The removal of both water and inorganic ions seems to be coupled and dependent on the active transport of sodium and chlorine (so-called NaCl pump).

Radiographic opacification of the gallbladder requires a minimum bile iodine concentration of 0.3%.[1,3] In oral cholecystography this iodine concentration is usually only attained after the bile is concentrated in the gallbladder.

The normal gallbladder is quite impermeable to highly ionized and water-soluble organic substances, such as taurocholate, BSP, iodipamide, and glucuronides, but is much more permeable to weakly ionized and lipid-soluble substances, such as unconjugated bilirubin.[61] A calculous gallbladder with mucosal thickening and fibrosis has a decreased permeability.[60] Failure to visualize the gallbladder in this condition during oral cholecystography, provided that intestinal absorption and biliary excretion of the contrast agent are normal and the cystic duct is patent, is likely caused by the gallbladder's inability to reabsorb water.[62] In acute cholecystitis an increased permeability to highly ionized substances is found, but there is no change in the absorption of lipid-soluble unconjugated bilirubin.[60] Nonvisualization of the gallbladder in this condition, assuming again that there is no other interfering pathology, results likely from the absorption of the contrast agent in both conjugated and unconjugated form. Although the absorption of the glucuronide of iopanoic acid has been documented,[63] it seems reasonable to assume that glucuronidized oral cholecystographic agents are deconjugated in acute cholecystitis by bacteria capable of producing the appropriate enzyme. In their original form, more lipid-soluble oral cholecystographic agents can then be absorbed from even a normal or only slightly inflamed gallbladder wall.

GASTROINTESTINAL AND RENAL EXCRETION

Oral cholecystographic contrast agents are mostly discharged from the common bile duct into the duodenum as glucuronides, whereas the IV cholangiographic agents are not metabolized and are excreted unchanged. Glucuronides of oral cholecystographic agents and IV cholangiographic agents are poorly absorbed from the intestinal tract and consequently excreted in the stool.[3] The amount of oral cholecystographic agents that is deconjugated by bacterial glucuronidase appears negligible, thereby preventing extensive enterohepatic recirculation of these agents.[28] Sixty-five percent of the oral dose of iopanoic acid is recovered from the feces within 5 days after administration in humans, and the rest is excreted with the urine.[29,64] The amount excreted in the urine decreases exponentially (each day about half of the amount of the preceding day is excreted), so that by the fifth day only traces of the contrast agents are found in the urine.[65] Nearly all of an oral cholecystographic contrast agent in the urine is in conjugated form, suggesting that back diffusion of the conjugate into the blood occurs.[28,30]

Urinary excretion of the biliary contrast agents occurs primarily by glomerular filtration, with only a small fraction being eliminated by tubular secretion.[3,30] There is a linear relation between the urinary excretion rate and the plasma concentration of the contrast agent.[30,35] This implies that, when the maximum capacity of the biliary excretion of the contrast agents is approached, a further increase in the plasma concentration or dose of the agent has little effect on the radiographic visualization of the biliary system and results only in additional renal excretion. The biliary transport maximum of contrast agents is reduced in both liver parenchymal disease and biliary obstruction. In these conditions the fraction of the contrast agent excreted in the urine increases drastically and approaches 100% in the most advanced stages, such as hepatic coma or complete common bile duct obstruction.[55,56]

TOXICITY

Biliary contrast agents are considerably more toxic than urographic contrast agents.[66] Severe systemic reactions associated with the use of oral cholecystographic contrast agents are rare, however, because their slow and inefficient intestinal absorption results in only relatively low plasma concentrations. Transient diarrhea is the most common side effect of these compounds. Intravenous cholangiographic contrast agents are approximately ten times more toxic than urographic contrast agents as reflected by their LD50 and the reported death rate of 1 in 5,000 examinations with iodipamide.[66] Systemic reactions involving the cardiovascular, pulmonary, and central nervous system are qualitatively similar to those produced by angiographic/urographic contrast agents and may be allergic, idiosyncratic, or toxic in nature. They have been discussed in great detail in previous chapters.

Biliary contrast agents differ from urographic contrast agents in two important features that are likely to account for their greater toxicity: (1) they have a considerable affinity to a variety of proteins,[14,25] and (2) they are taken up and temporarily accumulated in certain excretory cells, e.g., hepatocytes and epithelia of the renal tubules. After administration of biliary contrast agents a dose-dependent transient elevation of liver enzymes and serum bilirubin can be found.[3,67] Binding of these contrast agents to both lipoproteins in the cell membrane and different intracellular enzymes might explain the temporary impairment of hepatic function.

The most common serious adverse effect of biliary contrast agents, however, is exerted on the kidneys. Over 50 cases of acute renal failure have been reported after the administration of biliary contrast agents.[68,69] Different mechanisms are implicated in the development of renal failure. A direct toxic effect of the contrast medium on the tubular epithelium has been postulated.[70] Another hypothesis suggests obstructive renal failure caused by crystal deposition within the renal tubules.[71] Crystals might represent either the precipitated cholecystographic agent itself or another precipitated substance normally found in the urine, but increasingly excreted by the kidneys in the presence of a biliary contrast agent. One such substance is uric acid.[72–74] Uricosuria caused by different cholecystographic agents is of the same magnitude as with drugs currently used for the treatment of hyperuricemia, e.g., probenecid. A third hypothesis implicates hypotension as a major factor in the development of acute renal failure resulting in ischemic tubular necrosis.[75] A fall in systemic blood pressure indeed occurs after the administration of biliary contrast agents.[76,77] However, in a more recent experimental investigation, ischemic tubular necrosis could be induced by the administration of biliary contrast agents in the absence of systemic hypotension,[78] suggesting that the primary insult may occur on the small renal vessels and glomeruli. Whatever the etiology of acute renal failure after the administration of biliary contrast agents might be, it seems to be very important that the patient takes abundant fluids during this procedure and that excessive contrast material dosages are avoided, particularly in the presence of hepatic and/or renal disease.

REFERENCES

1. Joffe H, Wachowski TJ: Relation of density of cholecystographic shadows of the gallbladder to the iodine content. *Radiology* 1942;38:43–46.

2. Edholm P, Jacobson B: Quantitative determination of iodine in vivo. *Acta Radiol* 1959;52:337–346.

3. Sperber I, Sperber G: Hepatic excretion of radiocontrast agents, in Knoefel PK (ed): *International Encyclopedia of Pharmacology and Therapeutics*. Oxford, Pergamon Press, 1971, pp 165–235.

4. Hoey GB, Wiegert PE, Rands RD Jr: Organic iodine compounds as x-ray contrast media, in Knoefel PK (ed): *International Encyclopedia of Pharmacology and Therapeutics*. Oxford, Pergamon Press, 1971, pp 23–131.

5. Peterhoff R: Cholecystography with the sodium salt of iopanoic acid. *Acta Radiol* 1956;46:719–722.

6. Holmdahl KH, Lodin H: Absorption of iopanoic acid and its sodium salt. *Acta Radiol* 1959;51:247–250.

7. Goldberger LE, Berk RN, Lang JH et al: Biopharmaceutical factors influencing the intestinal absorption of iopanoic acid. *Invest Radiol* 1974;9:16–23.

8. Berk RN, Loeb PM. Contrast material for oral cholecystography, in Miller RE, Skucas J (eds): *Radiographic Contrast Agents*. Baltimore, University Park Press, 1977, pp 195–221.

9. Taketa RM, Berk RN, Lang JH, et al: The effect of pH on the intestinal absorption of Telepaque. *Am J Roentgenol* 1971;114:767–772.

10. Nelson JA, Moss AA, Goldberg HI, et al: Gastrointestinal absorption of iopanoic acid. *Invest Radiol* 1973;8:1–8.

11. Erlinger S: Physiology of bile secretion and enterohepatic circulation, in Johnson LR (ed): *Physiology of the Gastrointestinal Tract*, ed 2. New York, Raven Press, 1987, pp 1557–1580.

12. Bates TR, Gibaldi M, Kanig JL: Solubilizing properties of bile salt solutions. *J Pharm Sci* 1966;55:191–199.

13. Jackson MJ: Drug transport across gastrointestinal epithelia, in Johnson LR (ed): *Physiology of the Gastrointestinal Tract*, ed 2. New York, Raven Press, 1987, pp 1597–1621.

14. Lang JH, Lasser EC: Binding of roentgenographic contrast media to serum albumin. *Invest Radiol* 1967;2:396–400.

15. Kimbel KH, Heinkel K, Börner W: Die Ausscheidung des Nasalzes des dipinssäurebis-(2,4,6-trijod-3-carboxy-anilid) bei der ratte. *Arzneimittel-Forsch* 1956;6:225–227.

16. Burgener FA, Ciaravino V, Fischer HW: Failure of iosulamide to enhance hepatic tumors in rats. *Invest Radiol* 1982;17:46–49.

17. Ashworth CT, Sanders E: Anatomic pathway of bile formation. *Am J Pathol* 1960;37:343–355.

18. Fischer HW: The excretion of iodipamide, relation of bile and urine outputs to dose. *Radiology* 1965;84:483–490.

19. Song CS, Beranbaum ER: The role of serum albumin in hepatic excretion of iodipamide. *Invest Radiol* 1974;9:324.

20. Gonzalez MC, Sutherland E, Simon FR: Regulation of hepatic transport of bile salts. Effect of protein synthesis inhibition on excretion of bile salts and their binding to liver surface membrane fractions. *J Clin Invest* 1969;63:684–694.

21. Accatino L, Simon FR: Identification and characterization of a bile acid receptor in isolated liver surface membranes. *J Clin Invest* 1976;57:496–508.

22. Levi AJ, Gatmaitan Z, Arias IM: Two hepatic cytoplasmic protein fractions, Y and Z, and their possible role in the hepatic uptake of bilirubin, sulfobromophthalein and other anions. *J Clin Invest* 1969;48:2156–2167.

23. Arias IM: Transfer of bilirubin from blood to bile. *Sem Hematol* 1972;9:55–70.

24. Bissell DM: Formation and elimination of bilirubin. *Gastroenterology* 1975;69:519–538.

25. Sokoloff J, Berk RN, Lang JH, et al: The role of the Y and Z hepatic proteins in the excretion of radiographic contrast materials. *Radiology* 1973;106:519–523.

26. Orme ML'E, Davies L, Breckenridge A: Increased glucuronidation of bilirubin in man and rat by administration of antipyrine (Phenazone). *Clin Sci Mol Med* 1974;46:511–518.

27. Herzog RJ, Nelson JA, Staubus AE: Saturation kinetics of iopanoate in dogs with intact enterohepatic circulation before and after phenobarbital induction. *Invest Radiol* 1976;11:32–38.

28. McChesney EW: The biotransformation of iodinated contrast agents, in Knoefel PK (ed): *International Encyclopedia of Pharmacology and Therapeutics*. Oxford, Pergamon Press, 1971, pp 147–163.

29. McChesney EW, Hoppe JO: Observations on metabolism of iodopanoic acid. *Arch Int Pharmacodyn Ther* 1954;99:127–140.

30. Cooke WJ, Mudge GH: Biliary and urinary excretion of iopanoic acid in the dog. *Invest Radiol* 1975;10:25–34.

31. O'Maille ERL, Richards TG, Short AH: Acute taurine depletion and maximal rates of hepatic conjugation and secretion of cholic acid in the dog. *J Physiol (London)* 1965;180:67–79.

32. Preisig R, Cooper HL, Wheeler HO: The relationship between taurocholate secretion rate and bile production in the unanesthetized dog during cholinergic blockade and during secretin administration. *J Clin Invest* 1962;41:1152–1162.

33. Wheeler HO, Ross ED, Bradley SE: Canalicular bile production in dogs. *Am J Physiol* 1968;214:866–874.

34. Berk RN, Loeb PM, Ellzey BA: Contrast materials for intravenous cholangiography, in Miller RE, Skucas J (eds): *Radiographic Contrast Agents*. Baltimore, University Park Press, 1977, pp 223–250.

35. Loeb PM, Berk RN, Feld GK, et al: Biliary excretion of iodipamide. *Gastroenterology* 1975;68:554–562.

36. Nelson JA, Staubus AE, Riegelman S: Saturation kinetics of iopanoate in the dog. *Invest Radiol* 1975;10:371–377.

37. Berk RN, Loeb PM, Cobo-Frenkel A, et al: The biliary and urinary excretion of iopanoic acid: pharmacokinetics, influence of bile salts, and choleretic effect. *Radiology* 1976;120:41–47.

38. Berk RN, Loeb PM, Cobo-Frenkel A, et al: The biliary and urinary excretion of sodium tyropanoate and sodium ipodate in dogs: pharmacokinetics, influence of bile salts and choleretic effects with comparison to iopanoic acid. *Invest Radiol* 1977;12:85–95.

39. Berk RN, Loeb RM, Cobo-Frenkel A, et al: Saturation kinetics and choleretic effects of iodoxamate and iodipamide. *Radiology* 1976;119:529–535.

40. Goergen T, Goldberger LE, Berk RN: The combined use of oral cholecystopaque media and iodipamide. *Radiology* 1974;111:543–548.

41. Goresky CA, Haddad HH, Kluger WS, et al: The enhancement of maximal bilirubin excretion with taurocholate-induced increments in bile flow. *Can J Physiol Pharmacol* 1974;52:389–403.

42. Burgener FA, Fischer HW: The effect of bilirubin on biliary iodipamide excretion in the dog. *Invest Radiol* 1980;15:162–167.

43. Moss AA, Amberg JR, Jones RS: Relationship of bile salts and bile flow to biliary excretion of iopanoic acid. *Invest Radiol* 1972;7:11–15.

44. Burgener FA, Fischer HW: The effect of sodium taurocholate on biliary iodipamide excretion in the dog. *Gastroenterology* 1976;71:475–478.

45. Boyer JL, Scheig RL, Klatskin G: The effect of sodium taurocholate on the hepatic metabolism of sulfobromophthalein sodium (BSP): the role of the bile flow. *J Clin Invest* 1970;49:206–215.

46. Alpert S, Mosher M, Shanske A, et al: Multiplicity of hepatic excretory mechanisms of organic anions. *J Gen Physiol* 1969;53:238–247.

47. O'Maille ERL, Richards TG, Short AH: Factors determining the maximal rate of organic anion secretion by the liver and further evidence on the hepatic site of action of the hormone secretin. *J Physiol (London)* 1966;186:424–438.

48. Barnhart JL, Combes B: Effect of theophylline on hepatic excretory function. *Am J Physiol* 1974;227:194–199.

49. Berk RN, Goldberger LE, Loeb PM: The role of bile salts in the hepatic excretion of iopanoic acid. *Invest Radiol* 1974;9:7–15.

50. Vonk RJ, Jekel P, Meijer DKF: Choleresis and hepatic transport mechanisms. II. Influence of bile salt choleresis and biliary micelle binding on biliary excretion of various organic anions. *Naunyn Schmiedebergs Arch Pharmacol* 1975;290:375–387.

51. Ritt DJ, Combes B: Enhancement of apparent excretory maximum of sulfobromophthalein sodium (BSP) by taurocholate and dehydrocholate. *J Clin Invest* 1967;46:1108–1109.

52. Delage Y, Dumont M, Erlinger S: Effect of glycodihydrofusidate on sulfobromophthalein transport maximum in the hamster. *Am J Physiol* 1976;231:1875–1878.

53. Forker EL, Gibson G: Interaction between sulfobromophthalein (BSP) and taurocholate. The kinetics of transport from liver cells to bile in rats, in Paumgartner G, Preisig R (eds): *The Liver. Quantitative Aspects of Structure and Function*. Basel, Karger, 1973, pp 326–335.

54. Burgener FA: Intact animal models of normal and abnormal biliary excretion, in Milne ENC (ed): *Models and Techniques in Medical Imaging Research*. New York, Prager Scientific, 1983, pp 395–416.

55. Burgener FA, Fischer HW, Kenyon TD: Biliary excretion of iodipamide and iodoxamate in normal and common bile duct-obstructed dogs. *Invest Radiol* 1978;13:255–263.

56. Burgener FA, Fischer HW: Biliary excretion of iodipamide and iodoxamate in dogs with hepatic dysfunction induced by oral administration of dimethylnitrosamine. *Invest Radiol* 1979;14:502–507.

57. Burgener FA: Biliary iodipamide and iodoxamate excretion as function of basal bile flow in normal, common bile duct obstructed and liver-damaged dogs. *Fortschr Röntgenstr* 1981;134:40–43.

58. Feld GK, Loeb PM, Berk RN, et al: The choleretic effect of iodipamide. *J Clin Invest* 1975;55:528–535.

59. Burgener FA, Fischer HW, Adams JT, et al: Pharmaco-cholangiography with anticholinergic drugs in the dog. *Br J Radiol* 1976;49:769–775.

60. Rose RC: Absorptive functions of the gallbladder, in Johnson LR (ed): *Physiology of the Gastrointestinal Tract*, ed 2. New York, Raven Press, 1987, pp 1455–1468.

61. Ostrow J: Absorption of organic compounds by the injured gallbladder. *J Lab Clin Med* 1971;78:255–264.

62. Graham EA: The story of the development of cholecystography. *Am J Surg* 1931;12:330–335.

63. Berk RN, Lasser EC: Altered concepts of the mechanism of nonvisualization of the gallbladder. *Radiology* 1964;82:296–302.

64. Schroder JS, Rooney D: Excretion of 3-(3-amino-2,4,6-triiodophenyl)-2-ethyl-propanoic acid (Telepaque) by man. *Proc Soc Exp Biol Med* 1953;83:544–546.

65. McChesney EW, Banks WF Jr: Urinary excretion of three oral cholecystographic agents in man. *Proc Soc Exp Biol Med* 1965;119:1027–1030.

66. Ansell G: Adverse reaction to contrast agents. *Invest Radiol* 1970;5:374–384.

67. Fischer HW, Burgener FA: Fractionated dose administration schedule for cholecystography. *Invest Radiol* 1974;9:24–31.

68. Cholecystography and renal failure. *Lancet* 1971;7716:146–147.

69. Ansari Z, Baldwin DS: Acute renal failure due to radio-contrast agents. *Nephron* 1976;17:28–40.

70. Fink HE Jr, Roenigk WJ, Wilson GP: An experimental investigation of the nephrotoxic effects of oral cholecystographic agents. *Am J Med Sci* 1964;247:201–216.

71. Setter JG, Maher JF, Schreiner GE: Acute renal failure following cholecystography. *JAMA* 1963;184:102–110.

72. Mudge GH: Uricosuric action of cholecystographic agents. *N Engl J Med* 1971;284:929–933.

73. Postlethwaite AE, Kelley WN: Uricosuric effect of radiocontrast agents. *Ann Intern Med* 1971;74:845–885.

74. Burgener FA, Fischer HW: The uricosuric effect of Bilopaque (sodium tyropanoate). A contribution to the problem of nephrotoxicity of oral biliary contrast agents (in German). *Fortschr Röntgenstr* 1972;117:685–688.

75. Teplick JG, Myerson RM, Sanen FJ: Acute renal failure following oral cholecystography. *Acta Radiol Diagn* 1965;3:353–368.

76. Saltzman GF, Sandstrom KA: The influence of different contrast media for cholegraphy on blood pressure and pulse rate. *Acta Radiol* 1960;54:353–360.

77. Köhler R, Holsti LR: Biligrafin-forte and solu-biloptin in a comparative trial. *Acta Radiol* 1962;57:103–112.

78. Burgener FA, Fischer HW: Nephrotoxicity of sodium iopanoate in hydrated and dehydrated dogs. *Invest Radiol* 1978;13:247–254.

Chapter 21

Cholangiography and Pancreatography

Robert G. Gibney and H. Joachim Burhenne

Despite major advances in cross-sectional imaging techniques, such as ultrasonography, computed tomography, and magnetic resonance imaging, cholangiography and pancreatography remain as essential tools for the investigation of suspected biliary tract and pancreatic disease.

The history of cholangiography dates back to 1921 when Burckhardt and Müller obtained radiographic opacification of the gallbladder and bile ducts by direct percutaneous transhepatic injection of Kollargol (a silver-containing contrast agent) and air into the gallbladder.[1] This elegant study showed the diagnostic potential of cholangiography and led to the development of several other important cholangiography-related techniques and subsequently to the technique of pancreatography.

CHOLANGIOGRAPHY

Opacification of the intra- and extrahepatic portions of the biliary tract with contrast agents may be accomplished by a variety of radiologic and operative or endoscopic techniques, depending on the clinical situation. A list of these techniques is presented in Table 21-1.

Direct Cholangiography

Operative cholangiography was first performed by Mirizzi in 1932[2] and is now widely used for demonstration of the anatomy and pathology of the bile ducts at the time of biliary surgery and for exclusion of biliary calculi during cholecystectomy. Postoperative T-tube cholangiography is a routine procedure typically performed 7–10 days after common bile duct exploration. Cholangiography also facilitates

Table 21-1 Cholangiography Techniques

DIRECT CHOLANGIOGRAPHY

Percutaneous cholangiography
- Percutaneous transhepatic cholangiography[4]
- Percutaneous cholecystocholangiography[5]

Operative cholangiography[6,7]
- Primary: Before common duct exploration
- Secondary: After common duct exploration

Postoperative T-tube cholangiography[8]

Cholangiography during nonoperative stone extraction[3]

Endoscopic retrograde cholangiography[9]

Other techniques
- Retrograde cholangiography after choledochoenterostomy[10]
- Transjugular cholangiography[11]
- Other instrumental reflux cholangiography[12]
- Cholangiography via sinus or fistula opacification

INDIRECT CHOLANGIOGRAPHY

Intravenous cholangiography[13]

---

nonoperative biliary stone extraction, which may be performed through a mature T-tube tract 4 to 5 weeks after surgery.[3]

Percutaneous cholangiography is usually performed with a fine needle (22- or 23-gauge) using a transhepatic approach to the intrahepatic bile ducts.[4] When conventional transhepatic cholangiography has proved technically difficult, cholangiography may also be accomplished by means of percutaneous transhepatic opacification of the gallbladder.[5] Following either technique, a drainage catheter may be inserted percutaneously for decompression of the biliary tract. Endoscopic retrograde cholangiography is generally preferred when pancreatography is also required, when ampullary carcinoma is suspected, and when an endoscopic sphincterotomy is planned for treatment of choledocholithiasis.

Contrast Agents

Iodinated contrast agents, barium sulfate, and air can be used for direct cholangiography.

Iodinated Contrast Agents. Iodinated agents are the contrast agents of choice for all techniques of direct cholangiography. These are sterile aqueous solutions of diatrizoate sodium and/or meglumine (Hypaque, Winthrop Laboratories and Renografin, E.R. Squibb) or of iothalamate meglumine (Conray, Mallinckrodt).

The concentration of contrast agent that is used varies according to the clinical circumstances. The anatomy of the bile ducts, for instance, is best studied using full-strength contrast agents (at least 60% contrast concentration, which is approximately 30% iodine concentration) because a small volume of contrast will provide maximum opacification of the bile ducts. A dilute solution of 30% contrast (approximately 15% iodine concentration) is generally used if small biliary calculi are suspected, as in T-tube cholangiography.[14] A low-kV technique (70–75 kVp) is then used to maximize stone detection. Further contrast dilution to a 15% concentration of contrast (approximately 7.5% iodine concentration) is required for stone visualization if the bile ducts are prominently distended (common bile duct >10 mm in diameter) or if the gallbladder is being studied.

Another recommended technique uses full-strength contrast agents (at least 60% contrast concentration) with high kV (100–110 kVp) for both operative and T-tube cholangiography.[15,16] This technique has significant advantages for ductal penetration, stone detectability, less motion unsharpness, and overall high image quality (Fig. 21-1). A further advantage is reduced radiation doses to both patient and operator because of the high-kV technique. This technique can be used with all forms of direct cholangiography except where there is significant duct dilation when a reduction in contrast concentration is advisable. Whichever technique is used (full-strength contrast agent with high kVp or dilute contrast agent with low kVp), the great majority of calculi will be demonstrated. If one technique produces equivocal results, however, we recommend that both techniques be applied for maximum accuracy.

The contrast agent should be warmed to body temperature before administration in order to dissolve any crystals that may be in the solution. Care must be taken to avoid injection of air bubbles, which may be confused with calculi. An initial injection of 3-5 ml of contrast, followed by a further 5-10 ml of contrast, is usually sufficient for operative cholangiography. Introduction of contrast for all forms of cholangiography should be performed slowly without undue pressure. For this reason, we prefer a gravity infusion technique of contrast introduction, rather than syringe injection of contrast, for both operative and T-tube cholangiography. An injection pressure higher than a 20-30 cm column of water above the level of the ampulla of Vater usually results in ampullary spasm and may cause cholangiovenous reflux (see the discussion on adverse reactions). This may be the technique of choice, however, if filling of the intrahepatic bile ducts cannot be obtained with slow infusion of contrast.

Air. A spontaneous air cholangiogram is seen after sphincterotomy with a patent biliary-enteric anastomosis or if the papilla is patulous. Visualization of pneumobilia does not, however, exclude a partial biliary obstruction. Air may also be introduced as a contrast medium (pneumocholangiogram). This is particularly helpful if other contrast media are contraindicated. In practice, however, air is seldom used in this manner.

Barium Sulfate. Barium sulfate has been used successfully for both operative and T-tube cholangiography. Its use has been advocated in patients with a previous reaction to iodinated contrast agents.[17,18] The bile ducts tolerate barium sulfate

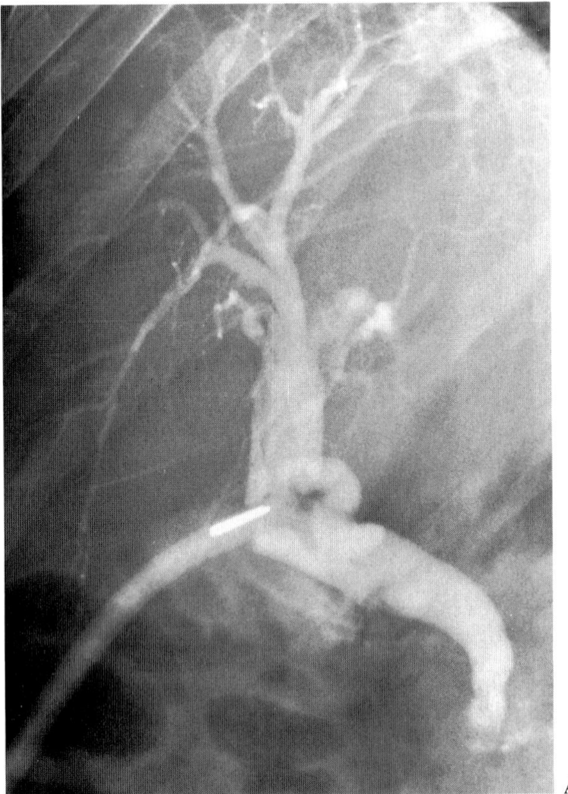

Figure 21-1 (*A*) Low-kVp T-tube cholangiogram and (*B*) high-kVp study from same patient. Note that tube in bile duct and stone are much better seen on high-kVp study. *Source*: Reprinted with permission from "Optimal Cholangiographic Technique for Detecting Bile Duct Stones" by WM Thompson et al in *American Journal of Roentgenology* (1986;146:537–541), Copyright © 1986, American Roentgen Ray Society.

well, and saline irrigation can be applied for subsequent washout. A 45% w/v solution of barium suspension in isotonic saline is sterilized in a standard autoclave and stored at room temperature. Barium sulfate causes no ampullary spasm resulting from irritation or hypertonicity, a phenomenon that may occur with the use of iodinated contrast agents.

Retrograde barium cholangiography during an upper gastrointestinal (GI) radiologic study is the method of choice to evaluate the patency of biliary-enteric anastomoses.[10] There is no evidence that such reflux cholangiography with barium gives rise to cholangitis. This technique is, however, contraindicated if either a blood-bile fistula or bowel perforation is suspected.

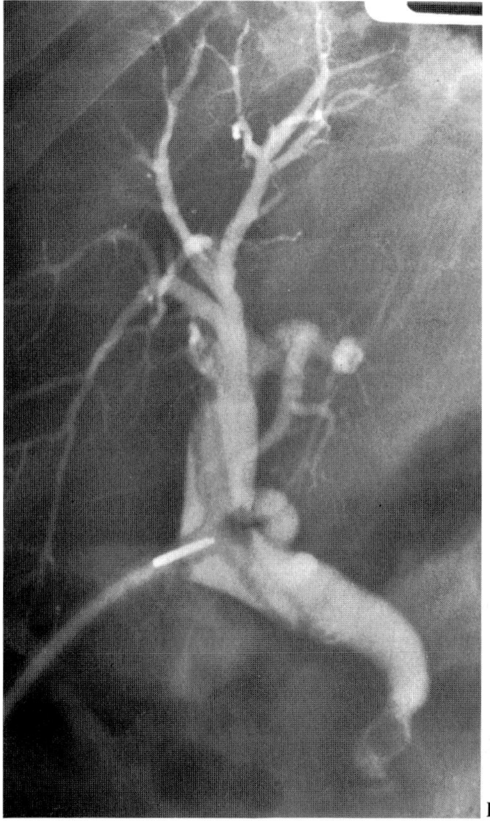
B

Adverse Reactions

Operative and T-tube cholangiography carry little hazard. When large volumes of contrast are used for direct cholangiography, some diarrhea is usually experienced when as much as 40 ml of iodinated contrast is injected. Overdistention of the ducts because of increased injection pressure may produce local pain, nausea, and vomiting. It may also force the contrast medium back through the bile canaliculi into the venous circulation, producing the so-called cholangiovenous reflux. A hypersensitivity reaction to the contrast agent is therefore a potential problem, but has been very rarely observed. If there is a history of previous serious contrast reaction, either barium or the low osmolar iodinated contrast agents may be safely used for cholangiography. Because the bile is frequently colonized with bacteria after biliary tract surgery, cholangiovenous reflux may also lead to bacteremia and even septic shock.[19] A significant reduction in adverse reactions results from use of a low-pressure gravity infusion technique for introduction of contrast for T-tube cholangiography.[19]

Spasm of the sphincter of Oddi may result from high pressure injection of contrast agents, administration of such analgesics as fentanyl citrate and morphine sulfate, and from operative manipulation or endoscopic cannulation of the sphincter. It can cause difficulties in radiologic interpretation and can predispose to cholangiovenous reflux. Such spasm may be overcome by intravenous injection of glucagon.[20]

Some reports have related pancreatitis after biliary tract surgery and even fatal pancreatic necrosis to operative cholangiography.[21,22] Reflux of contrast material into the pancreatic duct occurs in 9–33% of operative cholangiograms.[23,24] However, two large studies have concluded that operative cholangiography does not induce postoperative pancreatitis and that there is no relationship between pancreatitis and reflux of contrast into the pancreatic duct.[23,25] The major predisposing factors to the development of pancreatitis after biliary tract surgery are common bile duct exploration and a preoperative history of pancreatitis.

In the presence of biliary obstruction, cholangitis and septicemia are significant risks after both transhepatic and endoscopic retrograde cholangiography. For this reason, prophylactic antibiotic coverage should be administered before cholangiography in all suspected cases of biliary obstruction. Overfilling of the bile ducts should be avoided, and the obstruction should be relieved promptly by percutaneous, endoscopic, or surgical drainage.[26] Bile peritonitis is rarely a problem with fine-needle transhepatic cholangiography, but if percutaneous gallbladder puncture is performed, a transhepatic access route to the gallbladder is recommended to reduce the risk of bile leaking from the liver into the peritoneal cavity.[5]

Indirect Cholangiography—Intravenous Cholangiography

Indirect cholangiography may be accomplished by intravenous (IV) injection of contrast agents, which are excreted in the bile at a rate and concentration adequate for radiographic visualization of the bile ducts; this technique is IV cholangiography (IVC).

There has been a steady decline in the use of IVC over the past decade. Advances in the ability to diagnose biliary tract disease by ultrasonography, computed tomography, and direct cholangiography, together with reports of an unacceptably high frequency of nondiagnostic and inaccurate IVCs, have led to recommendations that the technique be abandoned.[27–29] Indeed, a recent survey of North American teaching hospitals, which we carried out in 1987, indicated that it already has been virtually abandoned in these hospitals (unpublished data). In Europe, however, some centers perform a routine preoperative IVC as an alternative to operative cholangiography during cholecystectomy.[30]

Contrast Media

Following its introduction in 1954, iodipamide meglumine (Cholografin, E.R. Squibb) was the only contrast agent available for use for IVC. Other chemically related compounds were subsequently developed; all are meglumine salts of a dimer composed of two triiodinated benzoate rings, each with a different interconnecting radicle group between the benzoate rings. The agents favored today are iodoxamate and iotroxinate (Cholovue, E.R. Squibb, and Biliscopin, Schering, respectively).

Technique (using Iotroxinate)

Before performing IVC, liver function tests should be performed to ensure that the serum bilirubin level is not elevated above 3-4 mg/dl, which would make biliary opacification unlikely. Elevation of serum bilirubin less than this level may result in reduced opacification of the bile ducts.

The preparation for an IVC includes a routine bowel cleansing and an overnight fast. Oral fluid intake is encouraged to maintain optimal hydration. An initial radiograph of the right upper quadrant is followed by an IV infusion of 50 ml of iotroxinate meglumine (iodine concentration 108 mg/ml) over 10-12 minutes (approximately 2 drops/sec). The infusion technique is preferred to an IV injection because it maintains a lower plasma concentration, resulting in less renal excretion of the contrast and enhanced biliary excretion.[31] The lower plasma concentration also reduces the risk of toxicity and side effects.[32]

The patients are examined supine in an approximately 20° right posterior oblique position. Abdominal compression may be applied, as for IV urography. A well-collimated radiograph of the right upper quadrant is repeated 10 to 20 minutes after the end of the infusion. A low-kV technique (60–70 kVp) is used to maximize the radiographic contrast. If the biliary tree is visualized, a series of linear tomograms is performed at 1-cm intervals. If there is no biliary tract visualization, conventional radiographs are repeated at 20-minute intervals up to 2 hours. Serial tomograms are obtained whenever the biliary tree is visualized.

Adverse Reactions

The toxicity associated with the use of iodipamide is another important factor that has led to the decline in the use of IVC. Both minor and major adverse reactions occur more frequently than with conventional urographic contrast agents. Minor reactions included nausea, vomiting, and a variety of skin reactions. The frequency of such reactions varied from 4.1–23.9%.[26] Major reactions included severe skin rashes, cardiorespiratory complications, hepatic and renal failure, hypotension, anaphylaxis, and death (1 death in 5,000 examinations with iodipamide compared with 1 in 40,000 urographic procedures).[33]

IVC is contraindicated in patients with IgM gammopathy and significant hepatic or renal failure. Furthermore, IVC contrast agents should not be used within 24-48 hours of administration of IV urographic or cholangiographic contrast agents or oral cholecystographic contrast agents. Failure to observe this rule may significantly increase the risk of serious morbidity and mortality.[26]

Less information is available about the incidence of adverse reactions to newer IV cholangiographic agents, such as iotroxinate and iodoxamate. Fewer patients have been examined, but studies have so far reported considerably less adverse reactions to these agents, especially when they are infused slowly.[30,34]

PANCREATOGRAPHY

For many years, direct pancreatography was possible only at the time of surgery on the pancreas. It is now much more common to perform nonoperative pancreatography, usually under endoscopic guidance, but occasionally by direct percutaneous opacification of the pancreatic duct (Table 21-2).

Techniques

Endoscopic cannulation of the ampulla of Vater was first accomplished in 1968, 3 years after peroral cannulation of the ampulla was first achieved under fluoroscopic guidance for the purposes of direct cholangiography and pancreatography.[12,38] Since that time, endoscopic retrograde pancreatography, frequently combined with endoscopic retrograde cholangiography (ERCP), has become a valuable tool for the investigation of pancreatic and hepatobiliary disease. ERCP is a combined endoscopic/radiologic procedure. Its endoscopic aspects are beyond the scope of this textbook, but have been well described elsewhere.[9] The radiologist must provide careful fluoroscopic monitoring with good quality radiographic equipment to guide the endoscopist and to obtain the best possible radiographs.

Percutaneous pancreatography is seldom indicated, but may be performed using a fine needle with sonographic or CT guidance in selected cases of pancreatic

Table 21-2 Pancreatography Techniques

Endoscopic retrograde pancreatography[9]
Operative pancreatography[35]
Percutaneous pancreatography[36,37]
Reflux pancreatography during cholangiography

duct dilation when opacification cannot be accomplished by endoscopy (Fig. 21-2).[36,37]

Contrast Agents

Sterile aqueous iodinated contrast agents are used exclusively for pancreatography. Unlike cholangiography where dilute contrast is frequently used, full-strength contrast agents (60%–76% contrast concentration, 30%–38% iodine concentration) are used for pancreatography. In patients with known hypersensitivity to iodinated contrast agents, low-osmolar contrast agents are recommended.

As with cholangiography, the contrast agent is first warmed to body temperature in a water bath. When the catheter has been passed through the orifice of the papilla, a slow injection of 2-3 ml of contrast is made under fluoroscopic observation. Sufficient contrast is then introduced to opacify the pancreatic ducts while carefully avoiding opacification of the pancreatic acini (acinarization) by overinjection (Fig. 21-3). It has been demonstrated that contrast material is absorbed

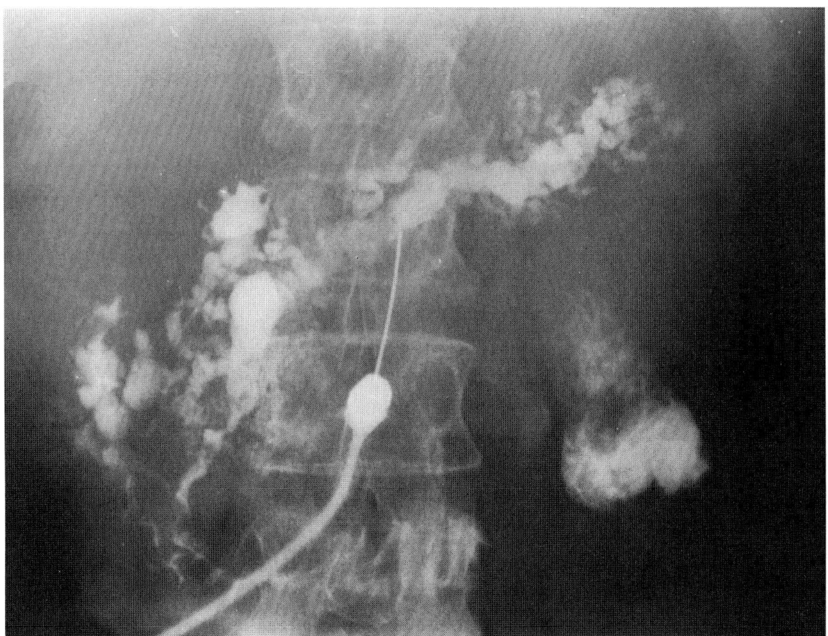

Figure 21-2 Percutaneous pancreatogram in a patient with chronic pancreatitis and stenosis of the main pancreatic duct close to the ampulla. The dilated pancreatic ducts contain calculi, and a pancreatic head pseudocyst is opacified.

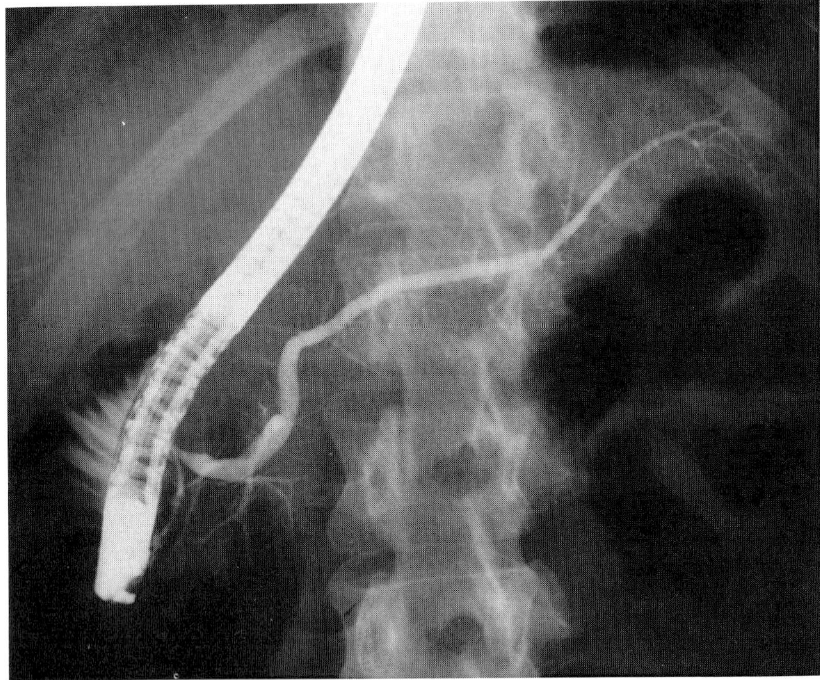

Figure 21-3 Normal endoscopic retrograde pancreatogram.

into the bloodstream by the pancreatic ductal epithelium. This absorption is not infrequently sufficient to permit renal opacification following pancreatography.[39]

Adverse Reactions

Pancreatitis is the most common adverse reaction following pancreatography. It develops in 0.5%–1.3% of patients following endoscopic retrograde pancreatography (ERP), but is typically mild and self-limiting.[26] Clinical pancreatitis is associated with epigastric pain and raised serum and urinary amylase levels. It must be distinguished from asymptomatic hyperamylasemia and hyperamylaseuria, which occur transiently in many patients following pancreatography. The exact causes of post-ERP pancreatitis are uncertain, but overdistention of the pancreatic ducts, pancreatic duct manipulation, and acinar rupture may all play a role. In patients where acinarization is accompanied by urographic visualization, there is a high risk of pancreatitis.[40] There is no convincing evidence that the use of low osmolar contrast agents results in a decrease in either asymptomatic hyperamylasemia or pancreatitis.[41,42]

The other major complication occasionally associated with pancreatography is development of pancreatic sepsis and pancreatic pseudocysts. These complications have an incidence of 0.3%, with a mortality rate that may be as high as 20%.[43] If opacification of a pancreatic pseudocyst is observed during pancreatography, the pseudocyst should not be completely opacified and prophylactic antibiotics should be administered after the procedure. Doing so reduces the incidence of postprocedure infection. Similarly, patients with pancreatic duct obstruction are at risk of developing pancreatic sepsis if contrast is injected beyond the site of obstruction.

The complication rate following endoscopic retrograde pancreatography can be minimized by careful fluoroscopic monitoring while contrast is injected slowly to prevent overfilling the pancreatic duct or complete opacification of pancreatic pseudocysts or obstructed ducts.

REFERENCES

1. Burckhardt H, Müller W: Versuche über die Punktion der Gallenblase und ihre Röntgendarstellung. *Dtsch Z Chir* 1921;162:168–197.

2. Mirizzi PL: La cholangiografia durante las operaciones de las vias biliares. *Bol Trab Soc Buenos Aires* 1932;16:1133.

3. Burhenne HJ: Percutaneous extraction of retained biliary tract stones: 661 patients. *AJR* 1980;134:888–898.

4. Mueller PR, vanSonnenberg E, Simeone JF: Fine-needle transhepatic cholangiography: indications and usefulness. *Ann Intern Med* 1982;97:567–572.

5. Teplick SK, Haskin PH, Paviledes CA, et al: Percutaneous transcholecystic cholangiography: experimental study. *AJR* 1985;144:1059–1063.

6. Jolly PC, Baker JW, Schmidt HM, et al: Operative cholangiography: a case for its routine use. *Ann Surg* 1968;168:551–565.

7. Schulenburg CAR: Operative cholangiography: 1000 cases. *Yearbook Radiol* 1970;242.

8. Burhenne HJ: Operative and postoperative biliary tract. in Margulis AR, Burhenne HJ (eds): *Alimentary Tract Roentgenology*, vol. 2, ed 2. St. Louis, CV Mosby, 1983, p 1772.

9. Stewart ET, Vennes JA, Geenen JE: *Atlas of Endoscopic Retrograde Cholangiopancreatography*. St. Louis, CV Mosby, 1977.

10. Bilbao MK, Dotter CT: Reflux cholangiography in sphincteroplasty or enterobiliary anastomosis. *Radiology* 1975;115:585–588.

11. Rösch J, Antonovic R, Dotter CT: Transjugular approach to the liver, biliary system and portal circulation. *AJR* 1975;125:602–608.

12. Rabinov KR, Simon M: Peroral cannulation of the ampulla of Vater for direct cholangiography and pancreatography. *Radiology* 1965;85:693–697.

13. Burhenne HJ, Murray JB: Intravenous cholangiography, in Margulis AR, Burhenne HJ (eds): *Alimentary Tract Roentgenology*, vol. 2, ed 3. St. Louis, CV Mosby, 1461–1478.

14. Ashmore JD, Kane JJ, Pettit HS, et al: Experimental evaluation of operative cholangiography in relation to calculus size. *Surgery* 1956;40:191–196.

15. Thompson WM, Halvorsen A, Gedgaudas RK, et al: High kVp versus low kVp for T-tube and operative cholangiography. *Radiology* 1983;146:635–642.

16. Thompson WM, Halvorsen A, Foster WL, et al: Optimal cholangiographic technique for detecting bile duct stones. *AJR* 1986;146:537–541.

17. Wolfman NT, Short WF: T-tube cholangiography with barium after reaction to iodinated contrast medium. *JAMA* 1975;232:523–524.

18. Serance SR, Dagradi AE: Endoscopic retrograde cholangiography using barium sulphate. *Am J Gastroenterol* 1978;69:57–62.

19. Dellinger EP, Kirshenbaum G, Weinstein M, et al: Determinants of adverse reaction following postoperative T-tube cholangiogram. *Ann Surg* 1980;191:397–403.

20. Dyck WP, Janowitz HD: Effect of glucagon on hepatic bile secretion in man. *Gastroenterology* 1971;60:400–404.

21. Boles ET: Postoperative pancreatitis. *Arch Surg* 1956;73:710–718.

22. Hershey JE, Hillman FJ: Fatal pancreatic necrosis following choledochotomy and cholangiography. *Arch Surg* 1955;71:885–889.

23. Bardenheier JA, Kaminski DL, Willman VL: Pancreatitis after biliary tract surgery. *Am J Surg* 1968;116:773–776.

24. Millbourn E: Klinische studien über die choledocholithiasis. *Acta Chir Scand* 1941;86(suppl 65):1–310.

25. Vernava A, Andrus C, Herrmann VM, et al: Pancreatitis after biliary tract surgery. *Arch Surg* 1987;122:575–580.

26. Ott DJ, Gelfand DW: Complications of gastrointestinal radiologic procedures. II. Complications related to biliary tract studies. *Gastrointest Radiol* 1981;6:47–56.

27. Rholl KS, Smothers RL, McClennan BL, et al: Intravenous cholangiography in the CT era. *Gastrointest Radiol* 1985;10:69–74.

28. Goodman MW, Ansel HJ, Vennes JA, et al: Is intravenous cholangiography still useful? *Gastroenterology* 1980;79:642–645.

29. Eubanks B, Martinez CR, Mehigan D, et al: Current role of intravenous cholangiography. *Am J Surg* 1982;143:731–733.

30. Alinder G, Nilsson U, Lunderquist A, et al: Preoperative infusion cholangiography compared to routine operative cholangiography at elective cholecystectomy. *Br J Surg* 1986;73:383–387.

31. Fuchs WA, Preisig R: Prolonged drip infusion cholangiography. *Br J Radiol* 1975;48:539–544.

32. Ansell G, Tweedie MCK, West CR, et al: The current status of reactions to intravenous contrast media. *Invest Radiol* 1980;6:32–39.

33. Ansell G: Adverse reactions to contrast agents, scope of problem. *Invest Radiol* 1970;5:374–391.

34. Taenzer V, Volkhart V: Double blind comparison of meglumine iotroxamate, meglumine iodoxamate, and meglumine ioglycamate. *AJR* 1979;132:55–58.

35. Howard JM, Short WF: An evaluation of pancreatography in suspected pancreatic disease. *Surg Gynecol Obstet* 1969;129:319–324.

36. Cooperberg PL, Cohen MM, Graham M: Ultrasonographically guided percutaneous pancreatography: report of two cases. *AJR* 1979;132:662–663.

37. Matter D, Bret PM, Bretagnolle M, et al: Pancreatic duct: ultrasound guided percutaneous opacification. *Radiology* 1987;163:635–636.

38. McCune WS, Shrob PE, Moscovitz H: Endoscopic cannulation of the ampulla of Vater: a preliminary report. *Ann Surg* 1968;167:752–756.

39. Sable RA, Rosenthal WS, Siegel J, et al: Absorption of contrast medium during ERCP. *Dig Dis Sci* 1983;28:801–806.

40. Roszler MH, Campbell WL: Post-ERCP pancreatitis: association with urographic visualization during ERCP. *Radiology* 1985;157:595–598.

41. Hamilton I, Lintott DJ, Rothwell J, et al: Metrizamide as contrast medium in endoscopic retrograde cholangio-pancreatography. *Clin Radiol* 1982;33:293–295.

42. Hannigan BF, Kelling PWN, Slavin B, et al: Hyperamylasemia after ERCP with ionic and non-ionic contrast media (letter). *Gastrointest Endosc* 1985;31:109–110.

43. Bilbao MK, Dotter CT, Lee TG, et al: Complications of endoscopic retrograde cholangio-pancreatography: a study of 10,000 cases. *Gastroenterology* 1976;70:314–320.

Part V

Myelographic Agents

Chapter 22

Water-Insoluble Myelographic Agents

Sven E. Ekholm and Jack H. Simon

Before the introduction of computed tomography (CT) and magnetic resonance imaging (MRI), conventional radiographic evaluation of the spinal cord and the subarachnoid space (SAS) required the injection of a contrast agent because of the limited contrast between soft tissues (intervertebral disc and spinal cord) and cerebrospinal fluid (CSF) using conventional x-ray film technique.

Gas was the first contrast agent used in the subarachnoid space to allow visualization of the spinal cord and intracranial structures.[1–3] The commonly used gases (air, oxygen and nitrous oxide) had the advantages of being nontoxic to the central nervous system (CNS) and were spontaneously absorbed after the examination. Gas was used mainly for localization of lesions, but was inadequate for definition of fine structures, such as nerve roots and blood vessels. Gaseous contrast agents were also associated with severe side effects due to disturbed pressure balance. The need for a positive contrast agent became evident.

LIPIODOL

As with many other discoveries, the application of the first positive myelographic contrast agent, Lipiodol, was the result of an accident.[4] Lipiodol is an iodized poppy seed oil containing 40% (w/v) iodine, which was used previously as a therapeutic agent in the epidural space for treatment of sciatica. During such a therapeutic procedure Lipiodol was accidentally injected into the SAS without any sign of adverse reactions. At the time, Sicard and Forestier also noticed that the viscous, CSF-immiscible Lipiodol flowed freely in the SAS when the patient was tilted head downward. The potential of this discovery became immediately apparent, and without further testing Lipiodol came into popular clinical use as the myelographic agent of choice.

Six years after the first report describing the application of Lipiodol, Odin et al. reported severe adverse reactions from Lipiodol. These included lower extremity weakness, bladder dysfunction, and meningeal irritation.[5] These side effects could be reduced be refinement of the oily contrast agents.[5] Lipiodol remained a popular contrast agent for myelography despite this report, but widespread popularity was not achieved until the publication of herniated disc syndrome in 1934,[6] and the diagnosis of the herniated disc using Lipiodol in 1936.[7] However, the high viscosity of Lipiodol made it far from an ideal diagnostic agent. It was non-cohesive and formed large, irregular globules, which resulted in relatively poor delineation of nerve roots and small tumors within the spinal canal.

PANTOPAQUE

In the early 1940s a group of researchers at the University of Rochester developed a new, stable agent for use through the entire spinal canal known as iophendylate (Pantopaque).[8,9] Iophendylate is a sterile mixture of ethyl esters of iodophenylundecanoic acids containing 30.5% firmly bound iodine. It has an oily appearance and is normally clear or pale yellow in color. When exposed to sunlight iophendylate becomes discolored and should be discarded. The specific gravity of Pantopaque at 20° C is 1.260, and its viscosity ratio is only about 1/17 that of Lipiodol at 37° C. Compared to Lipiodol, Pantopaque flows more freely, is more cohesive, and is easier to remove from the SAS after completion of an examination.[10]

Still not an ideal contrast agent, the standard iodine concentration of Pantopaque results in a very high contrast density, which reduces the detailed visualization of intraspinal structures. Moreover, the relatively high viscosity compared to CSF and the lack of miscibility with CSF prevent wide filling of nerve root sleeves by Pantopaque. However, these same factors make Pantopaque relatively easy to use. In particular, in the absence of dilution of Pantopaque with CSF, rapid examinations are not necessary, as with the water-soluble contrast agents. Moreover, the relatively high surface tension and specific gravity of Pantopaque simplify its positioning within the spinal canal.

Technique

In the majority of patients Pantopaque is introduced into the SAS in the lumbar region at the L2-3 level or below. After the examination is completed, Pantopaque is typically removed through the injection needle. It may, in some cases, be necessary to introduce a second needle at L5-S1 to facilitate a more complete removal of the contrast agent. Pantopaque can be injected through a cisternal or

C1-2 level puncture, for example, to locate the upper extent of a block. The contrast agent necessarily is then left in situ.

The optimal volume of Pantopaque injected into the SAS varies depending on the area to be examined, size of patient, and volume of SAS. In cases of suspected block it is usually sufficient to use small volumes (2–3 ml) to locate and outline the inferior aspect of the lesion. Commonly, in studies that reveal a block by tumor, this small amount of contrast medium is not removed at the end of examination, which allows re-examination without repuncture to evaluate subsequent treatment. Generally, however, a much larger volume is used to obtain an optimal demonstration of an abnormality within the spinal canal or for screening for suspected lesions. The optimal volume of contrast also varies widely among investigators.[10] For lumbar examinations, 9–24 ml of Pantopaque is often used; for thoracic or complete myelograms, 18–36 ml, and for examinations of the cervical region, 12–18 ml. In some patients the use of relatively large volumes has been recommended to optimize diagnostic accuracy.[11,12]

Pantopaque injected into the SAS will remain for a prolonged time unless it is removed. Ramsey et al. estimated the clearance to be in the range of 0.5-1 ml a year unless the surface area of the contrast medium is increased.[13] Other investigators claim a lower rate of resorption, which eventually takes place in only about 10% of all patients.[14] The standard practice is to remove as much Pantopaque as possible with the expectation of reducing the risks of late adverse reactions, the most important of which is chronic arachnoiditis. In view of experimental evidence supporting a relationship between chronic arachnoiditis and Pantopaque myelography, it seems reasonable to assume that the incidence of these inflammatory changes will be reduced by minimizing the contact time. There are, however, no data available that directly implicate prolonged contact time in the pathogenesis of chronic arachnoiditis. It is of interest that in centers in the United Kingdom it has been a common practice to use a smaller volume of Pantopaque (8 ml or less) with no attempt made to remove the Pantopaque after the examination.

Complications

Considering its wide use, Pantopaque is generally well tolerated and, although not entirely devoid of side effects, the incidence of serious reactions is low. When injected into the SAS, Pantopaque commonly causes a modest CSF pleocytosis and an elevation of total CSF protein, particularly gamma globulin.[15] These changes may be associated with meningeal reactions, such as headache, fever, malaise, and meningismus. In most cases the symptoms are mild.[16,17] Serious but rare reactions associated with Pantopaque may develop with a wide spectrum of symptoms, including such widely different symptoms as altered mental status, myoclonus, muscle spasm, cranial nerve palsies, and seizures.[16–20] In most patients the milder acute reactions remit within 2 or 3 days, but in a few instances they are followed by more serious, delayed reactions that may become chronic. Yet, the subacute and chronic reactions are not always preceded by acute reactions.

A number of these reactions have similarities to a classical hypersensitivity reaction. Luce et al. described two patients with a history of hay fever. Both patients were pretested and had a negative test to intradermal Pantopaque.[21] Subsequent myelograms were uneventful with no adverse reactions during the next few days. In both patients a small amount of the contrast medium was left in the SAS. Nine days after the myelogram one of the patients developed an aseptic meningitis and a simultaneous flare reaction in the skin at the site of the previous sensitivity testing. The other patient had a similar delayed reaction, with symptoms beginning 30 days postmyelography. In both patients the meningeal reaction cleared in several days, followed by improvement in the skin reaction. These reactions, as well as a number of reports of aseptic eosinophilic meningitis, which usually starts 10-30 days after uneventful myelography, are examples of unpredictable complications that show similarities with idiosyncratic reactions. Some of the previously described acute reactions could also be included in this group.

A more well-known complication of Pantopaque myelography is the chronic meningeal reaction. This has been reported as an incidental finding in as many as 67% of all patients, most often without any symptoms.[22] This chronic reaction, commonly called adhesive arachnoiditis, causes symptoms often undistinguishable from those of the underlying disease in a small percentage of patients.[23] The same chronic reaction can, however, also be found after surgery or a diagnostic lumbar puncture alone.[24] When chronic arachnoiditis develops the symptoms are usually not apparent until several months after myelography. The symptoms are most often related to lumbar or sacral root dysfunction,[25,26] but can, in severe cases, appear as a progressive myelopathy or hydrocephalus.[27–29] It has been postulated, but is not generally accepted, that the more severe reactions to Pantopaque are related to demyelinating diseases, such as multiple sclerosis.[29,30]

As mentioned above, the severity of the chronic reactions may be related to contact time with the Pantopaque. Although not proven, this hypothesis is supported to some extent by experiments in animals where there appears to be an association between severe reactions and the retention of large amounts of Pantopaque.[31] In animal studies an increased meningeal reaction has also been associated with blood in the SAS at the time of Pantopaque myelography.[31–35] Increased toxicity is also found when Pantopaque and other iodized oils are emulsified. In one study, animals injected with emulsions of small oil particles all died within 10 minutes to 2 weeks.[36]

Contraindications

There are few contraindications to Pantopaque myelography. These include any contraindications for lumbar puncture, a "bloody" spinal tap, which has been associated with intravasation and subsequent lung embolization,[37] and may also increase the risk of arachnoiditis,[31] and a well-documented iodine reaction. Pantopaque probably should be avoided in a patient with recent lumbar punctures which would increase the risk of an epidural injection. If acute or chronic meningeal reactions occur, an effort should be made to remove the residual contrast medium. Corticosteroids may be given as a brief systemic treatment for symptomatic relief.[38]

Despite these adverse reactions, Pantopaque, when properly used, has a very low neurotoxicity. With the exception of chronic arachnoiditis, the number and frequency of major side effects are very low. Its use in daily practice has, however, diminished markedly recently because of its diagnostic limitations compared to the new, nonionic, water-soluble contrast agents. The water-soluble contrast agents are also useful in CT of the spinal canal where Pantopaque has no application because of its density and poor mixing with CSF. In MRI, Pantopaque may even have a negative impact because it is becoming clear that residual contrast can be misinterpreted as small metastatic lesions.

REFERENCES

1. Dandy WE: Roentgenography of the brain after the injection of air into the spinal canal. *Ann Surg* 1919;70:397–403.

2. Jacobaeus HC: On insufflation of air into the spinal canal for diagnostic purposes in cases of tumors in the spinal canal. *Acta Med Scand* 1921;55:555–564.

3. Lindgren E: Myelography with air. *Acta Psychiatr Neurology* 1939;14:385–388.

4. Sicard JA, Forestier JE: Methode generale d'exploration radiologique par l'huile iode's (Lipiodol). *Bull Soc Med Hop (Paris)* 1922;46:463–468.

5. Odin M, Runstrom G, Lindblom A: Iodized oils as an aid to the diagnosis of lesions of the spinal cord and a contribution to the knowledge of adhesive circumscribed meningitis. *Acta Radiol* 1929;7(suppl):1–85.

6. Mixter WJ, Barr JS: Rupture of intervertebral disc with involvement of spinal canal. *N Engl J Med* 1934;211:210–215.

7. Hampton AO, Robinson JM: The roentgenographic demonstration of rupture of the intervertebral disc into the spinal canal after the injection of Lipiodol: with special reference to unilateral lumbar lesions accompanied by low back pain with "sciatic" radiation. *AJR* 1936;36:782–803.

8. Strain WH, Plati JT, Warren SL: Iodinated organic compounds as contrast media for radiographic diagnosis. I. Iodinated aracyl esters. *J Am Chem Soc* 1942;64:1436–1440.

9. Steinhausen TB, Dungan CE, Furst JB, et al: Iodinated organic compounds as contrast media for radiographic diagnoses. III. Experimental and clinical myelography with ethyl iodophenylundecylate (Pantopaque). *Radiology* 1944;43:230–235.

10. Shapiro R: *Myelography*, ed 4. Chicago, Year Book Medical Publishers, 1984.

11. Heinz RE, Brinker RA, Taveras JM: Advantages of a less dense Pantopaque contrast material for myelography. *Acta Radiol Diagn* 1966;5:1024–1031.

12. Kieffer SA, Peterson HO, Gold LHA, et al: Evaluation of dilute Pantopaque for large-volume myelography. *Radiology* 1970;96:69–74.

13. Ramsey GH, French JD, Strain WH: Iodinated organic compounds as contrast media for radiographic diagnosis. IV. Pantopaque myelography. *Radiology* 1944;43:236–240.

14. Wende S, Schliack H: Zur Frage von Pantopaque-Spatschaden. *Nervenarzt* 1961;32:415–416.

15. Ferry DJ, Gooding R, Standefer JC, et al: Effect of Pantopaque myelography on cerebrospinal fluid fractions. *J Neurosurg* 1973;38:167–171.

16. Davies FL: Effect of unabsorbed radiographic contrast media on the central nervous system. *Lancet* 1956;2:747–748.

17. Kieffer SA, Binet EF, Esquerra JV, et al: Contrast agents for myelography: clinical and radiological evaluation of Amipaque and Pantopaque. *Radiology* 1978;129:695–705.

18. Taren JA: Unusual complication following Pantopaque myelography. *J Neurosurg* 1960;17:323–326.

19. Mayer WE, Daniel EF, Allen MB: Acute meningeal reaction following Pantopaque myelography. *J Neurosurg* 1971;34:396–404.

20. Sinclair DJ, Ritchie GW: Morbidity in post-myelogram patients: a survey of 100 patients. *J Can Assoc Radiol* 1972;23:278–283.

21. Luce JC, Leith W, Burrage WC: Pantopaque meningitis due to hypersensitivity. *Radiology* 1951;57:878–881.

22. Johnson AJ, Burrows EH: Thecal deformity after lumbar myelography with iophendylate (Myodil) and meglumine iothalamate (Conray 280). *Br J Radiol* 1978;51:196–202.

23. Shaw MDM, Russell JA, Grossart KW: The changing pattern of spinal arachnoiditis. *J Neurol Neurosurg Psychiatr* 1978;41:97–107.

24. Epstein BS, Epstein JA: The siphonage technique for the removal of Pantopaque following myelography. *Radiology* 1972;103:353–358.

25. Hurteau EF, Baird WC, Sinclair E: Arachnoiditis following the use of iodized oil. *J Bone Joint Surg* 1954;36A:393–400.

26. Ward W, Matheson M, Gonski A: Three cases of granulomatous arachnoiditis after myelography. *Med J Aust* 1976;2:333–335.

27. Mason MS, Raaf J: Complications of Pantopaque myelography: case report and review. *J Neurosurg* 1962;19:302–311.

28. Jensen F, Reske-Nielsen E, Ratjen E: Obstructive hydrocephalus following Pantopaque myelography. *Neuroradiology* 1980;18:139–144.

29. Gupta SR, Naheedy MG, O'Hara RJ, et al: Hydrocephalus following iophendylate injection myelography with spontaneous resolution: case report and review. *Comput Radiol* 1985;9:359–364.

30. Kaufmann P, Jeans WD: Reactions to iophendylate in relation to multiple sclerosis. *Lancet* 1976;2:1000–1001.

31. Bergeron RT, Rumbaugh CL, Fang H, et al: Experimental Pantopaque arachnoiditis in the monkey. *Radiology* 1971;99:95–101.

32. Howland WJ, Curry JL, Butler AK: Pantopaque arachnoiditis: experimental study of blood as a potentiating agent. *Radiology* 1963;80:489–491.

33. Howland WJ, Curry JL: Experimental studies of Pantopaque arachnoiditis. *Radiology* 1966;87:253–257.

34. Howland WJ, Curry JL: Pantopaque arachnoiditis: experimental study of blood as a potentiating agent and corticosteroids as an ameliorating agent. *Acta Radiol Diagn* 1966;5:1032–1041.

35. Howland WJ, Curry JL, Bell RO: Experimental studies of Pantopaque arachnoiditis. II. Laboratory Studies. *Radiology* 1966;87:257–258.

36. Jaeger R: Irritating effect of iodized vegetable oils on the brain and spinal cord. *Arch Neurol Psychiatr* 1950;64:715–719.

37. Aspelin P, Lester J: Pantopaque pulmonary embolism following myelography. *Neuroradiology* 1977;14:43–44.

38. Junck L, Marshall WH: Neurotoxicity of radiological contrast agents. *Ann Neurol* 1983;13:469–484.

Chapter 23

Water-Soluble Myelographic Agents

Sven E. Ekholm and Jack H. Simon

The limitations of the oil-based contrast agents were recognized soon after their initial description, leading to efforts to develop nontoxic, water-soluble agents. Nine years after the introduction of Lipiodol for myelography, the first report describing a water-soluble contrast agent for myelography was published.[1] This contrast agent, sodium methiodal, had some definite advantages when compared to the oil-based agents. Its lower viscosity and density and improved miscibility with cerebrospinal fluid (CSF) allowed finer detail to be detected. Moreover, methiodal was spontaneously absorbed, which eliminated the need for removal following each investigation. However, sodium methiodal was very irritating. This necessitated spinal anesthesia and hindered its use above the lumbar region. Meglumine iothalamate, introduced in 1964,[2] could be used without supplementary spinal anesthesia, but was too toxic for use above the lumbar region, as was meglumine iocarmate.[3]

A major improvement in contrast agent design occurred in 1972 with the introduction of the first nonionic, water-soluble contrast medium, metrizamide (Amipaque). This contrast agent was the result of a suggestion by Almén that nonionic contrast agents would have a lower osmolality than ionic agents with an identical iodine concentration.[4] Metrizamide also proved to be far less neurotoxic than the ionic agents and was less likely to induce arachnoiditis.[5] With this contrast agent there was finally a marked improvement in diagnostic quality of subarachnoid investigations, and metrizamide became the first water-soluble contrast agent generally used for investigation of the entire subarachnoid space (SAS).

However, increased use of metrizamide, especially in high doses, soon made it apparent that this new agent was not completely free of neurotoxic properties.[6,7] Most of the adverse reactions associated with metrizamide are minor, such as headache (reported in 21–68% of patients)[8] and nausea (reported in 25–40% of patients),[9,10] but there have also been a significant number of more serious reactions, such as mental disturbances, cortical blindness, aphasia, encephalopathy, and seizures.[6,7,11–15]

However, the idea of a nonionic, water-soluble contrast agent was firmly and rapidly established. Metrizamide has been followed by a number of both monomeric and dimeric nonionic agents. Most of these have been developed for vascular use, but two, iopamidol (Isovue) and iohexol (Omnipaque), have already been registered in the United States for intrathecal use and will probably replace metrizamide for routine clinical application. Despite the substantial improvements in toxicity, no contrast agent has yet been produced that is entirely inert.[16]

PHYSIOLOGIC SYSTEMS

In this chapter, the physiologic systems of the central nervous system (CNS) that are believed to be affected by water-soluble contrast agents are reviewed, as are the hypotheses for contrast agent neurotoxicity.

Blood-Brain Barrier

To understand the origin of the adverse reactions that can develop after myelography, it is important to examine the unique environment of the CNS. The CNS seems to require a very stable chemical environment to function properly. This stability is accomplished normally through a variety of control mechanisms.[17,18] The concentration of several ions and nonelectrolytes in the CSF and the extracellular space of the CNS differs from that in serum.[19] Most significant is a lower concentration of protein, glucose, and potassium and a higher concentration of magnesium. The composition of CSF thus largely depends on specific transport mechanisms,[20] and many water-soluble solutes in the blood, such as contrast agents, either do not pass beyond the blood-brain barrier (BBB) or enter only slowly.[6]

Contrast agents injected into the SAS bypass the blood-brain barrier, diffuse through the extracellular spaces of the brain, and may then interfere with the delicate balance necessary to maintain normal neural function.[21]

Normal CSF Production

Water-soluble contrast agents mix with CSF and are transported out of the SAS along with CSF. This absorption into the blood is largely regulated by the production rate of CSF. Most of the CSF formation takes place in the ventricular system of the brain, and the major portion is normally derived from the choroid plexus. Experiments with isolated choroid plexus preparations indicate that 80% or more of the CSF is produced from this source alone.[22,23] The parenchyma of the CNS is probably the principal source of extrachoroidal CSF production.[24,25]

The mechanism of CSF production is complex and the current explanations are largely speculative.[26–28] It is believed that an ultrafiltrate of plasma passes through the capillary endothelium of the choroid plexus into the surrounding interstitial tissue as a result of hydrostatic pressure. This filtration can occur because the capillaries of the choroid plexus lack tight junctions. Further transport, at least for larger molecules, is hindered by the tight junctions between the epithelial cells covering the choroid plexus.[29] The transport of ions from the interstitial tissue to the CSF is partially dependent on intracellular carbonic anhydrase and an active sodium-potassium pump. The transfer of chloride ions may either be coupled to this pumping process, or there may be a separate transport system.[20,30] It is assumed that the resultant local osmotic forces are responsible for the movement of water. This concept is supported by the negligible difference in the osmolalities of plasma, plasma ultrafiltrate, and CSF.[17]

Interference with this complex system of CSF production could result in a secondary effect on the clearance of contrast agents and other substances from the CSF. Intravascular injection of certain drugs—for example acetazolamide[31] and furosemide[32,33]—has been shown to reduce CSF production markedly in animal experiments and in humans.[34] Other factors that may interfere with CSF production are rapid changes in serum osmolality caused by fluid injection and changes in the acid-base balance of serum. Rapid increases in serum osmolality in animals result in reduced CSF formation, whereas an acute drop in osmolality has the opposite effect.[35] Based on studies in an animal model it has been suggested that dehydration may decrease the absorption rate of CSF and contrast agents after myelography, thereby increasing the risk of toxic effects from the contrast agent.[36] With respect to disturbances in the acid-base balance, alkalosis is the only known factor that has any significant influence by causing a decrease in CSF production.[37]

Effect of Contrast Agents on CSF Production

Contrast agents may also interfere with CSF production. Harnish et al.[38] were able to demonstrate a reduction in CSF production after intravenous (IV) injection of an ionic contrast agent and later also with a nonionic contrast agent, iohexol.[39] The choroid plexus is known to be innervated by both adrenergic[40] and cholinergic[41] fibers. Stimulation of the adrenergic fibers diminishes CSF flow by about 30% acutely, and denervation results in an increased flow by approximately 30%.[40] Cholinergic innervation has the opposite effect, with stimulation causing an increased CSF flow by as much as 100% without changes in blood flow to the choroid plexus.[42] In an invertebrate preparation, Marder et al.[43] demonstrated that metrizamide acts both as an antagonist of cholinergic transmission and an inhibitor of the enzyme acetylcholinesterase. Dawson[44] showed that both ionic and nonionic contrast agents exerted a relative inhibitory effect on cholinesterase, which was least pronounced with iohexol. This effect may be the cause of the reduction in CSF flow after an IV injection of contrast agents.[38]

Absorption of CSF/Contrast Agents

The absorption of CSF into the blood is the result of a hydrostatic gradient, with the absorption rate relatively linear over a wide physiologic range.[34,45,46] When the pressure is raised above normal physiologic levels the resistance to flow also seems to diminish.[47] The absorption of CSF under normal physiologic conditions has until recently been regarded as occurring through the arachnoid granulations and villi alone. Indications of a supplementary lymphatic drainage in earlier investigative work have been largely ignored.[48,49] The arachnoid villi and granulations are ideally situated to drain CSF from the SAS into the major dural sinuses. The importance of these structures in CSF drainage has been well established.[48–50] Smaller arachnoid granulations also exist along the spinal nerve roots.[51] The mechanisms accounting for CSF drainage through the granulations and villi are controversial; some proponents favor a transendothelial vacuolization process[52] (a closed channel theory), whereas others favor an open, endothelial-lined channel system.[27,53]

Recent studies have shown that CSF is also normally drained via prelymphatic pathways through the cribriform plate to the cervical lymph nodes.[54,55] It seems reasonable, although not yet proven, that this kind of lymphatic drainage may also take place in other regions, including the spinal canal. Bradbury et al. demonstrated that a substantial amount of CSF is absorbed via prelymphatic pathways through the cribriform plate and that the magnitude of the flow is subject to hydrostatic and postural factors. This pathway may be important, particularly for larger molecules, but the rate of recovery of molecules below 5,000 daltons is insignificant in lymph.[55] This finding suggests that the lymphatic drainage pathway is normally only of minor significance for the absorption of water-soluble contrast agents, such as metrizamide, iohexol, and iopamidol, which have a molecular weight less than 1,000 daltons.

CSF Circulation

An adult male has about 150–200 ml of CSF in the SAS, which is replaced approximately three times a day. The CSF produced in the ventricles is continuously transported out of the ventricular system by pulsatile volumetric changes of the brain and the vessels of the choroid plexus that correspond with the cardiac cycle.[56] CSF is also propelled toward the fourth ventricle and out of the ventricular system by currents produced by the ependymal cilia.[57] There is a downward gravitational flow of CSF along the spinal cord. CSF reaches the lumbar sac in 60–90 minutes as measured by tracers injected into the cisterna magna.[58] There are also some indications for cranial-directed circulation of lumbar CSF. Labeled albumin injected into the SAS could be detected in the basal cisterns after approximately 1 hour and reached the intracranial cavity in about 12 hours.[59] Studies of CSF absorption after lumbar myelography in rabbits proved, however, that most of the contrast medium is absorbed in the lumbar area and very small amounts reach the cisterna magna.[60–62] Because there are some differences between humans and rabbit in CSF flow dynamics, a different absorption pattern might be expected in humans. It has been shown that a significant proportion of the contrast agent is absorbed in the lumbar region after lumbar myelography. In normal radioiodinated human serum albumin (RIHSA) cisternography, however, after the tracer is injected into the lumbar SAS it is distributed over the cerebral SAS in 10-12 hours.[63] Because the biologic half-life of metrizamide injected intrathecally in humans has been calculated to be approximately 4 hours, it thus seems reasonable to believe that a large portion of the contrast agent is absorbed in the spinal region.[64] Moreover, if there had not been a significant absorption of contrast in the lumbar region, the toxicity of the early ionic agents, such as sodium methiodal, would have made these agents obsolete for lumbar myelography as well.

Extracellular Space and Contrast Media Diffusion

In contrast to the blood-brain barrier,[65,66] there apparently is no barrier between the CSF and brain. The cells of the pia mater and ependyma are joined by gap junctions that allow the free exchange of water-soluble substances between the CSF and the extracellular space of the CNS.[29,65,67,68] There are a few exceptions. For example, the epithelial cells covering the choroid plexus are joined by tight junctions similar to those seen in the endothelial cells of the blood-brain barrier.[29] Aside from such exceptions, this lack of a barrier allows water-soluble substances, such as contrast agents injected into the SAS, to enter the extracellular space of the CNS and come into direct contact with the neurons and glia. Penetration of metrizamide into the CNS has been demonstrated in animal experiments,[69–71] as well as in patients.[72] Experimental data indicate that penetration is by simple diffusion[71] and that there is no difference between metrizamide and ionic contrast agents.[70] The higher toxicity of ionic contrast agents is therefore unrelated to differences in the depth of penetration or the rate of diffusion. However, factors that may influence the parenchymal concentration of a contrast agent are important because a direct relationship has been demonstrated between cerebral contrast agent concentration and complication rate following intrathecal administration of metrizamide.[73,74] Such a relationship could also be seen with iohexol where the incidence of headache and nausea increased when more than 17 ml (180 mg I/ml) was used.[75] Fenstermacher et al.[71] found a slight difference in the distribution of metrizamide in the neural tissue of rabbits compared to an extracellular marker (EDTA) and raised the suspicion of some cellular entry or binding.

Myelographic studies in rabbits by Ekholm et al. also suggested some retention of metrizamide in the neural tissue,[61] whereas iohexol seemed to follow a simple diffusion model.[62] Similarly, data from a perfusion study comparing metrizamide and iohexol demonstrated a smaller distribution volume for metrizamide when the spinal canal was perfused over a 4-hour period with an artificial CSF-containing contrast agent.[76] The lower distribution volume of metrizamide,[71,76] as well as the possible retention of metrizamide,[61] could be related to a membrane binding or cellular uptake. However, intracellular localization of metrizamide has never been shown,[69] and it has been considered unlikely that metrizamide, even in high doses, will diffuse into brain cells.[77] This difference between metrizamide and iohexol might thus result in a shorter neuron contact time, provided that the two contrast agents have identical absorption rates from the CSF.

NEUROTOXICITY

The major problem with water-soluble contrast agents is their neurotoxicity. The oil-based contrast agents differ in this respect because most remain in the subarachnoid space and are not in direct contact with the neurons. The oil-based contrast agents also do not cause any major disturbance in the delicate balance between the extra- and intracellular ions and osmotic forces.

The conventional ionic contrast agents are toxic to neural tissue. They can induce epileptiform activity when applied to cortical tissue,[78] most likely as a result of the free salt-forming ions and not their osmolality.[79] However, hyperosmolality can affect CSF composition as shown by Maly and Fex[21] who demonstrated that injection of hyperosmolar solutions of iohexol or metrizamide into the SAS of rabbits resulted in an increase of calcium and magnesium ions relative to sodium and potassium ions. Metrizamide has a similar although less pronounced effect as an isotonic solution. Isotonic iohexol was completely inert in this aspect. This relative increase in calcium and magnesium ions may be the origin of the depressive effect associated with hyperosmolality.[80]

In vitro experiments have also demonstrated a marked depressive effect of hypertonicity on glucose metabolism.[81] The depression of electrical activity from hypertonic iohexol[80] could be the result of a shift in the magnesium gradient in neurons as a result of an energy-depleted magnesium pump. The effect of hypertonicity on CSF composition has not been studied in humans.

Nonionic metrizamide possesses certain neurotoxic qualities that are not explained by hyperosmolality. In vitro electrophysiologic experiments have shown that metrizamide causes early excitatory changes in electrical activity and a mild late inhibitory effect from hypertonicity.[80] Iohexol (300 mg I/ml) caused only inhibition, which was more marked when the contrast agent concentration, and subsequently the osmolality of the CSF/contrast agent solution, was increased. Ionic contrast agents with their higher osmolality caused more pronounced depressive as well as excitatory effects. All of these effects were reversible with rinse-out within 30 minutes.[80]

The origin of the neural excitatory effect of metrizamide has not been definitely established. However, the presence of 2-deoxy-D-glucose (2-DG), a known inhibitor of glucose metabolism[82] in the metrizamide molecule, led Bertoni et al to suggest and demonstrate a competition between glucose and metrizamide for hexokinase in vitro.[83] Simon et al. subsequently demonstrated that there was no enzyme inhibitory effect from less neurotoxic nonionic contrast agents, including iohexol and iopamidol.[84]

Because glucose is the main source of energy in the CNS, the brain is extremely vulnerable to factors that interfere with glucose metabolism.[85,86] This vulnerability is heightened by the inability of the brain to store large amounts of glucose as glycogen. Consequently, the brain is dependent on an almost continuous supply of glucose.[87] However, hexokinase inhibition in vivo requires that metrizamide or an inhibitory derivative be transported intracellularly because hexokinase is an intracellular enzyme. To date, intracellular localization of metrizamide has not been proven.[69,77]

In vitro experiments have shown that metrizamide added to neural tissue slices depresses carbon dioxide production.[88] This depression indicates either a metabolic block or reduced glucose utilization secondary to reduced neural activity. Such a disturbance of metabolism is not seen with iohexol[81] or iopamidol.[89] There is, however, an effect from hyperosmolality,[81] which might be expected to be less significant in vivo because water would rapidly diffuse into the hyperosmotic areas.

Further suggestions that metrizamide neurotoxicity might be the result of an effect on glucose metabolism come from two types of deoxyglucose autoradiography experiments. Bech et al. demonstrated a generalized decrease in cerebral glucose utilization in rats exposed to subarachnoid metrizamide.[90] Simon et al. were able to demonstrate a localized decrease in cerebral glucose utilization using 14C-deoxyglucose autoradiography in rabbits in regions of locally high concentrations of metrizamide.[91] Animals were placed in a decubitus position, with the test drug layered over the dependent hemisphere for 3-6 hours; the opposite hemisphere acted as an internal control. In the same study, iohexol had no inhibitory effect.

Glucose is transported across the blood-brain barrier and intracellularly by means of a stereospecific membrane carrier.[65] This carrier can be competitively inhibited by a structural analog, such as 2-DG, but it will not transport the L-form of glucose.[19,87,92–94] The idea of a competitive inhibition of intracellular glucose transport by metrizamide remains an attractive hypothesis. However, more recent experiments suggest that metrizamide neurotoxicity may not be explained simply as an effect on glucose transport and/or metabolism. Isotopic-labeled glucose transport studies in synaptosomes and tissue slices have not shown a simple competitive inhibition pattern that might be expected for direct interference with transport mechanisms.[95] Continued experiments on rat brain synaptosomes have, however, shown significant alterations in both calcium and transmitter fluxes caused by metrizamide, as well as other nonionic contrast agents.[96]

CLINICAL APPLICATION

Although the myelographic technique is not the subject of this chapter, some general principles are discussed that should help minimize adverse reactions. The neurotoxicity problems noted with metrizamide continue to be a problem with the new, second-generation nonionic contrast agents, although the frequency and magnitude of all complications have been reduced and most of the severe complications are extremely rare. Several cautions seem advisable even with these newer contrast agents. For example, patients with known seizure disorders, alcoholism, severe chronic obstructive lung disease, chronic uremia, or known hypersensitivity to iodinated contrast media should probably not be candidates for examination even with these contrast agents unless it is absolutely necessary and appropriate actions have been taken to reduce the risks. The same is true for patients treated with neuroleptics or any other drug that lowers the threshold for seizure. Such drugs should be discontinued if possible for at least 48 hours before and 24 hours after myelography. If seizures arise after myelography, they can be managed initially with IV diazepam (Valium). Elderly and diabetic patients seem to be at relatively increased risk.[97] Similarly, patients with chronic arachnoiditis[5] might be considered a high-risk group due to their impaired absorption of contrast agents. Finally, any contraindication for diagnostic lumbar puncture is also a contraindication for myelography.

The patient should have solid foods withdrawn the night before examination, but a generous liquid diet is important to avoid dehydration. Dehydration may increase both the incidence and severity of adverse reactions as a result of decreased CSF production. If the oral intake is insufficient, fluid should be given parenterally to keep the patient well hydrated. Premedication is not needed, but in the extremely anxious patient it can be helpful to give 5–10 mg diazepam intramuscularly before the examination.

The total amount of contrast agent used should be limited to the minimal amount required to obtain a diagnostically complete study. Current recommendations continue to emphasize limiting total dose to less than 3 g I in the adult patient. Neurotoxicity has been related both to the amount and the localization of the contrast agent. For example, metrizamide located outside the occipital lobes has caused transient cortical blindness.[98] Aphasia has been noted in patients placed in the left lateral decubitus position, which resulted in higher metrizamide concentrations along the left cerebral hemisphere.[99] In a high dose study of iohexol, where a maximum of 4.5 g I was injected for lumbar myelography, there was an increase in the frequency of adverse reactions at the highest dose levels. Despite the high dose levels, however, there were no major adverse reactions.[75] In children, it is recommended that the dose should not exceed 100 mg I/kg body weight, but the size of the CNS may be a more appropriate criterion for dose selection than weight.[100]

At the completion of the examination the patient is typically confined to bed for 18–24 hours. The head should be elevated 30–45 degrees for the first 8 hours to minimize intracranial transport of contrast agent. The patient is then placed flat or the head slightly elevated to reduce the risk of CSF leakage. Restroom privileges are allowed after the initial 8 hours. It has been questioned whether this rigid postmyelography regime is really necessary. In fact, one study showed fewer adverse reactions in ambulatory patients compared to those confined to bed.[101]

As a final impression, it seems that the new nonionic contrast agents presently used for myelography are excellent diagnostic tools with an acceptable toxicity. By taking the appropriate precautions discussed above with regard to contrast toxicity, myelography should be a relatively safe procedure. Further improvements in toxicity may be expected with some of the new dimeric, nonionic water-soluble contrast agents under development and testing. If the neurotoxic properties of these molecules are not greater than those of the monomers, their increased molecular size should be beneficial by reducing diffusion into the CNS with resultant lower tissue concentration. The lower osmolality of these contrast agents, when comparing iodine concentration, should also be beneficial.

REFERENCES

1. Arnell S, Lidstrom F: Myelography with Skiodan (Abrodil). *Acta Radiol* 1931;12:287–288.

2. Campbell RL, Campbell JA, Heimburger RF, et al: Ventriculography and myelography with absorbable radiopaque medium. *Radiology* 1964;82:286–289.

3. Gonsette R, André-Balisaux G: A new hydrosoluble and resorbable contrast material for myelography and ventriculography. Presented at the XII International Congress of Radiology, Tokyo, Oct 9-11, 1969.

4. Almén T: Contrast agent design: Some aspects on the synthesis of water soluble contrast agents of low osmolality. *J Theor Biol* 1969;24:216–226.

5. Haughton VM, Eldevik OP: Complications from aqueous myelographic media: experimental studies, in Sackett JF, Strother CM (eds): *New Techniques in Myelography*. New York, Harper and Row, 1979, pp 184–194.

6. Junck L, Marshall WH: Neurotoxicity of radiological contrast agents. *Ann Neurol* 1983; 13:469–484.

7. Ekholm SE: Adverse reactions to intravascular and intrathecal contrast medium, in Preger L (ed): *Iatrogenic Diseases,* vol 1. Boca Raton, FL, CRC Press, 1986, pp 115–136.

8. Skalpe IO: Adverse effects of water soluble contrast media in myelography, cisternography and ventriculography: a review with special reference to metrizamide. *Acta Radiol* 1977; 355(suppl):359–370.

9. Sackett JF, Strother CM, Quaglieri CE, et al: Metrizamide—CSF contrast medium. *Radiology* 1977;123:779–782.

10. Bergstrom K, Mostrum U: Technique for cervical myelography with metrizamide. *Acta Radiol* 1977;355(suppl):105–109.

11. Gonsette RE: Cervical myelography with a new resorbable contrast medium: Amipaque. *Acta Neuro Belg* 1976;76:283–285.

12. Amundsen P: Metrizamide in cervical myelography: survey and present state. *Acta Radiol* 1977;355(suppl):85–97.

13. Dugstad G, Eldevik P: Lumbar myelography. *Acta Radiol* 1977;355(suppl):17–30.

14. Nickel AR, Salem JJ: Clinical experience in North America with metrizamide. *Acta Radiol* 1977;355(suppl):409–416.

15. Budny JL, Hopkins LN: Ventriculitis after metrizamide lumbar myelography. *Neurosurgery* 1985;17:467–468.

16. Lamb J, McAllister V, Nelson M, et al: A prospective comparison of iotrolan, iohexol and iopamidol for lumbar myelography. *Acta Radiol* 1986;369(suppl):524–527.

17. Katzman R, Pappius HM: *Brain Electrolytes and Fluid Metabolism*. Baltimore, Williams & Wilkins, 1973.

18. Cutler RWP: Neurochemical aspects of blood-brain-cerebrospinal fluid barriers, in Wood JH (ed): *Neurobiology of Cerebrospinal Fluid*. New York, Plenum Press, 1980, pp 41–51.

19. Fishman RA: *Cerebrospinal Fluid in Diseases of the Nervous System*. Philadelphia, WB Saunders, 1980.

20. Wright EM: Transport processes in the formation of the cerebrospinal fluid. *Rev Physiol Biochem Pharmacol* 1978;83:1–34.

21. Maly P, Fex G: CSF cation changes following subarachnoid injection of iohexol and metrizamide in rabbits. *Acta Radiol Diagn* 1984;1:65–71.

22. Miner LC, Reed DJ: Composition of fluid obtained from choroid plexus tissue isolated in a chamber in situ. *J Physiol (London)* 1972;227:127–139.

23. Welch K: Secretion of cerebrospinal fluid by choroid plexus of the rabbit. *Am J Physiol* 1963; 205:617–624.

24. Cserr HF, Cooper DN, Suri PK, et al: Efflux of radiolabeled polyethylene glycols and albumin from rat brain. *Am J Physiol* 1981;240:F319–F328.

25. Rosenberg GA, Kyner WT, Estrada E: Bulk flow of brain interstitial fluid under normal and hyperosmolar conditions. *Am J Physiol* 1980;238:F43–F49.

26. Davson H: *Physiology of the Cerebrospinal Fluid*. London, J & A Churchill, 1967.

27. Welch K: The principles of physiology of the cerebrospinal fluid in relation to hydrocephalus including normal pressure hydrocephalus, in Friedlander WJ (ed): *Current Reviews. Advances in Neurology*. New York, Raven Press, 1975, pp 247–332.

28. Wood JH: Physiology, pharmacology and dynamics of cerebrospinal fluid, in Wood JH (ed): *Neurobiology of Cerebrospinal Fluid*. New York, Plenum Press, 1980, pp 1–16.

29. Brightman MW: The intracerebral movement of proteins injected into blood and cerebrospinal fluid of mice, in Lajhta A, Ford DH (eds): *Progress in Brain Research*. Amsterdam, Elsevier, 1968, pp 19–37.

30. Segal MB, Pollay M: The secretion of cerebrospinal fluid. *Exp Eye Res* 1977;25(suppl): 127–148.

31. Tschirgi RD, Frost RW, Taylor JL: Inhibition of cerebrospinal fluid formation by carbonic anhydrase inhibitor, 2-acetylamino-1,3,4-thiadiazole-5-sulfonamide (Diamox). *Proc Soc Exp Biol Med* 1954;87:373–376.

32. Reed DJ: The effect of furosemide on cerebrospinal fluid flow in rabbits. *Arch Int Pharmacodyn* 1969;178:324–330.

33. Buhrley LE, Reed DJ: The effect of furosemide on sodium-22 uptake into cerebrospinal fluid and brain. *Exp Brain Res* 1972;14:503–510.

34. Cutler RWP, Page L, Galicich J, et al: Formation and absorption of cerebrospinal fluid in man. *Brain* 1968;91:707–720.

35. Hochwald GM, Wald A, Malhan C: The sink action of cerebrospinal fluid volume flow. *Arch Neurol* 1976;33:339–344.

36. Eldevik OP, Haughton VM, Sasse EA: The effect of dehydration on the elimination of aqueous contrast media from the subarachnoid space. *Invest Radiol* 1980;15:155–157.

37. Oppelt WW, Maren TH, Owens ES, et al: Effects of acid-base alterations on cerebrospinal fluid production. *Proc Soc Exp Biol* 1963;114:86–89.

38. Harnish PP, DiStefano V: Decreased cerebrospinal fluid production by intravenous sodium diatrizoate. *Invest Radiol* 1984;19:318–323.

39. Harnish PP, DiStefano V: Pharmacological action of radiographic contrast media. Reduced cerebrospinal fluid production in the dog. *J Pharmacol Exp Ther* 1985;232:88–93.

40. Lindvall M, Edvinsson L, Owman C: Sympathetic nervous control of cerebrospinal fluid production from the choroid plexus. *Science* 1978;201:176–178.

41. Edvinsson L, Nielsen KC, Owman C: Cholinergic innervation of choroid plexus in rabbits and cats. *Brain Res* 1973;63:500–503.

42. Haywood JR, Vogh BP: Some measurements of automatic nervous system influence on production of cerebrospinal fluid in the cat. *J Pharmacol Exp Ther* 1979;208:341–346.

43. Marder E, O'Neil M, Grossman RI, et al: Cholinergic actions of metrizamide. *AJNR* 1983; 4:61–65.

44. Dawson P: Contrast media and enzyme inhibition. I. Cholinesterase. *Br J Radiol* 1983; 56:653–656.

45. Katzman R, Hussey F: A simple constant-infusion manometric test for measurement of CSF absorption. I. Rationale and method. *Neurology* 1970;20:534–544.

46. Mortensen OA, Weed LH: Absorption of isotonic fluids from the subarachnoid space. *Am J Physiol* 1934;108:458–468.

47. Davson H, Hollingsworth G, Segal MB: The mechanism of drainage of the cerebrospinal fluid. *Brain* 1970;93:665–678.

48. Key EAH, Retzius MG: *Studien in der Anatomie des Nervensystems und des Bindegewebe*. Stockholm, Samson & Wallin, 1875.

49. Weed LH: Studies on cerebrospinal fluid. No. III. The pathways of escape from the subarachnoid spaces with particular reference to the arachnoid villi. *J Med Res* 1914;31:51–91.

50. Welch K, Friedman V: The cerebrospinal fluid valves. *Brain* 1960;83:454–469.

51. Kido DK, Gomez DG, Pavese AM, Jr, et al: Human spinal arachnoid villi and granulations. *Neuroradiology* 1976;11:221–228.

52. Tripathi RC: Ultrastructure of the arachnoid matter in relation to outflow of cerebrospinal fluid. A new concept. *Lancet* 1973;2:8–11.

53. Lee BCP, Gomez DG, Potts DG, et al: Passage of Amipaque (metrizamide) through the arachnoid granulations. *Neuroradiology* 1979;17:185–190.

54. Foldi M, Csillik B, Zoltan OT: Lymphatic drainage of the brain. *Experimentia* 1968; 24:1283–1287.

55. Bradbury MWB, Westrop RJ: Factors influencing exit of substances from cerebrospinal fluid into deep cervical lymph of the rabbit. *J Physiol* 1983;339:519–534.

56. Bering EA Jr: Introductory remarks for conference on the cerebrospinal fluid and the extracellular fluid of the brain. *Fed Proc* 1974;33:2061–2063.

57. Cathcart RS 3rd, Worthington WC Jr: Ciliary movement in the rat cerebral ventricles: clearing action and direction of current. *J Neuropath Exp Neurol* 1964;23:609–618.

58. DiChiro G, Hammock MK, Bleyer WA: Spinal descent of cerebrospinal fluid in man. *Neurology* 1976;26:1–8.

59. DiChiro G: Movement of the cerebrospinal fluid in human beings. *Nature (London)* 1964; 204:290–291.

60. Golman K, Wiik I, Salvesen S: Absorption of a non-ionic contrast agent from cerebrospinal fluid to blood. *Neuroradiology* 1979;18:227–233.

61. Ekholm SE, Foley M, Kido DK, et al: Lumbar myelography with metrizamide in rabbits—An investigation of contrast media penetration and absorption. *Acta Radiol Diagn* 1984;25:517–522.

62. Ekholm SE, Foley M, Morris TW, et al: Neural tissue uptake and clearance of iohexol following lumbar myelography in rabbits. *Acta Radiol Diagn* 1985;26:331–336.

63. Henriksson L, Voigt K: Age-dependent differences of distribution and clearance patterns in normal RIHSA cisternograms. *Neuroradiology* 1976;12:103–107.

64. Speck U, Schmidt R, Volkhardt V, et al: The effect of position of patient on the passage of metrizamide (Amipaque), meglumine iocarmate (Dimer-X) and ioserinate (Myelografin) into the blood after lumbar myelography. *Neuroradiology* 1978;14:251–257.

65. Bradbury MWB: *The Concept of a Blood-Brain Barrier*. Chichester, NY, Wiley, 1979.

66. Katzman R: Blood-brain-CSF barriers, in Siegel GJ, Albers RW, Katzman R, Agranoff BW (eds): *Neurochemistry,* ed 2. Boston, Little, Brown & Co, 1976, pp 414–428.

67. Brightman MW, Reese TS: Junctions between intimately opposed cell membrane in the vertebrate brain. *J Cell Biol* 1969;40:648–677.

68. Oldendorf WH, Davson H: Brain extracellular space and the sink action of the cerebrospinal fluid. *Arch Neurol* 1967;17:196–205.

69. Golman K: Distribution and retention of 125-I-labelled metrizamide after intravenous and suboccipital injection in rabbit, rat and cat. *Acta Radiol* 1973;335(suppl):300–311.

70. Sage MR, Wilcox J, Evill CA, et al: Brain parenchyma penetration by intrathecal ionic and nonionic contrast media. *AJNR* 1982;3:481–483.

71. Fenstermacher JD, Bradbury MWB, duBoulay G, et al: The distribution of 125-I-metrizamide and 125-I-diatrizoate between blood, brain and cerebrospinal fluid in the rabbit. *Neuroradiology* 1980;19:171–180.

72. Drayer BP, Rosenbaum AE: Metrizamide brain penetrance. *Acta Radiol* 1977; 355(suppl):280–293.

73. Caille JM, Guibert-Trainer F, Howa JM, et al: Contamination encephalic par la metrizamide apres myelographie. *J Neuroradiol* 1980;7:3–12.

74. Ekholm S, Fischer H: Neurotoxicity of metrizamide. *Arch Neurol* 1985;42:24–25.

75. Simon J, Ekholm SE, Kido DK, et al: High-dose iohexol myelography. *Radiology* 1987;163:455–458.

76. Holtas S, Morris TW, Ekholm SE, et al: Penetration of subarachnoid contrast medium into rabbit spinal cord. Comparison between metrizamide and iohexol. *Invest Radiol* 1986;21:151–155.

77. Gjedde A: The blood-brain barrier is impermeable to metrizamide. *Acta Neurol Scand* 1982; 66:392–395.

78. Gonsette RE: Biologic tolerance of the central nervous system to metrizamide. *Acta Radiol* 1973;335(suppl):25–44.

79. Hershkowitz N, Bryan RN: Neurotoxic effects of water-soluble contrast agents on rat hippocampus. Extracellular recordings. *Invest Radiol* 1982;17:271–275.

80. Bryan RN, Centeno RS, Hershkowitz N, et al: Neurotoxicity of iohexol: a new nonionic contrast medium. *Radiology* 1982;145:379–382.

81. Ekholm S, Reece K, Foley M, et al: The effect of iohexol on glucose metabolism compared with metrizamide. *Invest Radiol* 1984;19:574–577.

82. Gjedde A: Modulation of substrate transport to the brain. *Acta Neurol Scand* 1983;67:3–25.

83. Bertoni JM, Schwartzman RJ, van Horn G, et al: Asterixis and encephalopathy following metrizamide myelography: investigations into possible mechanisms and review of the literature. *Ann Neurol* 1981;9:366–370.

84. Simon JH, Ekholm SE, Morris TW, et al: Further support for the glucose hypothesis of metrizamide toxicity. *Invest Radiol* 1987;22:137–140.

85. Guroff G: Carbohydrates, in Guroff G (ed): *Molecular Neurobiology*. New York, Marcel Dekker, 1981, pp 284–311.

86. Sokoloff L: Circulation and energy metabolism of the brain, in Siegel GJ, Alber RW, Katzman R, Agranoff BW (eds): *Neurochemistry*, ed 2. Boston, Little, Brown & Co, 1976, pp 388–413.

87. Crone C: General properties of the blood-brain barrier with special emphasis on glucose, in Cserr H, Fenstermacher JD, Fencl V (eds): *Fluid Environment of the Brain*. New York, Academic Press, 1975, pp 33–46.

88. Ekholm SE, Reece K, Coleman JR, et al: Metrizamide—a potential in vivo inhibitor of glucose metabolism. *Radiology* 1983;147:119–121.

89. Ekholm SE, Morris TW, Fonte D, et al: Iopamidol and neural tissue metabolism. A comparative in vitro study. *Invest Radiol* 1986;21:798–801.

90. Bech L, Diemer NH, Gjedde A: The effect of metrizamide on regional brain glucose metabolism in the rat. *Acta Neurol Scand* 1984;69:249.

91. Simon JH, Ekholm SE, Morris TW, et al: The effect of subarachnoid metrizamide and iohexol on cerebral glucose metabolism in vivo. *Acta Radiol* 1986;369(suppl):549–553.

92. Bachelard HS: Specificity and kinetic properties of monosaccharide uptake into guinea pig cerebral cortex in vitro. *J Neurochem* 1971;18:213–222.

93. Cutler RWP: Neurochemical aspects of blood-brain-cerebrospinal fluid barriers, in Wood JH (ed): *Neurobiology of Cerebrospinal Fluid.* New York, Plenum Press, 1980, pp 41–51.

94. Sokoloff L, Reivich M, Kennedy C, et al: The (14-C) deoxyglucose method for the measurements of local cerebral glucose utilization: theory, procedure, and normal values in the conscious and anaesthetized albino rat. *J Neurochem* 1977;28:897–916.

95. Ekholm SE, Morris TW, Fonte D, et al: Effects of contrast media on neural tissue glucose uptake. Abstract 112, AUR 25th Annual Meeting, Charleston, SC, 1987.

96. Morris TW, Ekholm SE, Marinetti G, et al: Contrast media induced alterations of calcium and dopamine uptake in rat brain synaptosomes. Abstract 110, AUR 25th Annual Meeting, Charleston, SC, 1987.

97. Steiner E, Simon JH, Ekholm SE, et al: Neurologic complications in diabetics after metrizamide lumbar myelography. *AJNR* 1986;7:323–326.

98. Smirniotopoulos JG, Murphy FM, Schellinger D, et al: Cortical blindness after metrizamide myelography. *Arch Neurol* 1984;41:224–226.

99. Butler MJ, Cornell SH, Damasio AR: Aphasia following pluridirectional tomography with metrizamide. *Arch Neurol* 1985;42:39.

100. Pettersson H, Fitz CR, Harwood-Nash DCF, et al: Adverse reactions to myelography with metrizamide in infants, children and adolescents. I. General and CNS effects. *Acta Radiol Diagn* 1982; 23:323–329.

101. Sykes RHD, Wasenaar W, Clark P: Incidence of adverse effects following metrizamide myelography in non-ambulatory and ambulatory patients. *Radiology* 1981;138:625–627.

Chapter 24

Myelographic Agents in Computed Tomography

Sven E. Ekholm and Jack H. Simon

Although computed tomography (CT) is about 100 times more sensitive than conventional x-ray techniques in detecting attenuation differences, in selected patients non-contrast-enhanced CT may be inadequate for defining anatomic structures of interest. In 1974, Greitz and Hindmarsh suggested that the combination of intrathecal metrizamide with CT could be used to improve the morphologic evaluation of the subarachnoid space (SAS) in the skull base region.[1] The CT equipment available at that time, however, limited its usefulness, and conventional film techniques, using either gas and/or water-soluble contrast media, remained the methods of choice. Conventional radiographic cisternography using metrizamide was reported by Grepe in 1975.[2] This technique was subsequently refined in the following years by narrowing the distance from puncture site to the area of interest using lateral C1-2 or suboccipital approaches[3] and by applying complex motion tomography.[4] A large variety of applications of combined subarachnoid contrast x-ray film techniques have now been described, including use in myelography,[5] cerebellopontine angle cisternography, and ventriculography.[6–10]

Metrizamide's neurotoxicity, however, severely limited its use in conventional film technique. Its position had to be carefully controlled to avoid dilution, yet high local concentrations increased the likelihood of adverse reactions. It was, for example, almost impossible to obtain a bilateral study of the internal auditory canals in one examination. With refinements in CT technology, metrizamide and the second-generation nonionic contrast agents, however, can now be used routinely to examine the spinal SAS in its entirety, for ventriculography, and for cisternography when routine CT of the head is insufficient.

METHODS

CT Metrizamide Cisternography (CT-MC)

Technical improvements in CT equipment now permit excellent radiologic studies using metrizamide cisternography (CT-MC). The superb delineation of normal and abnormal structures in the SAS and definition of the margins of the brain and spinal cord make this technique preferable to conventional film techniques in most patients. The required amount of iodine for CT-MC has been estimated to be between 2–10 mg/ml (final concentration) in the CSF.[1,11] This is accomplished using between 2 and 10 ml of a nonionic contrast agent at a concentration of 180–200 mg I/ml. Because these doses are not entirely benign it is important to minimize the amount used intracranially. In our experience, using the latest generation of CT scanners, a total of about 5 ml contrast agent at about 180 mgI/ml is usually adequate for anatomic delineation of any area of interest, provided that a correct injection technique is used. For CSF circulation studies, twice this volume has been recommended.[12]

Various techniques for introduction of the contrast agent have been described.[11,13–15] It is helpful to verify a proper instillation of contrast in the SAS, preferably by using fluoroscopic control. Except for CSF circulation studies where the contrast agent is mixed and positioned in the cervicothoracic region,[16–17] all examinations of the intracranial structures are handled generally in the same way. After injection the needle is withdrawn and the patient tilted head down for rapid transport of the contrast agent to the basal cisterns. It is usually easier to maintain the contrast agent as a bolus if the patient is tilted while in the lateral decubitus position. The bolus can usually be followed by intermittent fluoroscopy until it reaches the cervical region after which the patient is turned prone or supine, depending on the major area of interest, with the head angled down in relation to the back. Better intracranial mixing between CSF and contrast agent is achieved by having the patient cough at that time. Coughing can sometimes also increase CSF leakage in patients with oto- or rhinorrhea, thus improving the likelihood for localization of the damaged area. The patient is expeditiously transferred to CT for completion of the examination. In patients with an indwelling ventricular shunt, it is sometimes desirable to inject the contrast medium through the shunt, a technique that is known as CT metrizamide ventriculography (CT-MV).

These techniques, CT-MC and CT-MV, are used in selected patients when conventional CT with and without intravenous (IV) contrast enhancement fails to establish a cause for hydrocephalus and when lesions, such as small tumors, are suspected within the SAS.[18] It is also possible to localize a lesion more precisely as being either within the intra- or extra-axial space, which may be important preoperative information.

As a tool in preoperative evaluation, CT-MC can also be used to (1) establish the precise location of CSF leakage or to study the flow of CSF; (2) establish the direct communication between a CSF density cyst and the surrounding ventricular or cisternal spaces, and (3) distinguish the freely movable ventricular cysts of cysticercosis when the larvae are alive, which differentiates them from, for example, an ependymal cyst. Cysts that are immobilized by adhesions may demonstrate contrast uptake 5 to 6 hours after introduction of contrast.[19] Delayed uptake would be less characteristic of arachnoid cysts.

CT Air Cisternography (CT-AC)

This technique, initially described by Sortland in 1979,[20] has proven to be sensitive in detecting small acoustic neuromas and causes very minimal discomfort. A total of 4-8 ml of air (or filtered CO_2) is injected by a lumbar approach. The air is allowed to ascend intracranially up to, but not beyond, the cerebellopontine angle (CPA) region by means of patient positioning. The CT examination is performed immediately with thin, overlapping axial slices while the head is kept in a horizontal position with the affected ear turned upward. In many patients it is possible to examine the opposite side by carefully rotating the patient.

As a contrast agent, air is excellent for CPA cisternography because it is compatible with very narrow collimation, which minimizes problems of partial volume averaging with the dense bone in this region. Air has been used in the SAS since 1919[21] and is to this day the only nontoxic contrast agent for neural tissue. It is completely absorbed after the examination, and the use of relatively small volumes in this procedure reduces the problems with CSF pressure disturbances encountered in conventional pneumoencephalography.

CT-Metrizamide Myelography (CT-MM)

The availability of CT for evaluation of spinal diseases has markedly changed the radiologic approach. Conventional CT of the spine has proven to be an accurate and sensitive technique, especially for the evaluation of disc disease and spinal stenosis in the lumbar region. However, most radiologists continue to use myelography as the primary diagnostic approach for the thoracic and cervical regions. Myelography is also the method of choice when the level of involvement cannot be precisely defined.

In selected patients, the diagnostic capabilities of CT are improved by the intrathecal injection of a contrast agent followed by CT-metrizamide myelography (CT-MM). CT-MM can be used as a primary technique (relatively low dose CT-MM) or after myelography. CT-MM may provide detail when the contrast agent has become too diluted for conventional x-ray or tomographic technique or to obtain additional information in selected regions after myelography. This combined approach provides a high degree of diagnostic accuracy in the evaluation of spinal lesions in children.[22] In adult patients myelography before CT-MM may be used to select a suspect area to be studied or to improve the diagnostic accuracy in complicated studies. CT-MM may preclude the need for a second puncture above a myelographic block by demonstrating the upper extent of the lesion or additional lesions when only a minimal amount of contrast medium passes the block.

One potential problem with CT-MM after myelography is the risk of increased contrast agent entering the intracranial subarachnoid space and thus increased toxicity. However, in a comparative study of patients who underwent CT-MM after cervical myelography versus those with cervical myelography alone, there was no significant difference in either the incidence or severity of adverse reactions between the two groups.[23]

When only a limited, predetermined area has to be examined, it is now common practice to use CT-MM as the primary method of examination, thereby reducing the amount of contrast agent injected into the SAS. CT-MM can, for example, be used in adult patients when there is a question of cervical disc disease or syringomyelia or in the patient at risk for contrast toxicity, such as patients with seizure disorders.

In a minority of patients with syringomyelia, there may be early visualization of the cystic cavity as a result of communication between the syrinx and the SAS. In all negative studies where a syringomyelia is still suspected, a delayed CT examination at 6-11 hours is advisable because delayed ''enhancement'' of the syrinx is often found, probably as a result of contrast agent diffusion.[24]

In children, CT-MM can often be used alone to evaluate certain types of lesions.[22] Pettersson and Harwood-Nash retrospectively compared myelography and CT-MM in 240 patients with a wide range of lesions and found that CT-MM frequently provided additional, important information compared to myelography alone.[22]

Because CT is far more sensitive than conventional filming, it is often advantageous to delay the CT-MM after myelography to allow the contrast concentration to decrease in the SAS. The half-life of metrizamide in the SAS and that of probably most of the newer, water-soluble agents is about 4 hours.[25] The time delay in most patients should then be 3–4 hours unless a marked dilution of the contrast agent occurs during myelography, in which case the time interval to CT is shortened. Before the CT scan the patient should be rotated to mix CSF and the contrast agent. When CT-MM is used as a primary procedure, the contrast agent volume should be about 5 ml at a concentration of 180-200 mg I/ml. Ideally, the injection should be given using fluoroscopic control and the contrast bolus positioned as close to the area of interest as possible. The patient is then transferred immediately to the CT scanner for examination.

CONCLUSION

The contrast agents described above are the same as the conventional water-soluble myelographic agents (see Chapter 23). Today, metrizamide has largely been replaced by less toxic agents, including iohexol and iopamidol. Although none of these agents is entirely inert, recent comparative studies suggest that when handled correctly they are relatively safe.

REFERENCES

1. Greitz T, Hindmarsh T: Computer assisted tomography of intracranial CSF circulation using a water-soluble contrast medium. *Acta Radiol Diagn* 1974;15:497–507.

2. Grepe A: Cisternography with the non-ionic water-soluble contrast medium metrizamide. A preliminary report. *Acta Radiol Diagn* 1975;16:146–160.

3. Corrales M, Tapia J: Encephalography with metrizamide. *Neuroradiology* 1977;13:249–254.

4. Roberson GH, Brismar J, Davis KR, et al: Metrizamide cisternography with hypocycloidal tomography: preliminary results. *AJR* 1976;127:965–967.

5. Shapiro R: *Myelography*, ed 4. Chicago, Year Book Medical Publishers, 1984.

6. Siqueira EB, Bucy PB, Cannon AH: Positive contrast ventriculography, cisternography and myelography. *AJR* 1968;104:132–138.

7. Wilkinson HA: Selective third ventricular catheterization for Pantopaque ventriculography. *AJR* 1969;105:348–351.

8. Oftedal SI: Intraventricular application of water-soluble contrast media in cats. *Acta Radiol* 1973;335(suppl):125–132.

9. Cronqvist S: Ventriculography with Amipaque. *Neuroradiology* 1976;12:25–32.

10. Skalpe I, Amundsen P: Clinical results with metrizamide ventriculography. *J Neurosurg* 1975;43:432–436.

11. Roberson GH, Taveras JM, Tadmor R: Computed tomography in metrizamide cisternography. Importance of coronal and axial views. *J Comput Assist Tomogr* 1977;1:241–245.

12. Hindmarsh T: Computer cisternography for evaluation of CSF flow dynamics. *Acta Radiol* 1977;355(suppl):269–279.

13. Drayer BP, Rosenbaum AE, Kennerdell JS, et al: Computed tomographic diagnosis of suprasellar masses by intrathecal enhancement. *Radiology* 1977;123:339–344.

14. Drayer BP, Rosenbaum AE, Reigel DB, et al: Metrizamide computed tomography cisternography: pediatric applications. *Radiology* 1977;124:349–357.

15. Roberson GH, Brismar J, Weiss A: CSF enhancement for computerized tomography. *Surg Neurol* 1976;6:235–238.

16. Hindmarsh T, Greitz T: Computer cisternography in the diagnosis of communicating hydrocephalus. *Acta Radiol* 1975;346(suppl):91–97.

17. Drayer BP, Rosenbaum AE, Higman HB: Cerebrospinal fluid imaging using serial metrizamide CT cisternography. *Neuroradiology* 1977;13:7–17.

18. Rosenbaum AE, Drayer BP: CT cisternography with metrizamide. *Acta Radiol* 1977;355(suppl):323–337.

19. Zee C-S, Tsai FY, Segall HD, et al: Entrance of metrizamide into an intraventricular cysticercosis cyst. *AJNR* 1981;2:189–191.

20. Sortland O: Computed tomography combined with gas cisternography for the diagnosis of expanding lesions in the cerebellopontine angle. *Neuroradiology* 1979;18:19–22.

21. Dandy WE: Roentgenography of the brain after the injection of air into the spinal canal. *Ann Surg* 1919;70:397–403.

22. Pettersson H, Harwood-Nash DCF: *CT and Myelography of the Spine and Cord*. Berlin/New York, Springer-Verlag, 1982.

23. Yu YL, duBoulay GH: Is there an increased risk of early side effects of metrizamide in post-myelogram computed tomography? *Neuroradiology* 1984;26:399–403.

24. Barnett HJM, Fox A, Vinuela F, et al: Delayed metrizamide CT observations in syringomyelia. *Ann Neurol* 1980;8:116.

25. Speck U, Schmidt R, Volkhardt V, et al: The effect of position of patient on the passage of metrizamide (Amipaque), meglumine iocarmate (Dimer-X) and ioserinate (Myelografin) into the blood after lumbar myelography. (abstracted). *Neuroradiology* 1978;14:251–257.

Part VI

Ultrasonography

Chapter 25

Contrast Agents in Ultrasonography

Robert M. Lerner, Raymond Gramiak, Michael Violante, and Kevin J. Parker

The initial clinical successes of ultrasound imaging were the direct result of a natural inherent contrast that allowed the discrimination of solid versus fluid-containing spaces. Subsequently, with the advent of gray-scale imaging, it became possible to determine internal features of solid organs so that changes in reflectivity pattern could indicate the presence of disease processes. Despite this inherent contrast some lesions, such as certain liver metastases, could not be readily recognized. Ultrasound imaging would benefit if contrast agents were available that could easily be delivered to the systemic circulation to change the reflectivity of normal tissues, to localize specifically in regions of disease, to serve as blood pool markers, or to identify physiologic processes. Although currently there is considerable clinical and investigative activity, especially in cardiac ultrasound, a great opportunity exists for the development of clinically useful agents for cardiac and abdominal imaging.

This chapter provides a historical perspective, a description of physical and physiologic considerations, and an overview of new agents and their applications.

HISTORY OF ULTRASOUND CONTRAST AGENTS

Our early experience with ultrasound contrast agents stemmed from a fortuitous echocardiographic study performed at cardiac catheterization during left heart injections of indocyanine green for calculation of cardiac output. An intense contrast effect filled the injected cavity and provided a marker that helped validate valvular and chamber anatomy in the nonanatomic presentation of M-mode.[1] Later we identified the pulmonic valve[2] using contrast and recently applied contrast to Doppler detection of left coronary artery blood flow.[3] It was evident in our early experiences that gas bubbles represented the contrast agent, that these were too large to pass through the capillaries of the pulmonary vascular bed, that some bubble-containing injections might be markedly delayed en route to the heart, and that others produced so much attenuation that information in deeper structures was destroyed.

Many agents, too extensive for individual comment, have been tested.[4] Those that consisted of injected gas, gas-fluid combinations, agitated fluids, or bubble-producing fluids, such as hydrogen peroxide[5] or ether,[6] all failed to produce left heart contrast. The injection of bubbles through a catheter wedged in a peripheral branch of the pulmonary artery was successful in achieving left heart contrast, but the need for cardiac catheterization and the potential for lung trauma[7] rendered this technique experimental. Ionic solutions certainly pass into the systemic circulation, but the very minimal contrast effect and the near toxic concentrations required preclude their use. Lipid emulsions and collagen preparations with particle sizes suitable for transpulmonary passage have failed to achieve usable contrast effects.[8,9]

PHYSICS/PHYSIOLOGY

In this section the presumed mechanisms of action and some of the requirements for ultrasound contrast are presented. In conventional radiology the role of contrast agents is to increase attenuation of the x-ray beam, but in ultrasound the role of contrast agents is to increase backscatter by changing the local acoustic properties. Gobuty has considered mechanisms of contrast enhancement for ultrasound agents composed of homogeneous or inhomogeneous fluids.[10] The mechanism of contrast production by homogeneous fluids relies on a change in the bulk acoustic impedance at an interface where a contrast agent has been delivered as compared to an adjacent region devoid of contrast. The resulting impedance mismatch enhances scattering only at the dissimilar region interface. Fluid contrast agents seem to be best suited for delineating luminal boundaries. If the regions where homogeneous fluid is delivered have a uniform concentration, they may not be expected to show significant volume scattering because of a lack of scattering centers. However, if the contrast agent remains intraluminal (even within small vessels) and does not permeate the extravascular spaces, the fine vascular network containing contrast would serve as scattering centers. Thus, volume type of scattering may potentially be achieved, but the weak impedance mismatch of current fluids results in a clinically inadequate level for imaging. Particulate-based contrast agents, on the other hand, may show increases in volume scattering, resulting in an increase in the overall reflectivity of regions containing these particles.

A theoretical analysis of ultrasound scattering from gas bubbles, solid particles, and tissue-like particles, such as red cell aggregates, has been presented by Lubbers.[11] It provides a useful perspective for potential ultrasound contrast agents. For a given particle size (much smaller than the ultrasound wave length), gas bubbles produce four orders of magnitude more scattering than solid (steel or glass) particles. Solid particles produce one order of magnitude more scattering than red cell aggregates, which probably scatter at levels similar to soft tissues. The apparent advantage of bubbles is undermined by the instability of small (micron-sized) bubbles in blood. Bubbles small enough to pass through the pulmonary capillary bed tend to collapse under surface tension forces.

Although Lubbers has provided a theoretical background for the ultrasonic visibility of contrast agent particles based on their acoustic impedance, size, and number, particle clumping may also influence backscatter, especially as a function of frequency. We speculate that clumping of particles with separations smaller than an acoustic wavelength produces coherent scattering; this phenomenon results in a component of scattering intensity proportional to the second power of the particle number, rather than the first power as for incoherent scattering.

In addition to the changes in backscattering derived from contrast, it is also important to consider changes of attenuation.[12] Ideally, an ultrasound contrast agent would influence the reflective properties of a region of interest without significantly altering sound attenuation. This is important because it is desirable to maintain the usual time-gain compensation (TGC) settings on conventional ultrasound equipment for uniform depiction of non-contrast-enhanced anatomic structures. Because ultrasonic scattering is a minor component of ultrasound attenuation in tissues (less than 10%), it is unlikely that a small increase in ultrasound scattering would cause a major change in sound attenuation.[13] However, gas and bubble contrast agents that cause vastly increased scattering and absorption may seriously limit depth penetration and produce an unacceptable "ring down" scattering artifact.

The requirements for ultrasound contrast may be summarized as follows. Delivery of an agent to the region of interest necessitates that the agent be able to pass the pulmonary capillary bed, have a predictable and sustained survival in the systemic circulation, and be free of embolic consequence in the end organ or the pulmonary capillary bed. Gravitational trapping of the injected intravascular agent due to either excessive or inadequate bouyancy needs to be avoided. Resulting echogenicity should be sufficiently changed compared to the background region for appreciable contrast to be achieved. Attenuation imparted by the contrast agent should be low enough so as not to alter the contrast enhancement and also should not obscure or limit deep penetration within an organ. The agent should be capable of accumulation in a specific organ system and either be excreted within a specific conduit by physiologic transport mechanisms or have sustained persistence in the vascular pool. The agent must be eliminated from the body after the study is completed. These criteria must all be met, in addition to the most important one of low toxicity.

One of the mechanisms recognized as responsible for delivery of contrast agents to specific target sites is macrophage activity. Macrophages, both fixed and wandering, have been identified as the receptors of particulate contrast agents, both solid and fluid. In the liver, relatively low dosages of perfluorocarbons have been successful in identifying metastases by increasing the echogenicity of the entire lesion or its rim. These locations suggest that wandering macrophages must be recruited because liver lesions are generally regarded as being free of macrophages. Higher dosages appear to be necessary for enhancement of liver background echogenicity, most likely the result of fixed macrophage pick-up. In the heart, fixed macrophages are not present, and those recruited for infarct imaging require higher dosage administration. Solid particles readily localize in liver macrophages, and their phagocytosis by wandering macrophages has not been utilized for ultrasonic contrast agent localization. It is clear that macrophages play an important role in ultrasound contrast dynamics and that investigation is required for a better understanding of dose variability, as well as for enhanced implementation of this physiologic process.

AGENTS/APPLICATIONS

Newer contrast agents include gas bubbles that are all nearly the same size and have enhanced longevity due to either a carrier vehicle or suspension in a viscous stabilizing liquid. The particulate agents include solid suspensions or liquid emulsions.

The applications of current interest include delineation of anatomic cavities, conduits, and abscesses; improved imaging of solid organs as in detection of liver metastases; and tagging of the blood pool for both luminal boundary delineation and lesion perfusion studies.

Cavities and Conduits

Localization of catheters or needles in anatomic spaces (ventricular shunt placement, pericardial effusion) and verification of pathologic nonanatomic cavities (ultrasound-guided abscess drainage) are areas where ultrasound contrast has been used, especially during interventional procedures.[14–16] These applications have been successful using agitated water as the contrast agent injected through a needle or catheter. It is doubtful that a more sophisticated contrast agent is needed for these simple uses. More precise studies of anatomic conduits for morphology and patency, such as in ultrasound hysterosalpingography and urinary tract evaluation (cystography and pyelography), could benefit from a reproducible agent.

SHU 454, an agent described in the section on vascular pool agents, has had some demonstrated successes in preliminary laboratory studies of the urinary tract.[17]

Early work with a variety of agents sought to render the normally gas-containing bowel a better sonic window by filling it with material having appropriate scattering and attenuation properties. Although this can be accomplished for the stomach, the inability to fill the remainder of the bowel has prevented it from being a primary sonographic target or window to distal structures, and little progress in this area is available to report at this time.[18]

Vascular Pool Agents

Vascular pool agents could enhance utilization of ultrasound by providing reflector sources within the vessels so that the speed and direction of blood flow become apparent during real-time imaging. Although it is occasionally possible to visualize slowly flowing blood in such venous structures as the inferior vena cava or the portal vein, this ability is thought to be related to the formation by red cell aggregates of large enough particles for significant ultrasound scattering compared to normal blood background. Exogenous injection of material that can be tracked would prove useful in demonstrating arterial and venous anatomy. The effect should be sustainable in the circulation for many minutes to complete an examination. Clearance should be on the order of minutes to hours so that the examination can be repeated.

A new generation of contrast agents is currently being developed and evaluated for potential passage through the pulmonary vascular bed. These may be classified as complex fluids (emulsions or suspensions) or precision microbubbles. Of the emulsions, perfluorocarbons are attractive candidates for the generation of ultrasound contrast because these materials possess a very high density (1.9 g/ml) and a low speed of sound (600 m/sec), both of which differ considerably from tissue. Experimentally these agents have been used as blood replacement fluids, have been associated with only minor reactions during acoustic contrast studies in patients, and seem generally safe according to long-term industrial exposure data. Intravenous injection in doses of 1.6 g/kg enhances the rim or the entire lesion in liver metastases from carcinoma of the colon, pancreas, or stomach. Lesions that did not demonstrate enhancement at this dosage could be recognized as filling defects after injection of 2.4 g/kg to enhance normal liver parenchyma. Usually an interval of 24 hours is required for visualization.[19] In the heart, direct injections into the coronary artery revealed excellent demonstration of myocardial perfusion,[20] while IV injection revealed recognizable low-level enhancement of normal myocardium.[21] In the presence of acute (2-day old) infarctions, doses of 20 g/kg produce enhancement of the entire infarct or of its rim.[22] These doses are rather high because the LD50 in rats is 26.6 g/kg. These perfluorocarbon agents may make it possible to visualize blood flow in vessels using real-time instrumentation. In addition, color-coded Doppler sensitivity should be enhanced by these agents.

Precision microbubbles probably represent the most exciting contrast agents under development for chamber and vessel delineation and for organ perfusion. Several methods are being investigated to prepare a suitable agent for minimally invasive, relatively simple, low-cost, and repeatable delineation of myocardial infarct size from 2-D echocardiograms at bedside. Already under intense study are suspensions of stable and small (less than 10-micron diameter) gaseous microbubbles produced by high-powered sonicators that induce acoustic cavitation in viscous solutions. Bubble persistence over 3 minutes has been reported.[23] Experimental studies in animals following direct injection into the left main coronary artery accurately delineated the region of poor perfusion in 45-minute and 5-hour occlusions.[24]

Microbubbles have been encapsulated in gelatin, albumin, or sugar. When injected into right and left atrial cavities, peak contrast as measured by a digital photometer was greater for these new agents than for the standard ones—blood, saline, and green dye. Pulmonary artery injections with the new agents, as well as with the standard ones, resulted in left heart opacification.[25] The future use of encapsulated gaseous agents is uncertain.

A promising new concept incorporates solid particles as carriers for microbubbles. The new agent SHU 454 (Schering) consists of a suspension of saccharide-stabilized microbubbles (median diameter 3 microns, 97% less than 6.5 microns). The agent is prepared immediately before its use and is stable for 5 minutes. Although this preparation does not cross the pulmonary capillary bed, peripheral venous injection leads to homogeneous and reproducible right heart chamber opacification.[26] This particle carrier-gas approach to an ultrasound contrast agent has promise, and with appropriate particle size and gas-stabilizing properties an agent capable of passing the pulmonary capillary bed may be developed.

Contrast enhancement of Doppler signals may extend their application to patients and organ systems presently marginally or inadequately examined by this technique. These contrast agents should increase the signal-to-noise ratio for the Doppler system by increasing scattering of particles within vessels, thereby leading to more accurate representation of the blood velocity. They should neither disturb the normal flow nor cause artifactual information from either settling or gravitational floating of particles that have different densities compared to blood. A good vascular flow marker, imaged by conventional B-scan equipment and capable of passing the pulmonary capillary bed, would markedly enhance the ability to extract functional data from the left ventricular cavity. The sometimes difficult-to-image endocardial surface could be readily delineated by this method, and automated border detection could provide excellent ventricular function data. The same type of technology would be useful for identification of clots, particularly those that are laminated to the heart wall and protrude little.

Perfusion defect imaging for the ultrasonic detection of myocardial infarction currently seems to require aortic or direct coronary artery injection. In ultrasound not related to the heart, perfusion imaging would be extraordinarily useful to demonstrate neovascularity of tumors and to study abdominal organs with compromised arterial flow. The contrast-enhanced backscattered power level may represent a method for estimating flow volume from which calculations of cardiac output, regurgitant fractions, and shunt flows could be made.

There is also a need for technology of this type in the evaluation of carotid artery disease. A leading indicator of the severity of stenosis is the peak systolic frequency shift as detected by continuous-wave (CW) Doppler. Velocity continues to increase with decreasing lumen size until the point is reached when velocity and flow volume begin to decrease until both reach zero at the point of total occlusion. Sampling of velocity in the descending limb of the velocity versus severity curve will suggest stenoses that are considerably less critical than those actually present. Because flow volume is also decreasing, an estimate of flow volume would be useful to identify the true nature of the stenosis. Some of these patients, when studied by CW Doppler, show no evidence of flow and are incorrectly diagnosed as suffering from a total occlusion, whereas an angiogram may reveal a trickle of flow and therefore the potential for surgical correction. It is evident that Doppler systems are not sufficiently sensitive to detect the compromised flow volumes in these patients and that contrast-enhanced reflectivity of blood could possibly improve Doppler ultrasound detection.

Intracoronary injections of uncontrolled-sized microbubbles in saline were used to show that certain flows detected near the pulmonary artery actually represent left coronary artery blood flow.[3] Features of diastolic waveforms were recognized and identified as typical of those recorded by other methods in coronary arteries. We had not been successful in detecting systolic flow in the left coronary artery until we studied a patient undergoing contrast coronary arteriography. An obvious early systolic flow component was demonstrated on the reflectivity-enhanced contrast study, whereas the control study made without contrast showed only a few wispy and unrecognizable signals. It seems clear that current ultrasound instrumentation cannot recognize certain important physiologic flows and would benefit from an enhanced signal from a circulating contrast agent.

Weak Doppler signals pose a serious imaging problem in the currently popular Doppler color flow imagers. These instruments process returning signals for B-scans and for Doppler blood flow images. The signal processing and system gain needs of these disparate imaging modes are considerably different. The backscattered signals from solid tissues in the B-scan image are probably 30 dB higher than the backscattered data from blood. The B-scan therefore uses a portion of the beam that is relatively narrow as compared to the 30 dB down portion required for Doppler. As a result, lateral resolution in the Doppler image is very poor, and processing techniques are needed to prevent the oversized color data from spilling across anatomic landmarks. When solid tissue boundaries are not present to contain flow, as in a jet in a cardiac chamber, the Doppler image will overrepresent the size of the flow pattern. Similarly, the overly wide beam in the Doppler makes slice thickness of the color flow detection considerably greater than the B-scan so that the duplex display will contain structures and flows originating from regions not common to both modalities. The current rash of cases of functional pulmonic insufficiency demonstrated by these instruments are in reality left coronary artery blood flow superimposed on the B-scan image of the pulmonary valve because the left coronary artery lies just under the pulmonary artery but outside the narrow B-scan plane.[3,27] Enhanced reflectivity of the circulating blood could minimize this disparity between B-scan and 2-D color flow imaging and probably simplify instrumentation, as well as make it more believable in displaying blood flows.

Solid Organs/Lesions

Ideal agents for imaging organs and/or lesions with ultrasound should show perfusion and such functional characteristics as localization to macrophages or specific cellular transport, i.e., renal excretion.

Solid particles of 10 microns or less consisting of fat emulsions,[8] gelatin spheres,[28] or collagen particles[9] have not been satisfactory because of low or poor contrast enhancement. In our laboratory, iodipamide ethyl ester (IDE) particles of uniform 1-micron diameter and with a high density (2.4 g/cc) were shown to accumulate rapidly in hepatic reticuloendothelial cells that are abundant in normal liver but are lacking in tumors and many other lesions.[29] About 95% of the particles are accumulated by the liver and 5% by the spleen, but none in any other organ or tissue. The bloodstream is cleared within 2 hours, and elimination from the organism occurs within 2 to 5 days. Imaging of rat livers with 7.5 MHz and 10 MHz instrumentation revealed good increase in backscatter and ready differentiation of normal from contrast-enhanced livers at safe levels.[30] The focus of current investigations is to determine the optimal particle size for clinical application in human livers and improved detectability with lower frequency ultrasound systems.

CONCLUSION

The contrast needs of cardiac, vascular, anatomic conduit, and solid organ imaging seem to be pointing toward a group of agents with related but differing properties. Transpulmonary passage will require particle sizes (solid, gaseous, or combined) smaller than the capillaries in order to provide contrast for chamber filling and clearance sequences, as well as for solid tissue perfusion. The strong echogenicity of microbubbles makes them particularly attractive for this purpose.

Perfusion needs of solid organ imaging and of Doppler enhancement will require a sustained increase in blood reflectivity sufficient to perform an extended clinical examination. The volume of injected particles/microbubbles for this purpose may be quite large and will probably create special problems in the control of particle/microbubble size. It will be especially important to ensure stability in the circulation to preclude embolization and to control the degree of reflectivity and attenuation within specified limits for quantitation of flow from backscatter.

The perfluorocarbons and the IDE particles both illustrate affinity of a contrast agent for a specific tissue, organ, or body system. Both types of particles seem to be concentrated in fixed macrophages in liver and spleen and may also be ingested by wandering macrophages responding to stimuli arising in neoplastic or inflammatory tissues. Clearly, much needs to be learned concerning the selective anatomic contrast localization offered by these agents. Research goals must include exploration of other mechanisms to produce additional tissue-specific agents and to accelerate and intensify the contrast effect. These are timely and difficult challenges to investigators involved in this endeavor. Conversion of these speculative expectations to clinical reality will certainly increase the utility of ultrasound in patient care. Although a commercial preparation is not yet available in the United States, there is activity among some of the drug companies to produce such a contrast agent for ultrasound.

REFERENCES

1. Gramiak R, Shah PM, Kramer DH: Ultrasound cardiography: contrast studies in anatomy and function. *Radiology* 1969;92:939–948.

2. Gramiak R, Nanda NC, Shah PM: Echocardiographic detection of the pulmonary valve. *Radiology* 1972;102:153–157.

3. Gramiak R, Holen J, Moss AJ, et al: Left coronary arterial blood flow: noninvasive detection by Doppler ultrasound. *Radiology* 1986;159:657–662.

4. Ziskin MC, Bonakdarpour A, Weinstein DP, et al: Contrast agents for diagnostic ultrasound. *Invest Radiol* 1972;7:500–505.

5. Xinfang W, Jiaen W, Hanrong C, et al: Contrast echocardiography with hydrogen peroxide. *Chin Med J* 1979;92:693.

6. Meltzer RS, Sartorius OEH, Lancee CT, et al: Transmission of ultrasonic contrast through the lungs. *Ultrasound Med Biol* 1981;7:377–384.

7. Serruys PW, Meltzer RS, McGhie J, et al: Factors affecting the success of attaining left heart echo contrast after pulmonary wedge injections, in Meltzer RS, Roelandt J (eds): *Contrast Echocardiography* The Netherlands, Martinus Nijhoff, 1982.

8. Fink IJ, Miller DL, Shawker TH, et al: Lipid emulsions as contrast agents for hepatic sonography: an experimental study in rabbits. *Ultrasonic Imaging* 1985;7:191–197.

9. Ophir J, Gobuty A, McWhirt RE, et al: Ultrasonic backscatter from contrast producing collagen microspheres. *Ultrasonic Imaging* 1980;2:67–77.

10. Gobuty AH: Perspectives in ultrasound contrast agents, in Parvez Z, Moncada R, Sovak M (eds): *Contrast Media: Biologic Effects and Clinical Application,* vol 3. Boca Raton, FL, CRC Press, 1987, pp 146–154.

11. Lubbers J, Van Den Berg JW: An ultrasonic detector for microgas emboli in a bloodflow line. *Ultrasound Med Biol* 1976;2:301–310.

12. Shung KK, Fei D, Yuang Y, et al: Ultrasonic characterization of blood during coagulation. *J Clin Ultrasound* 1984;12:147.

13. Lyons MA, Parker KJ: Absorption and attenuation. II. Experimental results. *IEEE. UFFC,* 1988;35:6.

14. Widder DJ, Simeone JF: Microbubbles as a contrast agent for neurosonography and ultrasound-guided catheter manipulation: in vitro studies. *Am J Radiol* 1986;147:347–352.

15. Chandraratna PAN, Reid CL, Nimalasuriya A, et al: Application of two dimensional contrast studies during pericardiocentesis. *Am J Cardiol* 1983;52:1120–1122.

16. Scatamacchia SA, Raptopoulos V, Davidson RI: Saline microbubbles monitoring sonography-assisted abscess drainage. *Invest Radiol* 1987;22:868–870.

17. Meyer-Schwickerath M, Fritzsch T: [Sonographic imaging of the kidney cavity system using an ultrasonic contrast medium.] *Ultraschall Med* 1986;7:34–36.

18. Rauch RF, Bowie JD, Rosenberg ER, et al: Can ultrasonic examination of the pancreas and gallbladder follow a barium UGI series on the same day? *Invest Radiol* 1983;18:523–525.

19. Mattrey RF, Strich G, Shelton RE, et al: Perfluorochemicals as ultrasound contrast agents for tumor imaging and hepatosplenography: preliminary clinical results. *Radiology* 1987;163:339–343.

20. Saito H, Ajima H, Suzuki M: Myocardial contrast echocardiography using artificial blood (Fluosol-DA): a comparison with left ventricular wall motion in the experimental ischemic heart. *J Cardiogr* 1984;14:677–688.

21. Matsuda M, Kuwako K, Sugishita Y, et al: [Contrast echocardiography of the left heart by intravenous injection of perfluorochemical emulsion.] *J Cardiogr* 1983;13:1021–1028.

22. Mattrey RF, Andre MP: Ultrasonic enhancement of myocardial infarction with perfluorocarbon compounds. *Am J Cardiol* 1984;54:206–210.

23. Feinstein SB, Ten Cate FJ, Zwehl W, et al: Two-dimensional contrast echocardiography. I. In vitro development and quantitative analysis of echo contrast agents. *J Am Coll Cardiol* 1984;3:14–20.

24. Sakamaki T, Tei C, Meerbaum S, et al: Verification of myocardial contrast two-dimensional echocardiographic assessment of perfusion defects in ischemic myocardium. *J Am Coll Cardiol* 1984;3:34–38.

25. Bommer WJ, Mason DT, DeMaria AN: Studies in contrast echocardiography: development of new agents with superior reproducibility and transmission through lungs. *Circulation* 1979;60 (suppl II):17.

26. Lange L, Fritzsch T, Hilmann J, et al: Right heart echo-contrast with anesthetized dog after I.V. administration of new standardized sonographic contrast agent. 3rd communication: comparison of various agents employed in contrast echocardiography. *Arzneimittelforsch* 1986;36:1037–1040.

27. Recusani F, Valdes-Cruz LM, Dalton N, et al: Detection of coronary flow by pulsed Doppler and color Doppler flow mapping and its differentiation from pulmonary insufficiency. *J Am Coll Cardiol* 1986;7:14A.

28. Ophir J, Gobuty A, Maklad N, et al: Quantitative assessment of in vivo backscatter enhancement from gelatin microspheres. *Ultrasonic Imaging* 1985;7:293–299.

29. Violante MR, Mare K, Fischer HW: Biodistribution of a particulate hepatolienographic CT contrast agent: a study of iodipamide ethyl ester in the rat. *Invest Radiol* 1981;16:40–45.

30. Parker KJ, Tuthill TA, Lerner RM, et al: A particulate contrast agent with potential for ultrasound imaging of liver. *Ultrasound Med Biol* 1987;13:555–566.

ACKNOWLEDGMENT

Mrs. Beverly Pollet is to be commended for her expert assistance during preparation of this manuscript, which was done in a cheerful and expeditious manner despite numerous revisions.

Part VII

Magnetic Resonance

Chapter 26

Contrast Agents in Magnetic Resonance Imaging

Hannu Paajanen and Martti Kormano

Proton magnetic resonance imaging (MRI) provides excellent inherent contrast and spatial resolution between most biologic tissues. Tissue contrast is a multi-faceted variable dependent upon physico-chemical tissue properties and upon operator-chosen variables for MRI signal acquisition. In certain regions of the body the endogenous contrast is insufficient for adequate diagnostic information, and the use of exogenous image contrast agents has been suggested. These agents can aid in the assessment of tissue perfusion, of blood-brain barrier (BBB) integrity, of gastrointestinal (GI) anatomy, and of renal function. MRI contrast agents can help define tumors and differentiate tumor constituents. Although the applications of MRI contrast media may sometimes parallel those of radiographic iodinated compounds, there are important differences in the principles of contrast manipulation, relationship of dose and effect, toxicology, and chemistry. These differences, the recent experimental and clinical results, and future uses of MRI contrast pharmaceuticals are reviewed in this chapter.

Contrast in proton MRI is a direct result of inherent differences in the observed MR signals from body tissues.[1] MRI contrast is determined by (1) *intrinsic* tissue factors, such as proton density, blood flow, and the T1 and T2 relaxation times, and (2) *external* factors, such as the strength of the external magnetic field and operator-chosen pulse sequences.[1] Much work has been performed to develop appropriate pulse sequences and imaging parameters—external factors—that will produce more contrast between adjacent tissues. One rationale for using exogenous contrast agents is that if the T1, T2, and proton density (H) of a lesion and the surrounding normal tissue are similar, or if different regions of the pathologic process (tumor versus edema) have similar MR parameters, the lesion

This work was supported by the Sigrid Juselius Foundation, Helsinki, Finland. Hannu Paajanen was the postdoctoral research fellow in Magnetic Resonance Contrast Media in the Contrast Media Laboratory, University of California, San Francisco (Head: Robert C. Brasch) during 1984–1985.

and normal tissue are isointense on a MR image. Because exogenous contrast agents virtually always localize differently, thereby altering differently the relaxation parameters, additional information can be obtained and low inherent contrast can be augmented by their use.

PRINCIPLES OF CONTRAST ENHANCEMENT

Proton density (H), flow (f(v)), T1, and T2 contribute to the image intensity of two adjacent tissues. Manipulation of physical tissue properties, such as temperature or viscosity, to alter relaxation times is not clinically feasible. There seems to be little need or opportunity to use pharmaceutical intervention to assess flow in blood vessels because rapidly moving blood provides enough inherent contrast in MRI.[2]

The intensity equation of the most common pulse sequence (spin echo) can be written as:[1]

$$I = Hf(v)(1\text{-}e^{\frac{-TR}{T1}})e^{\frac{-TE}{T2}} \qquad (1)$$

Spin echo (SE) signal intensity can be enhanced by increasing proton density (H), by decreasing T1, or by increasing T2. The important consequence of the intensity equation is that the relaxation times, T1 and T2, as well as sequence parameters, TR, TE, influence the signal intensity in an *exponential* manner, i.e., small variations of these factors cause relatively large changes in signal intensity. Proton density influences the signal intensity *linearly*. This means that the modification of relaxation times is more effective in increasing contrast enhancement than the modification of proton density.

The inherent variation of proton density between body soft tissues is reported to be up to 25%. Increasing the T1 contrast by altering proton density has been tried in animals with diuretics, lipid solutions, ethanol, and clomifene.[3] Because water is a major tissue constituent and is rich in protons, the relative hydration of tissues should influence contrast. Proton density can be easily varied in the GI tract. A water-filled stomach[4] can alter contrast in GI tract imaging. Changes in renal tissue hydration, through voluntary dehydration or administration of diuretics, have also been demonstrated.[5] Altering hydration within a tissue not only influences proton density but also changes relaxation times.[6] Not only the amount of tissue protons is important in influencing contrast but also the chemical environment of hydrogen. Protons in free water react differently to the MR signal than protons within fat, with fat protons having much shorter T1 and T2 times. Accordingly, the intensity signal from fatty tissue is higher than that of water. Oral administration of mineral oil[7] or plain liposomes[8] has been suggested for MRI of the GI tract or reticuloendothelial cells.

All these methods of influencing the image contrast through altering proton density have the disadvantage of lacking the sensitivity to cause large changes in image contrast at safe dosages. When compared to the effects of paramagnetic or ferromagnetic substances the responses elicited by altering proton density seem to be weaker. Therefore none of these methods has had any significant clinical value.

A more effective way to enhance MR contrast through intrinsic tissue parameters is to change the relaxation times T1 and T2 because of their exponential effect on signal intensity. Either paramagnetic or ferromagnetic substances are administered to change the tissue proton relaxation processes. The physical principles and imaging rationale of paramagnetic (*T1 contrast agent*) and ferromagnetic (*T2 contrast agent*) pharmaceuticals differ so much that these groups are considered separately.

Paramagnetic Relaxation Enhancement

Paramagnetic substances have a permanent magnetism that is due to spin moment. In the absence of an external magnetic field, the magnetic moments of paramagnetic substances are randomly aligned. When the external magnetic field is turned on, however, the magnetic moments align with the field and generate strong local magnetic fields that shorten both the T1 and T2 of neighboring protons. Opposite to paramagnetics, *diamagnetic* substances are typically repelled by and align against the stronger portion of the external magnetic field.

The mechanism by which paramagnetic agents provide contrast in MRI is fundamentally different from how radiographic contrast media work. X-ray contrast media are observed directly in radiographic images because of their ability to absorb x-rays. The paramagnetics, however, operate in an indirect fashion by altering the local magnetic environment (T1 and T2) of tissues. Figure 26-1 demonstrates the diamagnetic, paramagnetic, and ferromagnetic nature of matter.

Paramagnetic substances have one or more particles—protons, neutrons, or electrons—with a spin that is not canceled by another similar particle with an opposite spin. The magnetic moments of unpaired electrons are 657 times larger than the magnetic moments of the protons or neutrons; thus, local magnetic fields generated by unpaired electrons are much stronger, being as high as 10^4 gauss.[9] Therefore the chemicals that have unpaired electrons (transition or lanthanide metal ions, organic-free radicals) are the most effective paramagnetic contrast enhancers. When paramagnetic ions are added to water, the relaxation process of water molecules is more enhanced in the vicinity of the paramagnetic centers. Paramagnetic ions, through *dipole-dipole* interactions between the unpaired electrons of the paramagnetic species and the water protons, achieve the reduction in T1 and T2. Relaxation occurs through the tumbling of the paramagnetic-water complex.

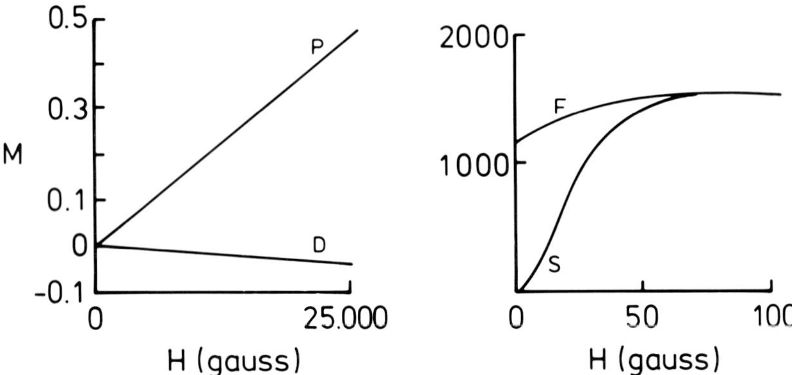

Figure 26-1 The behavior of diamagnetic, paramagnetic, superparamagnetic, and ferromagnetic materials in the external magnetic field. The induced magnetization (*M*, emu/g) of these materials is presented as a function of an applied magnetic field (*H*, *gauss*). Diamagnetic materials (*D*) have a small magnetic susceptibility when an external magnetic field is turned on. In comparison, paramagnetic agents (*P*) induce much more magnetization when the external field is on. Net magnetization of both these materials is linear and directly proportional to the strength of the applied field. In comparison, superparamagnetic (*S*) and ferromagnetic (*F*) materials induce *very large* magnetization in the external magnetic field. This is nonlinear, and is saturated even in weak magnetic fields. Note that superparamagnetic, paramagnetic, and diamagnetic materials do not possess any magnetization in the absence of an external field. Ferromagnetic agents, however, retain their magnetization when the external field is removed. *Source*: Modified from Cullity BD: *Introduction to Magnetic Materials*. Reading, MA, Addison-Wesley, 1972 and Saini S, et al: Ferrite particles: a superparamagnetic MR contrast agent for the reticuloendothelial system. *Radiology* 1987;162:211–216.

The relaxation process induced by paramagnetic ions is called *proton relaxation enhancement* (*PRE*).[10] The PRE depends on the magnitude of the local fluctuating fields induced by the spin moment of unpaired electrons and also on the time scale of the magnetic field fluctuation. This time dependence is characterized by the correlation time τc:

$$1/\tau_c = 1/\tau_r + 1/\tau_s + 1/\tau_m \tag{2}$$

where τ_r is the rotational correlation time, τ_s is the electron spin relaxation time, and τ_m is the water exchange time. Thus, the local magnetic field fluctuations can occur through rotation, translation, or chemical exchange. The rate of fluctuation is additive, and the fastest rate tends to dominate the others.

The use of paramagnetic ions as proton relaxation enhancers originates from the early studies of Bloch and co-workers.[11] The theoretical equations that describe the PRE in the presence of paramagnetics were developed further by Solomon and Bloembergen in the 1950s.[12,13] A detailed description of PRE is available elsewhere.[10,14] This discussion considers only those parts of the PRE theory that we feel are pertinent in developing the paramagnetic complexes as contrast agents for clinical MRI.

It is easiest to discuss changes in T1 or T2 as a result of paramagnetic contribution by considering the factor 1/T, which is additive and measures the *rate of relaxation*. The term *1/T* (i.e., relaxation rate) should not be confused with the term *relaxivity*, which means the effectiveness of the paramagnetic ion in changing relaxation. The relaxivity is determined as 1/T per molar concentration of paramagnetics.

Relaxation rates are *additive*, so that the observed tissue T1 after paramagnetic agent administration can be written:

$$1/T1_{obs} = 1/T1_i + 1/T1_p \qquad (3)$$

where $T1_{obs}$ is the observed relaxation time, Tl_i is the inherent tissue relaxation time without a paramagnetic effect, and $T1_p$ is the relaxation time of the paramagnetic agent. The same relationship holds true for T2. Because the relaxation rate of the paramagnetic agent is directly proportional to the concentration, we can write:

$$1/T1_{obs} = 1/T1_i + k\,(c) \qquad (4)$$

where k is a constant and (c) is the concentration of the paramagnetic ion. Because the PRE is additive, for the same concentration of paramagnetics the reduction of T1 or T2 is greater in tissues with inherent higher T1 or T2. This important phenomenon is illustrated in Figure 26-2. The effect of the paramagnetic contrast agent is always less in organs with a low inherent T1 and more in the tissues with a higher T1. Because tissue T2 is always shorter than T1, the absolute effect for T1 is greater than that for T2. It follows that paramagnetics influence T1 at lower concentrations than for T2, a factor that has practical importance in spin echo imaging.

Although the Solomon-Bloembergen equations appear to be complex, some simple conclusions can be drawn from them. The change in T1 or T2 produced by the paramagnetic ion is directly proportional to the concentration (c) of paramagnetic ion in the solution (equation 4), with a higher concentration providing a greater increase in the relaxation rates (or decrease in the relaxation times). Relaxation rates are also directly related to the second power of the effective magnetic moment of the paramagnetic ions (Table 26-1). The strength of the dipolar magnetic fields (i.e., the effective magnetic moment) caused by the unpaired electron is expressed in units of Bohr magnetons, as shown in Table 26-1. Metal ions with multiple unpaired electrons—for example, Gd^{3+}—tend to have higher magnetic moments, and thus they are more effective as contrast enhancers.

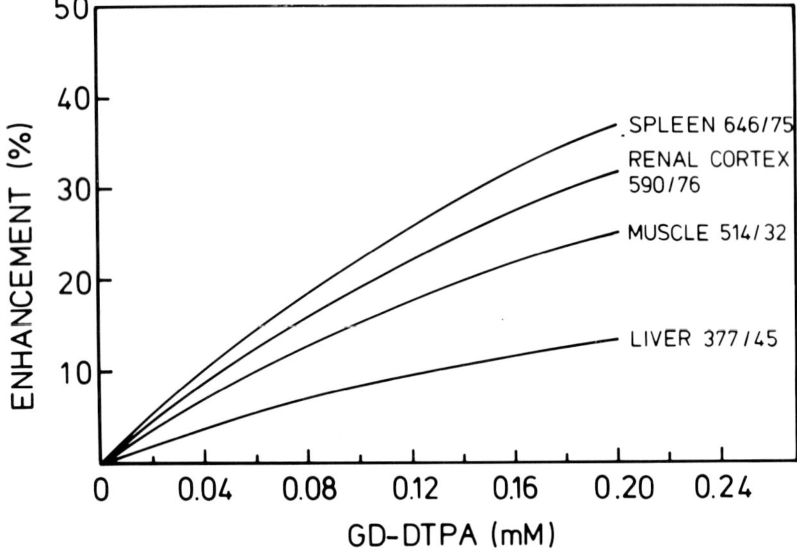

Figure 26-2 The contrast enhancement of tissues as a function of Gd-DTPA concentration; the computer-simulated effect of T1 and T2 (T1/T2) on tissue intensity. The contrast of tissues that have high inherent T1 and T2 is enhanced more than that of tissues with low inherent relaxation times. The curves are calculated using equations 1, 3, and 4 (SE 500/26). When the relaxivity of Gd-DTPA is known, it is possible to calculate the i/T1o at the known field strength. The values of T1/T2 are at 0.35 T. In vivo, however, tissue perfusion, amount of extravascular volume, and the like differ so much that the direct application of these kind of simulations is limited. For example, liver enhancement using Gd-DTPA is higher than expected on the basis of low inherent T1 and T2. *Source*: Reprinted with permission from ''Fast-Field-Echo MR Imaging with Gd-DTPA: Physiologic Evaluation of the Kidney and Liver'' by RI Pettigrew et al in *Radiology* (1986;160:561–563), Copyright © 1986, Radiological Society of North America Inc.

Relaxation rate (1/T) is also directly proportional to the electron spin quantum number. To produce a strong relaxation effect, the paramagnetic ion should have a high spin number (Table 26-1). Generally, the larger the number of unpaired electrons, the higher the spin numbers and the stronger the paramagnetic effect.

The distance between the paramagnetic ion and the protons that it is relaxing should be minimized because the PRE is inversely proportional to the sixth power of this distance. The increase in relaxation rate is proportional to the number of water molecules coordinated directly to the paramagnetic ion. For lanthanide aqua ions, usually up to ten water molecules are in close proximity (in the first hydration layer) to the paramagnetic center.[15] When the paramagnetic ion is chelated, the close access of water molecules is hindered, which decreases the relaxation rates. For example, the paramagnetic effect of Gd-DOTA is smaller by a factor of two

Table 26-1 Physicochemical Properties of Some Paramagnetic Metal Ions

Ion	*Electron Configuration*	*Number of Unpaired Electrons*	*Spin Quantum Number*	*Magnetic Moment (Bohr Magnetons)*
Ti^{3+},V^{4+}	$3d^1$	1	1/2	1.7–1.8
V^{3+}	$3d^2$	2	2/2	2.6–2.8
Cr^{3+},V^{2+}	$3d^3$	3	3/2	3.8
Co^{2+}	$3d^7$ (high spin)	3	3/2	4.1–5.2
Mn^{3+},Cr^{2+}	$3d^4$ (high spin)	4	4/2	4.9
Fe^{3+},Mn^{2+}	$3d^5$ (high spin)	5	5/2	5.9
Fe^{2+},Co^{3+}	$3d^6$ (high spin)	4	4/2	5.1–5.5
Cu^{2+}	$3d^9$	1	1/2	1.7–2.2
Eu^{3+}	$4f^6$	6	6/2	3.4
Gd^{3+}	$4f^7$	7	7/2	8.0

Source: *Introduction to Solid State Physics* by C Kittel, John Wiley and Sons Inc, © 1968.

than that of a free Gd $^{+3}$ ion. This decrease is partly due to the diminished number of water molecules in the first hydration layer. Koenig and co-workers have calculated that in the Gd-DOTA molecule approximately only one water molecule is in the first layer.[15]

Because relaxation rate (1/T) is also related to the external magnetic field strength, the ability of paramagnetics to induce relaxation improves at lower frequencies of the polarizing field. For example, small gadolinium chelates have two to three times stronger relaxivity at low field strengths.[15,16] The large proteins labeled with gadolinium exhibit five- to ten-fold greater relaxivities at low fields, but the peak of relaxivity is near 20 MHz.[17] With binding of the paramagnetic ion to a protein, the tumbling rate of the ion is reduced, resulting in increased correlation times and the increased relaxation rate. Extensive discussions of the frequency dependence of PRE have been published elsewhere.[15–19]

Ferromagnetic Relaxation Enhancement

Although virtually all biologic iron is paramagnetic, the clusters of iron ions create the collective domain with a magnetic moment 10–1000 times greater than the sum of the paramagnetic iron ions. This is due to magnetic ordering of the unpaired electron spins of the iron ions in the crystal.[20] In the clustered form iron is either *ferromagnetic* or *superparamagnetic* (Fig. 26-1). Particulate iron less than 300 å in size is superparamagnetic; larger particles are ferromagnetic. In this respect, the size of the particle is critical. Ferromagnetic substances also align with an external magnetic field, but tend to retain that alignment after the external field is removed. Superparamagnetic substances, similar to paramagnetic and diamagnetic agents, do not possess any magnetization in the absence of an external magnetic field.[20] When compared to paramagnetic materials, ferromagnetic and superparamagnetic substances have enormously large magnetic moments that are nonlinear.

Various investigators have proposed the use of ferromagnetic or superparamagnetic materials as contrast enhancers for MRI.[21–25] The fundamental difference between paramagnetics and ferromagnetics in regard to MRI is that paramagnetics enhance the proton signal (Fig. 26-3), whereas ferromagnetics or superparamagnetic material tend to destroy the signal, thus inducing negative contrast (Fig. 26-4).[21–25]

In tissue, even with extremely small doses, both substances produce local magnetic field gradients (inhomogeneities) that are several orders of magnitude greater than those produced by paramagnetic compounds.[23] For example, the net magnetic moment of a ferromagnetic magnetite particle is over 15 times greater

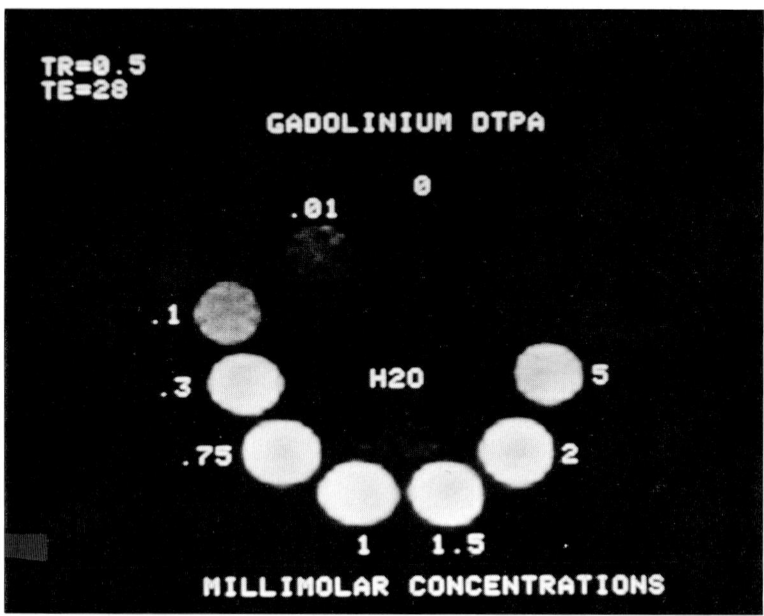

Figure 26-3 Different dilutions of Gd-DTPA (0-5 mmol) in test tubes (SE 500/28). Note that when the concentration of Gd-DTPA is increased, the image intensity is also increased at a certain level. The maximum intensity is achieved at the concentration of 1 mmol of Gd-DTPA; at higher levels, the image intensity turns darker again. The centrally located water test tube (H_20) is black on this image.

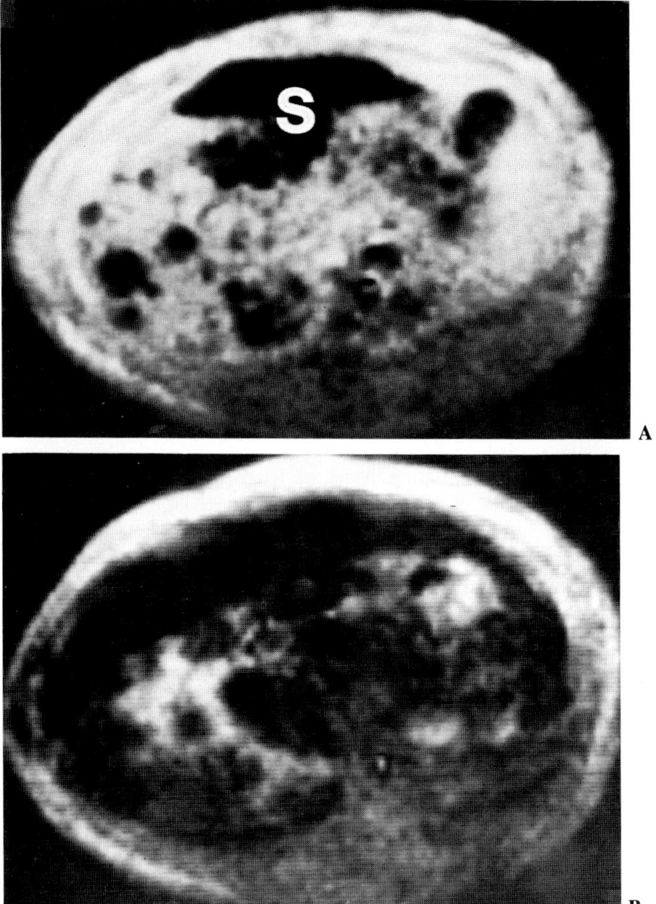

Figure 26-4 Ferromagnetic contrast material in the rabbit GI tract. The saturation recovery images (120/30) at the level of upper abdomen of rabbit before (*A*) and after (*B*) the administration of ferromagnetic particles into the stomach. Partially fluid-filled stomach (*S*), fat, and bowel content give a high signal on precontrast image. Ferromagnetic particles destroy the signal, and the bowel turns dark. *Source*: "Ferromagnetic Contrast Studies of Gastrointestinal Tract at 0.02 Tesla" by P Niemi, M Kormano, and H Paajanen, unpublished manuscript.

than the magnetic moment of a paramagnetic ferritine molecule of the same size.[25] Proton diffusion through these local field inhomogeneities results in extreme T2 shortening because of irreversible phase shifts and profound signal loss.[26] The decay of magnetization is multiexponential depending on particle size; thus, the ferromagnetic particles affect the proton relaxation times in a manner distinctly different from paramagnetics.[26] The effect on T1 is less pronounced.[22,24,25] For example, Olsson and associates reported that the T2 relaxivity for ferromagnetic magnetite was 800 1/smM, and the T1 relaxivity was 150 1/smM at 0.25 T.[24] The T2 effect could be further pronounced by entrapping magnetite into the starch matrix.[24]

According to Renshaw the proton relaxation rates are directly proportional to the number of ferromagnetic particles, to the second power of magnetic moments of the ferromagnetic particles, and to the viscosity of the solution.[25] Because the correlation time (τc) is assumed to be dominated by the translational diffusion of water molecules around the larger ferromagnetic particle, the T2 relaxation, which is dominated by translational diffusion, is more augmented than the T1 relaxation by ferromagnetic particles.[26]

OPTIMAL PULSE SEQUENCES

Paramagnetics

Imaging of tissues requires careful consideration of external instrument parameters (TR, TE, TI, slice thickness, etc.) versus the intrinsic tissue parameters (T1, T2, H). A decrease of T1 increases the image intensity (SE, IR), whereas a decrease of T2 produces the opposite effect (SE), assuming that the other factors are constant. The IR sequence produces images that are primarily dependent upon the T1 differences of the tissues.[27] The spin echo can be dominated by either *T1 or T2* relaxation, depending on the values selected for TR and TE.[1] Thus, when using appropriate software-controlled sequences and pulse parameters (TR, TI, TE), more T1- or T2-weighted images can be produced as desired.

When using paramagnetics, the shortening of T1 is desirable in order to increase the signal intensity of the lesion (Fig. 26-3). The shortening of T2 is competitive, causing a reduction in signal intensity when using SE sequence. Figure 26-5 shows schematically that better intensity augmentation is achieved by using a more T1-weighted sequence. For maximal paramagnetic contrast effect TR and TE should be as short as possible. However, unless flip angle imaging is used, a too short TR would ruin the signal-to-noise ratio because the recovery of magnetization is incomplete. A short TE is the most important pulse parameter in paramagnetic contrast enhancement.[27,28] Most commercial MR systems use SE imaging that has a minimum TE of 26-30 ms. Wolf et al. have obtained excellent liver images by using as short a TR/TE as 80/10 ms.[27] Theoretical calculations and experimental data[27–30] show that TE must be set to the minimal value available for both SE and IR sequences. In the IR sequence, TI should be between the T1 of the lesion before and after contrast enhancement.[29]

A variation of TR has relatively little effect on paramagnetic contrast enhancement unless TR is too long, because if the longitudinal relaxation is complete the T2 decay causes loss of signal difference between the enhanced and nonenhanced parts of the image.[27] However, a very short TR would ruin the MR signal.

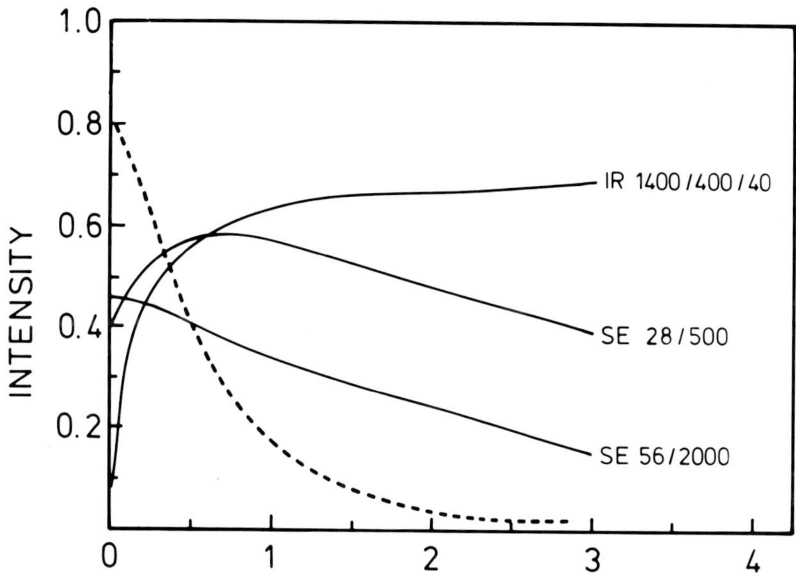

Figure 26-5 Kidney enhancement as a function of Gd-DTPA concentration (mmol); the simulated effect of different pulse sequences. The calculated curves show that the highest renal enhancement is seen when a highly T1-weighted sequence (IR 1400/400/40) is used. On SE 28/500 sequence, peak intensity is achieved at a renal concentration of approximately 1 mmol of Gd-DTPA; at higher levels, the intensity decreases again. This biphasic intensity curve is typical when using SE sequences and is due to the T2 effect. A T2-weighted SE sequence (SE 56/2000) results in a negative contrast enhancement. In comparison, the *dotted line* demonstrates the effect of ferromagnetic particles on the renal intensity. Ferromagnetics always destroy the intensity signal. All curves are calculated similarly as explained in Figure 26-1.

The above-mentioned guidelines are not generally applicable to fast imaging sequences.[31] For example, gradient-echo sequences utilize very short TR (40 ms), and image contrast is determined by the excitation pulse angle, which is varied from 10 to 90 degrees.[32] Similarly as in dynamic CT, five to ten images can be obtained per minute after paramagnetics administration.[31,32] Appropriate pulse parameters and tip angles for maximal contrast enhancement have to be determined.[31]

In summary, the tissue intensity response to paramagnetics is a complex biphasic curve; at low doses there is signal enhancement (T1 effect), but at higher doses of the paramagnetic substance signal loss occurs (T2 effect) (Fig. 26-5). An optimal tissue concentration of paramagnetic contrast agent is essential for maximal T1 decrease and maximal intensity enhancement; more is *not* better.

Ferromagnetics

Ferromagnetic particles induce a relaxation effect, which is preferentially more for T2 than T1. Optimal demonstration of the T2 effect requires the use of T2-weighted imaging. The effect is seen on T2-weighted images as loss of signal intensity, i.e., the tissue turns black. The objective is to destroy the signal of normal tissue, which accumulates ferromagnetic agent, and increase contrast between tissues that do not accumulate the ferromagnetic particles and those that do accumulate it (Fig. 26-4).

Ferromagnetic agents provide a simple monophasic effect of progressive signal loss (T2 effect) with increasing dose (Fig. 26-5). Although maximal lesion-to-normal tissue contrast occurs with heavily T2-weighted pulse sequences, T1-weighted sequences also demonstrate the effect of ferromagnetics.[33] A mixed T1-T2-weighted sequence may be even superior to a pure T2-weighted sequence because greater signal averaging occurs.[33]

TYPES OF CONTRAST AGENTS

Several types of pharmaceuticals, each with different advantages and disadvantages, have been proposed for MRI (Table 26-2). An ideal MR contrast enhancer should be pure, stable, nontoxic, water soluble, readily available, and inexpen-

Table 26-2 Potential Contrast Agents for MRI

PARAMAGNETIC METAL IONS
Transition metal series: Cr^{+2},Fe^{+2},Fe^{+3},Mn^{+2},Mn^{+3}
Lanthanide series: Eu^{+3},Gd^{+3}
CHELATES AND COMPLEXES
Gd-DTPA, Gd-DOTA, Cr-EDTA, Fe-EDTA, Gd-EDTA
Macrocyclic complexes
NITROXIDE SPIN LABELS
Pyrrolidine and piperidine derivates
Hypoxia-sensitive agents
TISSUES AND TUMOR-SPECIFIC CONTRAST AGENTS
Labels of monoclonal antibodies, porphyrins, liposomes, hydroxycolloids, bile avid agents, intravascular macromolecules
MOLECULAR OXYGEN AND PERFLUOROCARBONS
FERROMAGNETIC AGENTS
Magnetite

sive. It should have a potential to be conjugated to organ-specific biomolecules, and it should undergo efficient renal excretion. The substance should be nonreactive and nontoxic in vivo (no mutagenicity, teratogenicity, carcinogenicity, or immunogenicity).[8] Using these criteria as a guide, the major classes of MRI contrast agents can be evaluated.

Paramagnetic Metal Ions

Transition, lanthanide, and actinide metal ions contain unpaired electrons within electron orbitals, which make them paramagnetic. First-row transition metal ions—for example, iron, manganese, cobalt, nickel, copper—have partially filled 3^d orbitals. Because the magnitude of the relaxation rate is dependent on the magnetic moment, the most favored metal ions are Mn^{2+}, Cr^{3+}, and Fe^{3+}, which have five, three, and five unpaired electrons, respectively, in their 3^d orbitals (Table 26-1). Lanthanide metal ions have partially filled 4^f electron shells. There are seven 4^f orbitals, and the maximum spin is achieved when lanthanides contain seven electrons with unpaired spin. Gadolinium ion (Gd^{3+}) is the best known paramagnetic agent, with a spin quantum number of 7/2 (Table 26-1). Although the general rule of thumb is that the paramagnetic ions that have more unpaired electrons (higher magnetic moment) cause more efficient relaxation, there are some exceptions. The relaxation rate of the ion in an aqueous solution is not reliably predicted by its magnetic moment; significant and unexpected relaxation phenomena may result due to the effect of macromolecules, pH, and competition for binding.[34,35]

Although the transition and lanthanide series inorganic cations are powerful proton relaxers,[34] the unmodified metal ions are probably too toxic for human use. Typically they are metabolized, accumulate in the liver, and show slow biologic clearance.[8] Such diseases as fatty degeneration of liver, CNS demyelination, GI ulcerations, and renal and cardiovascular disorders may result from their use.[8]

Paramagnetic Metal Chelates and Complexes

Because of the inherent toxicity of free transition and lanthanide metal ions, chelated metal complexes have been proposed for MRI. The most commonly used chelators are ethylenediaminetetraacetic acid (EDTA) and diethylenetetramine pentaacetic acid (DTPA). Recently, new chelating complexes, such as gadolinium cryptelates and paramagnetic macrocyclic complexes, have been suggested.[36–38] Cryptelates are complexes of a lanthanide metal ion with amino acids of macrocyclic polyamines. The cryptelates, such as Gd DOTA, Gd NOTA, and Gd TETA, have been evaluated as MR contrast agents.[36] Another way to produce more tolerable paramagnetics is by using macrocyclic complexes, such as Mn (cyclam).[38] These complexes contain a metal ion encircled by a cyclic ligand that binds at four or five sites, leaving several sites available for interaction with water.[38]

The metal complexes are generally less effective in reducing T1 than the corresponding metals. Two reasons have been proposed for this difference: (1) The distance between the relaxing proton and unpaired electron is increased as a result of chelate, and the number of coordination sites around the paramagnetic ion (hydration sites) is diminished and (2) the formation of ligand bonds reduces the high-spin state of the paramagnetic ion.[39]

Different paramagnetic metal chelates have been tested in vivo in the search for an ideal MR contrast agent: Cr-EDTA,[40] Fe-EDTA,[41] Gd-DOTA,[42] Gd-EDTA,[41] Gd-DTPA,[43] and Mn-DTPA.[44] A major concern about the intravenous use of metal chelates is the potential for in vivo dissociation of the complexes, which would release toxic free metal ions. Therefore, a preparation of metal complexes having a high stability constant is required.[45,46] The chelates, such as EDTA, CDTA, and DTPA, when complexed with paramagnetic ions, remain unchanged in vivo to produce stable paramagnetic contrast agents.[45,46] It seems that the complexes of manganese, iron, and gadolinium do not undergo ligand substitution in vivo if their stability constants are approximately $>10^{16}$ for Mn and Gd and $>10^{22}$ for iron complexes.[45]

To date, Gd-DTPA and Gd-DOTA are most promising for human use.[36,37,43] They are rapidly distributed to the extravascular space, as are x-ray contrast media, and are excreted through the kidneys by glomerular filtration.[42,43]

Gd-DTPA is currently under clinical investigation in Europe and the United States. An increase in signal intensity can be produced in the kidney and liver by using as low a dose of Gd-DTPA as 0.05 mmol/kg.[47] The half-life of Gd-DTPA is 20 minutes, with predominant extracellular distribution and renal excretion.[43,48] Initial biodistribution of Gd-DTPA indicates maximum contrast enhancement of tissues between 15 and 25 seconds after IV administration and partial elimination within 70 seconds after administration.[49] Diagnostically relevant doses of 0.1–0.5 mmol/kg Gd-DTPA cause no demonstrable cardiovascular disorders nor changes in electrolyte concentration, liver function, or blood coagulation.[50] The enhancement of soft tissue signal can be demonstrated using various routes of administration, such as intravascular,[43] intrathecal,[51] oral,[52] pulmonary,[53] and intra-articular.[54]

Ferromagnetic Agents

Common examples of ferromagnetic materials are iron and magnetite (ferrites). Ferrites are crystalline iron oxides; their general formula is $Fe_2^{+3}O_3$x M^{+2}0 where M is a divalent metal ion, such as manganase or iron. Magnetite is a naturally occurring ferrite in which the metal ion is Fe^{+2}. The superparamagnetic or ferromagnetic iron compounds have been used for GI tract[21] and hepatobiliary imaging[22–24] and to label monoclonal antibodies for tumor imaging.[25]

Nitroxide Spin Labels (NSL)

There are a limited number of stable free radicals that have one or more sterically hindered unpaired electrons localized in a nitrogen-oxygen bond. Some paramagnetic NSL have prolonged stability at varying pH and temperature, long shelf life, chemical versatility for conjugation to biomolecules, and relatively low toxicity.[8] In addition, NSL are rapidly excreted through the kidneys at the rate of glomerular infiltration.[55,56] The major advantage of NSL is their tremendous chemical versatility; they have been attached as ''spin labels'' to a variety of biochemical substrates.[57] However, a relatively poor ability to relax protons limits the potential usefulness of NSL in clinical imaging.[55] Swartz and co-workers have developed an interesting approach of using nitroxides as hypoxia-sensitive MR contrast agents.[58] Despite extensive studies of distribution, clearance, metabolism, and toxicity of NSL, no clinically applicable spin label has been found so far.

Organ-Specific Contrast Agents

The chemical versatility of metal chelates and NSL allows for the development of organ-specific MR contrast media. One of the most exciting proposals is the paramagnetic labeling of *monoclonal antibodies*, which may distribute specifically to either normal organs or tumors. Initial reports demonstrated the feasibility of antimyosin antibody, bovine serum albumin, and IgG labeling by gadolinium and manganese complexes using the bifunctional chelate technique.[59,60] First experiments showed that protein-DTPA-metal ion conjugates included up to nine metal ions per protein molecule,[59,61] which yielded the superior water relaxation properties of macrocomplex compared to that of free metal. Unger and associates were able to attach 1.5 Gd atoms per anti-CEA monoclonal antibody, which was not sufficient for contrast enhancement of tumor.[61] After this preliminary and negative trial, more extensive labeling techniques and optimistic results have been published.[62,63] For example, a recent report by Japanese investigators indicates the feasibility of imaging myocardial infarction after the IV administration of antimyocardiac myosin monoclonal antibody labeled with Gd-DTPA.[63] By using polymeric chelate molecules, such as desferoxamine chelates of Fe^{+3} or DTPA chelates of Gd incorporated into the polylysine carrier, over 50 paramagnetic ions can be attached to one protein.[62,64,65]

Whether heavily labeled macromolecules are stable in vivo and whether their immunoreactivity is maintained remain to be clarified. An interesting approach is the use of ferromagnetic or superparamagnetic immunospecific MR contrast agents.[66] The rationale is that, by using ferromagnetic labels of monoclonal antibodies, the macromolecule is effective at subnanomolar concentrations, which should be sufficient for the binding to specific cell surface receptors needed to induce any change in tumor T1.[66] By labeling the antineuroblastoma monoclonal antibody with magnetite particles it was possible to decrease 58% of the T1 of tumor 24 hours after IV administration of monoclonals coupled with particles.[66] This type of particle, however, accumulates into macrophages in liver and spleen, which may limit its intravascular use.

Another possible approach in tumor imaging is to use paramagnetic *metalloporphyrins*. Porphyrins accumulate in liver, kidney, and tumorous tissues for unknown reasons.[67] Particularly promising is manganese-tetraphenylsulfonyl (MnTPPS), which remains stable in human plasma and has relatively high proton relaxation properties.[68] Upon IV injection into animals bearing experimental tumor, MnTPPS provided enhancement in several excised tissues, including kidney, liver, and tumor.[67] Figure 26-6 demonstrates our preliminary experiments using MnTPPS. The toxicity of paramagnetically labeled porphyrins may be, however, too high for human use.

Another application of macromolecules labeled with paramagnetic ions is with *blood pool MR contrast agents*.[69,70] Brasch and co-workers have used albumin-DTPA-Gd complex to enhance the intravascular MR signal. They were able to label the human serum albumin molecule with as many as 19 Gd-DTPA molecules.[69] This intravascular macromolecule seems to be useful for imaging myocardium, liver, brain and pulmonary tissue, i.e., the tissues with large blood perfusion.[69]

Liver- and spleen-specific agents include reticuloendothelial-avid hydroxycolloids and liposomes labeled with paramagnetics.[71,72] Liposomes are manmade lipid spheres composed of single or multiple lamellae separated by aqueous layers. They are rapidly removed from the circulation and deposited within the liver, spleen, bone marrow, and lungs. The process involves the phagocytic activity of macrophages. Thus, paramagnetics or ferromagnetics entrapped into liposomes enhance selectively the tissues that are rich in macrophage activity.[71,72] Liver- and bile-specific contrast agents include paramagnetic metal chelates that are taken up by hepatocytes and excreted into bile (Fe-EHPG, Gd-IDA)[73] and substances that accumulate into macrophages (magnetite).[22]

Molecular Oxygen

Molecular oxygen possesses two paired electrons with parallel spin and is thus paramagnetic. Young and co-workers administered 100% oxygen to five volun-

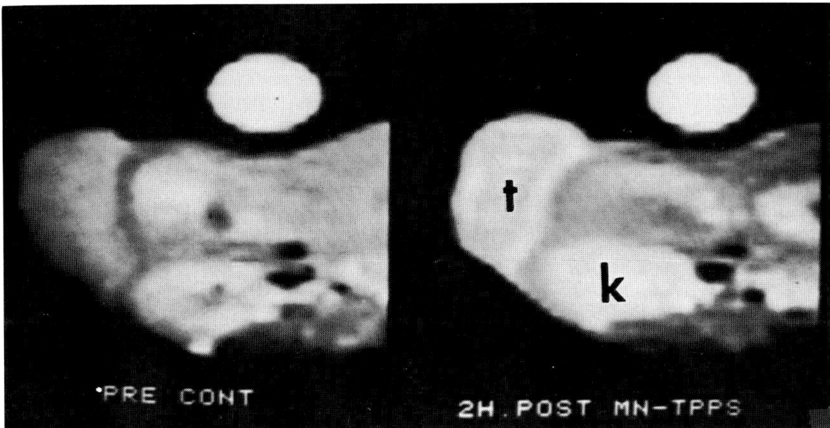

Figure 26-6 The contrast enhancement of sarcoma inoculated into nude mouse after metalloporphyrin Mn-TPPS4 injection. The porphyrin compound was administered at a dose of 0.08 mmol/kg, and a SE image (500/28) was obtained 2 hours later. The accumulation of Mn-TPPS4 in the tumor (*t*) is relatively specific, but enhancement is also seen in the kidney (*k*). *Source*: "Tumor Enhancement for MR Imaging: Comparison of Three Classes of Paramagnetic Contrast Media" by D Revel et al, Radiological Society of North America abstract 251, Chicago, 1985.

---

teers and observed the small increase in the MR signal from blood within the left ventricle (oxygen rich) as compared to that within the right ventricle.[74] The potential of perfluorinated blood and plasma substitutes, which have excellent oxygen solubility, to act as MR image contrast agents through increasing the oxygen concentration in organs is unclear.[75] It is also possible to enrich living tissues with oxygen-17 in the form of $H_2{}^{17}O$, which shortens the T2 times of tissues and does not affect T1 times. Perfusion of excised dog kidney with 10 ml of 5% $H_2{}^{17}O$ produced a significant change in renal intensity, which suggests the feasibility of using the oxygen-17 compounds as contrast agents in MR imaging.[76] However, the clinical usefulness of oxygen-derived contrast enhancers is yet to be determined.

PRECLINICAL AND CLINICAL APPLICATIONS

Central Nervous System

Paramagnetic small chelates, such as Gd-DTPA, cross the abnormal blood-brain barrier (BBB). Therefore, the mechanism of Gd-DTPA accumulation in the tumor tissue seems to be similar to that of x-ray contrast agents. Experimental studies of the damage to the BBB by hyperosmolar mannitol infusion clearly confirm this hypothesis.[77] Also, the damage to the blood-ocular barrier can be demonstrated by using Gd-DTPA-enhanced scanning.[78]

In the normal brain, paramagnetics cause a slight reduction in the gray/white matter contrast.[79] Brain areas where there is no BBB or an incomplete BBB, such as the pituitary gland, infundibulum, cavernous sinus, cranial nerves, and choroid plexus, also enhance after Gd-DTPA injection.[80,81] This ''physiologic'' enhancement should not confuse the interpretation of Gd-enhanced brain scans.

Contrast-enhanced MRI has been of particular value in defining the relationship of brain tumors and perifocal edema (Fig. 26-7).[82–84] Because the capillaries of neurinomas, adenomas, and meningiomas do not have normal function of the BBB, these tumors show consistent enhancement after Gd-DTPA administration. As in CT, the pattern of the enhancement may be central, patchy, diffuse, linear, or ring-like.[79,82,83] The effect of Gd-DTPA is best seen on a T1-weighted sequence; for example, IR 1400/400/30 or SE 500/30. Usually, the contrast enhancement lasts up to 50–60 minutes after injection.

Several nonenhanced MRI studies indicate that acoustic neurinomas are diagnosed with a high level of accuracy using T2-weighted MRI.[84] Gd-DTPA-enhanced MRI is, however, particularly useful for small intracanalicular tumors.[85] Intravenous administration of Gd-DTPA seems also to be useful in visualizing pituitary adenomas.[86] In general, normal pituitary tissue enhances promptly within 30 minutes after contrast injection, but adenomas enhance more slowly and persistently than the host pituitary.[86] Hence, the timing of postinjection imaging may be of critical importance in contrast-enhanced pituitary tumor imaging.

Tumors of higher malignancy grades generally show more contrast enhancement than tumors of lower grade.[79,83,87] For example, the increase of tumor intensity is almost twice as great after Gd-DTPA injection in high-grade gliomas than in low-grade gliomas.[79] More extensive clinical studies need to be done to clarify whether the separation of tumor and edema is always consistent and whether the correlation of contrast enhancement with tumor margins (and the degree of malignancy) after Gd-DTPA injection is reproducible.

In experimentally induced brain abscesses,[88,89] paramagnetics have been used successfully to define the disruption in the BBB and to differentiate lesions from surrounding edema. In experimental intracranial bacterial abscesses, Gd-DTPA clearly delineated the central pus, the surrounding capsule, and the edema of the abscess.[89] The Gd-DTPA enhancement was correlated with CT and microscopic studies. The increased signal around the abscess originated from the abscess capsule; the edema around the capsule did not enhance.[89]

Gadolinium-DTPA has also been tested in brain infarcts and multiple sclerosis. In one study, occlusion of the middle cerebral artery in cats indicated that Gd-DTPA causes enhancement of the lesion 16-24 hours after occlusion (39% increase in intensity).[90] The lesion enhancement then diminishes, being

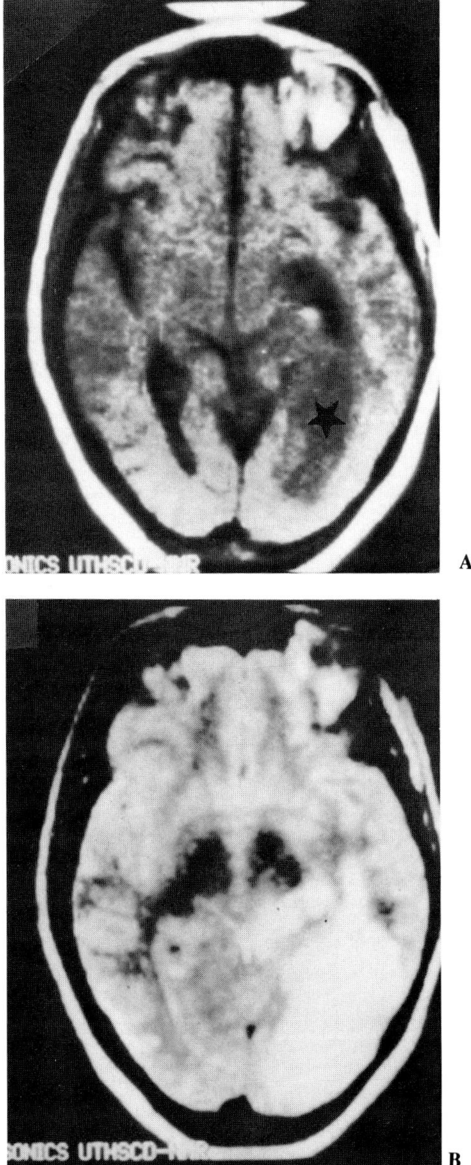

Figure 26-7 The contrast enhancement of glioblastoma in the left hemisphere after Gd-DTPA administration. On precontrast T1-weighted image (SE 500/28) the contour of the large tumor is hardly detected (*A*). On precontrast T2-weighted image (SE 2000/56) the tumor and surrounding edema are difficult to differentiate (*B*). After an injection of Gd-DTPA the margins of tumor are clearly seen on T1-weighted image (*C*).

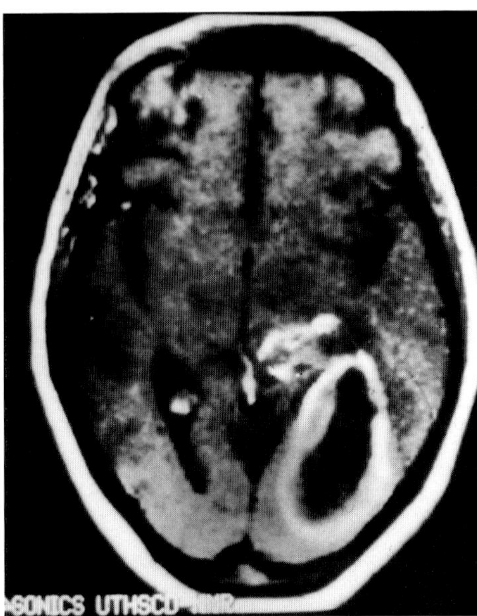

C

only 20% more than in normal white matter after 7 days. Acute cerebral ischemia (less than 16 hours after occlusion) did not enhance.[90] A preliminary clinical study of 11 patients with subacute or chronic brain infarcts confirmed that Gd-DTPA is useful in some patients.[91] Periventricular lesions and small, deep, white matter chronic infarcts, however, did not enhance. In general, T2-weighted sequences were most sensitive in demonstrating the infarcts. Gd-DTPA did increase the specificity of MRI.[91] Another preliminary report indicates that Gd-DTPA-enhanced MR imaging has the ability to separate active from inactive lesions in patients with multiple sclerosis.[92] In this study of 15 patients, contrast-enhanced MR was even more sensitive in showing enhancement than CT. Also, the ability of Gd-DTPA to delineate the lesion was due to the impaired BBB around multiple sclerosis plaques.[92]

Intraspinal tumors also accumulate Gd-DTPA.[93] The contrast agent was useful in outlining the tumorous masses, as well as in distinguishing intramedullary from extramedullary neoplasms.[93] Although in most cases of brain studies with paramagnetics the contrast is administered intravenously, intrathecal administration is also possible[94]. This technique may be of value in imaging abnormal cerebrospinal fluid (CSF) collections, intraspinal tumors, CSF rhinorrhea, and syringohydromyelia, and in studies of hydrocephalus and CSF flow dynamics.[94]

Although the early results using Gd-DTPA in different brain lesions are enthusiastic, some investigators have reported a more limited usefulness of contrast-enhanced MRI. In one study, the precontrast T2-weighted SE sequence was most sensitive in detecting brain lesions, showing 17 lesions out of 35 that were not seen on the post-Gd-DTPA images.[95] It was believed that Gd-DTPA may have value, particularly in the functional aspects of brain disease, such as the presence of perfusion of a lesion and an active break of the BBB.[95]

Cardiovascular System

Large blood vessels have a low signal intensity because of the rapidly flowing blood within them. The inherent contrast between large blood vessels and surrounding tissue diminishes in the microcirculation where the flow is slow.[2] Thus, paramagnetics have potential as perfusion markers in the microcirculation. Initial experimental studies used manganese salts to enhance successfully infarcted areas of the myocardium.[96] Experimental acute myocardial infarction produces a prolonged myocardial T1 and T2. Immediately after the intravascular injection of the paramagnetic agent, greater relaxation enhancement was seen in the normal myocardium than in the poorly perfused lesion.[97] Because of slow ''wash-in'' and ''wash-out'' of contrast medium from the infarcted tissue, the lesion remained enhanced for 5 minutes after Gd-DTPA injection.[97] In such a manner, the use of paramagnetics may improve the functional assessment of viable myocardium in acute cardiac ischemia. A recent study confirms that use of Gd-DTPA permits the detection of severe, resting myocardial flow reduction, but less severe flow differences are not readily detected.[98]

Detection of myocardial damage has been studied with infarct-avid monoclonal antibodies. The antibody labeled with Mn-DTPA and directed against cardiac myosin specifically localizes in the infarcted myocardium.[60] The complex was injected in a fairly large dose directly into the coronary arteries. Whether this experiment is reproducible using the IV route is unclear. Recently, Japanese investigators reported positive results in imaging infarcted canine heart tissue after IV injection of IgG-DTPA-Gd labeled with antihuman myosin antibody.[63]

If intravascular macromolecules, such as albumin, are tagged with paramagnetic agents, more specific information about the perfusion of different organs and tumors becomes available.[49,59,70] Human serum albumin can be paramagnetically labeled by binding up to 18 Gd-DTPA complexes per albumin molecule.[69] This technique enables effective MR imaging of the microcirculation of lung, heart, spleen, kidney, and brain tissue. Preliminary preclinical studies of the feasibility of albumin-Gd-DTPA complex in various ischemic disorders are in progress.[99] In a similar manner to the labeling of albumin with paramagnetics, red blood cells can be complexed with paramagnetics to enhance the vascular tree.[100]

Gastrointestinal Tract Imaging

Contrast enhancement of the GI tract to improve the differentiation of bowel from surrounding abdominal viscera is one of the most obvious applications of MRI contrast agents. Orally administered paramagnetic metal salts—for example, Gd-DTPA,[52] ferric chloride, ferric ammonium citrate,[4] Gd-oxalate,[101] ferrous sulphate heptahydrate, and iron dextran,[102] have been proposed for GI imaging. Clinical use of 1 mmol ferric ammonium citrate has permitted the delineation of the esophagus, stomach, duodenum, and the head of the pancreas, as well as the small bowel.[4]

The basic objective of gastrointestinal MRI is the differentiation of fat (bright) and air (dark) from bowel contents and tumors. Paramagnetic agents, such as Gd-DTPA and related compounds, increase the bowel intensity, which may cause difficulties in distinguishing tumors from abdominal fat. It is better to destroy the MR signal of normal gut by using ferromagnetic particles[103] or perfluorocarbon,[104] in the hope that tumors with a higher signal would be better visualized (Fig. 26-4).

Liver and Spleen

The liver and spleen MR contrast agents can be arbitrarily divided into three subgroups: *hepatobiliary agents*, *reticuloendothelial-avid* agents, and *nonspecific* contrast pharmaceuticals (nonspecific extracellular distribution).

Nonenhanced hepatic MRI cannot delineate well intra- or extrahepatic bile ducts. Therefore, contrast enhancement, achieved by administering a paramagnetic agent that is avidly extracted by the hepatobiliary system, may have clinical utility. For example, such agents as Gd-diethyl IDA and Gd di-isopropyl IDA,[46] Gd HIDA,[28] Fe- EHPG,[73] and its derivates have all been tested to alter MR characteristics of normally functioning liver. In addition, intravenous $MnCl_2$ loosely binds to serum proteins and is almost totally excreted into bile.[39] As mentioned earlier, $MnCl_2$ is, however, probably too toxic for human use. The bile-forming hepatocytes constitute almost 80% of the liver volume (Kupfer cells form only 2% of the total liver volume). Thus, the paramagnetic agents that are accumulated into hepatocytes and bile enhance hepatic MR signal very efficiently when compared to the macrophage-avid agents. Preliminary studies indicated that the liver signal could be increased up to 200% by IV administration of hepatobiliary-avid Fe(EHPG).[73] The acute toxicity of these compounds has, however, not been determined. Gd IDA[46] complexes and Gd-HIDA[28] have also been tested in experimental animals. All these agents are structural analogs of radioactive Tc IDA used presently in nuclear medicine. Another group of paramagnetic chemicals that may accumulate into bile are paramagnetic macrocyclic chelates. In rats, the IV injection of 16 μmol/kg of Mn (cyclam) caused the liver T1 relaxation rate to double at 15 minutes after injection.[38] No human studies using hepatobiliary paramagnetics have been performed so far.

The reticuloendothelial system (RES) consists of phagocytic cells in the liver, spleen, bone marrow, and lungs. The MR signal from the RES can be enhanced by using labeled liposomes.[72,105] Gd-DTPA incorporated into the lamellar phase of liposomes has been used to increase the T1 relaxivity of normal liver up to 110%.[105] Similarly, Mn-DTPA or $MnCl_2$[71] can be used to attach a paramagnetic label into liposomes and thus image liver, spleen, lungs, and bone marrow. Another group of agents incorporated into the RES are paramagnetic colloidal manganase sulfide[106] and gadolinium oxide.[107] Once again, these compounds cannot be applied to humans until their toxicity has been studied.

Recent studies have proposed the use of ferromagnetic or superparamagnetic particles as contrast agents for RES imaging.[21–25,108,109] Pure magnetite particles[21] or magnetite particles incorporated into albumin microspheres[108] are taken up almost exclusively by the RES following IV administration. Animal imaging and spectrometric data indicate that, in particular, liver, spleen, and pulmonary tissues are affected by ferromagnetics so that T2 and the signal intensity are reduced, resulting in "negative contrast."[21,108] The effect may last up to 6 months.[109] Soft tissue tumors inoculated into the liver are detected by using magnetite particles. For example, Widder noticed that the contrast of tumor and normal liver increases up to 200% after magnetite albumin administration.[108] Use of such ferromagnetic particles resulted in profound signal loss only from normal liver tissue but not from tumor tissue because tumors do not contain phagocytizing cells.[108,109] An interesting approach is to combine ferromagnetic particles and IV Gd-DTPA for liver cancer imaging. Using this "double-contrast" technique further increased the signal intensity of tumor and surrounding normal liver tissue;[110] Gd-DTPA accumulates in the tumor and increases tumor intensity, whereas ferrites decrease the normal liver intensity. Preliminary toxicity tests of ferromagnetic particles have shown a 20- to 250-fold margin of safety between the effective dose for imaging and a toxic dose.[108]

To date, plain Gd-DTPA has been ineffective as a contrast agent for hepatic tumor imaging because it is distributed both in normal liver and tumor interstitium. In fact, liver tumors may even be obscured after Gd-DTPA injection.[82] In one study of 23 patients with hepatic metastasis or hepatomas, the effect of Gd-DTPA was unclear. With some patients the contrast improved, while in others it became weaker.[111] Later, the same investigators published another study of 36 hepatic tumors and concluded that hemangiomas have a distinct pattern of contrast enhancement after Gd-DTPA injection: a diminished signal intensity on precontrast images, peripheral enhancement during the dynamic phase of contrast enhancement, and a complete fill-in on delayed images, similar to the pattern of hemangioma enhancement in CT.[112] Histologic correlation of hepatocellular carcinomas showed that the degree of contrast enhancement corresponded to tumor vascularity and that the peripheral halo corresponded to the fibrous capsular structure.[112] Modern fast imaging sequences have shown that the normal liver parenchyma is maximally enhanced within the first 2 minutes after Gd-DTPA injection.[113,114] After the initial vascular phase, Gd-DTPA is distributed into the interstitium, and the tumor-to-liver contrast begins to decrease in 3 to 8 minutes.[114] These findings clearly suggest the need for early dynamic imaging of hepatic tumors.

Other Applications

The primary benefit from the use of contrast agents in MRI is the addition of functional data to the superb morphologic information available in the images. Urography is a well-known example of the marriage of a functional and a morphologic study. For example, an acutely ischemic kidney following renal artery ligation may have the same MRI appearance as normally functioning kidneys. Paramagnetic agents that are administered intravenously and excreted through the kidneys, such as Gd-DTPA, provide a direct measure of renal function on the MR images.[113] In addition, the increased intensity of the kidneys, ureters, and bladder may help define contiguous abnormal masses.[115]

In the development of contrast agents a most obvious goal is better delineation of malignant tumors from normal tissue. In addition to monoclonals or porphyrins, the diagnostic role of Gd-DTPA has been evaluated in several types of animal and human tumors, including renal,[115] breast,[116,117] mediastinal,[118] pulmonary,[119] gynecologic,[120] and soft tissue.[121] In general, the usefulness of the Gd-DTPA type of agents lies in their ability to enhance the viable tissue around the tumor and thus differentiate the necrotic areas from edema.[116,122] However, it is impossible to differentiate the histologic type of tumor using Gd-DTPA; for example, sterile and bacterial abscesses, carcinoma, and hematoma may give similar patterns of enhancement.[116,122] All disease processes that increase the volume of extracellular space may potentially enhance after Gd-DTPA injection.

SUMMARY

The development of paramagnetic contrast agents offers an opportunity to expand the diagnostic yield of MRI and, in particular, to provide function-dependent images. The goal of future research will be to define agents that can be used safely and identify those having a unique distribution in the body so that specific tissues and functions can be highlighted.

REFERENCES

1. Wehrli FW, MacFall JR, Newton TH: Parameters determining the appearance of NMR images, in Newton TH, Potts DG (eds): *Modern Neuroradiology. Vol 2: Advanced Imaging Techniques*. San Anselmo, Clavadel Press, 1983, pp 81–117.

2. Mills CM, Brandt-Zawadzki M, Crooks LE, et al: Nuclear magnetic resonance: principles of blood flow imaging. *AJR* 1984;142;165–170.

3. Beall PT: Safe common agents for improved NMR contrast. *Physiol Chem Phys Med NMR* 1984;16:129–135.

4. Wesbey GE, Brasch RC, Goldberg HI, et al: Dilute oral iron solutions as gastrointestinal contrast agents for magnetic resonance imaging; initial clinical experience. *Magn Reson Imaging* 1985;3:57–64.

5. Hricak H, Crooks L, Sheldon P, et al: NMR imaging of the kidney. *Radiology* 1983;146:425–432.

6. Kundel HL, Schlakman B, Joseph PM, et al: Water content and NMR relaxation time gradients in the rabbit kidney. *Invest Radiol* 1986;21:12–17.

7. Newhouse JH, Brady TJ, Gebhardt M: NMR imaging: preliminary results in the upper extremities of man and the abdomen of small animals. *Radiology* 1982;142:246.

8. Brasch RC: Work in progress: Methods of contrast enhancement for NMR imaging and potential applications. *Radiology* 1983;147:781–788.

9. Mendonca-Dias MH, Gaggelli E, Lauterbur PC: Paramagnetic contrast agents in nuclear magnetic resonance medical imaging. *Sem Nuclear Med* 1983;13:364–376.

10. Burton DR, Forsen S, Karlstrom G: Proton relaxation enhancement (PRE) in biochemistry: a critical survey. *Progr NMR Spectroscopy* 1979;13:1–45.

11. Bloch F, Hansen WW, Packard M: The nuclear induction experiment. *Phys Rev* 1946;70:474–485.

12. Solomon I: Relaxation processes in a system of two spins. *Phys Rev* 1955;99:559–565.

13. Bloembergen N: Proton relaxation times in paramagnetic solutions. *J Chem Phys* 1957;27:572–573.

14. Dwek RA: Proton relaxation enhancement probes. *Adv Mol Relaxation Proc* 1972;4:1–53.

15. Koenig SH, Baglin CM, Brown RD III: Magnetic field dependence of solvent proton relaxation induced by Gd and Mn complexes. *Magn Reson Med* 1984;1:496–501.

16. Geraldes CFGC, Sherry AD, Brown RD, et al: Magnetic field dependence of solvent proton relaxation rates induced by Gd and Mn complexes of various polyaza macrocyclic ligands: implications for NMR imaging. *Magn Reson Med* 1986;3:242–250.

17. Lauffer RB, Brady TJ, Brown RD III, et al: 1/T1 NMRD profiles of solutions of Mn and Gd protein-chelate conjugates. *Magn Reson Med* 1986;3:541–548.

18. Gadian DG, Payne JA, Bryant DJ, et al: Gadolinium-DTPA as a contrast agent in MR imaging—theoretical projections and practical observations. *J Comput Assist Tomogr* 1985;9:242–251.

19. Brown MA: Effects of the operating magnetic field on potential NMR contrast agents. *Magn Reson Imaging* 1985;3:3–10.

20. Cullity BD: *Introduction to Magnetic Materials*. Reading, MA, Addison-Wesley, 1972.

21. Jacobsen T, Klaveness J: Magnetic particles as contrast media in MRI. Society of Magnetic Resonance in Medicine annual meeting (book of abstracts). 1985;2:868–869.

22. Saini S, Stark DD, Hahn PF, et al: Ferrite particles: A superparamagnetic MR contrast agent for the reticuloendothelial system. *Radiology* 1987;162:211–216.

23. Mendonca-Dias MH, Lauterbur PC: Ferromagnetic particles as contrast agents for magnetic resonance imaging of liver and spleen. *Magn Reson Med* 1986;3:328–330.

24. Olsson MBE, Persson BRB, Salford LG, et al: Ferromagnetic particles as contrast agent in T2 NMR imaging. *Magn Reson Imaging* 1986;4:437–440.

25. Renshaw PF, Owen CS, McLaughlin AC, et al: Ferromagnetic contrast agents: a new approach. *Magn Reson Med* 1986;3:217–225.

26. Hardy P, Henkelman RM: The mechanism of ferromagnetic contrast agents in proton relaxation. *Magn Reson Imaging* 1987;5:98.

27. Wolf GL, Joseph PM, Goldstein EJ: Optimal pulsing sequences for MR contrast agents. *AJR* 1986;147:367–371.

28. Greif WL, Buxton RB, Lauffer RB et al: Pulse sequence optimization for MR imaging using a paramagnetic hepatobiliary contrast agent. *Radiology* 1985;157:461–466.

29. Bydder GM, Young IR: MR imaging: Clinical use of the inversion recovery sequence. *J Comput Assist Tomogr* 1985;4:659–675.

30. Price AC, Runge VM: Optimization of pulse sequence in Gd-DTPA-enhanced magnetic resonance imaging, in Runge VM, Claussen C, Felix R, James AE (eds): *Contrast Agents in Magnetic Resonance Imaging*. Amsterdam, Excerpta Medica, 1986, pp 99–102.

31. Runge VM, Wood ML, Osborne MA, et al: Flash/Fisp—applications with Gd DTPA. *Magn Reson Imaging* 1987;5:95–96.

32. Matthei D, Frahm J, Haase A, et al: Regional physiological functions depicted by sequences of rapid magnetic resonance images. *Lancet* 1985;1:893.

33. Saini S, Hahn PH, Stark DD, et al: Contrast enhanced MRI of liver cancer using ferrite particles: pulse sequence and dose analysis. *Magn Reson Imaging* 1987;5:134.

34. Kang YS, Gore JC, Armitage IM: Studies of factors affecting the design of NMR contrast agents: manganese in blood as a model system. *Magn Reson Med* 1984;1:396–409.

35. Barnhart JL, Berk RN: Influence of paramagnetic ions and pH on proton NMR relaxation in biologic fluids. *Invest Radiol* 1986;21:132–136.

36. Knop RH, Frank JA, Dwyer AJ, et al: Gadolinium cryptelates as MR contrast agents. *J Comput Assist Tomogr* 1987;11:35–42.

37. Magerstadt M, Gansow OA, Brechbiel MW, et al: Gd(DOTA): an alternative to Gd(DTPA) as a T1,2 relaxation agent for NMR imaging or spectroscopy. *Magn Reson Med* 1986;3:808–812.

38. Jackels SC, Kroos BR, Hinson WH, et al: Paramagnetic macrocyclic complexes as contrast agents for MR Imaging: proton nuclear relaxation rate enhancement in aqueous solution and in rat tissues. *Radiology* 1986;159:525–530.

39. Wolf GL, Burnett KR, Goldstein EJ, et al: Contrast agents for magnetic resonance imaging, in Kressel HY (ed): *Magnetic Resonance Annual*. New York, Raven Press, 1985, pp 231–267.

40. Runge VM, Foster MA, Clanton JA et al: Contrast enhancement of magnetic resonance images by chromium EDTA: an experimental study. *Radiology* 1984;152:123–126.

41. Carr DH: The use of iron and gadolinium chelates as NMR contrast agents: animal and human studies. *Physiol Chem Phys Med NMR* 1984;16:137–143.

42. Josipowics N, Bonnemain B, Caille KH, et al: Pharmacokinetic studies of DTPA Gd and DOTA Gd in the rabbit. Society of Magnetic Resonance in Medicine annual meeting (book of abstracts). 1985;2:870.

43. Weinmann H-J, Brasch RC, Press W-R, et al: Characteristics of gadolinium-DTPA complex: a potential NMR contrast agent. *AJR* 1984;142:619–624.

44. Boudreau RJ, Frick MP, Levey RM, et al: The preliminary evaluation of Mn DTPA as a potential contrast agent for nuclear magnetic resonance imaging. *Am J Phys Imag* 1986;1:19–25.

45. Fornasiero D, Bellen JC, Baker RJ, et al: Paramagnetic complexes of manganase (II), Iron (III), and Gadolinium (III) as contrast agents for magnetic resonance imaging. *Invest Radiol* 1987;22:322–327.

46. Engelstadt B, Huberty J, White D, et al: Evaluation of gadolinium compounds potentially suitable for magnetic resonance using Gd-153 scintigraphy. *J Nucl Med* 1985;26:7–8.

47. Schorner W, Felix R, Laniado M, et al: Prufung des kernspintomographischen kontrastmittels Gadolinium-DTPA am Menschen. *Fortschr Rontgenstr* 1984;140:493–500.

48. Boudreau RJ, Burbidge S, Sirr S, et al: Comparison of the biodistribution of manganese-54 DTPA and gadolinium-153 DTPA in dogs. *J Nucl Med* 1987;28:349–353.

49. Schmiedl U, Moseley ME, Ogan MD, et al: Comparison of initial biodistribution patterns of Gd-DTPA and albumin-(Gd-DTPA) using rapid spin echo MR imaging. *J Comput Assist Tomogr* 1987;11:306–313.

50. Laniado M, Weinman HJ, Schorner W, et al: First use of GdDTPA/dimeglumine in man. *Physiol Chem Phys Med NMR* 1984;16:157–165.

51. Di Chiro G, Knop RH, Girton ME, et al: MR cisternography and myelography with Gd-DTPA in monkeys. *Radiology* 1985;157:373–377.

52. Kornmesser W, Laniado M, Hamm B, et al: The use of Gd-DTPA for gastrointestinal contrast enhancement in healthy male volunteers. *Magn Reson Imaging* 1987;5:136.

53. Montgomery AB, Paajanen H, Brasch RC, et al: Aerosolized gadolinium-DTPA enhances the magnetic resonance signal of extravascular lung water. *Invest Radiol* 1987;22:377–381.

54. Gylys-Morin VM, Hajek PC, Sartoris DJ, et al: Articular cartilage defects: detectability in cadaver knees with MR. *AJR* 1987;148:1153–1158.

55. Grodd W, Paajanen H, Eriksson UG, et al: Comparison of ionic and nonionic nitroxide spin labels for urographic enhancement in magnetic resonance imaging. *Acta Radiol* 1987;28:593–600.

56. Couet WR, Eriksson UG, Tozer TN, et al: Pharmacokinetics and metabolic fate of two nitroxides potentially useful as contrast agents for magnetic resonance imaging. *Pharm Res* 1984;5:203–209.

57. Dodd NJF: Spin label studies of cells, in Foster MA (ed): *Magnetic Resonance in Medicine and Biology*. Oxford, Pergamon Press, 1984, pp 66–91.

58. Swartz HM, Chen K, Pals M, et al: Hypoxia-sensitive NMR contrast agents. *Magn Reson Med* 1986;3:169–174.

59. Lauffer RB, Brady TJ: Preparation and relaxation properties of proteins labeled with paramagnetic metal chelates. *Magn Reson Imaging* 1985;3:11–16.

60. Brady TJ, Rosen BR, Gold HK et al: Selective decrease in T1 relaxation time of infarcted myocardium with the use of a manganase-labeled monoclonal antibody, antimyosin. Society of Magnetic Resonance in Medicine (book of abstracts). 1983.

61. Unger EC, Totty WG, Neufeld DM, et al: Magnetic resonance imaging using gadolinium labeled monoclonal antibody. *Invest Radiol* 1985;20:693–700.

62. Curtet C, Tellier C, Bohy J et al: Selective modification of NMR relaxation time in human colorectal carcinoma by using gadolinium-diethylenetriamine pentaacetic acid conjugated with monoclonal antibody 19–9. *Proc Natl Acad Sci* 1986;83:4277–4281.

63. Watanabe T, Yoshikawa K, Nishikawa J et al: Relaxation effect of contrast agent, anti-myocardiac myosin monoclonal antibody-diethylenetriamine pentaacetic acid-Gd. Society of Magnetic Resonance in Medicine (book of abstracts). 1987;2:928.

64. Shreve P, Aisen AM: Monoclonal antibodies labeled with polymeric paramagnetic ion chelates. *Magn Reson Imaging* 1986;3:336–340.

65. Manabe Y, Longley C, Furmanski P: High-level conjugation of chelating agents onto immunoglobulins: use of an intermediary poly(L-lysine)-diethylenetriamine pentaacetic acid carrier. *Biochemica Biophysica Acta* 1986;883:460–467.

66. Renshaw PF, Owen CS, Evans AE, et al: Immunospecific NMR contrast agents. *Magn Reson Imaging* 1986;4:351–357.

67. Lyon RC, Faustino PJ, Cohen JS, et al: Tissue distribution and stability of metalloporphyrin MRI contrast agents. *Magn Reson Med* 1987;4:24–33.

68. Chen C-W, Cohen JS, Myers CE, et al: Paramagnetic metalloporphyrins as potential contrast agents in NMR imaging. *FEBS letters* 1984;168:70–74.

69. Schmiedl U, Ogan M, Paajanen H, et al: Albumin labeled with Gd-DTPA as an intravascular, blood pool-enhancing agent for MR imaging: biodistribution and imaging studies. *Radiology* 1987;162:205–210.

70. Slane JMK, Lai C-S, Hyde JS: A proton relaxation enhancement investigation of the binding of fatty acid spin labels to human serum albumin. *Magn Reson Med* 1986;3:699–706.

71. Chilton HM, Jackels SC, Hinson WH, et al: Use of a paramagnetic substance, colloidal manganese sulfide, as an NMR contrast material in rats. *J Nucl Med* 1984;25:604–607.

72. Caride VJ, Sostman HD, Winchell RJ, et al: Relaxation enhancement using liposomes carrying paramagnetic species. *Magn Reson Imaging* 1984;2:107–112.

73. Lauffer RB, Greif WL, Stark DD, et al: Iron-EHPG as an hepatobiliary MR contrast agent: initial imaging and biodistribution studies. *J Comput Assist Tomogr* 1985;9:431–438.

74. Young IR, Clarke GJ, Bailes DR, et al: Enhancement of relaxation rate with paramagnetic contrast agents in NMR imaging. *J Comput Assist Tomogr* 1981;5:543–546.

75. Thomas SR, Clark LC Jr, Ackerman JL, et al: MR imaging of the lung using perfluorocarbons. *J Comput Assist Tomogr* 1986;10:1–9.

76. Hopkins AL, Barr RG: Oxygen^{-17} compounds as potential NMR T2 contrast agents: enrichment effects of H_20 on protein solutions and living tissues. *Magn Reson Med* 1987;4:399–403.

77. Runge VM, Price AC, Wehr CJ, et al: Contrast enhanced MRI: evaluation of a canine model of osmotic blood-brain barrier disruption. *Invest Radiol* 1985;20:830–844.

78. Frank JA, Dwyer AJ, Girton M, et al: Opening of blood-ocular barrier demonstrated by contrast-enhanced MR imaging. *J Comput Assist Tomogr* 1986;10:912–916.

79. Graif M, Bydder GM, Steiner RE, et al: Contrast-enhanced MR imaging of malignant brain tumors. *AJNR* 1985;6:855–862.

80. Kilgore DP, Breger RK, Daniels DL, et al: Cranial tissues: normal MR appearance after intravenous injection of Gd-DTPA. *Radiology* 1986;160:757–761.

81. Berry I, Brant-Zawadzki M, Osaki L, et al: Gd-DTPA in clinical MR of the brain. 2. Extraaxial lesions and normal structures. *AJR* 1986;147:1231–1235.

82. Carr DH, Brown J, Bydder GM, et al: Gadolinium-DTPA as a contrast agent in MRI: initial clinical experience in 20 patients. *AJR* 1984;143:215–224.

83. Felix R, Schorner W, Laniado M, et al: Brain tumors: MR imaging with Gd-DTPA. *Radiology* 1985;156:681–688.

84. Curati WL, Graif M, Kingsley DPE, et al: MRI in acoustic neuroma: a review of 35 patients. *Neuroradiology* 1986;28:208–214.

85. Vogl T, Bauer M, Hahn D, et al: MR-imaging of acoustic neuroma, plain and contrast enhanced studies. *Magn Reson Imaging* 1987;5:28.

86. Dwyer AJ, Frank JA, Doppman JL, et al: Pituitary adenomas in patients with Cushing's disease: initial experience with Gd-DTPA-enhanced MR imaging. *Radiology* 1987;163:421–426.

87. Claussen C, Laniado M, Schroner W, et al: Gadolinium-DTPA in MR imaging of glioblastomas and intracranial metastases. *AJNR* 1985;6:669–674.

88. Runge VM, Clanton JA, Price AC, et al: Evaluation of contrast-enhanced MR imaging in a brain-abscess model. *AJNR* 1985;6:139–147.

89. Grossman RI, Joseph PM, Wolf G, et al: Experimental intracranial septic infarction: magnetic resonance enhancement. *Radiology* 1985;155:649–653.

90. McNamara MT, Brant-Zawadzki M, Berry I, et al: Acute experimental cerebral ischemia: MR enhancement using Gd-DTPA. *Radiology* 1986;158:701–705.

91. Virapongse C, Mancuso A, Quisling R: Human brain infarcts: Gd-DTPA-enhanced MR imaging. *Radiology* 1986;161:785–794.

92. Grossman RI, Gonzalez-Scarano F, Atlas SW, et al: Multiple sclerosis: gadolinium enhancement in MR imaging. *Radiology* 1986;161:721–725.

93. Bydder GM, Brown J, Niendorf HP, et al: Enhancement of cervical intraspinal tumors in MR imaging with intravenous gadolinium-DTPA. *J Comput Assist Tomogr* 1985;9:847–851.

94. Rosen GM, Griffeth LK, Brown MA, et al: Intrathecal administration of nitroxides as potential contrast agents for MR imaging. *Radiology* 1987;163:239–243.

95. Brant-Zawadzki M, Berry I, Osaki L, et al: Gd-DTPA in clinical MR of the brain. 1. Intraaxial lesions. *AJR* 1986;147:1223–1230.

96. Brady TJ, Goldman MR, Pykett IL, et al: Proton nuclear magnetic resonance imaging of regionally ischemic canine hearts: effect of paramagnetic proton signal enhancement. *Radiology* 1982;144:343–347.

97. Wesbey GE, Higgins CB, McNamara MT, et al: Effect of gadolinium-DTPA on the magnetic relaxation times of normal and infarcted myocardium. *Radiology* 1984;153:165–169.

98. Johnston DL, Liu P, Lauffer RB, et al: Use of gadolinium-DTPA as a myocardial perfusion agent: potential applications and limitations for magnetic resonance imaging. *J Nucl Med* 1987;28:871–877.

99. Brasch RC: Developments in blood volume/perfusion contrast agents for magnetic resonance imaging. Evaluation of acute ischemia in the brain and heart. Presented at the Contrast Media 87 World Symposium, Monbazon, France, 1987.

100. Eisenberg AD, Conturo TE, Mitchell MR, et al: Enhancement of red blood cell proton relaxation with chromium labeling. *Invest Radiol* 1986;21:137–143.

101. Runge VM, Foster MA, Clanton JA, et al: Particulate oral NMR contrast agents. *Int J Nucl Med Biol* 1985;12:37–42.

102. Hall LD, Hogan PG: Paramagnetic pharmaceuticals for functional studies, in McCready VR, Leach MO, Ell P (eds): *Functional Studies using NMR*. Berlin, Springer-Verlag, 1986, 109–127.

103. Niemi P, Kormano M, Paajanen H: Ferromagnetic contrast studies of gastrointestinal tract at 0.02 Tesla. Submitted for publication.

104. Mattrey RF, Hajek PC, Gylys-Morin VM, et al: Perfluorochemicals as gastrointestinal contrast agents for MR imaging: preliminary studies in rats and humans. *AJR* 1987;148:1259–1263.

105. Kabalka G, Buonocore E, Hubner K, et al: Gadolinium-labeled liposomes: targeted MR contrast agents for the liver and spleen. *Radiology* 1987;163:255–258.

106. Chilton HM, Jackels SC, Hinson WH, et al: Use of a paramagnetic substance, colloidal manganase sulfide, as an NMR contrast material in rats. *J Nucl Med* 1984;25:604–607.

107. Burnet KR, Wolf GL, Schmacher HR, et al: Gadolinium oxide: a prototype agent for contrast enhanced imaging of the liver and spleen with magnetic resonance. *Magn Reson Imaging* 1985;3:65–71.

108. Widder DJ, Greif WL, Widder KJ, et al: Magnetite albumin micropheres: a new MR contrast material. *AJR* 1987;148:399–404.

109. Saini S, Stark DD, Hahn DF, et al: Ferrite particles: a superparamagnetic MR contrast agent for enhanced detection of liver carcinoma. *Radiology* 1987;162:217–222.

110. Weissleder R, Saini S, Stark DD, et al: Double contrast MR imaging of liver cancer. *Magn Reson Imaging* 1987;5:135.

111. Mano I, Yoshida H, Nakabayashi K, et al: Fast spin echo imaging with suspended respiration: gadolinium enhanced imaging of liver tumors. *J Comput Assist Tomogr* 1987;11:73–80.

112. Ohtomo K, Itai Y, Yoshikawa K, et al: Hepatic tumors: dynamic MR imaging. *Radiology* 1987;163:27–31.

113. Pettigrew RI, Avruch L, Dannels W, et al: Fast-field-echo MR imaging with Gd-DTPA: physiologic evaluation of the kidney and liver. *Radiology* 1986;160:561–563.

114. Saini S, Stark DD, Brady TJ, et al: Dynamic spin-echo MRI of liver cancer using gadolinium-DTPA: animal investigation. *AJR* 1986;147:357–362.

115. Laniado M, Claussen C, Kornmesser W, et al: Magnetic resonance imaging of renal tumors with gadolinium-DTPA, in Otto RC, Higgins CB (eds): *New Developments in Imaging*. New York, Thieme, 1986, pp 40–55.

116. Revel D, Brasch RC, Paajanen H, et al: Gd-DTPA contrast enhancement and tissue differentiation in MR imaging of experimental breast carcinoma. *Radiology* 1986;158:319–323.

117. Heywang SH, Hahn D, Schmidt H, et al: MR imaging of the breast using gadolinium-DTPA. *J Comput Assist Tomogr* 1986;10:199–204.

118. Hahn D, Nagele M, Seelos K, et al: Gd-DTPA contrast enhancement in conventional and fast MR-imaging of mediastinal masses. *Magn Reson Imaging* 1987;5:131.

119. Zeitler E, Kaiser W, Feyrer E, et al: Magnetic resonance imaging, with and without Gd-DTPA, of bronchial carcinoma, in Runge VM, Claussen C, Felix R, James AE (eds): *Contrast Agents in Magnetic Resonance Imaging*. Amsterdam, Excerpta Medica, 1986, pp 147–149.

120. Roth G: Experience with Gd-DTPA enhanced magnetic resonance imaging of gynecological tumors, in Runge VM, Claussen C, Felix R, James AE (eds): *Contrast Agents in Magnetic Resonance Imaging*. Amsterdam, Excerpta Medica, 1986, pp 167–169.

121. Pettersson H, Ackerman N, Kaude J, et al: Gadolinium-DTPA enhancement of experimental soft tissue carcinoma and hemorrhage in magnetic resonance imaging. *Acta Radiol* 1987;28:75–78.

122. Paajanen H, Grodd W, Revel D, et al: Gadolinium-DTPA enhanced MR imaging of intramuscular abscesses. *Magn Reson Imaging* 1987;5:109–115.

Part VIII

Miscellaneous Agents

Chapter 27

Arthrography

Richard W. Katzberg

In the past decade there have been significant improvements in the techniques of arthrography based on developments in contrast agents; new imaging modalities, such as computed tomography (CT), digital subtraction, and magnetic resonance (MR); and a better appreciation of fundamental abnormalities in joint pathology. With the advent of arthroscopy and MR performed without contrast agents, however, there has been an overall decline in the numbers of arthrographic studies performed nationwide. This chapter outlines the current trends in arthrography in the setting of newer methodologies and contrast agents and discusses the pharmacokinetics and morbidity of contrast agents together with the current developments in MRI contrast agents.

Arthrography is defined as a roentgenogram of a joint after the injection of a contrast agent that enhances the surfaces of the intra-articular structures. In general, three different procedures have been utilized in arthrography: (1) a negative contrast obtained with air or gas; (2) a positive contrast utilizing an iodinated opaque agent; and (3) a combination of these, i.e., the double-contrast technique.[1] The first reported arthrograms were performed in 1905 by Werndorff and Robinsohn, who used air in the knee joint.[2] Pneumoarthrography did not gain practical importance until the 1930s to 1940s when several detailed presentations on the subject were published.[3,4] Some authors attempted arthrography by outlining the menisci with gas produced within the joint as a manifestation of the vacuum phenomenon.[5,6] Pneumoarthrography utilizing injected room air or oxygen achieved only limited popularity because its accuracy was low. During the late 1940s, positive contrast arthrography gained popularity with improvements in the developments of iodinated contrast agents. Positive contrast arthrography was first popularized by Lindblom and is now a well-accepted procedure.[7] Double-contrast arthrography of the knee was introduced by Bircher in 1931.[8] This method was further refined by Andrew and Wehlin in 1960[9] and by Freiberger et al.[10] in 1966 and is now the standard technique for the examination of the knee joint.

TRENDS

In arthrography there is increasing utilization of low osmolality or nonionic contrast agents with or without epinephrine—often in association with injections of air for a double-contrast depiction of joint anatomy—and aided by more sophisticated imaging modalities, such as CT, multidirectional tomography, or digital subtraction technology. These methodologies improve tissue contrast, enabling an assessment of specific types of joint disorders. Examples of such beneficial techniques are CT-assisted knee and shoulder arthrography, tomographically assisted temporomandibular joint (TMJ) arthrography, and digital subtraction-assisted wrist and TMJ arthrography.

To assess the current and future status of arthrography, Hall surveyed 98 arthrographers, representing 92 departments, who performed or supervised a total of more than 33,000 arthrograms each year.[11] The number of arthrograms performed during the past 5–10 years decreased approximately 20%; 30% of the departments performed fewer arthrograms, and 15% performed more.

Knee arthrography accounted for 55% of all arthrograms performed. The number of knee studies performed during the past 5–10 years decreased 50% overall; of the respondents 85% performed fewer studies, 10% did the same number, and only 5% performed more arthrograms. All respondents mentioned arthroscopy as the major factor in this decline. Noninvasive MR imaging was thought by many to be an excellent technique for assessing the menisci and probably the cruciate ligaments. Many respondents thought that MR imaging, despite its high cost, eventually would replace both arthroscopy and arthrography as the technique of choice for assessing suspected internal knee injuries.

Shoulder arthrography accounted for 20% of the total arthrograms performed. Thirty percent of the respondents performed a greater number of these examinations in the last 5–10 years, whereas 10% performed fewer studies. The overall increase was thought to be related to a better orthopedic understanding of the syndrome of tendonitis/impingement/adhesive capsulitis. Most of the arthrographers were skeptical about a major role for sonography of the shoulder. They suggested that MR of the shoulder may not have a significant impact on the role of arthrography in the near future.

Arthrography of the temporomandibular joint accounted for almost 10% of all examinations in the survey. Use of this study had increased dramatically in recent years; few such arthrograms were performed 10 years ago. Although CT and MR have affected TMJ arthrography adversely, it was the opinion of many arthrographers that the overall underutilization of TMJ arthrography outweighed any impact of these imaging modalities and that in absolute terms the use of this examination would continue to increase. This accounted for 5–10% of arthrograms with recent increases.

Hip studies accounted for slightly more than 5% of all arthrograms. The utilization of both hip and wrist arthrography was felt to be related to trends in orthopedics. Elbow and ankle studies accounted for less than 5% of the total number of arthrograms in the survey. These examinations are primarily obtained to assess the presence of loose joint bodies or defects in the articular cartilage.

In summary, the use of arthrography has diminished only modestly over the past 5–10 years. Its continued popularity is based on very extensive clinical experience, an extremely low complication rate, and its low cost in comparison with arthroscopy and MR imaging.[1]

PHARMACOKINETICS OF WATER-SOLUBLE CONTRAST AGENTS

Contrast agents are necessary for imaging because small density differences in soft tissues do not allow optimal radiographic evaluation without enhancement. These agents should ideally be inert in every respect; however, significant toxicity related mainly to hyperosmolality does exist. Newer, low osmolality contrast agents are a promising step toward decreased toxicity.

Contrast agents available for arthrography depend upon iodine for their radiopacity and are water soluble, fully substituted, triiodinated, benzoic acid derivatives that are manufactured either as sodium or meglumine salts of either diatrizoate or iothalamic acid (Fig. 27-1). The theoretical ratio of iodine atoms to dissolved particles in solution for these agents is 1.5:1 because they are ionic and disassociate into two particles in solution, with the anion containing three iodine atoms and with a cation of sodium or methylglucamine (meglumine) containing no iodine atoms. Osmolalities range from 1200 to over 2000 mOsm/kg H_2O. The ionic monomers are small molecules at high concentrations in solution and are not metabolized.

Recent developments have reduced the high osmolality of these agents by changing the substituted side groups. This results in nonionic monomers in which the ratio of iodine atoms to dissolved particles in solutions is increased to 3:1 and the osmolality is more than halved. Nonionic monomers that have been recently developed include iopamidol and iohexol. A monoacidic dimer, ioxaglate (Hexabrix), is a ratio-3 contrast agent that has an ionized anion in solution with an osmolality of about 600 mOsm/kg H_2O at 320 mg I/ml. A nonionic dimer ratio-6 contrast agent, iotrol, has recently been developed that reduces the osmolality to 300 mOsm/kg of water at 300 mg I/ml, which is isotonic with blood.

Each molecule of conventional water-soluble contrast media disassociates in solution to one anion containing three iodine atoms and one cation, which is usually sodium or meglumine, that contains no iodine atoms. Dimerization

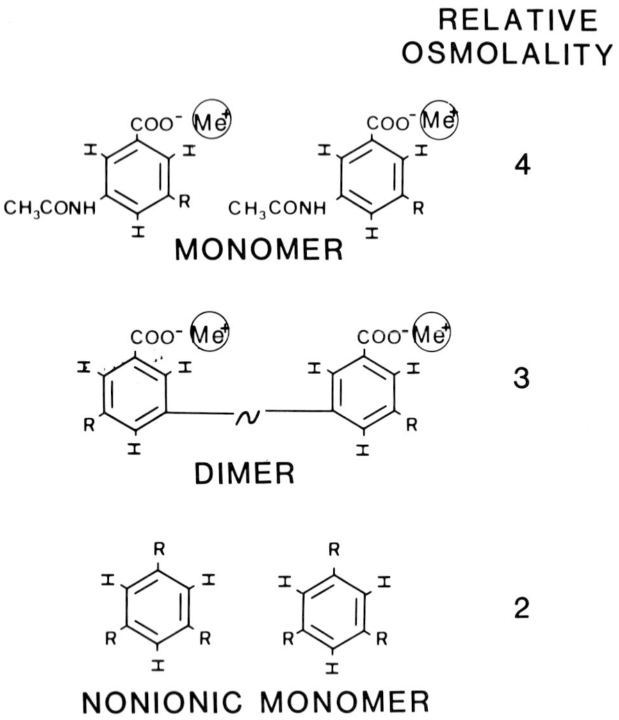

Figure 27-1 Relationship between the number of iodine atoms and the number of particles in the monomeric hypertonic contrast agents, the dimeric contrast agents, and the nonionic monomeric contrast agents.

---

reduces the osmolality by one-fourth, as each molecule disassociates into one anion containing six iodine atoms and two non-iodine-containing cations. The newer triiodinated nonionic contrast media have half the osmolality of the monomeric ionic contrast media with the same iodine content.

A rapid decrease of contrast density in the first 5 to 15 minutes after intra-articular injections of a contrast agent is a well-known phenomenon.[12–15] Diminution of contrast quality in arthrography is caused partly by direct resorption of the contrast agent through the synovial membrane and partly by dilution resulting from the osmotic effect of the contrast agent. It has been shown, for instance, that optimal contrast quality can be prolonged by adding epinephrine to the contrast agent, thereby reducing fluid movement across the highly vascular synovial membrane.[12,15]

The diffusion of the contrast agent through the synovial membrane seems to be largely dependent on the concentration gradient across the membrane and the molecular size of the contrast compound. Thus, one might anticipate a theoretical advantage for the dimeric contrast agents with molecular weights in the range of 1600 in comparison to either the ionic or nonionic monomers with molecular weights in the range of 800.[4,16]

An investigation in knee arthrography in dogs demonstrated that a dimer was superior to both an ionic and a nonionic monomer in terms of pharmacokinetic characteristics (Figs. 27-2 to 27-4) and radiographic quality (Table 27-1).[14] The dimer and the nonionic monomeric contrast agents were both superior to the ionic monomer in the radiographic evaluation, although the difference appeared to be more striking with the dimer than with the nonionic agent. A dimer consistently showed the highest total iodine values in the joint fluid. Because a dimer has both a larger molecular weight and a higher osmolality than a nonionic monomer, its larger molecular size seems to be a more important factor than its osmolality. The dimer draws more fluid into the joint than a nonionic agent because of its higher osmolality, but it tends to diffuse from the joint space at a slower rate because of its larger particle size. These two factors together result in nearly equal overall iodine concentrations of both agents in the joint fluid. The additional fluid drawn into the joint by the dimer apparently does not significantly hamper the radiographic quality of the arthrogram. This suggests that the rapid deterioration of radiographic quality is primarily due to the fast absorption rate of the contrast agent and is not due so much to the dilution of the contrast material. The total amount of

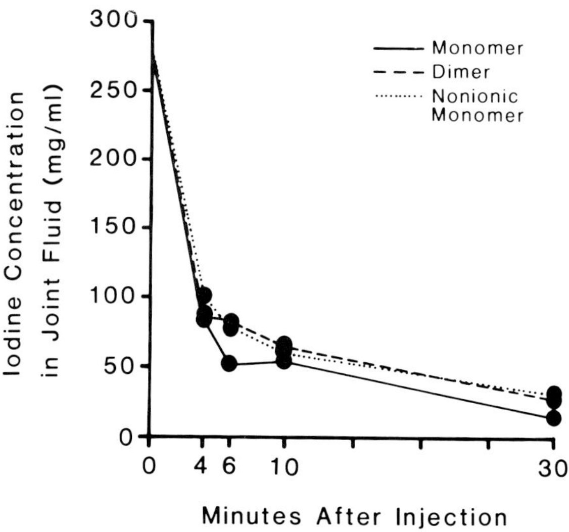

Figure 27-2 Relationship between iodine concentration in the knee versus time after injection of 4 ml of contrast material. Each point represents a summary of 6 data values. Note the very rapid decline in iodine concentration after the injection, which is a function of the absorption of contrast material and the movement of fluid into the joint space.

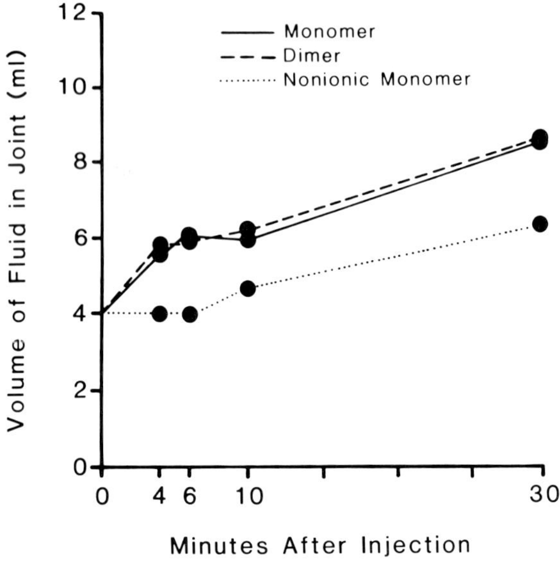

Figure 27-3 Volume of joint fluid (ml) versus time. The nonionic monomer creates less of a joint fluid flux into the joint space than do either the monomer or the dimer.

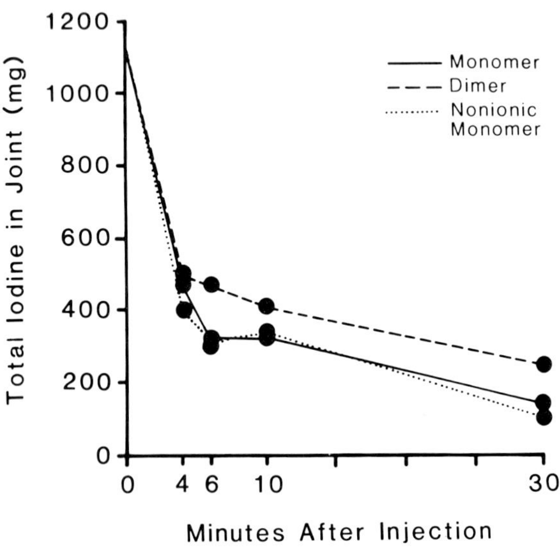

Figure 27-4 The important relationship between total iodine in the joint (mg) versus time after injection. Again note the rather exponential pattern to the decrease in total iodine in the joint versus time. The dimer is slightly superior than either the monomer or the nonionic monomer in the important parameter of total iodine with arthrography.

Table 27-1 Comparison of Contrast Agents in Knee Arthrography in Dogs (Values at 6 minutes post-injection)

	Total Iodine	*Joint Volume*	*Iodine Concentration*	*Radiographic Evaluation*
Monomer	− − −	+ + +	− − −	+
Dimer	− −	+ + +	− −	+ + +
Nonionic	− − −	+ +	− −	+ +
Monomer + EPI	−	+	−	+ + + +

fluid in the joint space is similar for the dimer and the monomer (Fig. 27-3). Most probably, this again results from the delayed diffusion of the dimer from the joint space compared to the faster diffusion of the monomer (Fig. 27-4). This disparity results in roughly equal osmotic forces for both agents.

A successful approach that decreases contrast material absorption from the joint space is the simultaneous intra-articular injection of epinephrine with the contrast agent.[12] Epinephrine slows the local circulation around the joint, which probably accounts for the delayed contrast absorption.

The effect of 0.3 ml of 1:1,000 epinephrine on iodine concentration, total iodine content, fluid volume, and radiographic quality was evaluated in knee arthrograms in dogs.[15] With the addition of epinephrine the iodine concentration and total iodine content were significantly higher initially and remained significantly higher than in controls over a 1-hour period (Fig. 27-5). The fluid volume in the knee was significantly lower with epinephrine. Both contrast absorption and dilution were major parameters in decreasing iodine concentration. With increasing time the effect of contrast absorption becomes the major factor in deterioration of arthrographic quality, having a two to three times greater effect on decreasing iodine concentration. When epinephrine was used, early radiographs were noted to be significantly better, and although the quality deteriorated rapidly in the controls, the enhanced quality with epinephrine persisted up to 1 hour. These findings confirmed those of Hall that epinephrine is a very useful adjunct in enhancing the quality of knee arthrography.[12] Similar findings have been noted in TMJ arthrotomography as well.[17]

MORBIDITY

Discomfort from arthrography performed with water-soluble contrast agents or double-contrast techniques occurs with a higher incidence than previously thought. A prospective study in 72 patients that examined immediate post pro-

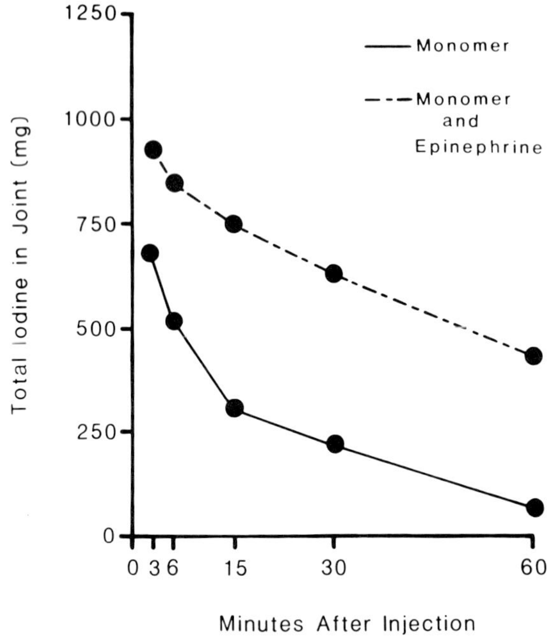

Figure 27-5 Relationship between total iodine content (mg) in the knee joint of the dog versus time comparing the monomer alone versus the monomer plus 0.3 ml of epinephrine. There is a significant difference in total iodine in the knee joint of the dog when utilizing the monomeric contrast agent in conjunction with epinephrine. Note also the rather exponential decrease in total iodine in the joint with the monomer alone, which is similar to that shown in Figure 27-4.

---

cedural discomfort and delayed post procedural discomfort after shoulder arthrography was reported by Hall et al. in 1981.[18] Three different techniques were compared using equal intra-articular volumes of positive contrast agent alone (13 ml), positive contrast agent (13 ml) diluted with 2% lidocaine, or positive contrast agent (3 ml) and 10 ml of air. All patients received 0.35 ml of 1:1,000 epinephrine injected intra-articularly with the positive contrast agent. Patients were asked to quantify their baseline shoulder discomfort on the basis of symptoms the day before and the day of arthrography. At the conclusion of the arthrogram the patients were again asked to quantify their discomfort using the same format.

Moderate or severe exacerbation of baseline shoulder discomfort, when evaluated 24–48 hours after arthrography, occurred in 74% of all patients. There was significantly less discomfort after double-contrast examinations than after other techniques, although the level of morbidity was still high. The incidence and severity of discomfort after shoulder arthrography were, thus, higher than previously reported or generally recognized.

These authors suggested that morbidity from shoulder arthrography is probably caused by a direct irritant effect of the positive contrast medium and/or the additional influx of fluid into an already distended joint, in response to the hyperosmolar contrast agent. The latter effect was consistent with the delayed onset of symptoms, which usually began 4–6 hours after the procedure, peaked in intensity 12 hours later, and, when severe, often lasted for several days. Smaller volumes of intra-articular contrast agent and/or the use of nonionic and polymeric contrast agents were suggested to reduce patient symptomatology.

A similar study evaluated morbidity with ionic and low osmolality contrast agents in TMJ arthrography.[17] Thirty-one patients who presented with TMJ pain and dysfunction and were suspected of having internal derangements were chosen for the study protocol. Either meglumine/sodium diatrizoate (60%, Renografin-60) or the monoacidic dimer (Hexabrix) was used, and the contrast agent vials were randomized and labeled sequentially. Patients were asked to rate their baseline TMJ discomfort on the basis of their symptoms on the day of arthrography and over the 24-hour period after the study. The arthrograms were rated for quality by three radiologists.

There were no statistical differences in the baseline discomfort between the two groups. The immediate discomfort experienced by those patients following the injection of Hexabrix was less than that experienced by those patients receiving Renografin-60 ($p < 0.05$).

There was no statistical difference in the amount of delayed (24 hour) discomfort between the patients in either the Hexabrix or the Renografin-60 group. Three patients (3/28 or 11%, one who received Hexabrix and two who received Renografin-60) had moderate exacerbation of pain over the 24-hour period following the procedure. The most common descriptions of the delayed discomfort after the procedure were soreness, stiffness, and/or mild headache.

Although the overall radiographic quality of the arthrograms was better for Hexabrix than Renografin-60 at the immediate and 10-minute periods, the differences were not statistically significant (Fig. 27-6). This study revealed that patients experienced maximal discomfort from TMJ arthrography with the initial joint filling and joint distention, which rapidly resolved over 10 minutes. The cause of pain in this study seemed to be joint distention and the use of a hypertonic medium. The rapid improvement that patients experienced after contrast injection is consistent with previous studies that showed rapid, almost exponential absorption of contrast agents from the joint resulting in decreased joint distention. The delayed exacerbation of patient discomfort following TMJ arthrography in 3 of 28 patients is most likely due to a direct chemical irritant effect of the contrast agent on the synovium.

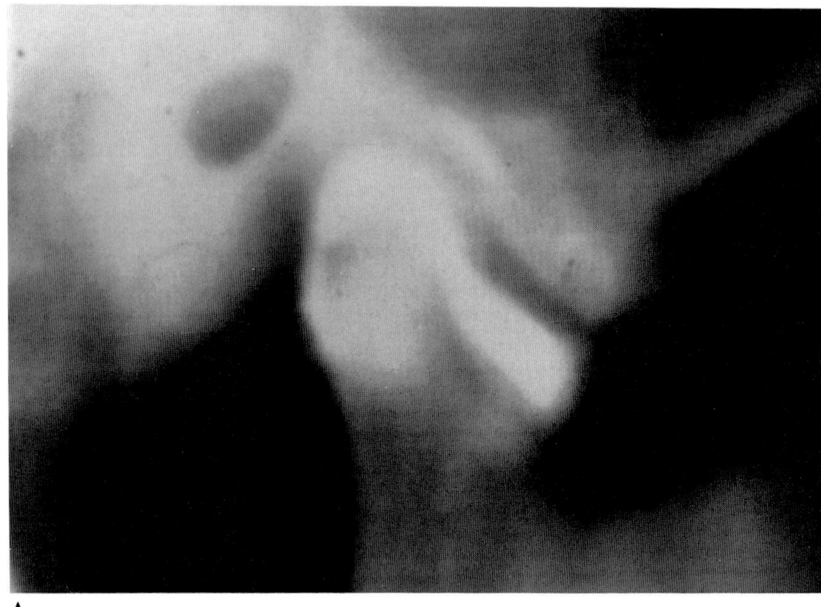

A

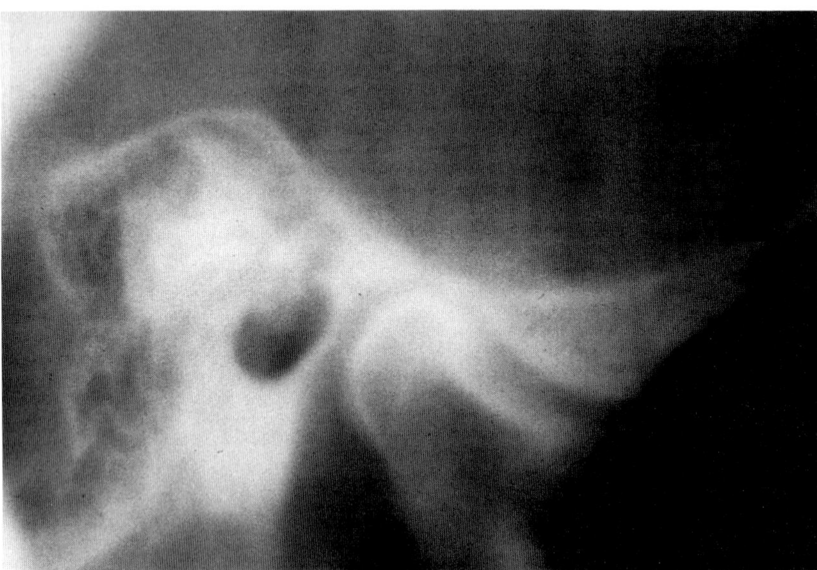

B

Figure 27-6 A TMJ arthrogram performed with (*A*) a low osmolality contrast medium, which shows increased opacification of the joint space at 5 minutes after the injection, versus the (*B*) lesser contrast opacification with the standard hypertonic contrast medium, meglumine diatrizoate (300 mg I/ml).

Fichtner and Weiss found transient inflammatory reactions in the synovial membrane after injection of 1 ml Urografin-76% or 1 ml 50% glucose in rat knee joints.[19] The changes were seen in biopsies taken 3 hours to 1 day after injection, and no changes were observed after 2 days. Reaction caused by the contrast agent was, however, less severe than that caused by the less concentrated glucose solution. Bodnya et al. injected 1 ml of a triiodinated contrast agent (diatrizoates in acid acetrizoates) in rabbit knee joints and found temporary and reversible inflammatory changes that could be observed after 2 hours, decreasing after 2 days and almost disappearing by the fourth day.[20] The response was milder when the contrast agent was first mixed with the same volume of 2% novocaine. Bjork et al. examined the knee joints of four rabbits after repeated injections of 0.5 ml of a dimer or saline and found no signs of joint damage.[16] Their animals were sacrificed 4–11 days after the injections.

Johansen and Berner found minimal inflammatory reactions in the synovial tissue of rabbit knee joints not only with various types of contrast agents but also with physiologic saline.[13] They found no correlation between the osmolality of the contrast agent and the severity of the inflammation. The authors concluded that joint distention can by itself cause synovitis. Human arthrography with air alone or with air and positive contrast agents causes a low-grade inflammatory process, consisting of eosinophil and round cell infiltration in the synovial membrane.

It has been demonstrated that, after the intra-articular injection of contrast agents, the synovium becomes edematous and hemorrhagic within 2 hours. Histologically, at 24 hours there is tissue eosinophilia and vascular congestion. The use of sodium-containing contrast agents may lead to a more severe reaction that includes contraction and adhesions of the synovium and greater tissue eosinophilia, as well as mast cell proliferation.[21] Sodium-containing agents may cause more pain, especially if injected inadvertently into the soft tissues, and therefore, the use of these agents should be avoided. Eosinophilia of the synovial fluid has also been demonstrated following arthrography.[22]

Arthrography can still be performed in patients with a previous history of contrast agent reaction.[23] One alternative is to perform an air-contrast examination or to perform the arthrogram with a nonionic contrast agent. Although pretreatment is currently recommended, it has not been shown conclusively to prevent a recurrent reaction.

As for severe contrast agent reactions, arthrography is considered extremely safe. Reactions to intra-articular contrast agents are very uncommon when compared to intravascular administration of these same agents. Newberg recently summarized this topic in conjunction with a questionnaire received from 57 radiologists who performed or supervised more than 126,000 arthrograms (Table 27-2).[23] No deaths were observed in this patient study group. The most common complication reported was a sterile chemical synovitis, which occurred in 150 patients. Vagal reactions occurred in 83 patients; six of them subsequently

Table 27-2 Complications of Arthrography in 126,000 Examinations

Complication	*Number of Patients*
Death	0
Severe reactions	
Hypotension	4
Vasomotor collapse and laryngeal edema	1
Air embolism	1
Vagal reactions	83
Hives	61
Cellulitis	1
Sepsis	3
Massive effusion	1
Severe pain	5
Sterile chemical synovitis	150

Source: Reprinted with permission from "Contrast Reactions to Arthrography" (pp 17–24) by AH Newberg in Syllabus for the Categorical Course on Diagnostic Techniques in the Musculoskeletal System, American College of Radiology, 1986.

experienced seizures, and another patient became apneic. Urticaria occurred in 61 patients. Several respondents noted that hives may occur up to several hours following the procedure. There were three instances of septic arthritis and one case of cellulitis following arthrography. There were only six severe contrast reactions: four hypotensive episodes, one vasomotor collapse with laryngeal edema that resolved, and one air embolism.

FUTURE DEVELOPMENTS AND MR CONTRAST AGENTS

MR has been shown useful in depicting the normal anatomy of the knee, as well as abnormalities involving the ligaments, menisci, and periarticular soft tissue structures. With its advantages of noninvasiveness, superior tissue contrast discrimination, adequate spatial resolution with surface-coil techniques, and ability to provide information about both intra- and extra-articular soft-tissue structures, MR is a promising diagnostic method for evaluation of joint structures. The major disadvantage at this time is the high cost of this modality in comparison to arthrography. Further refinements in techniques to improve soft tissue contrast of intra-articular structures suggests that MR will have a major impact on the use of knee arthrography.

The capability of MR to detect focal defects in the articular cartilage of the knee was investigated with intra-articular injections of saline or gadolinium-DTPA.[24] This study demonstrated that discrete cartilage lesions as small as 3 mm in diameter could be seen with T2-weighted sequences, regardless of other imaging parameters, such as number of signal averages, size of acquisition matrix, and slice thickness. The diagnostic capabilities increased, however, after the intra-articular injection of gadolinium-DTPA in a 500 μm concentration. Gadolinium-DTPA is of low toxicity and has a similar molecular size as the currently used water-soluble iodinated contrast agents. Whether there will really be a need for intra-articular contrast agents in conjunction with MR will require further study. Fat suppression MR techniques demonstrate superior depiction of cartilage over conventional spin echo techniques.[25] It is my opinion that MR of articular structures will develop successfully without the need for intra-articular contrast agents.

REFERENCES

1. Foote GA: Arthrographic contrast media, in Miller RE, Skucas J (eds): *Radiographic Contrast Agents*. Baltimore, University Park Press, 1977, pp 451–462.

2. Werndorff R, Robinsohn I: Ueber intraarticulare und interstitielle Sauerstoffinsufflation zu radiologisch-diagnostischen und Therapeitischen Zwecken. *Verhandl Dtsch Ges Orthop* 1905;4:9–11.

3. Mescham I, McGraw WH: Newer methods of pneumoarthrography of the knee with an evaluation of the procedure in 315 operated cases. *Radiology* 1947;49:675–711.

4. Sommerville EW: Air arthrography as an aid to diagnosis of lesions of menisci of knee. *J Bone Joint Surg* 1942;24:873–882.

5. Dittmar O: Der kniegelerks-Meniskus im Rontgenbilde. *Rontgenpraxis* 1932;4:442–445.

6. Gershon-Cohen J: Internal derangements of the knee joint: diagnostic scope of soft tissue roentgen examinations and the vacuum technique demonstration of the menisci. *AJR* 1945;54:338–347.

7. Lindblom K: Arthrography of the knee: a roentgenographic and anatomical study. *Acta Radiol* 1948;74(suppl):1–112.

8. Bircher E: Pneumoradiographie des knies und der anderen gelenke. *Schweiz Med Wschr* 1931;61:1210–1211.

9. Andrew L, Wehlin L: Double-contrast arthrography of the knee with horizontal roentgen ray beam. *Acta Orthop Scand* 1960;29:307–314.

10. Freiberger RH, Killoran PJ, Carodona G: Arthrography of the knee by double-contrast method. *AJR* 1966;97:736–747.

11. Hall FM: Arthrography: past, present and future. *AJR* 1987;149:561–563.

12. Hall FM: Epinephrine-enhanced knee arthrography. *Radiology* 1974;111:215–216.

13. Johansen JG, Berner A: Arthrography with Amipaque (metrizamide) and other contrast media. *Invest Radiol* 1976;11:534–540.

14. Katzberg RW, Burgener FA, Fischer HW: Evaluation of various contrast agents for improved arthrography. *Invest Radiol* 1976;11:529–533.

15. Spataro RF, Katzberg RW, Burgener FA, et al: Epinephrine-enhanced knee arthrography. *Invest Radiol* 1978;13:286–290.

16. Bjork L, Erikson U, Ingelman B: A new type of contrast medium for arthrography. *AJR* 1970;109:606–610.

17. Katzberg RW, Miller TL, Hayakawa K, et al: Temporomandibular joint arthrography: comparison of morbidity with ionic and low osmolality contrast media. *Radiology* 1985;155:245–246.

18. Hall FM, Rosenthal DI, Goldberg RP, et al: Morbidity from shoulder arthrography: etiology, incidence, and prevention. *AJR* 1981;136:59–62.

19. Fichtner HJ, Weiss JW: Zur frage der Gelenkschadigung durch Kontrastmittel. *Beitr Klin Chir* 1963;207:164–171.

20. Bodnya IF, Mikliaev YI, Yakovetz VV: [Response of the synovial membrane to the introduction into the joint cavity of liquid contrast medium] (In Russian). *Ortop Travmatol Protez* 1973;34:53.

21. Pastershank SP, Resnick D, Niwayama G, et al: The effect of water-soluble contrast media on the synovial membrane. *Radiology* 1982;143:331–334.

22. Hasselbacher P, Schumacher HR: Synovial fluid eosinophilia following arthrography. *J Rheum* 1978;5:173–176.

23. Newberg AH: Contrast reactions to arthrography, in *Syllabus for the Categorical Course on Diagnostic Techniques in the Musculoskeletal System*. American College of Radiology, 1986, pp 17–24.

24. Gylys-Morin VM, Hajek PC, Sartoris DJ, et al.: Articular cartilage defects: detectability in cadaver knees with MR. *AJR* 1987;148:1153–1157.

25. Totterman S, Szumowski J, Hornak JP, et al: MR fat suppression technique in the evaluation of normal structures of the knee (abstracted), in *Book of Abstracts: Society of Magnetic Resonance in Medicine*, 1987, p 128.

Chapter 28
Lymphography

Harry W. Fischer

Modern clinical radiographic visualization of certain lymphatic structures is based upon direct injection of contrast agents into a peripheral lymphatic. The visualization of lymphatics and lymph nodes by indirect technique (i.e., by injection of contrast materials into the tissues or body cavities and joints) is mainly of historical interest because it has not developed into a clinically useful technique. A major portion of the indirectly introduced contrast material remains at the injection site or enters the blood vessels, whereas only a minor portion enters the lymphatics.

The modern era of lymphography stems from the work of Kinmonth and coworkers.[1] A small amount of blue dye injected intradermally and subcutaneously into the web spaces between the toes passes into the lymphatics, coloring them greenish-blue within a short time. The subcutaneous lymph channels of the dorsum of the foot are then cannulated by direct cut-down on the visible blue-stained trunks. The contrast agent first used was a water-soluble one, of the group used for intravenous urography and for angiography. Currently, a fatty or oily contrast material is the agent of choice.

BLUE DYE

Description

The blue dye originally used clinically—Patent Blue V,[2] also known as Patent Blue Violet—was selected because earlier work had shown it to enter and clearly demonstrate the lymphatics[3] when it was injected into the tissues of experimental animals. The chemical structure of Patent Blue V, a diamino derivative of triphenylmethane dye, is shown in Figure 28-1. A blue dye, Alphazurine 2G, of closely related structure and also shown in Figure 28-1 has often been considered

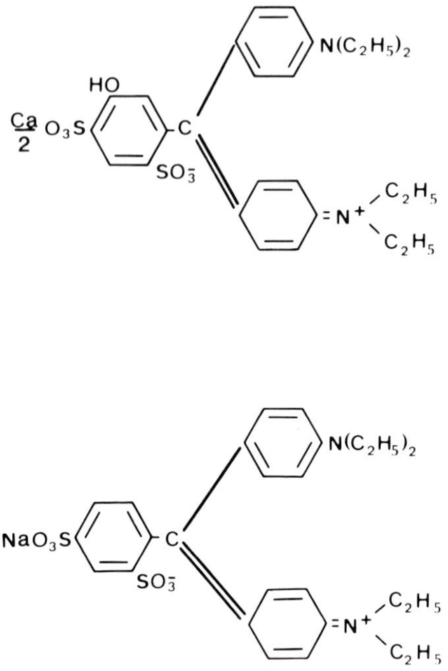

Figure 28-1 The chemical structure of the two triphenylmethane blue dyes used in visualization of lymph trunks for lymphographic injection. The compound 42051 (top) in the Color Index is officially termed Acid Blue 3 and is known as Patent Blue V or Patent Blue Violet. The compound 42045 (bottom) is termed Acid Blue I and is known as Alphazurine 2G. The two names are frequently used interchangeably, and commercial preparations may contain one or both compounds.

---

the same as Patent Blue V and has been used interchangeably and widely for lymphography.[4] An 11% water solution of Patent Blue V is said to be isotonic with body fluids. Other blue dyes that have been used are Evans Blue, Brilliant Blue FGG (FD and C Blue no. 1), direct sky blue, and red dye prontosil rubrum,[5] but none of these has had a wide or sustained usage.

Nuclear magnetic resonance studies on several samples of dyes from different manufacturers labeled Alphazurine 2G, Patent Blue Violet, and FD and C Blue no. 1 revealed three distinct compounds, with one mixture being labeled as all the same compound. Thin-layer chromatography has shown that the purity of the dyes varies from batch to batch.[6]

Patent Blue V, after interstitial injection, is absorbed from the tissues into the lymphatics and into the capillaries. The dye circulates through the body and is excreted in part through the kidneys. When larger volumes are injected, generalized but temporary bluish discoloration of the skin can occur, particularly in children. Bluish coloration of the urine is not unusual. The patient should be assured that these phenomena are not dangerous. The blue coloration of the skin in the region of the toes and foot and along the course of the lymph trunks of the extremity clears progressively and is essentially eliminated at about 2 weeks. Evans Blue is not as satisfactory in coloring the lymphatics as Patent Blue.[2] Direct sky blue dye colors the local tissue for an inordinately long period of time,[7] and the "tattoo effect" is objectionable to the patient and physician alike.

More recently a related blue dye has been introduced for the identification of the lymphatics by the same injection technique. This blue dye, the 2, 5 disulfonic acid isomer of Patent Blue, received approval from the Food and Drug Administration for clinical use after suitable demonstration of acute low toxicity in the rat and a successful clinical trial in 543 patients and volunteers.[8] The similarity of structural formula of this compound, officially named isosulfan blue (Lymphazurine), to Patent Blue Violet and Alphazurine 2G is shown in Figure 28-2.

Isosulfan blue for injection is formulated as a 1% solution based on the work of Gangolli et al. who found no tissue necrosis upon subcutaneous administration of triphenylmethane dyes in a 1% solution but widespread necrosis when in a 3% solution.[9] The compound was prepared in high purity as determined by high pressure liquid chromatography, which revealed 94.5% of the 2, 5 isomer and the remaining 5.5% as closely related isomers.[8] Phosphate was used as a buffer to achieve a final pH of 6.8–7.5.

Dose and Technique

The blue dye is injected subcutaneously in the web spaces between the toes or fingers. For the isosulfan blue, 1 to 3 ml of a 1% solution is recommended (1 ml per web space). Previously we used 0.1 ml of the 11% Patent Blue V or Alph-

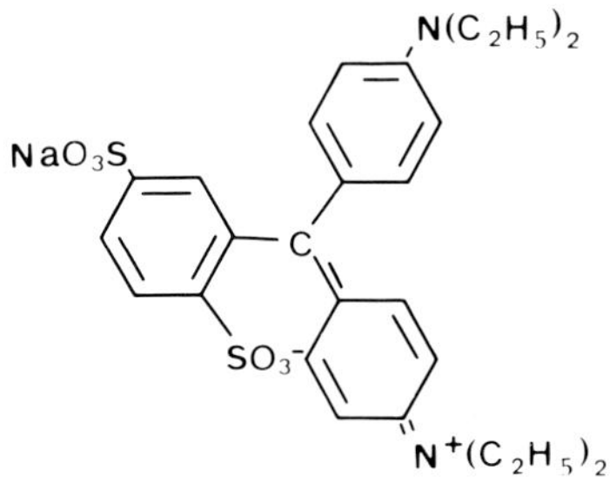

Figure 28-2 Structural formula of isosulfan blue. *Source*: Adapted with permission from "Use of Isosulfan Blue for Identification of Lymphatic Vessels: Experimental and Clinical Evaluation" by JI Hirsch et al in *American Journal of Roentgenology* (1982;139:1061–1064), Copyright © 1982, American Roentgen Ray Society.

azurine 2G diluted with 2 or 3 ml of fluid injected subcutaneously into the web spaces with massaging of the injected areas for a short time. The original blue had been diluted with 1% of a local anesthetic, but the decrease in discomfort to the patient was not great and the mixture of the blue dye with lidocaine resulted in some precipitation of dye particles.[10] The mixing of dye and local anesthetic has been suspected of causing reactions in excess of that with dye alone, but this is far from proven. However, because there is no clear advantage to using the dye-anesthetic mixture, it is no longer recommended.

For neck lymphography, 1 to 2 ml of the blue dye in a 1% solution is injected subcutaneously in three locations behind the ear.[11] For testicular lymphography, 1 ml of the dye solution is injected into the fold between the testis and epididymis under the lamina of the tunica vaginalis of the testis being exposed for removal.[12]

CONTRAST AGENTS

Description

The water-soluble, organic iodide monomeric contrast agents, in wide usage for angiography and urography, were originally used by Kinmonth[2] and others for direct lymphographic examination of edematous extremities. They can be injected with greater speed than the more viscous, oily Ethiodol. However, they diffuse rapidly from lymph channels and nodes, they do not adequately demonstrate nodes at a distance from the site of injection, and they only demonstrate the lymph structures very briefly.[13] With lymphography primarily being used now to visualize the lymph nodes for investigation of primary and secondary neoplasms, water-soluble contrast agents are unsatisfactory. Their use is now limited solely to investigation of certain cases of lymphedema if the radiologist so chooses.

The contrast agent in wide use for lymphography today is an iodized oil, known in the United States as Ethiodol and in European countries as Lipiodol Ultrafluid. The oil consists of the glyceryl esters of several fatty acids obtained from the natural poppy seed oil—namely, oleic, linoleic, palmitic, and stearic acids—that have been iodinated with the iodine entering at the double bonds of the unsaturated fatty acids, thereby converting linoleic acid to diiodosteric acid and oleic acid to monoiodosteric acid. The ethyl group is then substituted for the glycerol group (Fig. 28-3). The resultant iodinated ethyl esters of the fatty acids have a much lower viscosity than the higher viscosity glycerol esters, known commercially as Lipiodol. The viscosity of Ethiodol, nevertheless, is still quite high (55 centipoises at 20°C and 30 centipoises at 37°C), which accounts for the difficulty in injection through small needles, tubing, and lymphatic channels.[5] Ethiodol is yellow in the unopened ampule. A darkened color indicates decomposition, and the contrast agent should not be used. It should be protected from light when

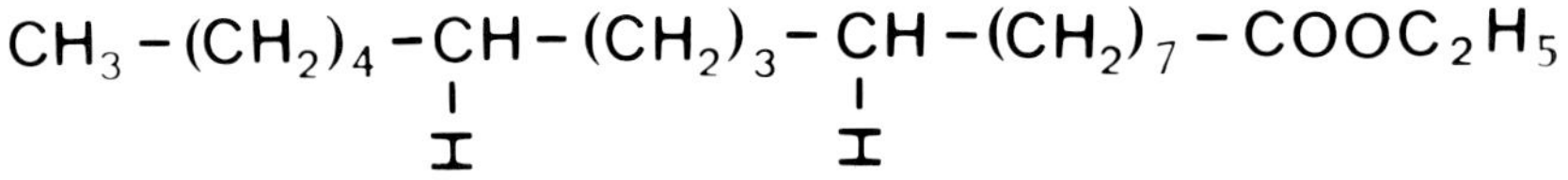

$$CH_3-(CH_2)_4-\underset{\displaystyle I}{\underset{|}{C}}H-(CH_2)_3-\underset{\displaystyle I}{\underset{|}{C}}H-(CH_2)_7-COOC_2H_5$$

Ethyl Diiodostearate

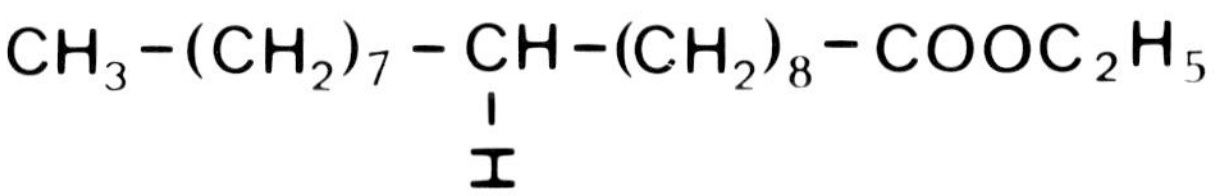

$$CH_3-(CH_2)_7-\underset{\displaystyle I}{\underset{|}{C}}H-(CH_2)_8-COOC_2H_5$$

Ethyl Monoiodostearate

Figure 28-3 The major components of Ethiodol or Lipiodol Ultrafluid are the ethyl esters of monoiodo- and diiodostearic acid shown here. Approximately 88% of this oily lymphographic agent is composed of these two compounds, the other 12% being uniodinated saturated and unsaturated components. The iodine content is 37%.

---

stored. When heated sufficiently and when in contact with air, iodine is liberated. The specific gravity is 1.280 at 15°C. The diiodo component is about 80% of the whole, the monoiodo is 12%, and the uniodinated saturated and unsaturated about 8%.[5]

Behavior of Ethiodol

When injected directly into a peripheral leg lymphatic, Ethiodol initially fills the lymphatics and nodes proximal to the junction of the thoracic duct with the subclavian vein. Radiographs taken during the injection and for a variable short time thereafter show both lymph vessels and nodes filled with the contrast agent, with some nodes later showing better filling. The contrast agent soon empties out of the lymph vessels, leaving the oily material in the nodes where it remains for many weeks and months and, occasionally, for a year or more. Studies in the dog at 3 days postlymphography showed that 23% of radioactively tagged [^{131}I] Ethiodol was retained in the nodes when a dose of 0.6 to 0.8 ml per kg of body weight was injected.[14] The customary human dose is 0.2 ml per kg of body weight, and because humans have more lymph nodes than the dog, the proportion retained in the nodes of an injected dose is probably much higher. Contrast that is not retained in the nodes passes into the central systemic veins via the thoracic duct. When the oily contrast material enters the veins, it passes through the right atrium and ventricle of the heart into the pulmonary artery and then into the lung capillaries, where it is trapped. The lung is a rather efficient initial filter, allowing little contrast material to pass through to the other organs at the dose levels used clinically. When the dose is larger, the lung functions as a less efficient filter and the amount reaching other organs increases. Ethiodol trapped in the lungs does not stay there unduly long. In dogs, at 3 days postlymphography after a 0.6 to 0.8 ml per kg dose, about 50% of the injected contrast material was in the lungs.[14] Some of the oily material is removed from the lungs by passing through the lung capillaries. Some also enters the macrophages in the alveolar spaces and there finds its way to the sputum. Some is broken down into its component parts and is metabolized.

Dosage

To visualize radiographically the inguinal, iliac, and aortic-abdominal nodes in an adult male, 6 to 8 ml of Ethiodol is injected into the lymph trunk on the dorsum of each foot. The adult female, generally of smaller size, needs only 4 to 7 ml per extremity. For children, depending on their size, 2 to 5 ml is sufficient. To visualize the axillary nodes, 2 to 4 ml is injected in the lymph trunks of the dorsum of each hand. For cervical lymphography 4 to 6 ml is recommended;[11] for testicular lymphography, 3 ml is sufficient for the injected side. The injection rate of lower extremity lymphography should be 0.1 to 0.2 ml/min. For injection of the arm or the cervical region, a slower rate of injection is advised to minimize extravasation from the delicate, thin-walled trunks. A rate of 0.5 ml/hr is recommended for testicular lymphography.[12]

Radiography

Leg and Testicular Injections

Radiographs are made at the conclusion of the injection and approximately 24 hours later. With the patient recumbent, anteroposterior (AP) and both posterior oblique views of the inguinal region, pelvis, and abdomen are recommended. A lateral view of the abdomen is optional. An AP view of the chest is optional to visualize the thoracic duct at the end of the injection. Image-intensified fluoroscopy is used to monitor the flow of contrast agent and to detect when it reaches the upper abdomen or lower thoracic duct.

Arm

Radiographs are made at the conclusion of the injection and 24 hours later of the axillary and supraclavicular regions in AP and both posterior oblique views.

Cervical

Radiographs are made at the conclusion of the injection and 24 hours later. AP and lateral views of the neck are obtained. Tomographic examinations and magnification techniques have been used at the discretion of the radiologist.

Prostate, Bladder

Direct injection of Ethiodol into the prostate for visualization of the lymphatic drainage of the bladder has been described. It is claimed to be more direct and less time consuming than rectal lymphography for assessing nodal metastases from the prostate.[15]

COMPLICATIONS

Complications, such as wound infections, delayed wound healing, and lymphangitis, are not considered to be caused by the contrast agent but by infection introduced at the time of lymphography.

Blue Dye

Adverse reactions to blue dye are infrequent, with an incidence of 1 per 700 examinations being reported in 32,000 surveyed lymphograms.[16] However, when a reaction occurs in a smaller series of patients, the incidence appears to the physician to be much higher.[17–19] Itching; nasal congestion; edema of eyelids, lips, and pharynx; urticaria; choking; bronchospasm; and cardiovascular collapse have occurred and have been treated successfully with epinephrine, hydrocortisone, and oxygen. Although a reaction to the blue dye may be confused with a reaction to the local anesthetic, later skin testing incriminated the dye rather than the anesthetic.[18,19] The role of impurities in the blue dye as a causative factor in adverse reactions has been raised[6] because samples are not necessarily pure.

Most lymphographers have had long experience with blue dye without having any patient reactions and have no objections to continuing its use. A number of experienced lymphographers, however, have found they do not need blue dye to identify the lymphatic for cannulation,[20] and it has been reported that not using blue dye neither adds to the time required for the examination nor increases the number of unsuccessful examinations.[21,22]

With the isosulfan blue dye, 5 of 543 patients (0.9%) had a mild allergic adverse reaction. Four of the five reactions occurred in patients who received the blue dye mixed with local anesthetic.[8]

Contrast Agents

The extravasation of oily contrast agent along the course of the lymph trunks or from the nodes is undesirable. Such extravasation decreases the amount of contrast agent available to visualize nodes and trunks and can result in an unsatisfactory diagnostic lymphogram. Ethiodol in the tissue produces a foreign body reaction that usually is not of consequence. Extravasation is more likely to occur when there is excessive speed or pressure of injection and where the lymphatics are more delicate and thin walled, as with arm, neck, and testicular injections. Extravasation from the nodes may be related to advanced disease weakening the node capsule.

A foreign body inflammatory reaction unavoidably occurs in lymph nodes that have been filled with Ethiodol, but this is not of clinical importance. A slight to moderate enlargement of the nodes can be recognized radiographically in the first few weeks after lymphography, followed by a gradual return of the nodes to normal size. The histologic changes result from the tissue's attempts to remove the foreign matter. After several months, giant cells and other inflammatory cells disappear, along with the oily droplets, so that no recognizable differences in node structure persist.

Patient sensitivity to the contrast agent, manifested as a pruritic skin rash, has been observed infrequently.[23] Iodine sialitis is also rare;[23] there has been one report of iodide goiter after lymphography.[24] Some reactions of this type are inevitable, because some iodine dissociates from the oil as indicated by studies with radioactively labeled oil[25] or by the instance of excretion of iodine in urine, when enough iodine is freed to opacify the bladder.[26] Therefore, a history of iodine sensitivity contraindicates lymphography. A history of allergy of any type should make the physician cautious, for it is known that allergic patients have a higher incidence of untoward reactions to other intravascular contrast agents. There has been little experience and very little enthusiasm for using a skin test or an oral iodine ingestion test to determine a patient's sensitivity to iodine before lymphography. Antihistamines and corticosteroids seem to be effective treatment for the dermatitic reactions to Ethiodol. Use of Ethiodol for lymphography results in a prolonged elevation of total serum iodine and protein-bound iodine.[27]

Lung and Brain Complications

The pulmonary complications of lymphography are the most significant and have accounted for almost all of the deaths reported. When an oily contrast agent is injected into the lymph trunks, any contrast not retained by the lymph nodes passes into the systemic venous circulation via the thoracic duct. Some contrast also enters the venous system by passage through lymphaticovenous communications normally, and when there is tumorous involvement of lymph nodes and obstruction to lymph flow, additional lymphaticovenous communications are established.

In many lymphographic examinations, a chest radiograph at the conclusion of the injection, and for the first few days postinjection, may not show any evidence of pulmonary oil emboli. However, emboli invariably occur, as can be demonstrated by studies with radioactively labeled contrast agents.[25,28] The great majority of patients tolerate these emboli without incident or symptoms, although the elevation of temperature that many patients experience in the immediate postinjection period has been ascribed to oil emboli in the lung. Despite the absence of symptoms, respiratory function studies have revealed a decrease in pulmonary diffusing capacity and pulmonary capillary volume. The maximum decrease is observed between 3 and 72 hours postinjection, with recovery occurring in 21 to 256 hours.[29] Up to 60% depression in function can occur. The decreased diffusion capacity is considered to be a result of decreased capillary blood volume; it correlates with the histologic evidence of embolization of many pulmonary capillaries by the oily globules.

Studies with radioactively labeled materials show that the maximum amount is in the lungs at 24 to 48 hours. Very shortly after, some of the stainable lipid material can be found in the interstitial tissues, and lipid material can be recovered from the sputum. Although some globules move onward through the capillaries, others are broken down in the lungs. The biologic half-life of radioactively labeled contrast agents in the lung is about 8 days.[30]

Although after lymphography the pulmonary oil embolization and resultant decrease in respiratory function are not productive of symptoms or outward disability, the complication can lead to morbidity and death. Patients who, before lymphography, have a significant loss of pulmonary diffusion capacity because of intrinsic lung disease are at greatest risk. Patients who are particularly susceptible are those who have had previous radiation to the lungs or whose lungs are involved by neoplasm.[23,31]

Embolization of oily globules of contrast agent to other organs, notably the brain, has produced morbidity and death in a number of patients. Blindness, hemiparesis, and coma have been observed.[32,33] Evidently the lung fails to filter or to trap the globules, possibly because of passage of droplets through small arteriovenous shunts in the normal or diseased lung or because the oil bypasses the lungs through a cardiac septal defect. Embolization to the brain is more likely to occur when the injected dose is large, but the explanation for this complication is not always readily apparent.

Occurrence of the uncommon significant pulmonary and the more rarely occurring cerebral embolism can be minimized by not injecting an excessive amount of the oily contrast agent.[16,34] In general, satisfactory radiographic information can be obtained by injecting substantially less than 10 to 15 ml or more per leg, which many first utilized as a dose. Seven to eight milliliters is now thought to be sufficient for smaller persons.

If it can be recognized that the nodes are fibrotic from prior radiation or other cause or if the nodes are largely replaced by tumor, then lower doses are appropriate because these nodes are likely to retain less contrast medium. Many lymphographers limit the amount injected by stopping the injection when the iodized oil reaches the upper lumbar level; the amount already injected is sufficient to fill all the subdiaphragmatic lymphatic structures. For this purpose, radiography or use of image-intensifier fluoroscopy is essential.

By the use of check films or fluoroscopy shortly after the beginning of the injection, an injection of the oily contrast agent into a peripheral vein that has mistakenly been cannulated can be avoided; if such an injection is recognized the injection can be terminated immediately.

Inhalation anesthesia and surgical procedures should be avoided or deferred in the immediate postlymphography period when lung function is most depressed. For the patient who has decreased pulmonary function because of lung disease and for whom the information to be derived from a lymphogram is clearly needed, exacerbation of lung dysfunction can be minimized by injecting one lower extremity and then not injecting the other until after an interval of 4 days. The patient should be able to tolerate the two separate lesser episodes of embolization.[5] If the injection volume can be held to a low level, even patients with pulmonary dysfunction can have lymphography without further functional deterioration.[35]

The immediate postlymphography decrease in pulmonary function and the early respiratory distress symptoms experienced by some patients are attributable primarily to capillary blockage by the oil globules. In a few patients, however, a lung response consisting of edema and inflammatory changes develops after the capillary obstructive phase. This can best be termed a chemical pneumonitis for it is considered to be a reaction to the oil or its components. It is evidently similar to the secondary or chemical phase of endogenous fat embolism that occurs when the embolized neutral fats are hydrolyzed to the more irritating fatty acids, which then damage the blood vessel endothelium and alveolar membranes, causing hemorrhage and exudation. The chemical pneumonitis may occur shortly after lymphography, or it may be delayed many days.[32] The chemical phase is characterized by fever, cough, and sputum, which is sometimes bloody, and minor to severe respiratory distress. Oxygen desaturation occurs. Rarely, enough blood is lost through the lungs to produce anemia. During the initial embolic phase, radiographically the lungs appear normal, or a fine generalized, diffuse stippled pattern is detectable. In the chemical phase, diffusely scattered or localized patchy infiltrates are found. Some of these infiltrates may become confluent and be labeled as infarcts, but more properly they should be considered as a pneumonia. Histologically there is a marked cellular reaction of histiocytes and foreign body giant cells.

Hepatic Oil Embolism

This complication occurs rarely and only when severe lymphatic obstruction is present in upper common iliac-lower aortic-abdominal nodes along with complete obstruction of both proximal common iliac veins and the distal inferior vena cava. The obstruction has almost always been caused by extensive node metastases. The Ethiodol reaches the liver by a circuitous route. The contrast agent, unable to flow beyond the iliac and aortic-abdominal lymphatics, shunts into the iliac veins through lymphaticovenous communications. Because it cannot flow past the obstructed veins, it travels via collateral circulation into the hemorrhoidal venous plexus and from there into the portal venous system and eventually the liver.[36] Hepatic oil embolization does not produce symptoms. Ultrasound scans, plain radiographs, and CT scans may show findings that have been described as unique.[37]

Although spread of tumor is a theoretical complication, there is no clinical evidence to show that this occurs. In the experimental animal, no significant changes in the barrier function of lymph nodes as a result of the oily contrast agent have been demonstrated.[38,39]

Other Complications

Hypotensive reactions have been reported. Fever is so common postlymphography that it is not considered to be a complication. It is usually low grade and does not last longer than 48 hours, and most often less. Headache, nausea, and vomiting are transient and probably nonspecific in origin.

Treatment of Complications

The initial lung embolic phenomena usually require no treatment because they are largely asymptomatic. Occasionally the patient experiences dyspnea, tachypnea, tachycardia, and cyanosis. Oxygen is indicated and, if signs of right heart failure occur, digitalis should be given. Hypotension must be treated.

With a chemical pneumonitis, oxygen is indicated for the hypoxia and respiratory distress. Intermittent positive pressure respiration is advocated. If anemia develops, transfusion may be necessary. With endogenous fat embolism, it is advisable to decrease the rate of hydrolysis of neutral fats and to slow the conversion to fatty acids that are considered responsible for the chemical phase. Ethyl alcohol, a satisfactory lipase inhibitor in endogenous fat embolism, presumably would also be of value in Ethiodol emboli. Dosage by mouth, 30 cc of 50% alcohol every 3 hours, is advised or, if necessary, 2,000 cc of 5% ethyl alcohol in

5% glucose intravenously. Evidence for the use of heparin is conflicting. Low molecular weight dextran intravenously is recommended to combat intravascular aggregation and clumping of formed elements of the blood. Calcium may be needed if the serum calcium falls because of binding and removal of calcium ions by the freed fatty acids.[5]

Cerebral oil embolism is treated similarly to lung embolism, with oxygen therapy being the cornerstone of therapy to maintain adequate oxygenation of the blood.

EXPERIMENTAL CONTRAST AGENTS

A new water-soluble contrast agent, iotasul, has been investigated for both direct and indirect lymphography. This contrast agent is a nonionic dimeric organic iodine-containing compound of higher molecular weight and lower osmolality than the previously investigated water-soluble organic iodides, such as the diatrizoate compounds, which are widely used as angiographic and urographic agents. When used in direct lymphography, because the new contrast agent has less tendency to spread through the walls of the vessels, longer visualization is afforded compared to the conventional water-soluble agents. When used in indirect lymphography in patients with lymphedema, iotasul provides more information by visualizing barely perceptible small vessels in the distal extremity. These radiographic assets, as well as evidence that no histologic changes are produced in lymph nodes, lymph vessels, lung, liver, and other organs and that only temporary changes in the area of injection are attributable to the contrast agent, are definite advantages of iotasul.[40–42]

Another new contrast agent, radiopaque perfluorocarbons, has been tested in animals and found to be of satisfactory radiopacity for a time in excess of several months. Biologic inertness was based on the lack of local and systemic effects of a deleterious nature.[43]

To our knowledge, no study has been completed showing that iotasul is as efficacious in revealing details of anatomy, specifically the demonstration of metastases, as Ethiodol. The use of perfluorocarbon in human lymphography has not been reported.

REFERENCES

1. Kinmonth JB, Harper RAK, Taylor GW: Lymphangiography by radiological methods. *J Fac Radiol* 1955;6:217–223.

2. Kinmonth JB: Lymphangiography in man: method of outlining lymphatic trunks at operation. *Clin Sci* 1952;11:13–20.

3. Hudack S, McMasters PD: Permeability of the wall of the lymphatic capillary. *J Exp Med* 1932;56:223–238.

4. *Colour Index*. ed 2, vol 3. Lowell, MA, Society of Dyers and Colourists, Bradford, Yorkshire, England, and American Association of Textile Chemists and Colourists.

5. Fischer HW: Contrast media, in Fuchs WA, Davidson JW, Fischer HW (eds): *Lymphography in Cancer*. New York, Springer-Verlag, 1969, p 20.

6. Hiranaka PK, Kleinman LM, Sokoloski EA, et al: Chemical structure and purity of dyes used in lymphograms. *Am J Hosp Pharm* 1975;32:928–930.

7. Smith JR, Dunlon EF, Protas JM, et al: Cannulation of the lymphatics of the lower extremity. *Surg Forum* 1958;9:811.

8. Hirsch JI, Tisnado J, Cho S-R, et al: Use of Isosulfan blue for identification of lymphatic vessels: experimental and clinical evaluation. *AJR* 1982;139:1061–1064.

9. Gangolli SD, Grasso P, Goldberg L: Physical factors determining the early local tissue reactions produced by food colourings and other compounds injected subcutaneously. *Food Cosmet Toxicol* 1967;5:601–621.

10. Newton DW, Rogers AG, Becker CH, et al: Evaluation of preparations of patent blue (alphazurine 2G) dye for parenteral use. *Am J Hosp Pharm* 1975;32:912–917.

11. Fuchs WA: Investigation techniques, in Fuchs WA, Davidson JW, Fischer HW (eds): *Lymphography in Cancer*. New York, Springer-Verlag, 1969, p 11.

12. Sayegh E, Brooks T, Sacher E, et al: Lymphangiography of the retroperitoneal nodes through the inguinal route. *J Urol* 1966;95:102–107.

13. Fischer HW: Lymphangiography and lymphadenography with various contrast agents. *Ann NY Acad Sci* 1959;78:799–808.

14. Koehler PR, Meyers WA, Skelley JF, et al: Body distribution of Ethiodol following lymphangiography. *Radiology* 1964;82:866–871.

15. Raghavaiash NV, Jordan WP Jr : Prostatic lymphography. *J Urol* 1979;121:178–181.

16. Koehler PR: Complications of lymphography. *Lymphology* 1968;1:117–120.

17. Biran S, Hockman A: Allergic reactions to patent blue during lymphangiography. *Radiol Clin Biol* 1973;42:166–168.

18. Kopp WL: Anaphylaxis from alphazurine 2G during lymphography. *JAMA* 1966; 198:668–669.

19. Mortazavi SH, Burrows BD: Allergic reactions to patent blue dye in lymphangiography. *Clin Radiol* 1971;22:389–390.

20. Maddison FE: Lymphatic cannulation without dye: technical notes. *Radiology* 1967;88:362.

21. Kapdi CC: Lymphography without the use of vital dyes. *Radiology* 1979;133:795–796.

22. Sigurjonsson K: Lymphography without the aid of vital dyes. *Lymphology* 1974;7:121–123.

23. Davidson JW: Lipid embolism to the brain following lymphography: case report and experimental study. *AJR* 1969;105:763–771.

24. Koutras DA, Sinaniotis CA: Iodide goitre following lymphography. *J Pediatr* 1973;83:83–84.

25. Patomaki LK, Verho S, Torsti R: Quantitative measurement in vivo of Lipiodol in lungs especially in connection with lymphography. *Nucl Med* 1974;13:227–235.

26. Wendth AF Jr, Moriarty DF, Cross UF, et al: Urinary bladder opacification following lymphangiography. *Radiology* 1968;91:762–763.

27. Jacobssen L, Saltzman GF: Effect of iodinated roentgenographic contrast media on butanol-extractable, protein-bound and total iodine in serum. *Acta Radiol* 1971;11:310–320.

28. Richardson P, Crosby EH, Bean HA, et al: Pulmonary oil deposition in patients subjected to lymphography. *Can Med Assoc J* 1966;94:1086–1091.

29. Gold WM, Youker J, Anderson S, et al: Pulmonary function abnormalities after lymphangiography. *N Engl J Med* 1965;273:519–524.

30. Fallat R: Pulmonary deposition and clearance of I^{131} labeled oil after lymphography in man: correlation with lung function. *Radiology* 1970;97:511–520.

31. Lee BJ, Nelson JH, Schwarz G: Evaluation of lymphangiography, inferior venacavography and intravenous pyelography on the clinical staging and management of Hodgkins disease and lymphosarcoma. *N Engl J Med* 1964;271:327–337.

32. Fuchs WA: Complications in lymphography with oily contrast media. *Acta Radiol* 1962;57:427–432.

33. Jay JC, Ludington LG: Neurologic complications following lymphangiography: possible mechanisms and a case of blindness. *Arch Surg* 1973;106:863–864.

34. Dolan PA: Lymphography: complications encountered in 522 examinations. *Radiology* 1966;86:876–880.

35. LaMonte CS, Lacher MJ: Lymphangiography in patients with pulmonary dysfunction. *Arch Intern Med* 1973;132:365–367.

36. Thornbury JR: Lymphatico-venous anastomoses involving the portal system: lymphographic changes in man, in *Progress of Lymphology II, Selected Papers from the Second International Congress of Lymphology*. Stuttgart, Thieme Verlag, 1968, p 105.

37. Lee S: Hepatic oil embolism following lymphangiography. *J Ultrasound Med* 1985;4: 357–359.

38. Blom JMH, Oort J: The effect of lymphography with Lipiodol ultrafluid on the barrier function of the lymph node. *Radiol Clin Biol* 1970;39:317–329.

39. Engzell V, Rubio C, Tjernberg B, et al: The lymph node barrier against Vx2 cancer cells before, during, and after lymphography: a preliminary report of experiments on rabbits. *Eur J Cancer* 1968;4:305–312.

40. Partsch H, Wenzel-Hora BI, Urbanek A: Differential diagnosis of lymphedema after indirect lymphography with iotasul. *Lymphology* 1983;16:12–18.

41. Siefert HM, Mutzel W, Schobel C, et al: Iotasul, a water soluble contrast agent for direct and indirect lymphography. Results in preclinical investigations. *Lymphology* 1980;13:150–157.

42. Wenzel-Hora BI, Kalbas B, Siefert HM, et al: Iotasul, a water-soluble (non-oily) contrast medium for direct and indirect lymphography. Radiological and morphological investigation in dogs. *Lymphology* 1981;14:101–112.

43. Long DM, Nielson MD, Multer FR, et al: Comparison of radiopaque perfluorocarbon and ethiofolin lymphography. *Radiology* 1979;133:71–76.

Chapter 29

Fistulography

Jovitas Skucas

A sinus is a tract extending from a focus of suppuration to the skin or mucous membrane through which pus can discharge. A fistula, in contrast, is a congenital or acquired passage leading from an abscess or hollow organ to either the skin or another hollow organ allowing the flow of either pus or secretions. An external fistula communicates with the skin, whereas an internal fistula does not. A fistula or sinus tract may drain pus, necrotic debris, or the contents of the gastrointestinal (GI), biliary, pancreatic, or urinary tracts. Occasionally these tracts may close spontaneously, although usually surgical intervention is necessary. In chronic sinuses there may be growth of epithelium into the tract, preventing closure.

Only external fistulas and sinuses or those that extend to easily accessible mucous membranes are considered here. Internal fistulas are best studied through the adjacent hollow organs using contrast agents appropriate for those structures. For example, biliary-enteric fistulas are usually detected with the appropriate barium examination; occasionally, such a fistula is suggested by hepatobiliary scintigraphy if activity in inappropriate tracts or loops of bowel is detected.[1,2] Likewise, at times endoscopic fistulography can be performed.[3]

A fistuloscope has been developed for external fistulas;[4] the instrument apparently is used similar to a catheter, with contrast injected through the fistuloscope once it has been advanced as far as possible.

Sinograms and fistulograms are widely performed procedures that provide useful information to the surgeon. However, since the article by Pendergrass and Ward in 1947,[5] only a limited number of publications dealing specifically with fistulography and sinus tract injection have appeared. An extensive review of fistulography was published in 1968 in German.[6]

CONTRAST AGENTS

The first mention of a contrast agent being used to outline a fistula was in 1897, when an iodine-glycerin compound was injected through a drain introduced into a pelvic fistula in a 17-year-old girl.[7] Shortly thereafter the heavy metals were introduced for sinus tract visualization. A bismuth paste was recommended in 1908.[8] Because of the resultant toxicity of the bismuth salts, including reports of pulmonary emboli,[9] iodized oil gradually replaced the heavy metals and was used with success for many years.[5,10,11]

Lipiodol, consisting of iodinated poppy seed oil, was the first organic iodine compound used.[12] In fact, the closely related Ethiodol (known as Lipiodol Ultra-fluid in Europe) was originally introduced as a contrast medium for sinus tracts, hysterosalpingography, and sialography.[13] The iodized oils were employed for years to study fistulas and sinus tracts.

Currently, most radiologists use the iodinated, water-soluble contrast agents. Not only are the water-soluble media readily adsorbed but they also mix with the retained contents of sinus tracts and abscesses and tend to outline the small ramifications of a sinus tract. A relatively dense but low viscosity contrast agent is preferred. Normally no dilution with water or saline is necessary. Especially with the smaller fistulous tracts, dilution may result in insufficient radiographic contrast for adequate visualization.

If there is a suspicion of communication with the spinal subarachnoid space, nonionic water-soluble agents are safer. If a bronchopleurocutaneous fistula is present, the ionic water-soluble agents in the lung parenchyma can lead to subsequent pulmonary damage. With such a fistula I prefer a relatively low concentration of barium sulfate or the nonionic water-soluble agents, although ionic water-soluble contrast media have been used in the past.[14,15] Aside from these relatively infrequent cases, the water-soluble iodinated agents remain today the contrast media of choice, with the ionic contrast media being preferred because of their lower cost.

Cold-vulcanized silicone rubber has been proposed for fistulography.[16] This is a rather viscous substance that is injected by a pump. After being mixed with a catalyst, the silicone rubber hardens in 3 to 5 minutes. It is claimed that the hardened, green-colored elastic rubber then serves as a guide to the surgeon.[16] The silicone rubber apparently is nontoxic and produces adequate radiographic contrast. This technique is currently rarely employed.

Currently, many abscesses are drained percutaneously, using a number of imaging modalities. A catheter is left in the abscess to allow for adequate drainage. Healing of the abscess cavity can be followed with serial sinograms. Generally, a full-strength water-soluble contrast agent designed for intravascular injection is employed for these studies (Fig. 29-1). The nonionic agents do not have any particular advantage in this indication, with the possible exception in the patient with an allergy to the ionic media.

TECHNIQUE

As stated by Gage and Williams, the approach used depends on the type of tract encountered.[11] One may explore a sinus tract to its full extent, explore a fistula

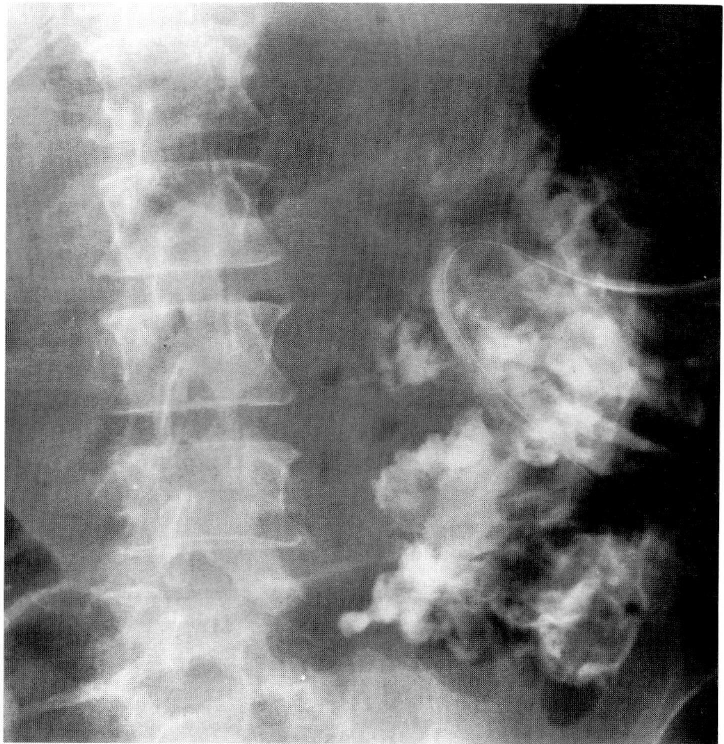

Figure 29-1 Left flank abscess. The abscess was localized with ultrasonography and a drainage catheter inserted. Initial injection of contrast outlines the extent of the abscess.

---

until the lumen of a viscus or duct is reached, or explore a cavity by instilling a limited amount of contrast material into the cavity and then use positioning to outline the full extent of the cavity.

The radiologic demonstration of the extent of a fistula quite often is incomplete because of poor technique. At times, because of incomplete filling, even large communicating cavities are not demonstrated. The most common mistake is simply not injecting enough contrast material, which can make it difficult to distinguish between large and small bowel or even an abscess cavity. A sufficiently tight seal is required at the sinus tract orifice to prevent backflow onto the patient's skin. Not only does backflow interfere with diagnostic interpretation but it also prevents the development of adequate pressure within the sinus tract. A simple method of creating a tight seal that works relatively well is to inflate the balloon on a Foley catheter and to have the patient hold the balloon tight against the sinus tract. Because the tip of the Foley catheter is within the sinus tract, backflow can be prevented by having the patient exert sufficient pressure upon the externally located balloon. The size of the Foley catheter selected depends upon the size of the tract. Some radiologists inflate the balloon within the fistula.[17]

A blunt conical rubber nozzle can be used in place of the Foley catheter to occlude the orifice.[18] A further refinement is to use a suction cup and achieve a tight seal through low pressure. Several different types of such "vacuum" cups have been described in the literature.[19–21]

If adequate occlusion of a sinus tract orifice is not possible with the above devices, a flexible catheter can be threaded into the tract for varying lengths. Some radiologists first introduce a guide wire to aid passage of a catheter.[17] A catheter sufficiently deep inside a fistula or sinus tract may not need a tight external seal; injected contrast preferentially outlines the passages adjacent to the catheter tip first.

At times, a sinus tract may be barely visible. A thin polyethylene catheter having an outside diameter approximately the same as that of the sinus tract can be threaded deep into the sinus cavity. If the catheter is considerably smaller than the tract, contrast will leak alongside the catheter and adequate pressure will not be maintained.

A simple method of maintaining a tight seal is to use a plastic drape that adheres to the patient's skin surrounding the fistula in a similar manner to surgical sterile drapes. A hole is then punctured through the drape over the fistula and a catheter advanced into the fistula (I. Brolin, personal communication, 1975). There may be sufficient pressure from the adhesive drape to prevent spillage of the contrast material.

The amount of pressure needed for an adequate examination is difficult to gauge. Excessive pressure may result in the breakdown of recently formed adhesions or the spread of a localized focus of infection. At times, the contrast material will flow by gravity. At other times, especially if small tracts are involved, considerable pressure is necessary. Filling of sinus tracts in the thorax may be aided by deep respirations.

If an external fistula communicates with bowel, generally more diagnostic information is obtained if fistulography is done before the appropriate barium study. Often, a barium enema or a small bowel examination does not demonstrate the fistulous communication. Even if seen, the fistulous connection to the bowel simply cannot be localized adequately.

If there are several sinus tracts present, generally all should be injected to outline the underlying anatomy. Leaving the catheters in their respective sinus tracts aids in orientation and identification.

Fistulography and sinus tract injections should be performed under fluoroscopic control. Radiography in two projections is essential for adequate localization. Oblique positions can be helpful. They can be accomplished either by having the patient turn on the fluoroscopic table, or if a transverse angulating table, such as the Siemens Orbiskop or Toshiba Gyroscope, is available, the tube and its corresponding image intensifier can simply be rotated around the stationary patient.

One technique for evaluation of acute penetration wounds of the abdominal wall was introduced by Cornell et al. in 1966.[22,23] A no. 4 French catheter is inserted through the skin wound into the subcutaneous tissues. The skin edge is secured around the catheter with purse-string sutures, and the contrast agent is injected through the catheter (Fig. 29-2). The catheter is then clamped, the patient rotated from side to side, and abdominal films obtained. Although Cornell and co-workers recommended sodium diatrizoate (50%), other nontoxic water-soluble agents have also been used. They recommended adding approximately 1 cc of methylene blue to the contrast as a possible aid to the surgeon. Based on preliminary results with 112 patients, they found that failure to demonstrate penetration of contrast into the peritoneal cavity is strong evidence that the peritoneum has not been punctured. There were no complications in their series. Specifically, no tissue necrosis or infection was encountered. This technique is rarely performed today, with computed tomography having largely supplanted it.

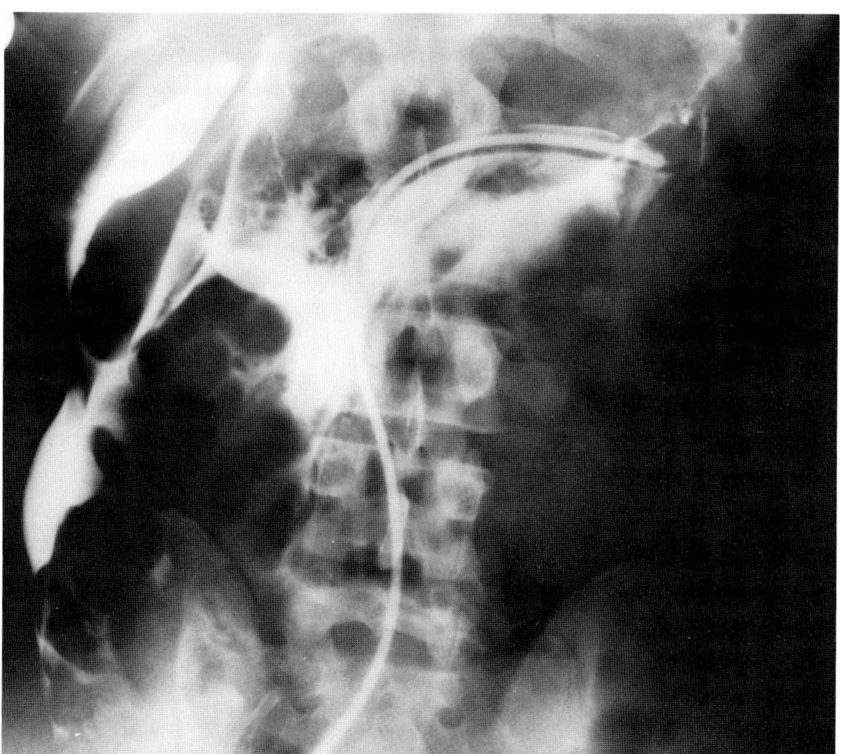

Figure 29-2 Sodium diatrizoate 50% was injected through a catheter inserted in a patient's stab wound. There is a layering of contrast around the gallbladder and several loops of bowel, indicating a peritoneal perforation.

If surgical excision of a fistula is contemplated, a catheter can be left in place to guide the surgeon.[24] In general, however, injection of methylene blue provides more information to the surgeon about the various fistulous ramifications.[25]

At times there is a spontaneous closure of a fistula after the injection of contrast. Rosemeyer discusses four such patients.[26] He used Urografin 60% as the contrast agent. Dell also found cessation of drainage after the injection of Lipiodol.[12] No explanation for this phenomenon is apparent.

COMPLICATIONS

There are relatively few complications associated with fistulography. If the injection pressure is too great, a focus of suppuration can be spread. There is also a possibility of intraperitoneal spread (Fig. 29-3) or intravasation (Fig. 29-4). As a

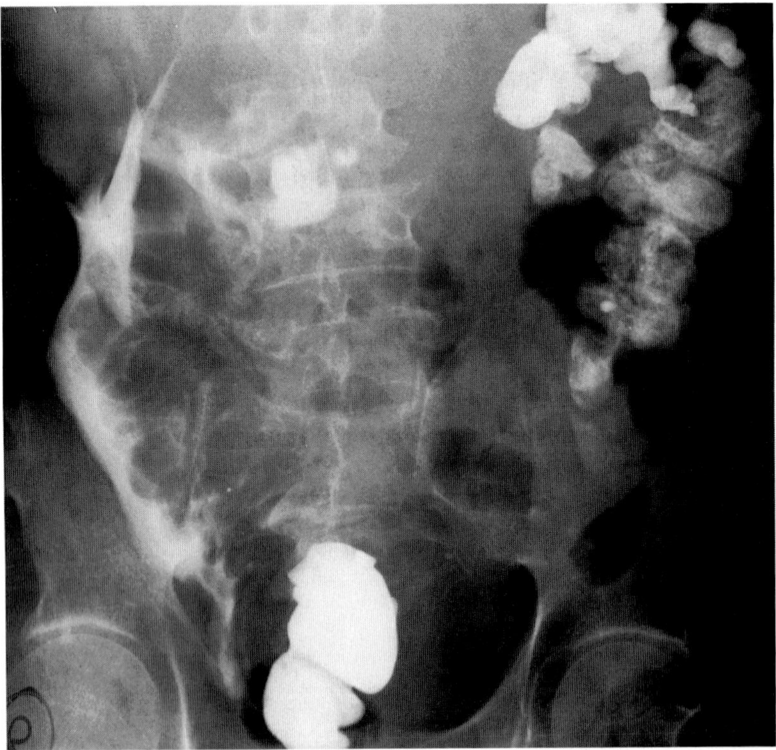

Figure 29-3 An abscess drainage was attempted under ultrasound guidance. A water-soluble iodinated contrast agent was injected to check catheter position. Contrast is spilling into the peritoneal cavity. The residual barium within bowel is from an earlier, nonrelated study.

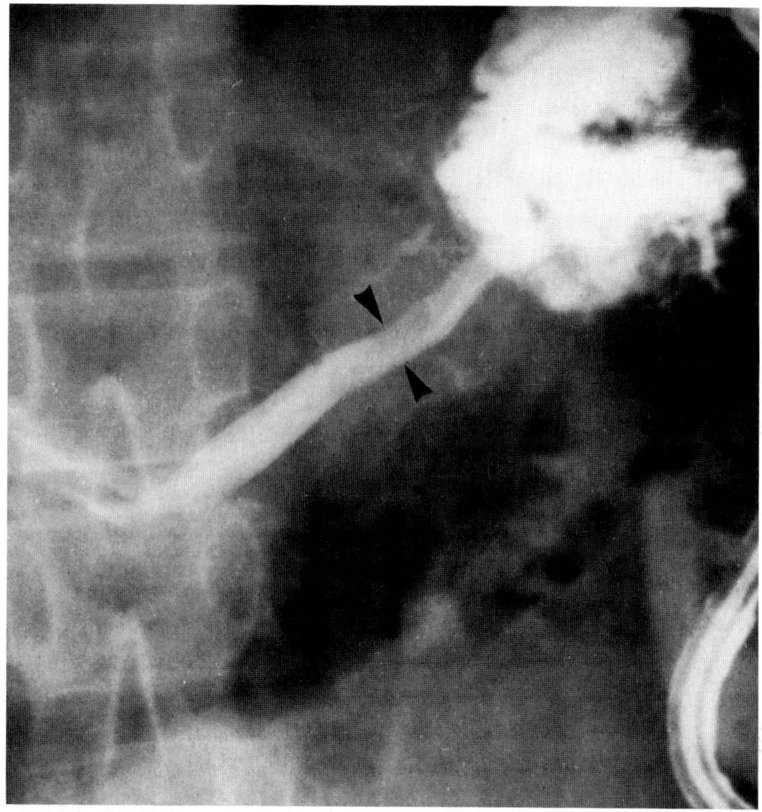

Figure 29-4 A fistula developed in the right upper quadrant after drainage of a pancreatic pseudocyst. It was draining small amounts of necrotic debris. Injection of contrast reveals free communication with the splenic vein (*arrows*). There was no bleeding after termination of the procedure. A fistulogram 7 days later revealed a subcapsular hematoma with no communication with the splenic vein.

result, the intravascular water-soluble contrast agents are preferred in most instances because they are readily absorbed from body cavities.

REFERENCES

1. White M, Simeone JF, Muller PR: Imaging of cholecystocolic fistulas. *J Ultrasound Med* 1983; 2:181–185.

2. McPherson GAD, Lavender JP, Collier NA, et al: The role of HIDA scanning in the assessment of external biliary fistulae. *Surg Gastroenterol* 1984;3:77–80.

3. Hoff G, Heldaas J: Pancreatic pseudocyst and pancreatico-duodenal fistula diagnosed by endoscopic fistulography. *Endoscopy* 1981;13:221–222.

4. Strekalovsky VP: New methods of endoscopic diagnosis of digestive tract diseases in the Soviet Union. *Endoscopy* 1982;14:135–138.

5. Pendergrass RC, Ward WC: Roentgenographic demonstration of sinuses and fistulas. *Radiology* 1947;57:571–577.

6. Werner H: Die Röntgendiagnostik der angeborenen und erworbenen Fisteln, in Diethelm L (ed): *Handbuch der Medizinischen Radiologie*, vol 8, *Röntgendiagnostik der Weichteile*. Berlin, Springer-Verlag, 1968, pp 522–605.

7. Graff H: Beitrag zum diagnostischen Wert der Röntgenstrahlen. *Fortschr Röntgenstr* 1897; 1:229.

8. Beck EG: Eine neue Methode zur Diagnose und Behandlung von Fistelgängen. *Zbl Chir* 1908; 35:555–557.

9. Leb A: Lungenembolie nach Fistelfullung mit Beck'scher Wismutpaste. *Bruns' Beitr Klin Chir* 1923;128:515–520.

10. Dyes O: Kontrastfüllung von Fistelgängen. *Der Chirurg* 1931;3:18–25.

11. Gage HC, Williams ER: The radiological exploration of sinus tracts, fistulae and infected cavities. *Radiology* 1943;41:233–248.

12. Dell JM Jr: Demonstration of sinus tracts, fistulas, and infected cavities by Lipiodol. *Am J Roentgenol* 1949;61:223–231.

13. Fischer HW: Contrast media, in Fuchs WA, Davidson JW, Fischer HW (eds): *Lymphography in Cancer*. New York, Springer-Verlag, 1969, p 13.

14. Schreer I, Birzle H: Die Bedeutung der Fistelfüllung in der chirurgischen Röntgendiagnostik. *Bruns' Beitr Klin Chir* 1972;219:560–564.

15. Trahan TJ: Bronchopleurocutaneous fistula: a case report. *J Am Osteopath Assoc* 1974; 73:995–998.

16. Stern W: A new method for demonstrating fistulas of the soft tissues. *J Int Coll Surg* 1965;44: 668–673.

17. Metges PJ, Silici R, Kleitz C, et al: La fistulographie. A propos de 126 examens. *J Radiol* 1980; 61:57–59.

18. Miller WT, Sullivan MA: Roentgenologic demonstration of sinus tracts and fistulae. *Am J Roentgenol* 1969;107:812–817.

19. Brekkan A: Fistulography with vacuum suction. *Acta Radiol* 1962;57:77–80.

20. Brekkan A, Axen O: Roentgen examination of enterostomy openings with use of vacuum suction. *Radiology* 1968;91:385–386.

21. Severini A: A new type of fistulograph. *Radiology* 1972;104:434–435.

22. Cornell WP, Ebert PA: Penetrating wounds of the abdominal wall: a new diagnostic technique. *Am J Roentgenol* 1966;96:414–417.

23. Cornell WP, Ebert PA, Greenfield LJ, et al: A new nonoperative technique for the diagnosis of penetrating injuries to the abdomen. *J Trauma* 1967;7:307–314.

24. Stanford W, Reuben CF, Flemma RJ, et al: Management of long-standing cardiocutaneous fistulas after resection of left ventricular aneurysms. *J Thorac Cardiovasc Surg* 1980;79:789–792.

25. Herrara HR, Mijangos J, Weiner RS: Mediastinal cutaneous fistulae arising as a complication of major cardiac surgery: the value of sinography and injection of methylene blue in their radical excision and immediate repair. *Br J Plast Surg* 1983;36:421–424.

26. Rosemeyer R: Beobachtungen über Spontanverschlüsse von Fisteln nach Kontrastdarstellung. *Arch Orthop Urfall-Chir* 1973;76:242–247.

Chapter 30

Bronchography

Anthony House

The first successful bronchogram was carried out by Jackson in 1918 by insufflating dry bismuth subcarbonate through a bronchoscope.[1] Over the next few years a number of contrast agents were tried, and in 1922 iodized poppy seed oil (Lipiodol) was introduced by Sicard and Forestier,[2] this being the first practical bronchographic agent.

Although the iodized oils remained in general use for many years, it soon became evident that Lipiodol was not without hazard. In 1945 Dormer et al.[3] added sulfonamide to Lipiodol, and a number of combinations of these substances were subsequently produced for bronchography. In 1948 Morrales and Heiwinkel combined a water-soluble opaque medium (Iodopyracet) with sodium carboxymethyl cellulose (CMC), which was marketed as Umbradil-viskos-B.[4,5] In 1953 a radiopaque medium of low solubility in water called Dionosil (N-propyl ester of iodopyracet) was introduced by Tomich et al.[6] This medium was suspended in either an aqueous solution of CMC (Dionosil aqueous) or in arachis oil (Dionosil oily). In the 1950s a number of investigators, including Nelson et al.,[7] described barium sulfate bronchography in humans. The bronchographic agent, Hytrast, a combination of iopydol and iopydone crystals with CMC, was described by Gildenhorn et al. in 1962.[8] Inhalation bronchography using the chemically inert metallic powder, tantalum, was evaluated by Nadel et al. in 1968.[9]

More recently, the use of perfluorocytlbromide, a contrast medium that is rapidly eliminated from the lungs, has been reported.[10] Inhalation bronchography using calcium ioglycamic acid, a water-soluble iodine-containing contrast agent in powder form, has been tried.[11] In 1984, the use of the nonionic contrast agent, metrizamide, was reported in infants.[12]

With the development of each bronchographic contrast agent, extensive animal and human investigations have been performed. Difficulties have arisen in relating the findings of animal experimentation to humans because of anatomic, physiologic, and bronchographic technical differences. Only by animal work, however, has it been possible to carry out controlled observations on toxicity and tissue damage.

BRONCHOGRAPHIC QUALITY AND SIDE EFFECTS

The bronchographic quality and possible side effects of the media depend mainly on their physical, pharmacologic, and physiologic properties. Ideally the contrast agent should be (1) capable of giving a good contrast radiograph, (2) capable of good adherence to the bronchial walls with maintenance of the bronchogram until the appropriate radiographs have been obtained, (3) miscible with bronchial secretions, (4) capable of filling or outlining even the small bronchi uniformly without filling the alveoli, (5) capable of giving consistently diagnostic studies, (6) easy to administer, (7) physiologically innocuous and pharmacologically inert, and (8) promptly and completely eliminated from the lung.[13,14]

Practical considerations include the interrelated variables of contrast elimination, bronchographic technique, and pulmonary status of the patient. In humans, most of the contrast in the trachea and the large and medium-sized bronchi is quickly eliminated after bronchography with the aid of gravity and coughing. Although gravity and coughing no doubt assist in the clearing of small bronchi and bronchioles as well, the mucociliary activity there is considered to be the most important mechanism, probably aided by the bronchial smooth muscle.[15–17] Mucus is produced mainly by the mucous glands, which extend as far peripherally as the small bronchi, and to a lesser degree by the goblet cells, which, with the cilia, extend to the terminal bronchioles.[15] Because of the absence of the mucociliary mechanism peripheral to this level, contrast tends to be retained here.[15,18–20] The retained contrast media are considered in the main to be phagocytosed, with the contrast containing macrophages migrating to the mucociliary level of the bronchial tree where they are removed. Or, they may enter the lymphatics to be carried to the lymph nodes or be retained in the alveoli or adjoining lung interstitium.[19,21–26]

With alveolar filling of the bronchial tree and the resulting delay in elimination and retention of the contrast, the likelihood of clinical and tissue reactions increases.[20,27–30] Excessive peripheral filling is more likely if the viscosity of the contrast medium is low, which may be an inherent property of the medium or the result of overdilution or over warming of the medium before bronchography.[31–33] Irritability of the bronchial tree produced by coughing is undesirable, as the exaggerated inspiratory effort that occurs before the expulsive phase of the cough causes the contrast to be sucked peripherally.[16] Irritation may be caused by a combination of the type of contrast used, patient preparation, and efficiency of anesthesia. Excessive bronchial secretions prohibit even distribution of the anesthetic agent.[34] Postural drainage before the procedure, especially when there is sputum production,[34,35] premedication with atropine to control mucus production, administration of codeine phosphate to suppress temporarily the central cough reflex, and mild sedation are useful.[34,36]

General anesthesia with control of the cough reflex is almost universally used in bronchography of infants and children.[37–39] Topical anesthesia may be used in older children and is almost always used in adults,[40–41] it being the safer mode although control of bronchial irritability is more difficult. Aerosol anesthesia, if available, produces more uniform and effective anesthesia.[11,35,42–44] Although lidocaine in general is used for topical anesthesia, it is not without risks. A safer and excellent anesthetic preparation is Forestier's solution (1% cocaine, phenol, and adrenalin). Because of its cocaine content, however, its use is prohibited in some countries, including the United States.

Transcricothyroid bronchography is not often employed because of the possibility of damage by the needle or contrast injection in the soft tissues of the neck.[45,46]

Too much contrast will lead to significant peripheral filling. Fluoroscopy during the examination helps control the amount of medium used, as well as the depth of contrast penetration. Postbronchographic postural drainage is of help in cleansing the lungs.[15] Peripheral overfilling with contrast retention is more likely in patients with chronic inflammatory or obstructive pulmonary disease because of difficulties in obtaining good anesthesia and impairment of the normal bronchial cleansing mechanism.[20,27,30,47,48] A number of investigators have shown that there is a temporary impairment of pulmonary function after bronchography of more than 20% for unilateral bronchography and over 30% if bilateral bronchography is performed.[49–53] This impairment is probably due to temporary interference with lung surfactant.[54] If there is prebronchographic impairment of pulmonary function, then only unilateral bronchography is desirable.

All of the commonly used media have a certain degree of toxicity, partly inherent in the contrast agent and partly a reflection of bronchographic technique. Improved patient preparation, the use of unilateral bronchography when appropriate, and fluoroscopy have helped decrease toxicity. Worldwide reviews published in 1958[40] and 1967[41] indicate that the mortality from bronchography is low. These surveys were in close agreement, with 18 fatalities being reported in over 100,000 bronchograms in the 1967 survey. Approximately half of the deaths resulted from the local anesthetic, primarily because of administration of excessive amounts.[41] The death rate directly attributable to the contrast media—about 1 per 12,500 examinations—occurred mainly in children or in patients with limited respiratory reserve.[41]

BRONCHOGRAPHIC CONTRAST MEDIA

Iodinated Compounds

Iodinated oils were widely used for bronchography until the 1960s. Their relatively high iodine content resulted in excellent contrast, but acute reactions to iodine did occur. The low viscosity led to peripheral filling with loss of anatomic detail and contrast retention. Because of the oils' strong chemical bonding, contrast was often retained for months to years. Consequent complications, in particular pneumonia, delayed reactions to iodine, and granulomatous formation have resulted in these compounds no longer being used.[40,41,54–57]

The search for a media of acceptable viscosity and toxicity led to the development of the suspended iodine compounds, Dionosil and Hytrast. The latter produces excellent bronchographic contrast and has very good adhesive properties, resulting in uniform coating of the mucosa and air contrast appearance. It is more irritating than the Dionosil media, however, and if peripheral filling occurs then the retained crystals, which are only slowly eliminated, can initiate pneumonia. Postbronchographic pyrexia is not uncommon. Other side effects, such as headache, nausea, vomiting, as well as mild transitory episodes of iodinism, may also occur. Use of Hytrast has, in the main although not entirely, been abandoned.[19,27,30,58–61]

Dionosil is available in both an aqueous and an oily form. Because the aqueous form is more irritating, the oil is preferred. On standing, a supernatant layer of the oil separates, and the bronchogram can be improved if half of this layer is poured off. Also the media needs to be shaken well just before use and ideally should be brought to body temperature. The amount used should be kept to a minimum; 12 to 20 ml is usually required per adult lung. With this amount, a satisfactory air contrast coating is obtained. Any retained iodine undergoes fairly rapid hydrolysis, while the suspending agent is phagocytosed. Mild clinical reactions, including pyrexia, nausea, vomiting, and headache, occur, as well as a temporary reduction in pulmonary function. The incidence of these reactions is increased when peripheral filling occurs. Iodine reaction and pneumonia are very rare. Reactivation of pulmonary tuberculosis is also very rare, although it has been reported.[3,6,17,28,62–66]

Although inhalation bronchography using a powder form of Dionosil and, more recently, powdered calcium ioglycamic acid has been carried out, these agents have not come into general use. Water-soluble iodine media have also been investigated. The early compounds were unacceptable, but recent use of one of the nonionic agents, metrizamide, indicates that these media may have some application.[12,34,67]

Noniodinated Compounds

A number of noniodinated opaque media, primarily metallic compounds, have been tried. Most of these have been too toxic, with the exception of barium sulfate and tantalum.

Barium sulfate with CMC in suspension or an aerosol powder has been found to produce excellent bronchograms with good patient tolerance. Because of its inert nature, most bronchographic physiologic reactions are more benign than those caused by iodinated media. Iodine sensitivity reactions are eliminated. However, prolonged retention of barium particles can occur, especially in sites of pulmonary disease. Granulomas may develop, and the possibility of long-term significant fibrotic changes has not been excluded.[7,24,63,68–75]

Inhalation of fine tantalum powder, a very inert material, produces superb bronchograms. However, there have been concerns as to possible fibrotic reaction in areas of overfilling, and, as with barium sulfate, this substance is not in general use for bronchography.[18,22,69,76–82]

Perfluorocarbon compounds[10] have also been studied and successful bronchograms reported, but their use has remained at an investigatory level.

CURRENT STATUS

Since the initial introduction of bronchographic contrast agents, until the last decade, bronchography has been an important and widely used examination in the evaluation of pulmonary disease. However, by the 1970s a two-thirds or greater decrease in the number of bronchograms had occurred, coinciding with the development and subsequent widespread availability of fiberoptic bronchoscopy and percutaneous needle biopsy.[41,83] Therapeutic advances in treating infections also contributed to this decrease. With the development of high resolution computed tomography (CT) and its greater diagnostic sensitivity, such as in the detection of small lung tumors and confirmation of suspected bronchiectasis, a further decrease in usage has occurred. As a result, in many radiology departments bronchography is only occasionally performed. With this decrease in usage has come a loss of impetus for the future development of improved bronchographic media.

However, bronchography still has a role where the newer diagnostic techniques are not readily available. Even when available, bronchography has useful, although limited, application, such as in the accurate mapping of sites of bronchiectatic involvement when surgery is contemplated or in the evaluation of suspected congenital pulmonary anomalies in children.[38]

In a number of centers bronchography is carried out immediately following fiberoptic bronchoscopy.[84–87] It has been our experience that excellent bronchograms can be obtained utilizing such a technique. By eliminating the need for a "two-stage examination" such a procedure has high patient acceptance, as well as reducing costs and morbidity. It is particularly applicable when only the occasional bronchogram is carried out. With the conventional separate examination, both patient acceptance and very good anesthesia with resultant improvement in the bronchogram can be achieved by the use of aerosol inhalation of topical anesthesia.

Dionosil, in particular the oily component, remains the most widely used agent due to rapid clearing of the iodinated portion and the very low incidence of significant reaction. Aerosol bronchography has the major advantages of enabling an easy examination for both operator and patient and producing uniformity of contrast distribution. Thus, the development of a satisfactory aerosol contrast agent would, no doubt, lead to this becoming the agent and procedure of choice for bronchography.

REFERENCES

1. Jackson C: The bronchial tree—its study by insufflation of opaque substances in the living. *Am J Roentgenol* 1918;5:454–456.

2. Sicard JA, Forestier JE: Methode generale d'esploration radiologique par l'Luile iodee. *Bull Mem Soc Med Hop Paris* 1922;46:463–469.

3. Dormer BA, Friendlander J, Wiles FJ: Bronchography in pulmonary tuberculosis. Part III. *Am Rev Tuberc* 1945;51:62–69.

4. Morales O: Further studies with viscous Umbradil (Umbradil-viskos) *Acta Radiol* 1949;32:317–335.

5. Morales O, Heiwinkel H: A viscous water-soluble contrast preparation. *Acta Radiol* 1948;30:257–266.

6. Tomich EG, Basil B, Davis B: The properties of n-propyl 3:5-di-iodo-4-pyridone-N-acetate (propyliodone). *Br J Pharm* 1953;8:166–170.

7. Nelson SW, Christoforidis A, Pratt PC: Barium sulfate and bismuth subcarbonate suspensions as bronchographic contrast media. *Radiology* 1959;72:829–838.

8. Gildenhorn HL, Springer FB, Wang SK, et al: New contrast medium for bronchography. *Dis Chest* 1962;42:596–599.

9. Nadel JA, Wolfe WG, Graf PD: Powdered tantalum as a medium for bronchography in canine and human lungs. *Invest Radiol* 1968;3:229–238.

10. Liu MS, Long DM: Biological disposition of perfluoroctylbromide: tracheal administration in alveolography and bronchography. *Invest Radiol* 1976;11:479–485.

11. Strecker EP, Kraemer C, Reinbold WD, et al: Inhalation bronchography using powdered calcium ioglycamic acid. *Radiology* 1979;130:303–309.

12. Smith W, Franken EA: Metrizamide as a contrast medium for visualization of the tracheobronchial tree: its drawbacks and possible advantages. *Pediatr Radiol* 1984;14:158–160.

13. DiGuglielmo L: Radiocontrast agents for bronchography, in Knoefel PK (ed): *International Encyclopedia of Pharmacology and Therapeutics V*, section 76, vol 2. Oxford, Pergamon Press, 1971, pp 395–411.

14. Knoefel PK: The respiratory tract, in Knoefel PK (ed): *Radiopaque Diagnostic Agents*, Springfield, IL, Charles C Thomas, 1961, pp 77–83.

15. Holden WS: The behaviour of contrast medium in the bronchial tree. *Br J Radiol* 1957;30:530–536.

16. Holden WS, Ardran GM: Observations on the movement of the trachea and main bronchi in man. *J Faculty Radiol* 1956–1958;8–9:267–275.

17. Holden WS, Crone RS: Bronchography using Dionosil oils. *Br J Radiol* 1953;26:317–332.

18. Gamsu G, Weintraub RM, Nadel JA: Clearance of tantalum from airways of different caliber in man evaluated by a roentgenographic method. *Am Rev Resp Dis* 1973;107:214–224.

19. Light JP, Oster WF: A study of clinical and pathological reaction to the bronchographic agent Hytrast. *Am J Roentgenol* 1964;92:615–622.

20. Mounts RJ, Molnar W: The clinical evaluation of a new bronchographic contrast medium. *Radiology* 1962;78:231–233.

21. Felton WL: The reaction of pulmonary tissue to Lipiodol. *J Thorac Surg* 1953;25:530–542.

22. Friedman PJ, Tisi GM: 'Alveolarization' of tantalum powder in experimental bronchography and the clearance of inhaled particles from the lung. *Radiology* 1972;104:523–535.

23. Hellström B, Holmgren HJ: The reaction of the lung on bronchography with viscous Umbradil (Umbradil-viskos B) (ASTRA), Umbradil (ASTRA) and carboxymethyl cellulose. *Acta Radiol* 1949;32:471–485.

24. Nelson SW, Christoforidis AJ, Pratt PC: Further experience with barium sulfate as a bronchographic contrast medium. *Am J Roentgenol* 1964;92:595–614.

25. Paterson JLH: An experimental study of pneumonia following the aspiration of oily substances; lipoid cell pneumonia. *J Pathol Bacteriol* 1938;46:151–164.

26. Záková N, Svoboda M: Morphological changes in the lungs following bronchography with barium sulfate. *Acta Univ Carol (Med)* 1965;11:125–135.

27. Cabrera A, Pickren JW, Sheehan R: Crystalline inclusion pneumonia following the use of Hytrast in bronchography. *Am J Clin Pathol* 1967;47:154–159.

28. Lang EK: A comparative study of febrile reactions to Hytrast, aqueous Dionosil and oily Dionosil in bronchography. *Radiology* 1964;83:455–459.

29. Light JP, Oster WF: Clinical and pathological reactions to the bronchographic agent Dionosil aqueous. *Am J Roentgenol* 1966;98:468–473.

30. Rayl JE: Clinical reactions following bronchography. *Ann Otol Rhinol Laryngol* 1965;74:1120–1132.

31. Grainger RG, Catellino RA, Lewin K, et al: Hytrast: experimental bronchography comparing two different formulations. *Clin Radiol* 1970;21:390–395.

32. LeRoux BT, Duncan JG: Bronchography with Hytrast. *Thorax* 1964;19:37–43.

33. Peck ME, Neerken AT, Salzman E: Clinical experience with water-soluble bronchography compounds. *J Thorac Surg* 1953;25:234–245.

34. Fennessy JJ: Bronchography. *Otolaryngol Clin N Am* 1973;6:579–600.

35. Nelson SW, Christoforidis AJ: Bronchography in diseases of the adult chest. *Radiol Clin N Am* 1973;11:125–152.

36. Dure-Smith P, Freundlich IM: Bronchography: the rational use of pre-medication and local anesthesia. *J Can Assoc Radiol* 1971;22:199–200.

37. Brunner S: Bronchography during infancy and childhood. *Dis Chest* 1967;52:201–204.

38. Levy M, Glick B, Springer C, et al: Bronchoscopy and bronchography in children. *Am J Dis Child* 1983;137:14–16.

39. Wilson JF, Peters GN, Fleshman K: A technique for bronchography in children. *Am Rev Resp Dis* 1972;105:564–571.

40. Bronchography: Summary of a world-wide survey report of committee on broncho-esophagology. *Dis Chest* 1958;33:251–260.

41. Bronchography: A report of the committee on broncho-esophagology. *Dis Chest* 1967;51:663–668.

42. Kandt D, Schlegel M: Bronchologic examinations under topical ultrasonic aerosol inhalation anesthesia. *Scand J Resp Dis* 1973;54:65–70.

43. Martin BH, Israel J, Stovin JJ: Anesthesia and intubation for bronchography. *Radiology* 1972;104:536.

44. Nelson SW, Christoforidis AJ: An automatic inhalation-actuated aerosol anesthesia unit: a new method of applying topical anesthesia to the oropharynx and tracheobronchial tree. *Radiology* 1964;82:226–234.

45. Wright FW: Accidental injection of Dionosil into the neck during bronchography. *Clin Radiol* 1970;21:384–389.

46. Zucherman SD, Jacobson G: Transtrachcal bronchography. *Am J Roentgenol* 1962;87:840–843.

47. Wright FW: Bronchography with Hytrast. *Br J Radiol* 1965;38:791–795.

48. Wood PB, Nagy E, Pearson FG, et al: Measurement of mucociliary clearance from the lower respiratory tract of normal dogs. *Can Anaesthesiol Soc J* 1973;20:192–206.

49. Surprenant K, Wilson A, Bennett L, et al: Changes in regional pulmonary function following bronchography. *Radiology* 1968;91:736–741.

50. Christoforidis AJ, Nelson SW, Tomashefski JF: Effects of bronchography on pulmonary function. *Am Rev Resp Dis* 1962;85:127–129.

51. Kokkola K: Respiratory gas exchange after bronchography. *Scand J Resp Dis* 1972;53:114–119.

52. Motley HL, Tomashefski JF: Acute effects of Lipiodol instillation on respiratory gas exchange. *Am J Physiol* 1951;167:812.

53. Zavod WA: Functional pulmonary changes following bronchography. *Am Rev Tuberc* 1948;57:626–631.

54. Schürch SF, Roach MR: Interference of bronchographic agents with lung surfactant. *Resp Physiol* 1976;28:99–117.

55. Burns AJ: Lipiodol in bronchography—its disadvantages, dangers and uses. *Am J Roentgenol* 1933;30:727–746.

56. Robertson PW, Morle KDF: Delayed pulmonary complications of bronchography. *Lancet* 1951;1:387.

57. Roodboets AP, Swierenga J, Oesner AP: Transient pulmonary densities around retained Lipiodol. *Thorax* 1966;21:473–481.

58. Misener FJ, Quinlan JJ, Hiltz JE: Hytrast: a new contrast medium for bronchography. *JAMA* 1965;92:607–610.

59. Morley AR: Pulmonary reaction to Hytrast. *Thorax* 1969;24:353–358.

60. Palmer PES, Barnard PJ, Cushman RPA, et al: Bronchography with Hytrast. *Clin Radiol* 1967;18:94–100.

61. Webb WR, Fitts CT: Evaluation of Hytrast as a bronchographic medium. *Am Surg* 1963;29:491–495.

62. Björk L, Lodin H: Pulmonary changes following bronchography with Dionosil oily (animal experiments): *Acta Radiol* 1957;47:177–180.

63. Dunbar JS, Skinner GB, Wortzman G, et al: An investigation of effects of opaque media on the lungs with comparison of barium sulfate, Lipiodol and Dionosil. *Am J Roentgenol* 1959;82:902–926.

64. Holden WS, Cowdell RH: Late results of bronchography using Dionosil oils. *Acta Radiol* 1953;49:105–112.

65. Jain SK, Agarwal RL: Reactivation of a tuberculous lesion following bronchography. *Tubercle* 1980;61:105–107.

66. Walker HG, Ma H: Oily and aqueous propylidone (Dionosil) as bronchographic contrast agents. *J Can Assoc Radiol* 1971;22:148–153.

67. Atwell RJ, Pedersen RL: A water-soluble contrast medium for bronchography: report on clinical use. *Dis Chest* 1950;18:535–541.

68. Cember H, Hatch TF, Watson JA, et al: Pulmonary effects from radioactive barium sulphate dust. *AMA Arch Indust Hyg* 1955;12:628–634.

69. Edmunds LH Jr, Graf PD, Sagel SS, et al: Radiographic observations of clearance of tantalum and barium sulfate particles from airways. *Invest Radiol* 1970;5:131–141.

70. Erickson LM, Shaw D, MacDonald FR: Prolonged barium retention in the lung following bronchography. *Radiology* 1979;130:635–637.

71. Fite F: Granuloma of lung due to radiographic contrast medium. *AMA Arch Pathol* 1955;59:673–676.

72. Johnson PM, Montclair NJ, Benson WR, et al: Toxicity of bronchographic contrast media. *Ann Otol Rhinol Laryngol* 1960;69:1103–1113.

73. Nice CM, Waring WW, Killelea DE, et al: Bronchography in infants and children; barium sulfate as a contrast agent. *Am J Roentgenol* 1964;91:564–570.

74. Shook CD, Felson B: Inhalation bronchography. *Chest* 1970;58:333–337.

75. Willson JKV, Rubin PS, McGee TM: The effects of barium sulfate on the lungs. *Am J Roentgenol* 1959;82:84–94.

76. Bianco A, Gibb FR, Kilpper RW, et al: Studies of tantalum dust in the lungs. *Radiology* 1974;112:549–556.

77. Dilley RB, Nadel JA: Powdered tantalum: its uses as a roentgenographic contrast material. *Ann Otol Rhinol Laryngol* 1970;79:945–952.

78. Gamsu G, Platzker A, Gregory G, et al: Powdered tantalum as a contrast agent for tracheobronchography in the pediatric patient. *Radiology* 1973;107:151–157.

79. Hinchcliffe WA, Zamel N, Fishman NH, et al: Roentgenographic study of the human trachea with powdered tantalum. *Radiology* 1970;97:327–330.

80. Nadel JA, Wolfe WG, Graf PD, et al: Powdered tantalum—a new contrast medium for roentgenographic examination of human airways. *N Engl J Med* 1970;283:281–286.

81. Sailer R, Kissler B,Stauch G, et al: Pulverformige Kontrastmittel zur Bronchographie. Tierexperimentelle Untersuchungen mit Tantal, Wolfram und Hytrast. *Fortschr Röntgenstr* 1973;119:727–736.

82. Schlesinger RB, Schweizer RD, Chan TL, et al: Controlled deposition of tantalum powder in a cast of the human airways: applications for aerosol bronchography. *Invest Radiol* 1975;10:115–123.

83. Fraser RG: Editorial—bronchography. *J Can Assoc Radiol* 1972;23:236–243.

84. Lutch JS, Ryan KG: Bronchography combined with bronchoscopy. *Chest* 1979;75:108.

85. Ono R, Loke J, Ikeda S: Bronchofiberscopy with curette biopsy and bronchography in the evaluation of peripheral lung lesions. *Chest* 1981;79:162–166.

86. Simelaro JP, Marks B, Meals R, et al: Selective bronchography following fiberoptic bronchoscopy. *Chest* 1979;76:240–241.

87. Taber RE: Bronchography after bronchoscopy. *Ann Thorac Surg* 1984;37:264.

Chapter 31

Sialography and Dacryocystography

Heun Y. Yune

The techniques of sialography and dacryocystography developed similarly. Arcelin in 1912[1] was credited as the first person to visualize the duct systems of the submandibular gland using a bismuth emulsion. Ewing in 1909 used bismuth paste to outline a lacrimal sac abscess.[2] However, not until the introduction of oily contrast materials, such as Lipiodol, was contrast radiography of these structures popularized. In 1925, Wiskovsky[3] reported to the Czechoslovakian Society of Otolaryngology a method of visualization of Wharton's duct using Lipiodol. In the same year and shortly thereafter, Barsony,[4] Uslenghi,[5] and Carlsten[6] independently claimed that they had performed the first successful sialogram. Carlsten used Lipiodol, while Barsony used a 20% potassium iodide solution. Before the wide acceptance of Lipiodol, a liquid barium mixture and bismuth paste were also tried in dacryocystography.[7,8] In 1924 Bollack[9] first reported the use of Lipiodol for visualization of the lacrimal tract. Earlier in 1921, Sicard and Forestier (see Chapter 18, Hysterosalpingography) introduced Lipiodol as a radiographic contrast agent, and by the time it was applied to the salivary glands and the lacrimal tracts, it had already been found to be a suitable contrast agent for bronchography, myelography, and hysterosalpingography. Subsequently, other oily contrast materials, such as Pantopaque and Ethiodol, and various aqueous contrast agents, such as Sinografin, were tried.[10–17] Ethiodol today remains the contrast material of choice both in sialography and dacryocystography.

SIALOGRAPHY

Nettleship (or similar) lacrimal dilators are used to identify and gradually dilate the salivary duct opening (Stensen's or Wharton's). When the orifice is dilated to approximately one-half to two-thirds of a millimeter in diameter, a tapered-tip, clear Teflon tube or a sialogram cannula is inserted into the duct.[18] In the great majority of patients, French no. 4 Teflon tubing is the adequate tube size. Occasionally, however, the orifice may be so patent that French no. 4 tubing may not adequately seal the duct orifice around the tubing and injected contrast material may reflux into the oral cavity, preventing proper filling of the intraglandular duct system and the acinar spaces. A large caliber tubing is useful with these patients. Either using tapered-tip Teflon tubing or an olive-tip (or similar) sialogram cannula, the objective is to achieve a good seal of the salivary duct orifice around the tubing so that a full column opacification of the duct system and the acinar spaces is obtained without resorting to undue speed and pressure of contrast injection.

Readers are referred to detailed descriptions of the technique.[18,19] A few major points of technical importance that will help improve the chance for high quality, low risk examinations are presented below. First, the examination is performed under fluoroscopic monitoring and the progression of the examination recorded on high quality spot films. Using fluoroscopy prevents insufficient or excessive contrast material injection and makes the reasons for incomplete visualization of the duct system or parts of the gland parenchyma more readily apparent. Second, to avoid artifactual "filling defects" in the contrast column within the duct, precautions should be taken to avoid introducing air bubbles during the contrast injection. The injection should be performed continuously under low pressure. The syringe loaded with the contrast agent and the injection tube or cannula with an interposing stopcock should be preassembled and all air purged from the entire assembly before introducing the tube into the duct. Intermittent injection can result in breaking and segmentation of the oily contrast agent, which may mimic a diseased duct and/or filling defects. When using an oily contrast agent, low pressure, continuous injection is more easily accomplished with a glass syringe than a plastic syringe. High injection pressure increases the chance of complications.

In the past, some authors have advocated the injection of contrast material until the patient complains of a sensation of distention and some pain, which will ensure contrast filling of the acinar spaces.[20–27] This technique is called "distension sialography." However, with this method, minor changes in the duct system can be completely overshadowed, especially when fluoroscopic monitoring is not used. It also increases the chance of ductal or acinar rupture and escape of contrast material into the salivary gland parenchyma or surrounding tissues.

At the termination of the contrast injection, the stopcock is turned off, and frontal, lateral, and oblique overhead radiographs may be obtained for visualization of finer anatomic details. When the films obtained are satisfactory, the injection tube is withdrawn from the duct, and a sialogogue is administered. After waiting for a few minutes, the overhead radiographs are repeated in standard positions (frontal, lateral, and oblique projections). A rough estimation of salivary production in the gland and highlighting of diseased portions of the gland parenchyma interspersed with normal areas are depicted on these "postevacuation" radiographs.

Conventional sialography is less frequently performed since the advent of ultrasonography, computed tomography, and magnetic resonance imaging for the study of mass lesions of major salivary glands. However, its indication in suspected salivary duct obstruction and parenchymal damage secondary to chronic inflammation remains unchanged.

DACRYOCYSTOGRAPHY

The initial steps of probing the lacrimal canaliculus and the introduction of a tapered-tip Teflon tube are identical to those for sialography except that, as the lacrimal canaliculus is smaller and shorter than the salivary duct, it must not be probed more than several millimeters in depth and more caution must be used so that it is not traumatized during probing. The catheter tip may be inserted only a few millimeters and certainly no more than 5 mm in depth. The injection tube is anchored to the skin of the cheek with adhesive tape. The injection tube assembly and the connection to a syringe preloaded with contrast agent are again identical to the preparation for sialography. Under fluoroscopic monitoring, spot film radiographs are obtained with the patient's head turned to various projections, including frontal, lateral, and oblique, during the contrast agent injection.[28] If the lacrimal tract is patent, a continuous injection will ensure that an uninterrupted column of contrast fills the lacrimal sac through the lacrimal canaliculus and the nasolacrimal canal and spills into the inferior nasal meatus.[29] Usually, the inferior canaliculus is initially cannulated unless it is obstructed or otherwise so severely diseased that it cannot be used. The contrast agent injection usually refluxes through the superior canaliculus while simultaneously draining through the nasolacrimal canal.[30]

A preliminary preparation of the lacrimal sac is desirable when performing dacryocystography. Some authors advocate the use of a local anesthetic,[12,31–33] but in most instances the patients who need dacryocystography have partial or complete obstruction of the lacrimal tract, thus reducing the chance of achieving anesthesia of the lacrimal tract. Gentle manipulation of probes in the lacrimal canaliculus really does not require preliminary local anesthesia.[34] Expression of the lacrimal sac content by gentle finger pressure or by irrigation through injection tubing before the contrast agent is injected is desirable.[31,35–37] Standard overhead radiographs are not necessary when good quality spot films that can resolve images of a structure in the range of 1-mm diameter are available.[38]

CONTRAST AGENTS

As mentioned above, the contrast material of choice is Ethiodol. The history of its development, the chemical composition, and physical characteristics are discussed in Chapter 18.

In sialography, a low viscosity, oily contrast agent, such as Ethiodol, is preferred over water-soluble contrast materials because the latter opacify the acinar structures too quickly and also infuse into the salivary gland parenchyma beyond the acinar spaces, resulting in loss of finer radiographic detail. Infusion of the iodinated contrast agent beyond the acinar space also increases the chance of so-called iodine parotitis. The water-soluble contrast agents are indicated only when an oily contrast agent cannot opacify the acinar structures. The acinar structures should normally be opacified with Ethiodol; when there is no opacification, to verify the presence of diffuse acinar scarring or damage, a water-soluble contrast agent may be used.[39] There is no particular brand of aqueous contrast agent that is preferred in such an instance.

In dacryocystography, again Ethiodol is better suited than the aqueous contrast agents. If an aqueous contrast medium is to be used, those with a viscosity-enhancing agent are preferable. Without proper viscosity, the flow of the contrast agent is too rapid if the lacrimal tract is patent, and it will be difficult to obtain proper visualization of fine structural details, such as the lacrimal canaliculi, the interior of the lacrimal sac, and the nasolacrimal duct. The viscosity-enhancing substances added to the aqueous contrast agents are polyvinylpyrrolidone and carboxymethylcellulose.

Use of a high viscosity, nonionic, low osmolality aqueous contrast agent, such as Amipaque, coupled with magnification and subtraction technique, has been described.[40] The nonirritating nature of such an agent is considered to be the basis of increased patient comfort and safety. Hypothetically, if the duct system is injured and injected contrast material should leak into the surrounding interstices, pre-existing chronic inflammatory changes could readily evolve into foreign body granuloma if nonabsorbable oily contrast material or viscosity-enhancing additives were used. The same concern applies both in dacryocystography and sialography.

The proper dosage to be used in each individual patient is determined by fluoroscopic monitoring and the detection of acinar space opacification.[18] Using the above-mentioned techniques, the usual dosage of contrast agent for parotid sialography is 0.7 to 1.0 cc.[20] Even in a very large gland the dose seldom exceeds 2.0 cc.[27,41] In the submandibular gland, a slightly lesser amount is sufficient. In dacryocystography, more than 1.5 cc of contrast agent is seldom required to obtain several spot films while the contrast is being injected continuously.

COMPLICATIONS

Except for the so-called iodine parotitis, complications associated with sialography and dacryocystography, especially when oily contrast agents are used, are primarily related to a faulty examination technique. The contrast agent used plays a secondary role.

In both procedures, because preliminary probing and dilation of a small duct and canal structure are necessary, accidental perforation through the wall of the duct may occur, especially when it is diseased or scarred. When such an injury to the duct occurs, the injected contrast agent can easily infiltrate the surrounding tissues.[42] Contrast infiltration can also result from an excessively high pressure of injection. In such a case, when the contrast agent is an oily material, it can conceivably result in foreign body granuloma reactions.[43,44] These, however, are usually scattered rather than confined in a localized space, and the amount that infiltrates is so small that long-lasting tissue changes are rarely encountered.

The real nature of iodine parotitis is not fully understood. The major salivary gland seems to be sensitive to a high concentration of iodine, and when this occurs, the salivary gland becomes edematous and very painful.[45]

When the oily contrast agent is retained in cavitary spaces, such as in the destroyed acini in Sjögren's disease, or in an abscess cavity, it can trigger a foreign body reaction. Clinically, however, this seldom is a problem. The salivary gland that is extensively destroyed and scarred, producing very little saliva and thus unable to wash out the injected contrast agent,[39] will experience more symptoms and signs from the underlying disease than from a foreign body reaction.

With an obstructed nasolacrimal duct, the injected contrast material may be temporarily retained in the dilated lacrimal sac. It, however, can be easily expressed back by gentle finger pressure.

In sialography, when the injection tubing tightly seals the duct opening, the contrast material injected, added to the retained saliva, raises the pressure inside the duct system and the acini, resulting in visible swelling of the salivary gland. Such swelling may develop rapidly. This acute swelling, however, is not often associated with pain, although patients almost invariably report a subjective sensation of tightness in the area of the salivary gland. Pain from the acute swelling without iodine parotitis is seen in approximately 10% of these patients and subsides quickly upon removal of the injection tubing and administration of a sialogogue. Swelling of the gland may remain even after the pain subsides. The pain associated with iodine parotitis is usually much more severe and lasts longer.

REFERENCES

1. Arcelin F: Radiographie d'un calcul salivaire de la glande sublinguale. *Lyon Méd* 1912; 118:769–773.

2. Ewing AE: Roentgen ray demonstration of the lacrimal abscess cavity. *Am J Ophthalmol* 1909; 26:1–4.

3. Wiskovšký B: Sialodochografie. *Zbl Hals-Nasen Ohrenheilkd* 1926;8:320.

4. Barsony T: Idiopathische sternongang dilatation. *Klin Wschr* 1925;4:2500–2501.

5. Uslenghi JP: Nueva technica para la investigacion radiologica de las Glandulas salivaires. *Rev Soc Argent Radiol Electrol* 1925;1:4.

6. Carlsten DB: Lipiodolinjektion in den ausführungsgand der speicheldrüsen. *Acta Radiol* 1926; 6:221–223.

7. Campbell DM, Carter JM, Doub KP: A method for roentgen ray demonstration of the nasolacrimal passageways. *Am J Roentgenol* 1922;9:381–387.

8. Van Gangelen G: Die Röntgenuntersuchung der tranewege. *Acta Otolaryngol* 1921;2:391–397.

9. Bollack J: Sur l'exploration radiographique des voies lacrymales par l'injection d'huile iodes. *Ann Ocul* 1924;161:321–335.

10. Brands T: *Diagnose und Klinik der Erkrankungen der grossen Kopfspeicheldrüsen*. Berlin, Urban und Schwarzenberg, 1972.

11. Hettwer KJ, Folsom TC: The normal sialogram. *Oral Surg* 1968;26:790–799.

12. Agarwal ML: Dacryocystography in chronic dacryocystitis. *Am J Ophthal* 1961;52:245–251.

13. Demorest BH, Milder B: Dacryocystography. I. The normal lacrimal apparatus. *Arch Ophthalmol* 1954;51:180–195.

14. Domke K: On the demonstration of the nasolacrimal tract with falitrash B. *Dtsch Gesundh* 1967;22:2144.

15. Malik SRK, Gupta AK, Chateyee S, et al: Dacryocystography of normal and pathological lacrimal passages. *Br J Ophthalmol* 1969;53:174–179.

16. Sargent EN, Ebersole C: Dacryocystography: the use of Sinografin for visualization of the nasolacrimal passage. *Am J Roentgenol* 1968;102:831–839.

17. Zizmor J, Lombardi G: *Atlas of Orbital Radiology*. Birmingham, AL, Aesculapius Publishing, 1973.

18. Yune HY, Klatte EC: Current status of sialography. *Am J Roentgenol* 1972;115:420–428.

19. Yune HY: Sialography and dacryocystography, in Miller RE, Skucas J (eds): *Radiographic Contrast Agents*. Baltimore, University Park Press, 1977, pp 485–492.

20. Blady JV, Hocker AF: Sialography, its technique and application in the roentgen study of neoplasm of the parotid gland. *Surg Gynecol Obstet* 1938;67:777–787.

21. Blady JV, Hocker AF: The application of sialography in non-neoplastic diseases of the parotid gland. *Radiology* 1939;32:131–141.

22. Einstein RA: Sialography in the differential diagnosis of parotid masses. *Surg Gynecol Obstet* 1966;122:1079–1083.

23. Kimm HT, Spies JW, Wolfe JJ: Sialography: with particular reference to neoplastic diseases. *Am J Roentgenol* 1935;34:289–296.

24. Ollerenshaw RGW, Rose SS: Radiological diagnosis of salivary gland disease. *Br J Radiol* 1951;24:538–548.

25. Osmer JC, Pleasants JE: Distention sialography. *Radiology* 1966;87:116–118.

26. Pendergrass EP, Schaeffer JP, Hodes PJ: *The Head and Neck in Roentgen Diagnosis,* ed 2. Springfield, IL, Charles C Thomas, 1956.

27. Rubin P, Holt JF: Secretory sialography in diseases of the major salivary glands. *Am J Roentgenol* 1957;77:575–598.

28. Lloyd GAS, Jones BR, Welham RAN: Intubation macrodacryocystography. *Br J Ophthalmol* 1972;56:600–603.

29. Pettit LH, Coin CC: Dacryocystography. *Radiol Clin N Am* 1972;10:129–142.

30. Hartmann E, Gilles E: *Roentgenologic Diagnosis In Ophthalmology*. Philadelphia, JB Lippincott, 1959, pp 166–177.

31. Aakhus T, Bergaust B: Dacryocystography in obstruction of the lacrimal passages. *Acta Radiol Diagn* 1969;8:369–375.

32. Iba GB, Hanafee WN: Distention dacryocystography. *Radiology* 1968;90:1020–1022.

33. Nahata MC: Dacryocystography: in disease of lacrimal sac. *Am J Ophthalmol* 1964; 58:490–493.

34. Rodriguez HP, Kittleson AC: Distension dacryocystography. *Radiology* 1973;109:317–321.

35. Campbell W: Radiology of the lacrimal system. *Br J Radiol* 1964;37:1–26.

36. Demorest BH, Milder B: Dacryocystography. II. The pathological lacrimal apparatus. *Arch Ophthalmol* 1955;54:410–421.

37. Law FW: Dacryocystography. *Trans Ophthalmol Soc UK* 1967;87:395–407.

38. Street DF, Howell MH: An alternative radiographic technique for macrodacryocystography. *Br J Radiol* 1967;40:235–236.

39. Chisholm DM, Blair GS, Low PS, et al: Hydrostatic sialography as an index of salivary gland disease in Sjögren's syndrome. *Acta Radiol Diagn* 1971;11:577–585.

40. El Gammal T, Brooks BS: Amipaque dacryocystography—biplane magnification and subtraction technique. *Radiology* 1981;141:541–542.

41. Blair GS: Hydrostatic sialography: an analysis of a technique. *Oral Surg* 1973;36:116–130.

42. Castren JA, Korhonen M: Significance of dacryocystography in lacrimal drainage system infections. *Acta Ophthalmol* 1964;42:188–192.

43. Lilly GE, Cutcher JL, Steiner M: Radiopaque contrast mediums: effect on dog salivary gland and subcutaneous tissue. *J Oral Surg* 1968;26:94–98.

44. Mandel L, Baurmash H: Radiopaque contrast solutions for sialography. *J Oral Ther Pharm* 1965;2:73–80.

45. Talner LB, Lang JH, Brasch RC, et al: Elevated salivary iodine and salivary gland enlargement due to iodinated contrast media. *Am J Roentgenol* 1971;11:380–382.

Part IX

Pediatric Agents

Chapter 32

Pediatric Use of Contrast Agents

E.A. Franken, Jr., Yutaka Sato, and Wilbur L. Smith

INTRAVASCULAR CONTRAST AGENTS

General

In general, the same types of agents are used for angiography and angiocardiography in the pediatric age group as are used with adults. Too, as the predominant route of excretion is via the kidneys, the same drugs are utilized for excretory urography. In this section the specific problems of the use of intravascular agents in infants and children are described. Their applications in individual examinations are presented in later sections.

Although the ideal intravascular contrast agent would have high radiopacity and no toxicity, drugs now available vary in iodine content, toxicity, and cost effectiveness. Clinical toxicity of intravascular contrast agents has components of chemical, osmolar, and idiosyncratic effects, all of which are of importance in infants and children. Chemotoxicity affects predominantly the nervous, respiratory, and cardiovascular systems and the kidneys. In each of these systems toxicity of contrast agents to children is evident in the same fashion as in adults. Seizures may occur after entry of contrast agents into the brain; they are particularly apt to occur in neonates with previous disease affecting the central nervous system.[1–3] Pulmonary hemorrhage occurs with massive overdosage of contrast material in infants.[4] Pulmonary edema is occasionally seen in infants with doses accepted as clinically safe.[5–7] Cardiovascular toxicity is discussed below in the section on angiocardiography. Renal function can be adversely affected by contrast material, although the exact incidence is disputed.[8–10]

Osmolar effects of contrast agents can be considerable, particularly in the newborn. For example, intravenous (IV) injection of 5 cc/kg of a traditional ionic contrast agent with an osmolality about five times that of serum will raise serum osmolality by 5% and circulating blood volume by 20%.[5] These increases are caused by entrance of fluid into the vascular space that must occur when such hypertonic fluid is given. The end result is hypertonic dehydration, the degree of which varies considerably. Such osmolar effects are thought to contribute to the occasional patient with acute pulmonary edema in response to administration of contrast material.

A subject seldom mentioned in the discussion of contrast agents is "acute osmole poisoning."[11] With rapid shifts in serum osmolality and blood volume, compensatory changes must occur in other organ systems. In the central nervous system compensatory effects can be measured in cerebrospinal fluid and cerebral venous pressures, both of which respond to sudden loss of brain fluid. The result is a situation promoting cerebral hemorrhage. As clinical manifestations of this phenomenon may be subtle, evaluation of "osmole poisoning" is seldom part of contrast agent investigations.

Allergic-type reactions are found with contrast agents in children, as well as in adults. As major causes of morbidity and mortality they are probably of much less importance than are chemotoxicity and hypertonicity.[12] Mechanisms of allergic-type reactions are considered in Chapter 9.

Are children more or less apt to suffer from contrast agent toxicity? From anecdotal reports it seems that accidental administration of inordinate amounts of contrast is more common in infants.[4] Similarly, pulmonary edema and complications of hypertonicity occur with some frequency in infancy, probably owing to the large amounts of drug given as compared to body surface area and blood volume.[5]

The overall incidence of reactions to IV contrast agents is about 5% in all age groups, with severe reactions in 0.05% and death in 1 in 40,000 examinations.[13] In a past survey of members of the Society of Pediatric Radiology,[14] there were five severe reactions and no deaths in 12,000 excretory urograms. These results are approximately what would be expected for all age groups. Although most deaths from contrast agents occur in 50- to 70-year-old patients,[15] the relative risk per individual in each age group—adult versus child—is not known.

In the past most contrast agents used in infants and children were salts of diatrizoate or iothalamate, containing 28–38% iodine (Hypaque 50%, Renografin 60%, Conray).[14] Agents with higher concentrations of iodine, such as Renografin 76%, Hypaque-M 75%, and Angio-Conray, have been used in angiocardiography to minimize subsequent dilution in the cardiac chambers. The quality of examinations and side effects have been related only to the amount of contrast used, with the possible exception that meglumine salts are less toxic.[16] These drugs are still used, particularly in children beyond infancy. In North American patients younger than 1 year of age, they are rapidly being replaced by the new low osmolar agents. As has been detailed elsewhere in this book, the new low osmolar nonionic water soluble agents are of two general types: (1) nonionic monomers containing three atoms of iodine per molecule and (2) ionic dimers containing six atoms of iodine per molecule and therefore three atoms of iodine per particle.[17] Metrizamide, iohexol, and iopamidol are nonionic monomers; ioxaglate (Hexabrix) is an ionized dimer. Owing to the increased iodine per particle in solution and certain surface adherence characteristics, osmolality is reduced by an approximate factor of 3 when compared to the traditional contrast agents.

The low osmolality contrast agents are less toxic than the traditional agents both clinically and in the laboratory, producing less physiologic derangement.[18,19] They are particularly valuable in the neonate, in whom water balance is critical, and in the child with congenital heart disease who is at risk for left ventricular failure. They offer considerable advantages in chemotoxicity and osmolality, but probably will not affect the incidence of allergic-type reactions. They are currently much more expensive than the traditional agents. At this time the cost effectiveness of these new agents in all situations has yet to be determined. There seems to be general agreement that their safety in the neonate and others at particular risk for contrast agent toxicity justifies their use in those groups.

Excretory Urography

Let us first consider how IV contrast agents are excreted and those anatomic and physiologic differences in the infant that affect excretion.[20] A bolus of contrast is rapidly dispersed throughout the vascular system and the extracellular space. As the contrast material reaches the kidney, it enters the urinary filtrate almost exclusively through glomerular filtration. Water is absorbed from the urinary filtrate, and the contrast material in the nephron becomes more concentrated and ultimately is visible on the radiograph. Thus, the quality of an excretory urogram depends upon the dose of iodine administered, the degree of dilution of the drug in body fluids, and the glomerular filtration rate (GFR) and tubular resorption of the kidney.

There are important differences in many systems in the newborn infant that affect the excretory urogram adversely. First, the extracellular space is relatively greater, allowing greater dilution of administered contrast material. Second, renal blood flow and the GFR are substantially reduced in the neonatal kidney. Lastly, there is reduced renal concentrating ability because of relative tubular insufficiency. So, given a specific amount of contrast material for excretory urography in the neonate, the intravascular level of contrast is reduced by dilution, less enters the urinary filtrate because of the reduced GFR, and less water is absorbed from the nephron. All these differences approach the adult standards as the infant ages, albeit at variable rates. In general, at 6–8 weeks of postnatal age there are sufficient changes so that good excretory urography is possible.

The traditional contrast agents—diatrizoates and iothalamates—are acceptable and effective for excretory urography in children. Unless there develops a policy for universal use of the new nonionic contrast agents for excretory urography in all patients, considerations of cost effectiveness might recommend continuation of use of these drugs for excretory urography in patients over 1 year of age. No matter which of these agents is used, the maximum dosage is of critical importance. There are no good prospective studies regarding the appropriate dosage of contrast for excretory urography in children.[21] Recommendations by various pediatric radiologists have ranged from 2 cc/kg up to 5 cc/kg.[22,23] As pulmonary edema may complicate examinations using 3 cc/kg, we prefer not to exceed 2 cc/kg, except in unusual circumstances.

At this time most reported series using the nonionic contrast agents in infants and children are from outside North America.[21,24–26] Although studies of small groups show little difference in side effects when compared to traditional agents, one large study indicates substantially less osmotic effect and other biochemical changes with the new contrast agents. Similar results have now been noted in North America.[27] The quality of urograms with these agents is as good and in some aspects better than with the traditional agents. The nephrogram is usually comparable with all contrast media, being dependent primarily on the amount of contrast delivered to and filtered by the kidney. As there is minimal osmotic diuresis with the new nonionic contrast agents, there may be less calyceal distention and filling of the ureter and bladder. Conversely, the lack of osmotic diuresis results in a greater concentration of iodine in the urine with a resultant better quality pyelogram. Almost all studies conclude that the overall quality of the excretory urogram is superior with the new agents.

In conclusion, it seems that the low osmolality contrast agents produce excretory urograms of superior quality in infants and children at less risk than the traditional agents. Unless cost constraints are overwhelming, they are the drugs of choice for excretory urography in the infant less than 1 year of age and in those with other risk factors: congenital heart disease with heart failure, renal disease, potential hypotension, and previous reaction to traditional contrast media. Although there are discernible differences among these new contrast agents, they are minimal. In our opinion, local custom and cost considerations are probably of more importance than the chemical differences of the various new contrast agents. As much less osmotic diuresis accompanies these contrast agents in excretory urography, the administered dose can be reduced. Cremin and Rhodes reduced the dose for urography from 3 cc/kg to 2 cc/kg and still found a 25% improvement in radiographic quality.[21]

Angiography

Systemic effects of the total amount of contrast material mentioned above are also appropriately considered in angiography elsewhere in the body. In general, the same agents used in adults are appropriate for children. The nonionic contrast agents seem to have particular advantages in arteriography and venography of the extremities because of reduced pain, and in neuroangiographic procedures, as discussed in Chapter 11. The specific agent and amount to be used for each type of angiographic procedure differ according to the area to be studied, as well as the age and size of patient.

As noninvasive modalities, such as computed tomography, ultrasound, and magnetic resonance, become more useful, use of angiography in the pediatric age group has decreased.[28] Current indications, in addition to neuroradiologic procedures, include patients with trauma, gastrointestinal (GI) bleeding, renovascular disease, portal hypertension, and primary liver tumors.[28,29] In some patients, IV digital subtraction angiography, with its elimination of arterial catheters, may suffice.[30]

Although the same angiographic contrast agents are used for children and adults, the amount administered must be modified according to the age and size of the patient. The total administered dose should rarely exceed 2 cc/kg. Recommended dosages of contrast material for angiography in children are given in Table 32-1.[29] These can be reduced substantially if digital subtraction techniques are used.[30]

Table 32-1 Recommended Amounts of Contrast Material for Angiography in Children

		Contrast Dose by Patient Weight		
Injection Site	*< 10 kg*	*10-20 kg*	*20-40 kg*	*> 40 kg*
Aorta (ml/kg)	1.5	1.5-1.2	1.2-1.0	0.9-0.8
Celiac axis (ml/kg)	1.5-1.3	1.3-1.1	1.1-0.9	0.8
Hepatic artery (ml/kg)	1.8	0.7	0.6	0.5
Splenic artery (ml/kg)	1.0	0.9	0.8	0.7
Superior mesenteric artery (ml/kg)	1.0	0.9	0.8	0.7
Inferior mesenteric artery (ml/kg)	0.7	0.6	0.5	0.4
Renal artery (ml)	1-2	3-5	5-7	7-9
Inferior vena cava (ml/kg)	1.5	1.5-1.2	1.2-1.0	0.9-0.8
Splenoportogram (ml/kg)	1.5-1.3	1.3-1.1	1.1-0.9	0.8

Source: Adapted with permission from "Pediatric Abdominal Angiography: Panacea or Passé" by AV Moore et al in *American Journal of Roentgenology* (1982;138:433–443), Copyright © 1982, American Roentgen Ray Society.

Angiocardiography

Cardiovascular effects of contrast agents can be substantial. Left ventricular end-diastolic pressure, pulmonary and systemic vascular resistance, coronary flow, and excitation levels for arrhythmia can all be affected adversely.[13,19,31,32] Most cardiac changes occur with intracoronary artery injections, but can be detected with aortic or even IV administration. Hypertonicity with increased circulating blood volume has an obvious effect on cardiovascular dynamics. Infants with congenital heart disease and actual or potential left ventricular failure are at greater risk than those with right heart lesions.[33]

In angiocardiography using conventional agents, most radiologists prefer contrast agents with a relatively high iodine content (usually 38%) because of subsequent dilution in the cardiac chambers. Viscosity of the contrast agent must be sufficiently low so that efficient injection of the material is possible; this requires warming to near body temperature. Meglumine salts of diatrizoate are generally less viscous than those of sodium, but are probably more toxic.[16] Renografin-76, which contains 4.48 mg of sodium/ml, can be used. Hypaque-M 75% has less viscosity but more sodium. Angio-Conray is another acceptable agent. Appropriate agents for coronary arteriography in adults are equally applicable to children.

The low osmolality nonionic contrast agents are of value in pediatric angiocardiography.[34,35] Although toxicity in an individual patient may be difficult to detect, larger series indicate less alteration of hemodynamics in those patients given the new contrast agents as compared to the traditional agents.[31,36] The diagnostic quality of the examinations is unchanged, but less clinical and biochemical deterioration occurs. Cumberland recommends that the low osmolality agents be used in angiocardiography for all children (particularly infants) and for those with congestive heart failure, hypertension, risk for hypotension, and severe coronary artery disease.[31]

There is far from universal agreement on the maximum amount of contrast agent to be used in pediatric angiocardiography. Enough contrast must be given and enough injections performed so that accurate diagnosis of the cardiac disease is achieved, while minimizing toxicity to the patient. Individual injections utilize 0.5 cc/kg to 1.0 cc/kg. Maximum doses for pediatric angiocardiography recommended by various authors include 2 cc/kg (Giammona et al.);[37] 4 cc/kg (children), 3 cc/kg (infants), 1.0–1.5 cc/kg (premature infants) (Stanger et al.);[16] and 5 cc/kg (Fox et al.).[38] It is understood that the risk of chemotoxicity, hypertonic dehydration, and pulmonary edema is increased as more contrast is given. A delay of 20–30 minutes between injections to allow for some of the contrast to be excreted and for osmolality to return to near normal is sometimes appropriate when large injections must be given for diagnosis.[16]

CONTRAST AGENTS FOR COMPUTED TOMOGRAPHY

Oral and Rectal Agents

Ideally, the gut lumen from stomach to rectum should be opacified during abdominal CT, with no associated interference of visualizing other intra-abdominal structures. Gut opacification is of particular value in infants in whom the paucity of intra-abdominal fat is a disadvantage for radiography. Evaluation of the pancreas, retroperitoneum, and mesentery is particularly difficult without contrast in the adjacent bowel. Problems in attempting to opacify the bowel include the difficulty in completely filling the duodenum, small bowel, and colon; inhomogeneity and streak artifacts; motion artifact; and achieving an adequate coating of the intestinal wall. A variety of contrast agents and techniques have been devised to overcome difficulties in contrast filling of the bowel. The multiplicity of methods reflects the lack of complete success of any single technique.

In the early years of body CT imaging, water-soluble iodine-containing agents in low concentration (e.g., 3% Gastrografin) were most frequently recommended. Their advantages include reasonable homogeneity without streak artifact and their theoretical ability to stimulate peristalsis and thus assure duodenal and distal small bowel filling. More recently a number of barium products have been made available for opacification of the gastrointestinal tracts. These preparations are used in a 1.0–1.5% w/v suspension of barium and are easily administered. Patient acceptance of the barium mixtures is much higher than of the iodine-containing contrast agents. Controlled studies comparing barium verses water-soluble agents show no differences in completeness of bowel opacification (including the duodenum), homogeneity of agent, and motion artifact.[39–41]

Most published guides for body CT in children recommend Gastrografin for gut opacification mixed with various flavoring agents.[42–44] Kaufman has advocated Hypaque in 1.5% concentration as producing decreased streak artifact, especially in the colon.[45] Others recommend barium.[46–48] Recent conversations with several pediatric radiologists indicate that the barium suspensions are now generally used.

Sodium ioxitalamate (Telebrix 38) is a water-soluble iodine-containing agent that has been used for bowel opacification in pediatrics.[49,50] Its attributes include a taste acceptable to children. The drug is unavailable in the United States and can be obtained on only a limited basis in Canada.

Techniques of administration of oral contrast agents to infants and children are quite variable. Published recommendations of several pediatric radiologists are listed in Table 32-2 where some factors (age grouping, timing of doses) are modified to allow consistency in comparing individual recommendations. It seems universally agreed that oral contrast agents in large quantities should be started at least 20 minutes before the examination. If colon opacification is needed, some

Table 32-2 Recommendations for Oral Contrast Agents*

	Berger, Kuhn, & Brusehaber[42]	*Cremin & Mervis*[44]	*Kaufman*[45+]	*Siegel*[55]	*Stanley*[43]
Agent and concentration	3% Gastrografin (or 1% barium)	3% Gastrografin	Hypaque in 1.5% solution	Barium	3% Gastrografin
Time of administration					
Routine	20-30 min PTE‡	2/3 30-60 min PTE, 1/3 10-20 min PTE	20-30 min PTE for upper abdomen or 45 min PTE for full abdomen	Routine to include colon	Upper abdomen, colon—1 hr PTE Lower abdomen—2-3 hrs PTE
If colon required	Auxiliary enema	Auxiliary enema	1 dose 12 hr PTE 1 dose 45 min PTE	1 dose evening 1 dose hr PTE	3-4 hr PTE & repeat hr PTE
Patient age and amount of contrast					
Neonates–6 mo	60-120 cc		Per neonatologist		
6 mo–1 yr	180 cc	100 cc	120 cc	90-120 cc	50-60 cc
2 yr–3 yr	270 cc	200 cc	240 cc		
5 yr–6 yr	360 cc	300 cc	360 cc	240 cc	250 cc
10 yr and over	540 cc	500 cc	480 cc	480 cc	600 cc

*Age grouping and timing of doses recommended by individual authors are modified for this table to allow consistency in comparing recommendations.

+Personal communication.

‡Prior to examination.

contrast needs to be given several hours in advance, as well as just before the examination.

A supplemental enema may be given if colon opacification is unsatisfactory after administration of oral contrast material; some recommend it routinely.[42,44] The amounts recommended in Table 32-3 generally fill the rectum, sigmoid, and portion of descending colon.

Intravenous Agents

Intravenous contrast agents are used almost universally to augment body CT examinations in children as in adults. Advantages of their use include better (1) delineation of abdominal organs, (2) determination of vascular anatomy, and (3) accentuation of differences between tumor and host tissues. Allergy to the contrast agent is an obvious contraindication to its administration. If liver metastases are suspected, consideration should be given for examination before and after contrast, as certain metastases may become isodense with normal hepatic parenchyma after contrast administration.

In the choice of IV contrast agent for body CT, the same factors mentioned above regarding intravascular contrast agents are pertinent. At the time of this writing, appropriate indications for the new low osmolar agents in body CT are preliminary. There is clear evidence that the vascular phase of the new agents is more intense, particularly immediately after injection.[51] This intensity may be related to reduced intravascular dilution with these drugs. The distribution volume and disposition phase of the low osmolar agents may also play a role.[52] Early

Table 32-3 Recommendations for Auxiliary Enemas for Colon Opacification*

	Berger, Kuhn, & Brusehaber[42]	*Kaufman*[45+]	*Cremin & Mervis*[44]
Agents and concentration	3% Gastrografin	Hypaque in 1.5% solution	2% Gastrografin
Age of patient and amount of contrast			
Neonate–6 mo	60 cc		
6 mo–1 yr	90 cc	100 cc–	50 cc
2 yr–3 yr	120 cc	200 cc	100 cc
5 yr–6 yr	180 cc	(age not	150 cc
10 yr and older	240 cc	specified)	200 cc

*Age grouping of individual authors modified to produce consistency in comparing recommendations.
+Personal communication.

studies on the clinical importance of these differences show that they have no appreciable effects.[53,54] We are unable to give specific recommendations regarding these drugs for body CT as there is currently insufficient information on their indications and cost effectiveness in children. Anecdotally we have noticed that sedated children are less likely to awaken if the nonionics are the administered contrast agents.

There is general agreement among pediatric radiologists as to types, dose, and route of administration of IV contrast material to infants and children for body CT.[42–45,47,51,55] Most use the same traditional agents available for urography. Although some[44] routinely administer 3 cc/kg (up to a maximum of 100 cc), more recommend 2 cc/kg (100 cc maximum).[43,45] The author's recommendation is to give 2 cc/kg in routine studies, but an additional 1 cc/kg can be given, usually as a second dose, in special situations. In the infant less than 1 year of age or in the presence of congenital heart disease, renal failure, or hypotension, the low osmolar agents seem a prudent choice. If a bolus technique is used to delineate vascular anatomy, one-half of the dose should be administered rapidly, scanning begun, and the other half given more slowly during the examination. If the lower abdomen is to be scanned, the bolus might be 40% of the total dose.[45]

If there is a need to study the bladder wall during CT of the pelvis, contrast material in a dose of 0.5 cc/kg can be given 20–30 minutes before the scan.[55] This eliminates a need for catheterization of the bladder. If the patient has a catheter already in place, saline can be instilled. Lesions along the bladder wall are often better seen because contrast may obscure a small tumor or inflammatory mass.

For craniocerebral CT, recommendations of Fitz are IV administration of contrast material at a dose of 3 cc/kg (120 ml maximum) given as a bolus, with immediate scanning.[46] A common practice among pediatric radiologists is to use a similar technique but to limit the contrast used to 2 cc/kg.

ALIMENTARY TRACT

General

In many studies of the alimentary tract in the pediatric age group the contrast agent of choice is identical to that of adults, but special modifications or techniques may be necessary. For instance, routine study of the small bowel in children is accomplished with barium preparations mentioned in other chapters. However, there are several instances where the technique of administration, preparation of the contrast agent, or substitution of more effective or less toxic agents is appropriate. Pediatric modifications of agents and techniques used in adults are mentioned here.

Certainly the safest contrast agent to use in the gut is room air. Circumstances in which this is the preferred agent include the neonate with suspected duodenal or pyloric obstruction. As there is no danger of aspiration, the only risk is perforation of the stomach or duodenum by rapid injection of too much air. A safe amount for most neonates is 30–50 cc given by nasogastric tube; if additional air is needed fluoroscopic monitoring is advisable. In the neonate with a gasless abdomen[56] or suspected ascites,[57] air outlining the location and size of the intestines is of value. Air insufflated through the rectum can be used to identify the colon in patients with questionable intestinal obstruction or ascites.[58,59]

Barium is used for most routine studies of the GI tract in infants and children. Its superior mucosal definition and resistance to dilution make barium the preferred agent. In general, the same multiple agents available for adults are used in children. Flocculation of most commercial barium products is a real problem in the stomach and small bowel of infants. Astley[60] attributes its cause to the increased mucus of the infant's stomach; Sellink[61] mentions the relative increase in lactic acid in small bowel. Sellink notes that as a barium suspension disintegrates its viscosity increases, making delineation of mucosal surfaces difficult. He proposes enteroclysis to overcome this problem because it delivers a large quantity of barium suspension to the small bowel over a short period of time.

In the esophagus and stomach any suitable barium suspension may be used in the child, but one must accept that mucosal coating will be less than that achievable in adults. In the small bowel Sellink's[61] and Ratcliffe's[62] recommendation for enteroclysis will produce better delineation of anatomy and abnormality than oral barium. Both indicate that rapid delivery of contrast into the distal duodenum or proximal jejunum with early filming of the bowel before flocculation occurs results in optimal radiographs of the pediatric small bowel. Ratcliffe recommends 50% w/v barium sulfate infused by gravity. He uses 75–100 cc in infants and small children, increasing the volume to 250 cc in an adolescent. Sellink uses a barium suspension with a specific gravity of 1.15, mentioning that a higher level is too dense and a lower one is apt to flocculate. He recommends a flow rate not to exceed 40 cc/min in infants; the rate should be reduced if there is too much pylorogastric reflux or in the presence of ileus. From 1–5 years of age the flow rate is increased to 60 cc/min and in older children to 80 cc/min. The maximum dose for an infant would be 200 cc, increasing to 400 cc at 1–2 years of age.

In European countries, enteroclysis is usually performed with prior sedation and drugs to stimulate peristalsis. North American radiologists are less enthusiastic about the use of sedatives and other drugs.

Double-contrast examination of the small bowel utilizing barium and methylcellulose has been used in children with results similar to those of adults.[63] We have no personal experience with this technique. The colon examination in children is most often performed by the single-column barium technique, rather than double-contrast methods, unless hematochezia or perhaps inflammatory bowel disease is being investigated. This is because the reasons for examination are more diverse, and the detail obtained with a double-contrast examination is not needed in most patients.

In infants in whom Hirschsprung's disease is suspected, there is a danger of fluid overload from absorption of water from the colon during a barium study. Addition of table salt, 1 tablespoon per gallon of barium suspension, results in a sodium concentration of approximately 70 mEq/L and obviates this threat.[64]

Barium is often retained for many days within the appendix after contrast studies. A potential complication mentioned in the past is barium retention in the appendix, with formation of a barolith and subsequent appendicitis.[65] The rarity of appendicitis after barium examinations makes the contrast an unlikely cause of this condition. Follow-up examinations to evaluate retention of appendiceal barium and prophylactic appendectomies for retained barium are therefore not justified.

In certain instances when there is potential gut perforation, barium is not an ideal contrast material. In the lungs its effects are primarily mechanical, but death can occur with massive aspiration.[66,67] Gastrografin and similar water-soluble agents are not unduly harmful when spilled into the pleura or peritoneum, but they are very hypertonic and toxic to the lungs. In recent years the newer low osmolar agents have been recommended as substitutes when barium is inappropriate.[58,66,68–70] They have the advantages of minimal pulmonary toxicity and low osmolality. Metrizamide (and probably most similar agents) is minimally absorbed from the normal gut, making prolonged visualization possible.[71] There is little effect on osmolality or blood volume after administration.[72] The low osmolar agents are safe and are the contrast agents of choice in GI examinations when barium is contraindicated.[66,68,70]

Inspissation of barium suspensions within the colon occurs infrequently with the current commercial mixtures available. It does remain a risk in the small and large bowel of the child with cystic fibrosis, where barium is a potential initiator of meconium ileus equivalent.[58,73] In these infants the use of ionic water-soluble agents or even the low osmolar agents should be considered, particularly if large quantities must be administered. Similarly, one should consider the use of these agents in examinations for prolonged ileus or to study a blind pouch of bowel.

Contrast study of the proximal pouch in the newborn infant with esophageal atresia is often needed preoperatively. For the same reasons cited above, we believe that classical water-soluble agents are contraindicated. Personal experience has been that, because aspiration is frequent, Dionosil can be harmful in such situations. The newborn's trachea is unable to accommodate aspiration of a plug of Dionosil, and partial airway obstruction ensues when this occurs. A barium suspension is the agent of choice to outline the blind proximal pouch of esophageal atresia. Of even greater importance are the amount of barium used and the route of administration. A catheter should be passed into the esophagus under fluoroscopic control and a minute amount (1 to 2 cc) of barium instilled. After confirmatory spot films are obtained, barium is removed from the pouch by suction. Oral administration of barium by the infant sucking a bottle and administration of a large quantity of contrast material by catheter are to be condemned, because these methods invariably result in aspiration.

Contrast study of the vomiting infant is frequently indicated if partial obstruction of the esophagus, gastroesophageal reflux, hiatus hernia, or gastric outlet obstruction is suspected. Hypertrophic pyloric stenosis, the most common indication for abdominal surgery in infancy, may require contrast investigation, particularly if physical examination or ultrasound investigation is negative or indeterminate. Although there are reports of using the newer low osmolar agents in these situations,[58] barium suspensions, used properly, are appropriate.[41] The most important part of the technique from a safety standpoint is to avoid significant aspiration. Thus, huge amounts of barium should not be used. After examination for gastric outlet obstruction, removal of most or all of the ingested barium by suction through a nasogastric tube is advisable.

Metrizamide[68] and other low osmolar agents are the contrast agents of choice in antegrade examination of the neonatal esophagus,[69] stomach, and bowel[68,73] when there is potential bowel perforation. They are relatively harmless to the peritoneal cavity and quite effective in demonstrating a perforation, as well as the specific changes of inflammation, edema, and obstruction in the gut.

Several recent series demonstrate the value of contrast enemas in the diagnosis of necrotizing enterocolitis in the neonate.[74–76] Because water balance in these small infants is critical, Gastrografin is probably not a useful agent, as the iodine concentration at isosmolar levels is low. Barium is a reasonable alternative as the above series demonstrate. Considering the substantial risk of perforation in these patients and the small amount of contrast needed, the low osmolar agents might be considered.

For many years barium enemas have been used with considerable efficacy and relative safety in the diagnosis and therapeutic reduction of ileocolic intussusception.[77] The barium enema procedure, in addition to its diagnostic value, has the ability to reduce intussusception in many instances, obviating the need for surgery. For whatever reasons, perforation of the colon during diagnostic and/or therapeutic barium enema in this condition is a real risk.[78–80] Perforation is generally associated with an area of necrotic bowel and may be proximal or distal to the intussusceptum. There may be multiple perforations. Those infants at risk for perforation tend to be younger, more acutely ill, and more often have associated bowel obstruction than other children with intussusception. If a contrast enema, diagnostic or therapeutic, is to be performed in such a patient, we recommend use of Gastrografin or another water-soluble agent, diluted to a level acceptable to the radiologist for fluoroscopic observation (we customarily dilute 2 parts water to 1 part Gastrografin). Others recommend the use of barium, citing its superior contrast and radiographic density and the relatively low risk of barium peritonitis with current methods of treatment.[78,81] In the infant who is not severely ill and has no bowel obstruction, most pediatric radiologists continue to perform diagnostic and therapeutic enemas for intussusception with barium. For many years rectal air insufflation with or without fluoroscopic control has been used in Japan, China, and Argentina as a therapeutic measure for intussusception. There is recent enthusiasm in North America and Australia for the use of air or other gas to reduce intussusception. Preliminary reports are encouraging.

Therapeutic Uses

The therapeutic use of water-soluble contrast agents in meconium ileus was first reported by Noblett in 1969,[82] and numerous subsequent reports attest to its efficacy. Meconium ileus occurs principally in neonates with cystic fibrosis and is characterized by distal small bowel obstruction in the first days of life. Intestinal obstruction in uncomplicated meconium ileus results from abnormal, sticky meconium that fills the distal ileum. Noblett noted that after a Gastrografin enema the infant might pass the meconium per rectum, alleviating intestinal obstruction and the necessity for surgery. Postulated mechanisms for successful removal of the meconium included effect of the surface-wetting agent in Gastrografin (Tween-80), osmotic diarrhea induced by the hypertonic contrast material, and mechanical effects of the enema. Noblett reported that relief of the obstruction required the retrograde reflux of Gastrografin into the dilated ileum containing abnormal meconium. Use of this technique or a modification thereof has produced a dramatic reduction in mortality of this disease.[83–88] At the same time some hazards have become apparent, and there is uncertainty regarding the mechanisms of effect of the enema.

Probably the greatest problem is recognition of the infant with complicated meconium ileus, i.e., accompanied by intestinal atresia, volvulus, perforation, or meconium peritonitis. Complications of this nature occur in about one-half of affected infants.[89] Free intraperitoneal air or fluid may accompany complicated meconium ileus. A failed therapeutic enema may be an additional indicator.

Intense dehydration accompanies Gastrografin enema in the neonate. Mechanisms are two-fold: (1) hypertonic contrast material in the gut creates a "third space" with influx of fluid into the colon and ileum, and (2) absorption of hypertonic Gastrografin into the vascular space produces an osmotic diuresis.[83] The result is a substantial reduction of cardiac output and blood volume. Effects are exaggerated when the enema is administered under anesthesia.[90] Close cooperation among the radiologist, neonatologist, and surgeon is necessary to assure maintenance of adequate hydration during the procedure.

Perforation of the colon may occur in uncomplicated meconium ileus spontaneously or as a result of enemas.[89] There is some evidence that Gastrografin[84] or other hypertonic contrast agents[85] may cause perforation, probably by the irritant effects of contrast material with long application time to abnormal bowel mucosa. The wetting agent, Tween-80, produces mucosal damage in animals only if the colon is dilated.[91]

Recommendations on the appropriate agent for therapeutic enema in meconium ileus must be tentative, as all the truly important factors in the procedure are unknown. Although it was assumed that reflux of contrast agent into dilated, meconium-filled ileum was necessary, radiologists have reported successful procedures in which contrast failed to enter the ileum.[88] Wood and Katzberg[86] note positive effects using Tween-80 in a 1–2% solution mixed with contrast material and water to produce an isotonic medium. Several investigators report success using less hypertonic contrast, such as Hypaque 25%–50%.[83–85] There is anecdotal evidence that the new low osmolar agents[86] or even a barium suspension[83,89] may suffice.

Our current thinking is that the need to produce an osmotic diarrhea with hypertonic agents refluxing contrast into the dilated ileum and the risk of the known irritant effects of contrast material on bowel mucosa are not fully defined. Nevertheless, the procedure is quite effective in relieving the intestinal obstruction of meconium ileus at a risk substantially less than that of surgery. Our feeling is that using Hypaque in a concentration of 25–40% has a known positive effect with minimal toxicity, given appropriate attention to hydration of the infant.

There are other conditions similar to meconium ileus in which water-soluble contrast agents may play a role. Meconium ileus equivalent occurs in patients with cystic fibrosis after the newborn period and is characterized by intestinal obstruction in an ileum filled with intraluminal putty-like material. Oral or rectal contrast agents are useful in relieving this condition.[92] In meconium plug syndrome and related disorders in premature infants, a plug of meconium in the colon is associated with bowel obstruction, and contrast enemas (even isotonic) may dislodge the plug.[93] Other uses include aiding liquefication of stool in severely obstipated children and even dislodgement of intestinal worms in subacute ascariasis.[94]

CYSTOURETHROGRAPHY

Cystography and micturating cystourethrography (CUO) maintain an important role in evaluation of the pediatric urinary tract. The apparent importance of the vesicoureteral reflux in urinary tract infection, assessment of the bladder and urethra in urinary obstruction, and evaluation of structural abnormalities are the usual reasons for performing this examination. The study is seldom indicated in the child with chronic abdominal pain, enuresis, failure to thrive, or similar vague clinical situations. Too, other modalities, such as nuclear medicine cystograms (which give less radiation and eliminate drug toxicity) or ultrasound,[95] which is even less invasive, can be substituted for the classical cystourethrogram in some instances.

Because the quality of the radiographic examination in CUO is dependent principally upon iodine concentration and thus radiopacity of the contrast agent, the choice of the appropriate agent should be based upon the degree of radiopacity necessary and its potential for adverse effects. In years past, barium preparations were sometimes utilized. Sodium iodide is rarely used in contemporary medical practice;[96] it is quite toxic. The newer lower osmolar contrast agents may prove valuable in the future. In North America most CUOs are performed using the traditional agents diatrizoate (Hypaque, Renografin), acetrizoate (Cystokon), or iothalamate (Cysto-Conray) in varying concentrations. Most studies of these agents have shown inconsequential differences in any of the agents' efficacy or toxicity. Nogrady and Dunbar object to the use of Cystokon because of the possibility of systemic absorption from the urinary tract.[97]

Controlled clinical studies on the most appropriate concentration of contrast agent are not available, as they would require more invasive evaluations, such as cystoscopy. It is known that inflammation of the bladder in animals is directly related to increasing concentration and amount of contrast material and that there are no appreciable differences among the various traditional water-soluble agents.[98] There is clinical evidence that cystographic contrast agents produce inflammation in the human bladder.[98–100] It is therefore reasonable to perform CUO with the lowest iodine concentration that still provides diagnostic information.

Although available packaged agents are in the range of 25%-30% concentration, CUOs of diagnostic quality can be performed using a concentration of 5%-10%.

Local or systemic allergic reactions can occur with introduction of contrast agents into the urinary tract because the contrast material is absorbed systemically from the bladder.[101] The subsequent renal excretion of absorbed contrast material should not be mistaken for ureteral reflux. McAlister et al.[100] have reported one instance of anuria following CUO; its etiology was uncertain.

Other complications of CUO relate to techniques in performance. There is both experimental[98] and clinical[100] evidence that gravity-drip administration of intravesical contrast material is better tolerated and safer than manual injection. An effective and safe height for introducing the contrast agent by gravity seems to be 60 cm. Infection of the urinary tract occurs in up to 6% of examinations.[102] Secondary systemic sepsis is a particular threat in the child with a dilated urinary tract.

How much contrast should be administered during CUO? The amount must vary according to the age of the patient, status of the urinary tract, and information desired from the examination. In the anuric neonate with suspected Potter's syndrome, 1 ml may suffice. In an older child with a flaccid neurogenic bladder several hundred milliliters are required. One must guard against a tendency to fill maximally a dilated urinary tract. For instance, in the presence of massive hydronephrosis and megacystis, only enough contrast material to assess the degree of dilation and anatomic abnormalities present should be given. In patients with a normal bladder in whom a voiding study of the urethra is needed, one must administer sufficient volume to assure prompt micturition. The older child and adult can voice that status. In the infant, clinical judgment of the physician is often required. Looking for spontaneous Babinski movement of the toes is helpful.

In summary, the authors' recommendations for CUO in the pediatric age group are that all of the currently widely used agents are satisfactory, with few differences among them. These include sodium acetrizoate (Cystokon), sodium diatrizoate (Hypaque), and meglumine/sodium diatrizoate (Renografin). The cost effectiveness of the newer low osmolar agents for cystography has yet to be determined. Of more importance than the specific agent is its concentration; a 10% solution provides images of diagnostic quality with low toxicity. The amount of the contrast agent used must be individualized. The route of administration should be by gravity-drip infusion.

BILIARY TRACT

There are now multiple methods to image the gallbladder and biliary tree—particularly, ultrasound, ERCP, and percutaneous transhepatic cholangiography—that are of greater value than oral cholecystography and IV cholangiography.[103] We have not performed an oral cholecystogram or classic IV cholangiogram in the pediatric section of this radiology department in the last 8 years. If by chance there were some reason to do so, the recommendations given in the previous edition of this book could be used.[104] For oral cholecystography Taybi[105] recommends (1) less than 13 kg body weight, 0.15 g/kg; (2) 13 to 25 kg, 2 g; and (3) over 25 kg, 3 g. For the rare IV cholangiogram, the authors use meglumine iodipamide (Cholografin meglumine) in a dose of 0.5 cc/kg up to a maximum dose of 20 cc.

In many European countries sodium ioglycamate (Biligram) is used for IV cholangiography and cholecystography, and it has some role in pediatric investigation.[106] The drug is unavailable in North America.

On occasion it is of value to opacify the bile ducts during CT examination. A preliminary report indicates that this is feasible with IV cholangiographic agents.[107] We have attempted this using IV Cholografin in infants, with variable success.

BRONCHOGRAPHY

Bronchography is seldom indicated or performed in North America. The reasons are multiple: (1) the procedure is uncomfortable and has a high morbidity; (2) bronchiectasis is no longer a major problem; and (3) alternate methods of diagnosis, particularly endoscopy and CT, are of greater value. Occasionally, bronchograms are necessary to outline congenital anomalies,[108] to define further abnormalities found at endoscopy, and to document fistulas. More often, contrast material arrives in the pediatric trachea and bronchi from aspiration or fistulas from the alimentary tract. The same comments on drugs for bronchography are therefore applicable to those GI examinations in which there is substantial risk that the contrast agent will enter the respiratory tract.

As bronchography requires close patient cooperation, most pediatric bronchograms are necessarily performed under general anesthesia.[109–111] This creates a problem in obtaining satisfactory bronchograms and increases the hazards of the procedure. Contrast material is best distributed to the bronchial tree by the suction effect of inspiration. With general anesthesia the positive pressure blow of the anesthetist must be utilized to some extent in respiration; therefore, distribution of the contrast agent may be compromised.[109] Segmental or lobar collapse occurs in up to 45% of pediatric bronchograms under general anesthesia;[112] both the anesthetic gas and a high oxygen concentration accelerate absorption of air distal to the partial bronchial obstruction induced by bronchographic material.[112,113] The hazard of pediatric bronchography is difficult to quantitate, but in one large series of adults and children, 18 fatalities occurred in children or others with compromised respiratory reserve.[111]

Problems of bronchographic agents are of two types: mechanical airway obstruction and effect on bronchial and alveolar tissues. The former is a particular problem in infants with their smaller airways and probably contributes to the relatively greater risk of bronchography in this age group. Oil-based agents, such as Dionosil, cause little tissue reaction but do produce considerable atelectasis and reduction in ventilation.[112]

Most barium preparations in use contain multiple additives, many of which are potentially toxic to respiratory tissues. Tracheally instilled commercial barium mixtures can cause serious problems, with an inflammatory response lasting for months.[114] Fatalities have been noted in infants aspirating barium during the course of GI examinations.[67]

Entry of the classic water-soluble iodine preparations (diatrizoate, iothalamate) into the respiratory system produces multiple deleterious effects. Clinically, hypoxia and respiratory distress are found.[115] Pathologically there can be intense alveolar edema and a considerable inflammatory response that can last for days.[116,117] The use of these agents is mentioned only to be condemned.

There has been recent interest in the use of the newer low osmolality contrast agents, particularly metrizamide, when contrast study of the tracheobronchial tree is indicated[108] or when aspiration is likely.[70,115] Alford et al.[118] demonstrated intense inflammation following intratracheal instillation of large quantities of metrizamide into experimental animals. Other experimental studies comparing metrizamide to more traditional bronchographic agents indicate that it causes considerably less reaction than the older drugs.[117] There is a report of persistence of metrizamide in the human lung for 5 months.[119] Clinical studies of metrizamide[108,115] and ioxaglate[70] indicate these are probably now the safest agents for use in the respiratory tract. However, because of its low viscosity and surface coating characteristics, metrizamide produces less than ideal delineation of the tracheal and bronchial mucosa. It rapidly "alveolizes" so immediate spot film recording of anatomy is necessary.[108]

In summary, bronchography is seldom indicated in infants and children. In those occasional instances where it must be performed, the authors recommend metrizamide as the least toxic drug, especially in the infant. It should also be used in GI investigations with a high potential for aspiration. In the older child or adolescent, an oily-contrast material, such as oily Dionosil, is recommended in those situations where delineation of bronchial anatomy is needed.

The amount of Dionosil used in pediatric bronchography must be reduced considerably as compared with that for adults. We follow the recommendation of Bell[109]: 1.25 cc per year of age for the right lung and 1.0 cc per year of age for the left lung. The pathologic side is generally examined first; most bronchographers do both sides during the same anesthetic episode.[110] After one side is done and the appropriate roentgenograms obtained, most of the material is removed by suction and/or coughing and the opposite lung is then studied. Obviously, both lungs should not be done consecutively in this fashion if there is a possibility of respiratory embarrassment.

REFERENCES

1. Junck L, Enzmann DR, DeArmond SJ, et al: Prolonged brain retention of contrast agent in neonatal herpes simplex encephalitis. *Radiology* 1981;140:123–126.

2. Junck L, Marshall WH: Fatal brain edema after contrast-agent overdose. *AJNR* 1986; 7:522–525.

3. Naheedy MH, Sakkubai N, Griffin AJ: Prolonged retention of contrast medium in an hypoxic neonatal brain. *Surg Neurol* 1983;20:369–372.

4. Kassner EG, Elguezabal A, Pochaczevsky R: Death during intravenous urography: overdosage syndrome in young infants. *NY State J Med* 1973;73:1958–1966.

5. Wood BP, Smith WL: Pulmonary edema in infants following injection of contrast media for urography. *Radiology* 1981;139:377–379.

6. McAlister WH, Siegel MJ, Shackelford GD: Pulmonary oedema following intravenous urography in a neonate. *Br J Radiol* 1979;52:410–411.

7. Berdon WE: Pulmonary edema in infants who receive contrast material. *Radiology* 1981;139:507.

8. Kashani IA, Higgins SS, Griswold W, et al: Renal function in children after large dose contrast medium angiocardiography. *Japan Heart J* 1984;26:451–456.

9. Dawson P: Contrast agent nephrotoxicity. An appraisal. *Br J Radiol* 1985;58:121–124.

10. Avner ED, Ellis D, Jaffe R, et al: Neonatal radiocontrast nephropathy simulating infantile polycystic kidney disease. *J Pediatr* 1982;100:85–87.

11. Kravath RE, Aharon AS, Abal G, et al: Clinically significant physiologic changes from rapidly administered hypertonic solutions: acute osmol poisoning. *Pediatrics* 1970;46:267–275.

12. Committee on Radiology, American Academy of Pediatrics: Water-soluble contrast material. *Pediatrics* 1978;62:114–116.

13. Spataro RF: Newer contrast agents for urography. *Radiol Clin N Am* 1984;22:365–380.

14. Gooding CA, Berdon WE, Brodeur AE, et al: Adverse reactions to intravenous pyelography in children. *Am J Roentgenol* 1975;123:802.

15. Shehadi WH: Death following intravascular administration of contrast media. *Acta Radiol Diagn* 1985;26:457–461.

16. Stanger P, Heymann MA, Tarnoff H, et al: Complications of cardiac catheterization of neonates, infants, and children: a three-year study. *Circulation* 1974;50:595–608.

17. Dawson P, Grainger RG, Pitfield J: The new low-osmolar contrast media: a simple guide. *Clin Radiol* 1983;34:221–226.

18. Wolf GL: Safer, more expensive iodinated contrast agents: how do we decide? *Radiology* 1986;159:557–558.

19. Dawson P: Chemotoxicity of contrast media and clinical adverse effects: a review. *Invest Radiol* 1985;20:S84–S91.

20. Franken EA, Smith WL, Smith JA: Excretory urography in the neonate. *Perinatal/Neonatal* 1979;3:15–16.

21. Cremin BJ, Rhodes AH: Contrast media in paediatric radiology. *Br J Radiol* 1983;56:779.

22. Cohen MD: Intravenous urography in neonates and infants. What dose of contrast should be used? *Br J Radiol* 1983;52:942–944.

23. Diament MJ, Kangarloo H: Dosage schedule for pediatric urography based on body surface area. *Am J Roentgenol* 1983;140:815–816.

24. Meradji M, Gershom EB: Excretory urography with four different contrast media: radiological and biochemical trials in 295 young infants. *Ann Radiol* 1984;27:199–206.

25. Stake G, Smevik B: Iohexol and metrizamide for urography in infants and children. *Invest Radiol* 1985;20:S115–S116.

26. Jorulf H: Iohexol compared with diatrizoate in pediatric urography. *Acta Radiol* 1983;S366:42–45.

27. Robey G, Reilly BJ, Carusi PA, et al: Pediatric urography: comparison of metrizamide and methylglucamine diatrizoate. *Radiology* 1984;150:61–63.

28. Afshani E, Berger PE: Gastrointestinal tract angiography in infants and children. *J Pediatr Gastroenterol Nutr* 1986;5:173–186.

29. Moore AV, Kirks DR, Mills SR, et al: Pediatric abdominal angiography: panacea or passe? *Am J Roentgenol* 1982;138:433–443.

30. Capitanio MA, Faerber EN, Gainey MA, et al: Digital subtraction angiography and its application in children. *Ped Clin N Am* 1985;32:1449–1460.

31. Cumberland DC: Low-osmolality contrast media in cardiac radiology. *Invest Radiol* 1984;19:S301–S305.

32. DiSessa TG, Zednikova M, Hiraishi S, et al: The cardiovascular effects of metrizamide in infants. *Radiology* 1983;148:687–691.

33. Sagy M, Aladjem M, Shem-Tov A, et al: The renal effects of radiocontrast administration during cardioangiography in two different groups with congenital heart disease. *Eur J Pediatr* 1984;141:126–129.

34. Carlsson EC, Rudolph A, Stanger P, et al: Pediatric angiocardiography with iohexol. *Invest Radiol* 1985;20:S75–S78.

35. Anthony CL, Tonkin ILD, Marin-Garcia J, et al: A double-blind randomized clinical study of the safety, tolerability and efficacy of Hexabrix in pediatric angiocardiography. *Invest Radiol* 1984;19:S335–S343.

36. Kunnen M, Van Egmond H, Verhaaren H, et al: Cardioangiography in children with iohexol, metrizoate and ioxaglate. *Ann Radiol* 1985;28:315–321.

37. Giammona ST, Lurie PR, Segar WE: Hypertonicity following selective angiocardiography. *Circulation* 1963;28:1096–1100.

38. Fox KM, Patel RG, Bonvicini M, et al: Safe amounts of contrast medium for angiocardiography in neonates. *Eur J Cardiol* 1977;5:373–380.

39. Chambers SE, Best JJK: A comparison of dilute barium and dilute water-soluble contrast in opacification of the bowel for abdominal computed tomography. *Clin Radiol* 1984;35:463–464.

40. Megibow AJ, Bosniak MA: Dilute barium as a contrast agent for abdominal CT. *Am J Roentgenol* 1980;134:1273–1274.

41. Garrett PR, Meshkov SL, Perlmutter GS: Oral contrast agents in CT of the abdomen. *Radiology* 1984;153:545–546.

42. Berger PE, Kuhn JP, Brusehaber J: Techniques for computed tomography in infants and children. *Radiol Clin N Am* 1981;19:399–408.

43. Stanley P: Computed tomographic evaluation of the retroperitoneum in infants and children. *J Comput Tomogr* 1983;7:63–75.

44. Cremin BJ, Mervis B: Paediatric abdominal computed tomography: the technique and use in neuroblastomas and pelvic masses. *Br J Radiol* 1983;56:291–298.

45. Kaufman RA: Liver-spleen computed tomography. A method tailored for infants and children. *J Comput Tomogr* 1983;7:45–57.

46. Fitz CR: Craniocerebral computed tomography. *Pediatr Radiol* 1983;13:148–149.

47. Berger PE: CT of the abdomen and retroperitoneum. *Pediatr Radiol* 1983;13:153–154.

48. Taylor S: CT scanning in paediatric oncology. *Radiography* 1985;51:81–83.

49. Ruijs SHF: A simple procedure for patient preparation in abdominal CT. *Am J Roentgenol* 1979;133:551–552.

50. Azouz EM, Hassell P, Nogrady MB, et al: Bowel opacification using "Telebrix 38" for CT scanning. *J Can Assoc Radiol* 1982;33:233–235.

51. Spataro RF, Fischer HW, Kormano M: Clinical comparison of Hexabrix, Iopamidol, and Urografin-60 in whole body computed tomography. *Invest Radiol* 1984;S372–S375.

52. Jensen LI, Dean PB, Nyman U, et al: Contrast media for CT: an analysis of the early pharmacokinetics. *Invest Radiol* 1985;20:867–870.

53. McClennan BL, Lee JKT, DiSantis DJ, et al: A double-blind clinical study comparing the safety and efficacy of Hexabrix and Renografin-76 in contrast enhanced computed body tomography. *Invest Radiol* 1984;S378–S384.

54. Berland LL: Double blind comparison of Hexabrix and Renografin-76 in computed tomography. *Invest Radiol* 1984;S376–S377.

55. Siegel MJ: Computed tomography of the pelvis. *Pediatr Radiol* 1983;13:154–155.

56. Tucker AS: Air contrast vs. metrizamide in neonatal gasless abdomen. *Am J Roentgenol* 1984;153:430–431.

57. Johnson JF, Phillips EL: Air as a gastrointestinal contrast agent for identifying pseudoascites in the newborn. *Am J Roentgenol* 1981;1247–1248.

58. Ratcliffe JF: The use of low osmolality water soluble (LOWS) contrast media in the pediatric gastrointestinal tract. A report of 115 examinations. Pediatr Radiol 1986;16:47–52.

59. Jorulf H: Ascites and other intraperitoneal fluid, in Franken EA (ed): *Gastrointestinal Imaging in Pediatrics*, ed 2. Philadelphia, Harper & Row, 1982, pp 410–424.

60. Astley R, French JM: The small intestine pattern in normal children and in coeliac disease. *Br J Radiol* 1951;24:321–330.

61. Sellink JL: Technique for the x-ray examination of the small intestine in babies. *J Belge Radiol* 1980;63:5–12.

62. Ratcliffe JR: The small bowel enema in children: a description of a technique. *Clin Radiol* 1983;34:287–289.

63. Hormann D: Doppelkontrastuntersuchung des kindlichen Dunndarms mit Bariumsulfat und Methylzellulose. *Radiol Diagn* (Berl) 1983;24:499–506.

64. Franken EA: *Gastrointestinal Radiology in Pediatrics*. Hagerstown, MD, Harper & Row, 1975.

65. Merten DF, Lebowitz ME: Acute appendicitis in a child associated with prolonged appendiceal retention of barium (Barium appendicitis). *South Med J* 1978;71:81–82.

66. Meradji M: Radiological appraisal to the upper digestive tract in infants and young children. *J Belge Radiol* 1980;63:25–32.

67. McAlister WH, Siegel MJ: Fatal aspirations in infancy during gastrointestinal series. *Pediatr Radiol* 1984;14:81–83.

68. Cohen MD, Weber TL, Grosfeld JL: Bowel perforation in the newborn and diagnosis with metrizamide. *Radiology* 1984;150:65–69.

69. Belt T, Cohen MD: Metrizamide evaluation of the esophagus in infants. *Am J Roentgenol* 1984;143:367–369.

70. Ratcliffe JF: The use of ioxaglate in the paediatric gastrointestinal tract: a report of 25 cases. *Clin Radiol* 1983;34:579–583.

71. Cohen MD: Prolonged visualization of the gastrointestinal tract with metrizamide. *Radiology* 1982;143:327–328.

72. Clarke E, Siegle RL: Effect of oral metrizamide on hematocrit and serum osmolality in the neonate. *Invest Radiol* 1984;19:599–600.

73. Fischer WW, Nice CM Jr: Barium impaction as a cause of small bowel obstruction in an infant with cystic fibrosis. *Pediatr Radiol* 1984;14:230–231.

74. Leonidas JC, Bhan I, Leape LL: Barium enema in suspected necrotizing enterocolitis: is it ever indicated? *Clin Radiol* 1980;31:587–590.

75. Negrette J, Ziervogel MA, Young DG, et al: Barium enema examination in neonates with suspected necrotizing enterocolitis. *Z Kinderchir* 1986;41:19–21.

76. Uken P, Smith W, Franken EA, et al: Use of the barium enema in the diagnosis of necrotizing enterocolitis. *Pediatr Radiology* 1988;18:24–27.

77. Ravitch M: *Intussusception in Infants and Children*. Springfield, IL, Charles C. Thomas, 1959.

78. Armstrong EA, Dunbar JS, Graviss ER, et al: Intussusception complicated by distal perforation of the colon. *Radiology* 1980;136:77–81.

79. Mahboubi S, Sherman NH, Ziegler MM: Barium peritonitis following attempted reduction of intussusception. *Clin Pediatr* 1984;23:36–38.

80. Humphry A, Ein SH, Mok PM: Perforation of the intussuscepted colon. *Am J Roentgenol* 1981;137:1135–1138.

81. Eklof O, Hald J, Thomasson B: Barium peritonitis. Experience of five pediatric cases. *Pediatr Radiol* 1983;13:5–9.

82. Noblett HR: Treatment of uncomplicated meconium ileus by Gastrografin enema: a preliminary report. *J Pediatr Surg* 1969;4:190–197.

83. Frech RS, McAlister WH, Tekenberg J, et al: Meconium ileus relieved by 40 percent water-soluble contrast enemas. *Radiology* 1970;94:341–342.

84. Leonidas JC, Burry F, Fellows RA, et al: Possible adverse effect of methylglucamine diatrizoate compound on the bowel of newborn infants in the meconium ileus. *Radiology* 1976;121:693–696.

85. Grantmyre EB, Butler GJ, Gillis D: Necrotizing enterocolitis after Renografin-76 treatment of meconium ileus. *Am J Roentgenol* 1981;136:990–991.

86. Wood BP, Katzberg RW: Tween-80/diatrizoate enemas in bowel obstruction. *Am J Roentgenol* 1978;130:747–750.

87. Nordshus T, Eriksson J, Langslet A: The use of iohexol in meconium obstruction in the newborn. *Fortschr Röntgenstr* 1986;144:358–359.

88. Martin DJ: Experiences with acute surgical conditions. *Radiol Clin N Am* 1975;13:297–329.

89. Donnison AB, Schwachman H, Gross RE: A review of 164 children with meconium ileus seen at the Children's Hospital Medical Center, Boston. *Pediatrics* 1966;37:833–850.

90. Maneksha FR, Betta J, Zawin M, et al: Intraoperative hypoxia and hypotension caused by Gastrografin-induced hypovolemia. *Anesthesiology* 1984;61:454–456.

91. Wood BP, Katzberg RW, Ryan DH, et al: Diatrizoate enemas: facts and fallacies of colonic toxicity. *Radiology* 1978;126:441–444.

92. McPartlin JF, Dickson JAS, Swain VAJ: The use of Gastrografin in the relief of residual and late bowel obstruction in cystic fibrosis. *Br J Surg* 1973;60:707–710.

93. Swischuk LE: Meconium plug syndrome: a cause of neonatal intestinal obstruction. *Am J Roentgenol* 1968;103:339–346.

94. Bar-Maor JA, de Carvalho JLAF, Chappell J: Gastrografin treatment of intestinal obstruction due to *Ascaris lumbricoides*. *J Pediatr Surg* 1984;19:174–176.

95. Schneider K, Jablonski C, Fendel H: Kontrastsonographie der harnwege im Kindesalter. *Ultraschall Med* 1986;7:30–33.

96. Talarico RD, Patel RC, Lavengood RW; Cystourethrography. Reversible and irreversible damage to bladder. *NY State J Med* 1979;79:2080–2082.

97. Nogrady MB, Dunbar JS: The technique of roentgen investigation of the urinary tract in infants and children, in HJ Kaufmann (ed) *Progress in Pediatric Radiology*, vol 3. Chicago, Yearbook Publishers, 1970, pp 3–50.

98. McAlister WH, Shackelford GD, Kissane J: The histologic effects of 30% Cystokon, Hypaque 25%, and Renografin-30 in the bladder. *Radiology* 1972;104:563–565.

99. Shopfner CE: Clinical evaluation of cystourethrographic contrast media. *Radiology* 1967;88:491–497.

100. McAlister WH, Cacciarelli A, Shackelford GD: Complications associated with cystography in children. *Radiology* 1974;111:167–172.

101. Currarino G, Weinberg A, Putnam R: Resorption of contrast material from the bladder during cystourethrography causing an excretory urogram. *Radiology* 1977;123:149–150.

102. Glynn B, Gordon IR: The risk of infection of the urinary tract as a result of micturating cystourethrography in children. *Ann Radiol* 1970;13:283–287.

103. Myllyla V, Paivansalo M, Pyhtinen J, et al: Sensitivity of ultrasonography in the demonstration of common bile duct stones and its ranking in comparison with intravenous cholangiography and endoscopic retrograde cholangiopancreatography. *Fortschr Röntgenstr* 1984;141:192–194.

104. Franken EA Jr: Pediatric use of contrast agents, in Miller RE, Skucas J(eds): *Radiographic Contrast Agents*. Baltimore, University Park Press, 1977, pp 495–505.

105. Taybi H: The biliary tract in children, in Margulis AR, Burhenne HJ (eds): *Alimentary Tract Roentgenology*, ed 2. St. Louis, CV Mosby, 1973, pp 1504–1530.

106. Fahr K, Oppermann HC, Willich E: Vergleichende Studie über die Anwendung eines neuen gallenkontrastmittels im kindesalter. *Mschr Kinderheilk* 1978;126:391–394.

107. Takahara T, et al: Evaluation of the intrahepatic bile duct after operation for congenital biliary atresia using CT scan with Biligrafin enhancement. *J Japan Soc Pediatr Surg* 1981;17:659–666.

108. Smith WL, Franken EA: Metrizamide as a contrast medium for visualization of the tracheobronchial tree: its drawbacks and possible advantages. *Pediatr Radiol* 1984;14:158–160.

109. Bell HE: Bronchography in children. *Arch Dis Child* 1967;42:55–56.

110. Brunner S: Bronchography during infancy and childhood. *Dis Chest* 1967;51:663–668.

111. Committee on Bronchoesophagology: Bronchography. *Dis Chest* 1967;51:663–668.

112. Robinson AE, Hall KD, Yokoyama KN, et al: Pediatric bronchography: the problem of segmental pulmonary loss of volume. I. A retrospective study of 165 pediatric bronchograms. *Invest Radiol* 1971;6:89–94.

113. Robinson AE, Hall KD, Yokoyama KN, et al: Pediatric bronchography: the problem of segmental pulmonary loss of volume. II. An experimental investigation of the mechanism and prevention of pulmonary collapse during bronchography under general anesthesia. *Invest Radiol* 1971;6:95–100.

114. Dunbar JS, Skinner GB, Wortzman G, et al: An investigation of effects of opaque media on the lungs with comparison of barium sulfate, Lipiodol and Dionosil. *Am J Roentgenol* 1959;82:902–907.

115. Wang J-Z, Kohda E: Congenital H-type tracheoesophageal fistula: 2 cases safely diagnosed with metrizamide. *Nippon Acta Radiol* 1985;45:1009–1016.

116. Rust RJ, Cohen MD, Ulbright TM: Clinical, radiographic and pathologic effects of Amipaque on rabbit lung. Comparison with barium and Gastrografin. *Acta Radiol* 1982;23:553–559.

117. McAlister WH, Askin FB: The effect of some contrast agents in the lung. An experimental study in the rat and dog. *Am J Roentgenol* 1983;140:245–251.

118. Alford BA, Dee P, Feldman P: The effects of Metrizamide on the lung. *Pediatr Radiol* 1983;13:1–4.

119. Oppermann HC, Willich E: Neonatal bowel contrast studies with metrizamide [Letter]. *J Pediatr Surg* 1984;19:331.

Appendix A
Intravascular Contrast Agents

Table 1 lists most of the ionic intravascular contrast agent formulations available as of 1986. They are listed in ascending iodine content. The osmolality figures were supplied by the various manufacturers, and small differences at similar iodine content should not be significant.

Table 2 lists the lower osmolality new intravascular contrast agent formulations. One of these is an ionic dimer (ioxaglate), and the other two are nonionic.

Source: Reprinted with permission from "Catalog of Intravascular Contrast Media" by HW Fischer in *Radiology* (1986;159:561–563), Copyright © 1986, Radiological Society of North America Inc.

Table A–1 Higher Osmolality Intravascular Contrast Agents

Generic Name	Percentage in Solution	Trade Name	Iodine, mg/ml	Osmolality, mOsm/kg	Viscosity 25°C	Viscosity 37°C	Size Availability
Iodamide meglumine	24	Renovue-Dip[*]	111	433	1.8	1.4	300 ml B
Iothalamate meglumine	30	Conray-30[†]	141	600	2.0	1.5	50 ml V 100 ml V 150 ml B 200 ml B 300 ml B
Diatrizoate meglumine	30	Hypaque 30%[‡]	141	633	1.92	1.43	100 ml B 300 ml B
Diatrizoate meglumine	30	Reno-M-Dip[*]	141	566	1.9	1.4	300 ml B
Diatrizoate meglumine	30	Urovist Meglumine DIU/CT[§]	141	640	1.9	1.4	300 ml B
Diatrizoate sodium	25	Hypaque 25%[‡]	150	696	1.55	1.17	300 ml B
Iothalamate meglumine	43	Conray-43[†]	202	1,000	3.0	2.0	50 ml V 100 ml V 150 ml B 200 ml B 250 ml B
Diatrizoate meglumine	60	Angiovist 282[§]	282	1,400	6.1	4.1	50 ml V 100 ml V 150 ml V

Iothalamate meglumine	60	Conray†	282	1,400	6.0	4.0	20 ml V 30 ml V 50 ml V 100 ml V 100 ml B 150 ml B 200 ml B
Diatrizoate meglumine	60	Hypaque 60%‡	282	1,415	6.16	4.10	20 ml V 30 ml V 50 ml V 100 ml V 100 ml in 200 ml B 150 ml in 200 ml B 200 ml in 200 ml B
Diatrizoate meglumine	60	Reno-M-60*	282	1,500	4.6	4.0	10 ml V 30 ml V 50 ml V 100 ml V 100 ml B 150 ml B
Diatrizoate sodium 8% meglumine 52%	60	Angiovist 292§	292	1,500	5.9	4.0	30 ml V 50 ml V 100 ml V
Diatrizoate sodium 8% meglumine 52%	60	MD-60†	292	1,539	6.2	5.0	30 ml V 50 ml V

Note.—B = bottle, V = vial.
* Trade name of E.R. Squibb, New Brunswick, N.J.
† Trade name of Mallinckrodt, St. Louis.
‡ Trade name of Winthrop-Breon Laboratories, New York.
§ Trade name of Berlex Laboratories, Cedar Knolls, N.J.

Table A–1 continued

Generic Name	*Percentage in Solution*	*Trade Name*	*Iodine, mg/ml*	*Osmolality, mOsm/kg*	*Viscosity* 25°C	*Viscosity* 37°C	*Size Availability*
Diatrizoate sodium 8% meglumine 52%	60	Renografin-60*	292	1,420	5.9	4.0	10 ml V 30 ml V 50 ml V 100 ml V 100 ml B
Diatrizoate sodium	50	Hypaque 50%‡	300	1,550	3.43	2.43	20 ml V 30 ml V 50 ml V 150 ml in 200 ml B 200 ml in 200 ml B
Diatrizoate sodium	50	MD-50†	300	1,522	3.2	2.4	30 ml V 50 ml V
Iodamide meglumine	65	Renovue-65*	300	1,558	8.7	5.7	50 ml V 300 ml B
Diatrizoate sodium	50	Urovist Sodium 300§	300	1,550	3.3	2.4	50 ml V
Diatrizoate sodium 29.1% meglumine 28.5%	57.6	Renovist II*	309	1,517	5.6	3.8	30 ml V 50 ml V
Iothalamate sodium	54.3	Conray-325†	325	1,700	4.0	3.0	30 ml V 50 ml V
Diatrizoate meglumine	76	Diatrizoate meglumine USP 76%*	358	1,980	15	9.2	50 ml V
Diatrizoate sodium 10% meglumine 66%	76	Angiovist 370§	370	2,100	13.8	8.4	50 ml V 100 ml V 150 ml B 200 ml B

Diatrizoate sodium 10% meglumine 66%	76	Hypaque-76‡	370	2,016	13.34	8.32	30 ml V 50 ml V 100 ml V 100 ml in 200 ml B 150 ml in 200 ml B 200 ml in 200 ml B
Diatrizoate sodium 10% meglumine 66%	76	MD-76†	370	2,140	14.7	9.1	50 ml V 100 ml B 150 ml B 200 ml B
Diatrizoate sodium 10% meglumine 66%	76	Renografin-76*	370	1,940	13.8	8.4	20 ml V 50 ml V 100 ml B 200 ml B
Diatrizoate sodium 35% meglumine 34.3%	69.3	Renovist*	371	1,900	9.1	5.7	50 ml V
Diatrizoate sodium 25% meglumine 50%	75	Hypaque-M, 75%‡	385	2,108	12.69	7.99	20 ml V 50 ml V
Iothalamate sodium	66.8	Conray-400†	400	2,300	7.0	4.5	25 ml V 50 ml V
Iothalamate sodium 26% meglumine 52%	78	Vascoray†	400	2,400	17.0	9.0	25 ml V 50 ml V 100 ml B 150 ml B 200 ml B
Diatrizoate sodium 30% meglumine 60%	90	Hypaque-M, 90%‡	462	2,938	34.7	19.50	50 ml V
Iothalamate sodium	80	Angio-Conray†	480	2,400	14.0	9.0	50 ml V

Table A–2 Lower Osmolality Intravascular Contrast Agents

Generic Name	*Percentage in Solution*	*Trade Name*	*Iodine, mg/ml*	*Osmolality, mOsm/kg*	*Viscosity*		*Size Availability*
					25°C	*37°C*	
Iohexol	38.8	Omnipaque*	180	411	2.81	2.05	10 ml V 20 ml V
Iopamidol	40.8	Isovue-M 200†	200	413	3.3‡	2.0	20 ml V 50 ml V
Iohexol	51.8	Omnipaque*	240	504	4.43	3.08	10 ml V 100 ml V 200 ml B
Iopamidol	61	Isovue 300†	300	616	8.8‡	4.7	20 ml V 50 ml V 100 ml B 200 ml B
Iohexol	64.7	Omnipaque*	300	709	10.35	6.77	10 ml V 30 ml V 50 ml V 100 ml V
Ioxaglate sodium 19.6% meglumine 39.3%	58.9	Hexabrix§	320	600	15.7	7.5	20 ml V 30 ml V 50 ml V 100 ml in 150 ml B 150 ml in 150 ml B 200 ml in 250 ml B
Iohexol	75.5	Omnipaque*	350	862	18.50	11.15	50 ml V 100 ml V 200 ml B
Iopamidol	76	Isovue 370†	370	796	20.9‡	9.4	20 ml V 50 ml V 100 ml B 200 ml B

Note.—B = bottle, V = vial.

* Trade name of Winthrop-Breon Laboratories, New York

† Trade name of E.R. Squibb, New Brunswick, N.J.

‡ Value determined at 20°C.

§ Trade name of Mallinckrodt, St. Louis.

Appendix B

Barium Sulfate: Basic Properties

Jovitas Skucas

Barium was recognized as a separate element in 1808 by Sir Humphrey Davy.[1] He named it after the Greek word for heavy (*barys*). Barium is a relatively common element that interacts readily to form a number of water-soluble and insoluble compounds. In nature, it is found only as a salt, with barium sulfate and barium carbonate being the most common compounds.

SIZE AND SHAPE

Barium sulfate is a white, crystalline powder with a molecular weight of 233.39. It has a specific gravity of 4.50 and is made up of rhombic crystals.[2] X-ray crystallographic studies show that a single crystal contains four molecules and forms a rhomboid with dimensions of 8.88, 5.45, and 7.15 Ångstroms.[3] Thus, even if these crystals are placed end to end along their greatest dimension, a single extremely fine ground or precipitated particle of barium sulfate, 0.1 μm in diameter, contains rough fragments or aggregates of large crystals or an irregular aggregate of more than 112 single small crystals. In no commercially available barium sulfate contrast agents are even the majority of particle sizes below 0.1 μm. The most widely used precipitated barium sulfate has an average particle size distribution of approximately 1 μm. Precipitated barium sulfate is available in *mean* particle sizes as small as 0.3 μm and larger than 12 μm. Barium sulfate particles between 0.3 and 0.8 μm are generally precipitated not as pure crystals; they are used as additives for various custom formulations, with most of the information being proprietary. The sizes between 3 and 12 μm are generally used in mixtures of "high density" products designed for double-contrast gastric studies. Larger particle sizes can be obtained by crushing mined barium sulfate nodules to the desired size.

One report of size distribution for four commercial barium sulfate preparations listed a range of 0.07–0.70 μm.[4] The smallest common size was 0.13 μm and the largest 0.36 μm. In another assay, the majority were above 0.55 μm in diameter, and about 35% were larger than 1.5 μm.[5] In 1977, the following particle sizes were published:[6] Topcontral, 0.2–1.6 μm; Barotrast, approximately 3.0 μm; and Micropaque, 0.1–1.5 μm. One barium product, designed for gastric coating, has barium particles up to 12 μm[7] or 18.8 μm[8] and even higher. This product is characterized by an extreme heterogeneity of the particle sizes,[7] although all barium contrast agents are nonuniform both in particle size and shape; even ''homogeneous'' preparations vary over a range of particle sizes.

Colloidal particles can form clear suspensions or scatter light and become opalescent. Barium sulfate preparations of the same concentration but different particle size have their maximum turbidity at particle sizes of 0.4–0.5 μm, a size where the particles are no longer colloidal;[9] barium sulfate suspensions in which the particles are larger than ''turbid'' size appear white and opaque to ordinary light. This appearance is due almost entirely to reflection, with these white barium sulfate suspensions having particles larger than the colloid range.

One variable in the stability of a suspension is particle size. However, the coating properties of a suspension are not improved if the particle size is reduced below 1 μm.[10] If ground too fine, the extremely small particle size of pigments can decrease the hiding power of paint. However, there is no one optimal size. Even in the paint industry where the problem is less complex, the optimal size of a particle depends upon the pigment concentration, film thickness, and type of application.[11] Diffraction studies have shown that the optimal particle size for the most favorable reflection of x-rays averages 1 μm. Radiographs in these studies become more blurred and fuzzy as the particle size is decreased.[12] Reduction of particle size below a certain value gives no advantage. Knoefel et al. found that barium sulfate particles of 0.3 μm did not visualize the gastric mucosa as well as did 0.5 μm particles.[13] Suspensions in which 38% of the particles were greater than 1.5 μm visualized the gastric mucosa better than 0.3 μm particle suspensions.

Particle size can be measured with electron microscopy. Because the barium sulfate particles are crystals, they should not deform during sample preparation.[4] If the particles are coated with additives, as most commercial barium sulfate preparations are, the coating may deform during the drying process. Aggregation into larger clumps may also be seen; this may be the result of sample preparation. A photocentrifuge is useful in measuring very fine particles below 0.3 μm. For more coarse sizes, sedimentation by gravity can be used. Because dilute suspensions are necessary to achieve free-fall conditions, the dilution of additives may introduce spurious findings. In the paint industry, x-ray contact microradiography has been used to resolve 0.5 μm and larger particle sizes.[14] Laser light scattering can measure particle sizes as small as the order of the wave length of the monochromatic light used.

The specific surface area of barium sulfate crystals can be determined by the adsorption of dry nitrogen gas.[15]

The characteristics of finely ground or precipitated barium sulfate depend upon both the degree of subdivision and the size distribution of the particles. Several samples may pass a maximum particle size test, but still vary widely in particle size distribution (Figs. B-1 and B-2).

A uniform grain size gives a maximum percentage of voids and the fewest points of contact. For a given crystal form, with crystals uniform in size, the percentage of voids is the same regardless of crystal size; any variation in particle size rapidly decreases the percentage of voids. Furthermore, a fine particle size has more points of contact per unit volume than a coarse size and thus has a greater tendency to cake. In contrast, a nonuniform particle size suspension can result in a more dense film per unit thickness because of better filling of the voids.

The term "density" as applied to barium sulfate suspensions means the proportion of volume that is occupied by barium sulfate particles. For example, barium

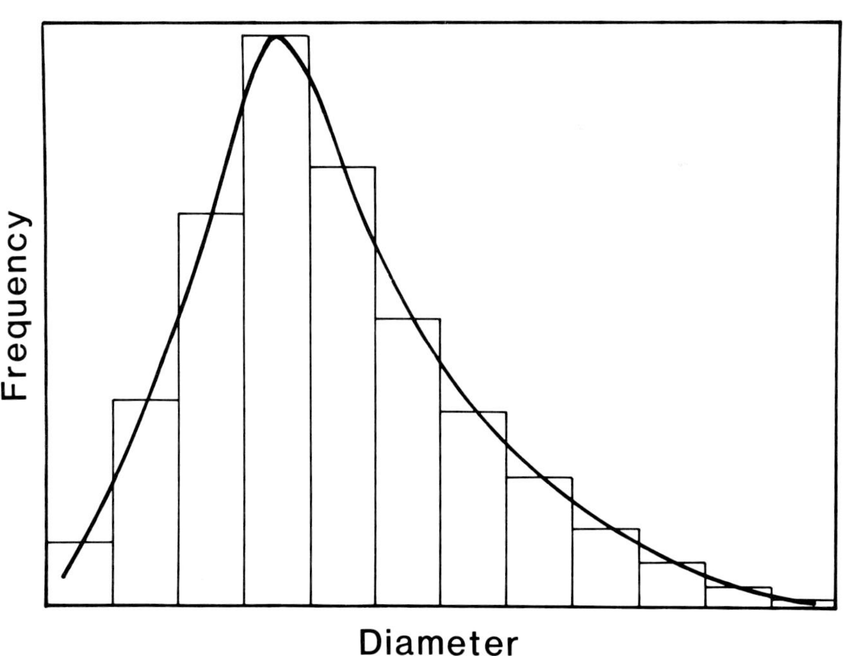

Figure B-1 Typical histogram obtained from a particle size counter. These counters can vary in accuracy considerably.

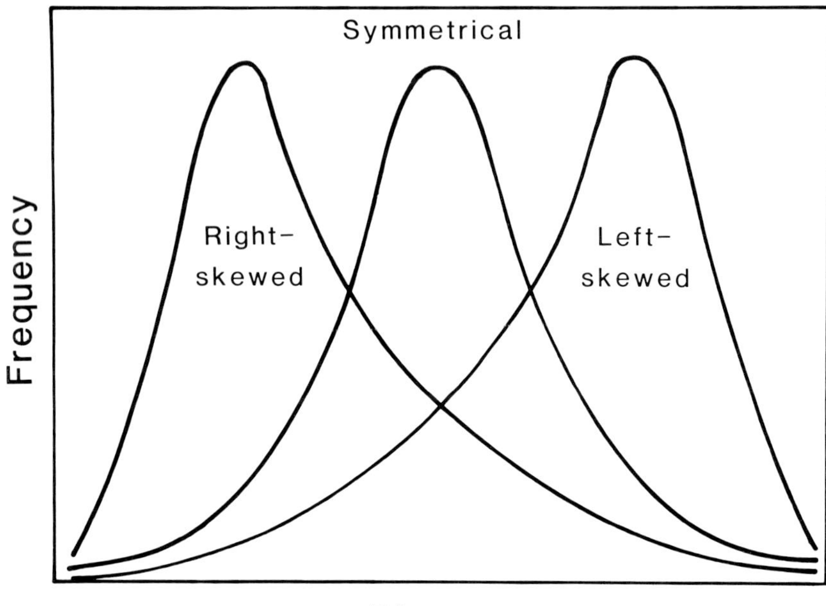

Figure B-2 Typical size distribution curves. Powders, such as barium sulfate, have right skewed distribution curves.

sulfate having 40% voids would be less dense than a mixture of only 25% voids. These voids can consist of air, water, additives, or any combination. The proportion of voids varies both in the "dry" powder state and after being mixed into suspensions. Even dry barium sulfate contains some moisture and wet barium sulfate some air. An analogy can be made to mortars of fine sand, where the density of mortar can range considerably. As in mortars, the densest barium sulfate particle combinations are obtained not with uniform particles but with mixtures. These mixtures are more dense when the coarse grains are in a proportion double to that of fine grains. Even denser mixtures are obtained with three different particle sizes.

In extremely fine powders the percentage of voids is increased through the adsorption of air. If the particles have adsorbed air cushions so that they do not come in contact with each other, the percentage of voids increases enormously as the size of the particles is decreased. This is true of fine powders even if they are not perfect spheres. Thus, fine moist sand occupies more space and weighs less per unit volume than dry sand because a film of water coats each grain of sand and separates it from the adjacent ones.

SEDIMENTATION

A suspension of pure barium sulfate will settle out under the influence of gravity in accordance with Stokes' law. Stokes determined the drag encountered by a sphere falling through a fluid as a result of gravity. The sedimentation rate is proportional to the square of the particle diameter and is inversely related to the viscosity of the suspending medium. Although the law does not apply to particles of various shapes, nevertheless, it is of value for comparative purposes. The closer the particles approach a smooth sphere, the more accurate are the determinations based on Stokes' law. Although barium sulfate particles are not spheres, for practical purposes they conform to this generalization in the ranges usually associated with contrast media. Another approximation is that Stokes' law does not readily apply to particles with a diameter of less than about 1 μm, such as colloids.[16] At very small sizes, the repulsive forces secondary to surface charges and particle collisions reduce the settling tendency.

The dilution ratio, which is the weight of liquid to solid, is also important in settling. For example, if a highly dilute barium sulfate suspension containing a *wide* particle size range is allowed to settle, the coarsest particles settle to the bottom at a comparatively rapid rate while the slower-settling fine particles remain on top. Later, a relatively slow, gradual clarification takes place. If exceedingly fine particles are present, there is no sharp demarcation between the settling solids and the supernatant liquid. All the particles have free movement and, except those of colloidal size, settle at a constant velocity in accordance with Stokes' law.

The settling rate can be studied as a function of time by weighing the amount of substance that has sedimented.[17] When increasing amounts of barium sulfate are added to a liquid and the suspension mixed and then allowed to settle, the following can be observed. First, a dilution is reached in which the fastest settling particles form into a portion and settle collectively. Second, at increasing concentrations this collective portion begins to form at progressively earlier periods until eventually a point is reached when the settling of particles is in a mass and no independent particle movement can be identified. This settling continues with a sharp separation between the barium sulfate and the supernatant liquid. Third, after a constant settling rate is achieved, there is a decrease in the settling rate if the concentration is increased further. This decrease results from retardation by particle interference. Beyond this point the flocs and particles are in intimate contact; further settling removes a portion of the interstitial liquid.

The particle size or bulkiness test for barium sulfate generally depends upon the sharp separation between the settling particles and the supernatant liquid. This sharp line is often *not* seen when additives are used. Different sedimentation rates can influence system stability significantly. Homogeneous particles of approximately 1 μm in size settle very slowly.

If a well-mixed 10% w/v dilution of relatively coarse, wide particle range barium sulfate is initially observed in a clear cylinder, it forms a homogeneous mass (Fig. B-3, phase A). After a short time, however, it assumes a flocculent structure. This structure forms several distinct zones, the clear zone 1 (described above) and a collective portion consisting of several parts (Fig. B-3, phase B). These different zones are seen most easily if a horizontal x-ray beam is used.[18] Reflected light from nonradiopaque substances in suspension may mask settling of the barium sulfate particles. The coarser, granular particles reach the bottom first. Immediately after this, and somewhat simultaneously, the more slime-like flocs nearest the bottom settle. They fill the interstitial spaces between the coarser particles and then build up, one upon another, in a zone of increasing depth (Fig. B-3, phases C and D). The individual flocs settle to a point where they rest directly upon one another. Further separation of liquid must come through liquid being pressed out of the flocs and the interstitial spaces.

In a rapidly settling suspension with a large particle size range, zone 1 in the earliest stages may be turbid because finely divided material remains in suspension. Later, this extremely fine material settles and the liquid clears. If the suspension contains an electrolyte, this zone may remain turbid for days. Without suspending agents, all the particles above true colloidal range will undergo slow compression (Fig. B-3, phase E). The sediment at the bottom of the suspension is formed in the final stage of settling. Particles will not settle beyond this point (Fig. B-3, phase F). The flow of liquid through the cake is inversely proportional to the fluid viscosity and the sediment thickness. Thus, the ability to redisperse the sediment from a highly viscous barium sulfate preparation is more difficult than from a relatively fluid preparation.

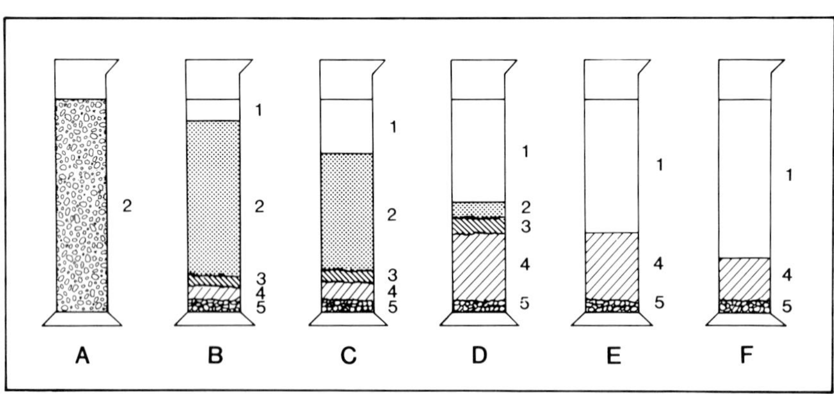

Figure B-3 Typical six-phase sedimentation for a suspension containing a wide particle size range. With the exception of zone 1, there are no sharp boundaries.

The clear zone 1 can contain colloidal particles because true colloidal suspensions (sols) are clear. With barium sulfate this would be an extremely dilute and temporary state because of its relative insolubility.

Both large uniform particles and agglomeration of particles make redispersion easier. As in dry powders, a large range in particle size can result in excessive settling and hard cake formation by filling in voids. The rate of cake formation is directly proportional to the ratio of solids to liquids in the suspension, up to the point of floc interference.

Ions of additives adsorbed on barium particles are most influential with particle sizes under 2 μm. In general, suspensions containing particle sizes larger than 0.1 μm will eventually settle out. Suspensions of particles ranging in size from 0.01 to 0.1 μm will not settle out, except after long standing. It is believed that *no* commercial barium sulfate medium has the majority of the particles in this size range.

There is considerable variation in the sedimentation rates of various commercial barium sulfate suspensions.[19] Some settle out in layers even before the examination is completed. Others form only a narrow zone 1 after settling for days. Extremely good barium suspensions with proper additives can perform superbly even in relatively low concentrations. These suspensions have enough particles and enough additives to produce homogeneous patterns even with a horizontal x-ray beam. The barium suspensions designed for CT tomography studies can have only 1–2 w/v% of barium, yet they do not settle readily. The key is to have sufficient amounts of the right type of additives to prevent settling. In these products the additives are proportionately greater than in conventional suspensions. Likewise, the larger particle, high density barium suspensions available for double-contrast gastric studies cannot simply be diluted and used for single-contrast small bowel and colon studies; they settle out and flocculate, resulting in suboptimal examinations. Other contrast media designed for single-contrast studies can be diluted considerably before significant settling occurs.

COLLOIDAL BARIUM SULFATE

When low concentrations of compounds insoluble in water, such as Prussian blue, are precipitated from weak solutions, the liquid becomes turbid. If the suspended material absorbs visible light the colloid suspension will be deeply colored. When a concentrated beam of light is passed through such a colloidal suspension, its path can be identified. The extremely small particles suspended in the liquid reflect light impinging on them, with the scattering being least for long light waves. This is the Tyndall effect of a colloidal suspension (sol).[20]

The colloidal range of dimensions is important for an understanding of barium sulfate suspensions. The International Union of Pure and Applied Chemistry defines a *colloid* as any material with at least one of its three dimensions between 1–1000 nanometers.[21] Presently, broad limits are used in the definition of a colloid, mainly because other physicochemical properties in addition to particle size are important. Strictly speaking, the term "colloid" should be used to define a physical system, although some colloid chemists refer to any finely divided phase as a colloid. Barium sulfate processors have extended the term further to include any barium sulfate contrast medium that contains colloidal additives. For example, one "colloidal barium sulfate" had particles so large that the term colloidal should not apply to it.[22]

Colloid chemistry is generally accepted to be the chemistry of surface properties. The surface forces are important factors in determining a suspension's properties. A *sol* is a colloidal system in which a liquid is the dispersion medium.

The settling of colloids under the influence of gravity is opposed by osmotic pressure and by Brownian movement. *Brownian movement* refers to a state of continuous motion of very small particles, first noted by Brown in a study of pollen suspended in water.[23] After a time the two opposing forces come to a sedimentation equilibrium. The distribution, even with same size particles, is not uniform; the concentration of particles increases toward the bottom of the sol.[24]

Studies on the crystalline barium sulfate salt have found that the distinction between crystalloids and colloids is not tenable and that a crystalline material can be made to assume a colloidal state under certain conditions;[25,26] for example, a low concentration of an extremely fine salt in modified aqueous liquids. Such suspensions have little x-ray absorption.

True colloidal sols of barium sulfate are often positively charged because of a strong preferential adsorption of the barium ion. They are stable in the presence of slightly adsorbed anions, such as sulfocyanate or iodide.[27] This explains why it is possible to obtain such a stable colloidal solution of barium sulfate by mixing solutions of barium sulfocyanate and cobalt sulfate.[26] In view of the stability of this colloid solution, one might expect to find the cobalt ion also strongly adsorbed by barium sulfate, because it would likewise have a stabilizing effect.

Many precipitation reactions may result in colloidal solutions. However, growth of the particle may result in sizes outside the colloid range. A low electrolyte count assists in the formation of a colloid.

Barium sulfate is too soluble in water (0.002 g/L) to give very fine crystals and form a hydrosol. Hydrosols may be obtained from chemical reaction by forming a substance within a fluid where the substance has practically no solubility. The greater the solubility of a substance, the larger are the crystals separating out from its supersaturated solution. Seemingly quite insoluble substances, such as barium sulfate or silver chloride, yield precipitates, the crystal-like nature of which can hardly be recognized. Nevertheless, their solubility is sufficiently great that stable aqueous colloidal solutions cannot be prepared.[28,29]

A colloid suspension (sol) of barium sulfate can be made with aqueous alcohol.[30] Barium sulfate is much less soluble in aqueous alcohol than in water. Alcohol solutions of sulfuric acid and barium acetate are used. Stable colloidal solutions of barium sulfate can be prepared by double decomposition, employing pure glycerol as the solvent.[31] A particularly stable variety was made using ethylate of barium and sulfuric acid. Weiser[31] prepared a stable negative solution of barium sulfate by adding a slight excess of 0.1 N solutions of sodium sulfate to 0.1 N solutions of barium chloride, using as a solvent 1 part water to 5 parts glycerol.

Another method for making a colloidal barium sulfate suspension was patented by Sanders for the sizing and filling of paper and textiles.[32] The suspension is prepared from a barium gluconate solution by the addition of a solution of hydroxylamine sulfate or anhydrous sodium sulfate. This preparation will stay in a stable suspension only if there is less than 12% of barium sulfate; at higher concentrations, the barium flocculates. Barium sulfate also forms a colloidal gel with the highly corrosive liquid, selenium oxychloride. When a solution of selenium oxychloride containing dissolved barium chloride is added to a like solution containing sulfuric acid, barium sulfate precipitates in a gelatinous form. This colloidal form of barium sulfate, however, immediately changes to the ordinary form and precipitates out when treated with water.[33]

Gengou prepared "colloidal" barium sulfate in the presence of sodium citrate.[34] Soxhlet reported that a suspension obtained by adding sulfuric acid to barium arabinate was milky after standing 4 years.[35] Gelatinous barium sulfate can be obtained by mixing concentrated solutions of barium sulfocyanate and manganese sulfate.[36] Relatively solid gels are produced by this mixing.

One exercise in colloid chemistry is to prepare a negative hydrosol of barium sulfate by precipitation in the presence of sodium citrate.[37] The particles are so finely divided that this weak sol looks like a true solution. Adsorption of the citrate ion accounts for the peptizing (dispersing) influence. If the solution is sufficiently acid to convert the sodium citrate into a poorly ionized citric acid, the citrate ion peptization is decreased, larger particles form, and the $BaSO_4$ precipitates.

There are no colloidal aqueous suspensions of barium sulfate available to the radiologist. Such a suspension would be either extremely dilute or a nearly solid gel; it would be unsuitable as a contrast agent except possibly for one of the newer imaging modalities where a very dilute contrast agent is needed.

EFFECT OF COLLOIDS

The type of interface between water and barium sulfate particles has considerable influence on the resultant suspension. This interface depends on the quantity and quality of colloidal substances that coat the particles and act as buffers between the barium sulfate particles and surrounding water. The influence of these coating materials is greatly out of proportion to their weight. As an example, an additive can completely coat the particles even though it comprises less than 1% of the total weight of the barium sulfate preparation. Some additives exert considerable influence even in amounts of 1 part per million.

When barium sulfate is dispersed in the presence of gelatin or some other protective colloid, it does not necessarily have the same properties it would possess if it were in a pure powdered form. The protective colloid forms a film around each group of particles, which often alters the barium sulfate properties and behavior.

Many substances in the colloidal size range, when suspended in water, are stable because of surface electric charges. In general, stability of a colloidal system is achieved when the electrostatic repulsion counterbalances the van der Waals attraction force. If the charges are neutralized by adding a substance with opposite surface electrical charges, the suspension is no longer stable and the substance precipitates. If the particles have no affinity for the solvent, the system is lyophobic (possessing an aversion to liquid).[38] If water is the suspending liquid, the system is hydrophobic. In general, a hydrophobic system is stable only if an appropriate stabilizer material is present.

Some colloidal solutions (sols) do not respond to neutralization of their surface charges. They have thermodynamic stability resulting from the sol's affinity for the solvent.[38] These solutions are formed when the colloid material and the solvent are placed in contact and are known as lyophilic colloids; if suspended in water, they are called hydrophilic.

There is some ambiguity in the definition of hydrophobic and hydrophilic sols. In a strict sense, a hydrophobic sol implies that there would be no wetting of the powder. Yet, many compounds that are wetted are still considered hydrophobic.

Electrical charge and hydration are the predominant factors that govern the stability of colloidal sols. Until recently, differentiation was made between hydrophobic and hydrophilic colloids depending on which of the two factors was predominant. Although this separation is sometimes useful, it is not always clear-cut. Both factors are important in many of the same colloids. In actual practice, intermediate properties are common; many colloidal sols rely for stability both on an interaction with the dispersion medium and on electrical charges. The relative importance of hydration and electrical charge varies with the product. One such example is clay suspended in river water, where it forms a negatively charged lyophobic colloidal system. When it encounters salt water, flocculation occurs and a delta begins to form at the river mouth.

A hydrophobic colloid can be coated with a hydrophilic colloid; the resultant surface is changed, and the product becomes hydrophilic and will no longer be affected by electrolytes. One is dealing here with surface layers of colloidal thickness, although the particles themselves are much larger. Barium sulfate particles can be coated in such a fashion. Several substances can be added on the surface.

A barium sulfate hydrophobic sol is of little use. It is sensitive to electrolytes in the dispersion medium and is generally unstable. If it is coated with a hydrophilic colloid, it is no longer as dependent on the presence of electrolytes and is quite stable even at high concentrations. As a result, most ''colloidal'' barium sulfate suspensions are made hydrophilic.

STABILITY

Electrostatic Effects

In general, colloid particles suspended in water carry an electric charge acquired by adsorption of ions, polyelectrolytes, or charged macromolecules.[39] In some, the particles introduce ions into solution. Adsorption of dipolar molecules can affect the distribution of charge. The solids do not always take on the same charge.

Barium sulfate may have either a negative or a positive charge.[4] Either ion may be adsorbed, depending upon whether the positive barium ion or the negative sulfate ion is first adsorbed in excess in the precipitating process.[40] Barium sulfate may be precipitated red in the presence of hydrous ferric oxide if an excess of sulfate ion has been adsorbed first. Its negative charge then causes it to adsorb the red, positively charged hydrous ferric oxide. The reverse is also true; if there is an excess of barium ion adsorbed there is little color in the precipitate. However, the supernatant liquid is colored red because the positive precipitate will not adsorb the positive ferric oxide.[37] Both precipitation reactions are carried out with the same reagents in the same liquid. The only difference is an excess quantity of one ion or the other. In general, barium sulfate has a positive charge in an acid solution and a negative charge in an alkaline solution. In water, the barium sulfate particles tend to have a slightly negative charge secondary to attached OH groups. Under some conditions certain solids in pure water adsorb either hydrogen ions or hydroxyl ions, thus promoting further ionization of the water and developing a charge on the particles. Furthermore, colloidal micelles may themselves ionize.[37]

Because the electrical charge of the system as a whole is nearly neutral, the dispersion medium must contain an almost equivalent charge of the opposite sign. These charges are carried by an excess of ions of one sign at the particle surface and an excess of ions of the opposite sign in solution. If a single particle is considered separately immersed in the liquid, it is surrounded by an electrical double layer. One layer is formed by the electrical charge on the surface of the particle. The charges of this layer consist of point charges concentrated around certain active spots on the crystal faces. When this particle, such as barium sulfate, is immersed in water with a small amount of salt, acid, or alkali dissolved, the ions in solution are attracted to the active points, thus forming a second outer layer that completes the "double layer" of ions.[41] The adsorption of charged polyelectrolytes can also add interface charges.

The electrical forces associated with a colloidal particle in water containing electrolytes thus come from two sources. One set arises either from the inner layer of ions that are strongly adsorbed on the surface or from ions located on the surface that are not fully saturated. The other source comes from the outer ions of the double layer that surrounds the particle. The number of outer ions around the particle may not neutralize completely the particle's charge. The hydration of the adsorbent surface, the adsorbed ions, and steric repulsion also play roles in the resultant electrical forces.

The charged nucleus with its bound charges from the dispersion medium, along with their accumulated and hydrated ions that are attracted and dispersed in their vicinity, form molecular aggregates in what constitutes the colloidal micelle.

The electrical charge of the outer layer extends over a variable distance from the particle surface and diffuses gradually with increasing distance. The "thickness" of this diffuse layer is of colloidal dimensions and is independent of particle size. In barium sulfate suspensions, the double layer thickness is generally small compared to the size of the particles. In general, the electrostatic potential decreases exponentially with distance.

The particle charge thus determines the stability of colloidal particles. Flocculation takes place, as a rule, when these charges are removed. Particles repel each other when they contain the same charge, but by the same token, they are sensitive to the action of electrolytes. A similar effect probably also applies to the much larger barium sulfate particles.[8] Many lyophobic sols flocculate when small amounts of electrolyte are added. Flocculation takes place when the double layer repulsion forces are reduced, the particles approach closely, and the van der Waals forces become important. The system stability is thus balanced by van der Waals attraction and electrical charge repulsion forces from the electrical double layer (Fig. B-4). It is believed that the attraction energy increases rapidly as two particles approach each other, whereas the repulsive energy changes more slowly with distance.[42] At very close distances, however, there is a marked increase in repulsion caused by the electrons surrounding the two particles. The sum of these forces results in an energy barrier that prevents particle adherence. This barrier varies with the size of the particles and their surface charge.

Polyelectrolytes form their charge by dissociation. They are hydrophilic colloidal-size molecules and contain large numbers of ionizable groups. Flocculation, titration curves, and other phenomena of these hydrophilic colloids are complicated because of the interaction among charged groups. For example, in a polyelectrolyte molecule with carboxyl groups, the dissociation of the first few groups gives the molecule a negative charge that hampers the dissociation of the

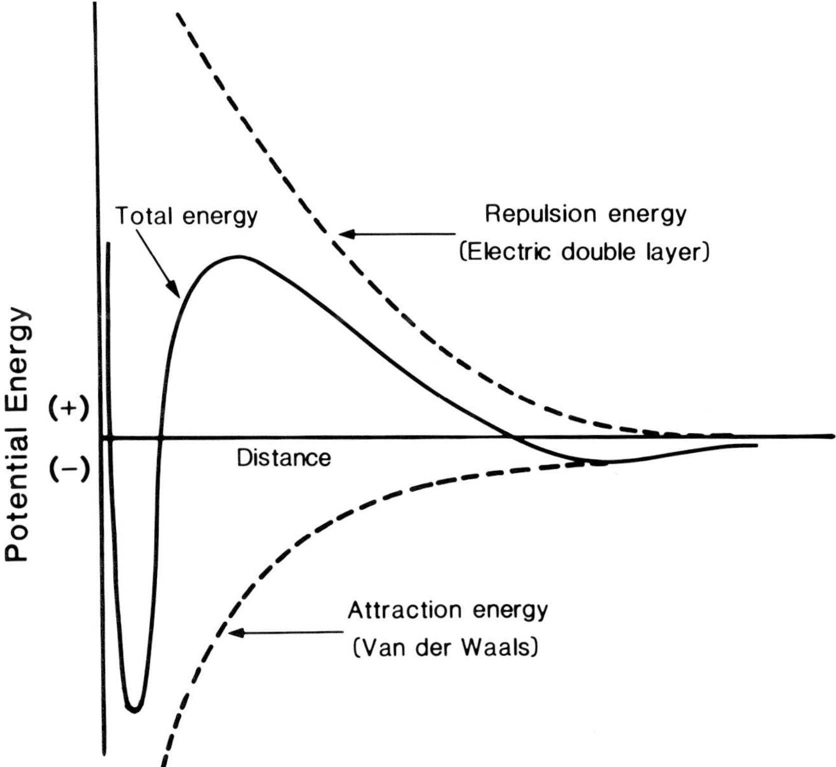

Figure B-4 The potential energy attraction and repulsion forces for two colloidal particles as they approach each other. Initially, there is an energy barrier against adhesion of the two particles; however, if this barrier can be overcome, the particles will attract each other.

---

following groups. The interaction between different charged groups is partly screened by the ionic atmosphere around each of them. In addition, polyelectrolytes can exert steric effects, discussed in the next section. If the polyelectrolyte is a long-chain molecule, such as polymethacrylic acid, its form in solution is a loose coil:[43]

$$
\begin{array}{ccccccccccc}
 & CH_3 & & CH_3 & & CH_3 & & CH_3 & & \\
 & | & & | & & | & & | & & \\
\text{etc.}— & C & —CH_2— & C & —CH_2— & C & —CH_2— & C & —CH_2— & \text{etc.} \\
 & | & & | & & | & & | & & \\
 & COOH & & COOH & & COOH & & COOH & &
\end{array}
$$

The coil increases with increasing charge as a consequence of growing mutual repulsion between the charged groups. The ion's strength also affects the degree of swelling of the coil. The distribution of ions around a polyelectrolyte molecule differs from the ordinary ionic double layer around a lyophobic colloid. Often, small ions penetrate within the polyelectrolyte molecule, and it is not always permissible to consider the charge as "smeared out."

The interaction of the double layers of both coated and uncoated particles in barium sulfate suspensions plays a major role in the characteristics of the final suspension. One example is the electroviscous effect shown by many hydrophilic colloids. The electrical charge on the hydrophilic sols increases viscosity. There is a considerable decrease in this high viscosity when small amounts of an electrolyte are added. The electrolyte neutralizes the potential barrier existing between particles and allows their mutual approach. The destruction of the potential barrier applies to different parts of the same macromolecule, allowing a folding together of the molecule to a less stretched form. The less stretched form is less viscous. The discharged sol does not coagulate because of the molecule's hydration. This hydration in turn can be reduced by dehydrating substances, and flocculation then occurs easily.

The rate of particle migration secondary to an electrical charge can be measured.[4,40] The electrophoresis of such small particles is similar to ionic migration in an electric field. However, because barium sulfate particles are outside the range of colloids, certain approximations must be made. The solid particles are assumed to be spheres. The electrical charge is usually assumed to be uniformly distributed throughout the surface. In practice, these conditions are not always applicable. The measured electrophoretic mobility is an average value, with the particles in a suspension having variable mobilities.

Several commercial devices are available to measure electrophoresis. Generally, the migration rate is converted to electrophoretic mobility by dividing the velocity by the field strength. The apparatus should be mounted horizontally so that vertical settling due to gravity is eliminated. The electrophoretic mobility is dependent on the adsorbed layer thickness.[44] It varies widely among different commercial barium sulfate preparations.[8,45] For at least one barium sulfate preparation, however, the mobility decreases with increased concentration (Fig. B-5).

The electrophoretic mobility is pH-dependent.[46,47] Commercial barium sulfate preparations show marked variation in electrophoretic mobility at low pH levels.[4,8] The variability is greatest below a pH of 4. Pure barium sulfate gives erratic results, confirming that the coating on the barium sulfate particles is the major factor in surface charge.

Steric Effects

An uncharged system containing colloid additives would undergo rapid coagulation unless there were repulsive forces to counteract the van der Waals attraction between particles. These repulsive forces can be electrostatic, as discussed in the previous section, or exist as steric stabilization.

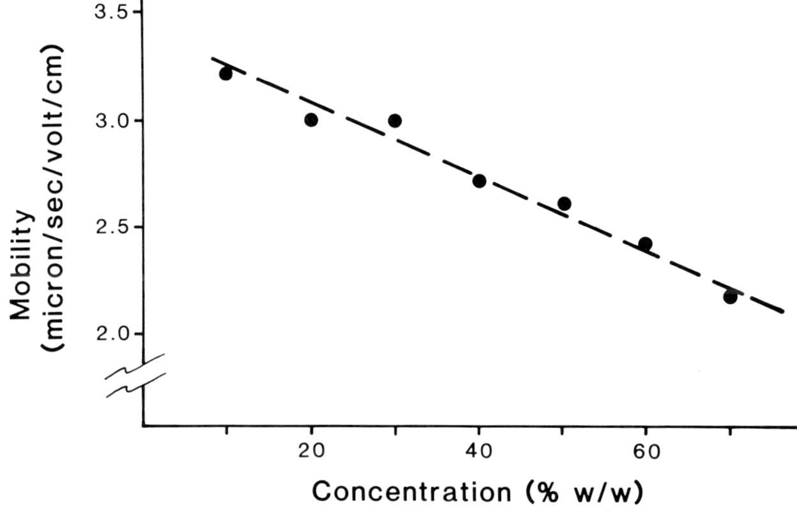

Figure B-5 Electrophoretic mobility as a function of concentration of Barosperse. *Source*: Adapted with permission from "Studies in Adherence on Contrast Media to Mucosal Surfaces" by SE Schwartz, HW Fischer, and AJS House in *Radiology* (1974;112:727–731), Copyright © 1974, Radiological Society of North America Inc.

The advantages of steric stabilization are its relative insensitivity to electrolytes, ability to maintain low viscosity at high particle concentration, and relative stability both in liquid and solid form. Steric stabilization is caused by adsorbed long-chain molecules repelling each other, resulting in repulsion between particles. This repulsion is caused by volume restrictions of the long-chain molecules and their associated osmotic effect. In general, better stabilization is achieved by using several polymers; thus, one polymer serves as an anchor on the particle, whereas another extends into and is soluble in the surrounding liquid.[48,49] The polymers can be ionic, such as gelatin, or nonionic.

FLOCCULATION

Negatively charged colloids can be precipitated by positively charged ions and other colloids. The precipitated particles can form a hard, compact and nonreversible mass by *coagulation* or, if the particles remain separated by a liquid film so that close contact by particles is not possible, by *flocculation*.[50] Flocculation is primarily a chemical process and results in a coarse precipitate. It should be differentiated from sedimentation, which results in homogeneous settling. Neither is flocculation the same as dilution. Use of a nonflocculating barium product allows detection of smaller amounts of dilution within the bowel, because clumping of barium sulfate particles in flocculation tends to mask the amount of dilution present.

Hydrophobic colloids can be flocculated by relatively small amounts of electrolytes that furnish polyvalent ions or change the pH of the system. The hydrophobic colloid is most unstable at the isoelectric point—the point at which the particle will not migrate to either electrode in an electric field. The negative charge, for example, on a hydrophobic barium particle may be reduced by contact with an ion of positive charge, arising from either the water supply or intestinal secretions, until it is practically zero. The colloid is then unstable and flocculates. This point is recognized by a failure to migrate in an electric field.

The degree of flocculation varies with the degree of dehydration produced by ions in the suspending medium. Ions react strongly with the polar water molecules. Small highly charged ions are more effective than large ions of low charge. However, large concentrations of ions are needed to produce flocculation of lyophilic sols.

It is the flocculation tendency, not the particle size, that determines the quality of the final suspension.[22] Flocculation can be decreased by keeping down the concentration of the coagulation electrolyte. For example, secondary aggregates of $BaSO_4$ become larger as the concentration of the coagulation electrolyte NH_4NO_3 steadily increases in the suspending liquid in the reaction of

$$Ba(NO_3) + (NH_4)_2SO_4 = BaSO_4 + 2NH_4NO_3$$

Particle flocculation can be prevented by adding strongly adsorbed protective agents, such as sugars or protective colloids. These agents form water-soluble films around the primary particles and prevent not only their primary growth but also their coalescence into large aggregates. Some of these protective colloids are sodium citrate, bentonite, glucose, gelatin, agar, tannin, gum arabic, casein, starch, biologic fluids, plant extracts, albumin, molasses, glycerol, invert and cane sugars, and various celluloses. These protective colloids interfere with precipitation of many insoluble compounds. Because of their small size they can cover the entire surface when used in small amounts. Thus, they protect the particles from the precipitating action of electrolytes. These agents also prevent close contact of individual barium sulfate particles, thus also helping prevent flocculation. However, because they are hydrophilic, they change the viscosity of the system. The usual concentration of these additives is from 0.5 to 1.0%, except if the additive is to be left with the finished product. Larger quantities may then be advantageous because they may make granulation easier or allow the dry product to be mixed in an extreme range of concentrations for various examinations. For example, a 20% w/v suspension would have only one-third as much suspending agent as a 60% w/v suspension. At the lower concentration, the amount of additive may fall below an effective range.

Organic ions can cause flocculation even at a low concentration, probably because of different absorbability on the colloid of the organic ions as compared with inorganic ions. As a result, the organic ions have an effect greater than expected from their valence. Thus, some barium sulfate compounds appear to be remarkably well suspended when first mixed but flocculate when added to the mucin and acid in the stomach.[8] The addition of gastric juice to barium sulfate USP increases flocculation up to ten times.[13] A small amount of barium sulfate appears to flocculate more than a large amount. With complete flocculation there is no barium sulfate left in suspension. Some radiologists, when confronted with flocculation, simply increase the amount of contrast used.

An apparent anomaly exists. Although gelatin, agar, and other polymers ordinarily increase the stability of colloidal solutions, small amounts of these substances may induce flocculation.[51] Minute quantities of gelatin added to barium sulfate suspensions cause rapid sedimentation. This sometimes simplifies the filtration and washing of barium sulfate. As the ratio of the gelatin is increased, it passes through the precipitation range into the protective or stabilized range.

One example of a commonly used protective agent shows how the protective colloid can influence barium sulfate suspensions. Sodium alginate has a molecular weight of 15,400 and contains many micelles. It undergoes an extreme degree of hydration and carries a weak negative charge with a pH of 7.7. It coats the finely divided barium sulfate particles and promotes deflocculation in as small a quantity as 0.66% of the dry barium sulfate powder. It is practical in any pH range from 4 to 10. Below a pH of 4, the alginic acid begins to precipitate. This precipitation explains the poor performance in some stomach examinations where there is high acidity. Hard water contains sufficient ions of calcium salts to give algin sols a higher viscosity than sols made in distilled water. Hard water can thus also modify the performance of the suspension and may influence the concentrations at which the barium sulfate is used. Algin sols are readily decomposed by bacterial action if allowed to stand for prolonged periods. Thus, preservatives may be needed for suspensions containing alginates.

The flow properties of barium suspensions, because of their normally high barium sulfate concentration, depend in part on the interaction between the solid particles. If the particles do not form aggregates and do not flocculate, the suspension will be fluid at high concentrations and the yield value will be low. If, on the other hand, there is a tendency to form aggregates, there is a high yield value and thixotropic behavior. In general, flocculation is promoted by gentle agitation, lack of a negative charge, and a nidus that acts as a nucleus for the formation of a floc.[10] It can be influenced by the presence of acid or mucin within bowel.

Flocculation can be modified by the addition of surface-active deflocculating agents, such as sodium carbonate, sodium silicate, tannic acid, aluminum stearate, potassium citrate, sodium citrate, potassium sodium tartrate, sodium phosphate, sodium hexametaphosphate, and such combinations as 0.1% each of sodium carbonate and sodium trisilicate. Sodium carboxymethylcellulose has been reported to produce deflocculation,[52] probably secondary to the addition of the hydroxyl groups that this substance contains. The salts of organic acids help produce stability. One in vitro study showed that sodium citrate gave the best suspension stability.[22] The effect of sodium citrate was accentuated by adding sorbitol. Other deflocculants, such as the carbopols and the lignosulfates—Maraspherse and the Daxads—have been used. The lignosulfates have a tendency to decompose upon storage in liquid suspension, forming hydrogen disulfide. These and other additives may become adsorbed on the barium particles and modify their interaction.

With such deflocculants as citrate, flocculation results if there is excess hydrochloric acid, sodium chloride, or gastric mucin. A sulfated polysaccharide, degraded carageenan, has been proposed as a deflocculant in the presence of dilute acid.[53]

The tendency to flocculate is less if the barium contrast agent is suspended in 0.4% saline, rather than water.[54] If the saline concentration is increased to 0.9%, however, the resultant coating appears "wet" and there is a loss of diagnostic quality.

By choosing suitable surface-active agents, the state of flocculation can be either increased or decreased. One increases flocculation to increase the yield value, apparent viscosity, and thixotropic behavior. Carefully controlled, these factors help achieve satisfactory coating and surface films.

If a very small particle size barium sulfate suspension is prepared, there is a large surface area per unit weight and very little sedimentation. Such a suspension, however, may produce excessive flocculation when mixed with gastric or small bowel secretions.[55] Small particle size by itself does not prevent flocculation. The large particle size products, in contrast, sediment rapidly. Even in the stomach rapid sedimentation is not always associated with good coating.

Deflocculation reduces viscosity and allows a higher concentration of barium sulfate particles. For example, 5% tannic acid has been used with 15% sodium hydroxide as a deflocculating agent in oil well drilling "muds" that consist largely of barium sulfate. The tannic acid permits higher concentrations of barium sulfate without an increase in viscosity.

The tannins yield anions that, in the ionized state, have the negative charge uniformly distributed along the extent of the molecule. Ford et al. have considered the action of tannins as "plating" the suspended particles.[56] They obtained semiquantitative correlations between the amount of tannin added and the amount necessary to yield maximum deflocculation; such maximum deflocculation corresponded to complete plating of all the particles present. Deflocculants are added to barium sulfate to counteract flocculating substances, such as sodium chloride and mucin. As a result, the addition of tannic acid to some barium preparations produces a better "mucosal" pattern, and various amounts of tannic acid had been used beginning in 1946.[57] The chief action of tannic acid is as a "peptizing" or deflocculating agent. Although some investigators credit the improvement with tannic acid to its precipitation of proteins,[58] it is believed that tannic acid makes poor or unprotected barium sulfate look better not because of any such protein precipitation, but because of its deflocculation action.

This is not to be construed as an endorsement for tannic acid. Tea and some wines contain small amounts that are believed to be relatively safe. In the 1960s, however, there were several reports of tannic-acid-induced liver damage, including some fatalities.[59,60] Subsequent work established that tannic acid is absorbed from the colon mucosa and can produce liver and kidney damage.[61–63] The damage seems to be dose dependent.[58,64] Modern barium formulations have eliminated a need for such substances as tannic acid. It is more effective with "pure" USP barium sulfate and not as useful with current commercial barium sulfate preparations.

VISCOSITY

The viscosity of a fluid is the ratio of stress to the shear rate. More simply, it is the resistance to flow. Fluidity is the opposite of viscosity. The viscosity varies with temperature; even for water, the viscosity is greater at lower temperatures.

Specific viscosity is the ratio of the viscosity of any substance compared to the viscosity of water. A substance that will deform permanently to a degree directly proportional to the force applied to it has true viscosity. Pure liquids, true solutions, and highly dilute suspensions have true viscosity and are called Newtonian fluids (Fig. B-6). In these fluids the stress and shear rate have a linear relationship. Such a liquid flows even when a very small force is applied. At times barium sulfate exhibits Newtonian flow. This occurs only in very low concentrations and before any significant settling has taken place. Many liquids containing colloidal dispersions have non-Newtonian flow.

If a minimum shear stress must be applied before the material starts flowing, it has plastic flow. This type of flow is seen with some dispersions of solids in liquids. It is believed that, with plastic flow, the individual particles are in close contact and that some bonds must be broken before flow can start.

The viscosity of some fluids decreases at higher shear rates; they are known as pseudoplastic fluids and possess properties intermediate between the true Newtonian fluids and plastic fluids (Fig. B-6). Their shear rate is a complex function. The relatively high apparent viscosity at the lower rate of shear is known as false

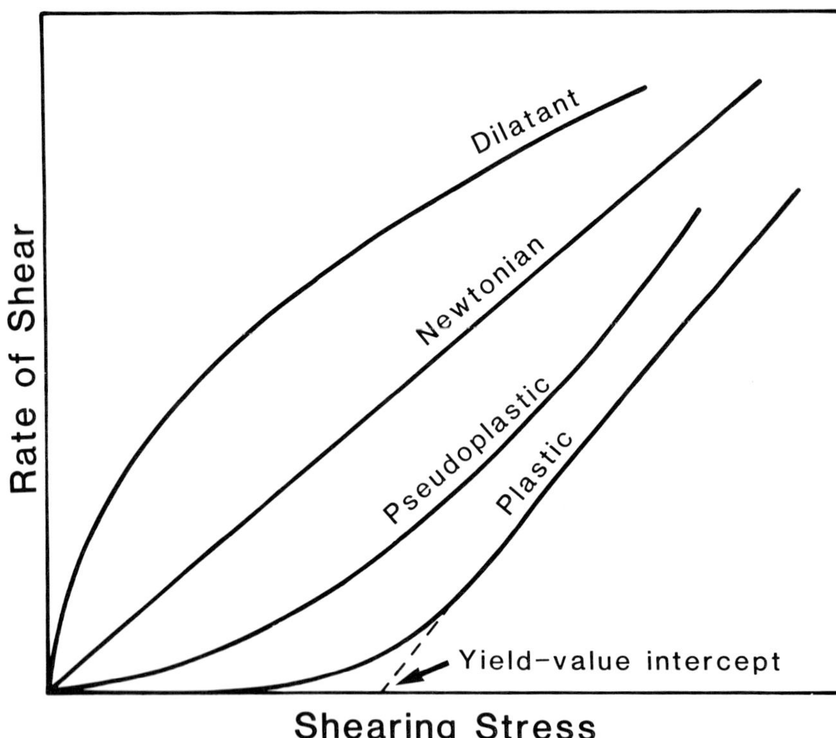

Figure B-6 Shear-stress curves for several suspensions. Most barium sulfate suspensions do not have Newtonian flow.

body. The gradually accelerated rate of flow as the stress increases is probably the result of a progressive orientation of the particles to parallelism with the lines of liquid flow and the consequent release of interlocked liquid. Pseudoplasticity is found in some colloidal dispersions, suspensions, and high-polymer solutions, such as hydroxyethylcellulose. Whether pseudoplasticity is advantageous in double-contrast barium examinations has been questioned by Russian investigators.[65]

Some highly viscous suspensions (gels) become fluid when shaken or made to flow fast, an effect produced by a breakdown of weak bonds (Fig. B-7). After being left undisturbed, the viscosity increases and a gel reforms. This property, called *thixotropy*, is useful in oil well drilling muds, some paints, and barium suspensions used for double-contrast studies.[8,66] Thus, when such a barium sulfate suspension with thixotropic properties is stirred or made to flow rapidly, the effective viscosity decreases. After introduction and application onto the mucosa, there is an absence of flow while radiographs are taken. During this state, the viscosity increases again and assists in preventing too much drainage of the

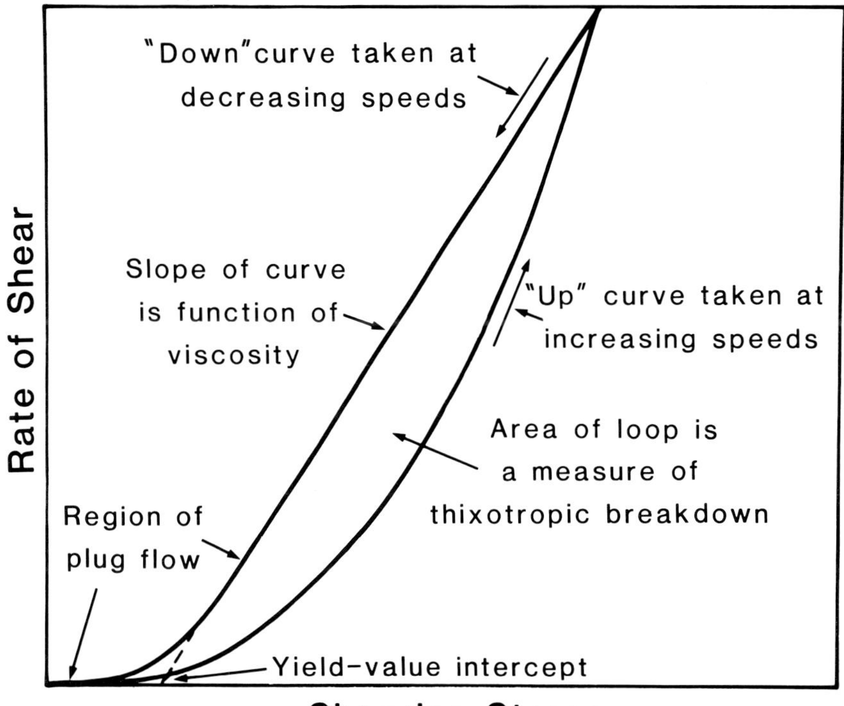

Figure B-7 Thixotropic flow as found in some barium suspensions used for double-contrast studies of the gastrointestinal tract.

---

suspension. If, however, the thixotropic "setting" of the suspension occurs too rapidly, flow lines and large polyp-like drops will be retained. This phenomenon may be accentuated by mucin and the dehydrating action of bowel.

Many commercial barium suspensions exhibit plastic or pseudoplastic flow (Fig. B-6). The yield point or minimum stress varies with different suspensions. The dispersion does not form its original shape immediately after stirring; rather, there is a delay. If the delay is short, the thixotropic flow may approach plastic or pseudoplastic flow. Many barium sulfate suspensions containing a methylcellulose derivative or similar compound exhibit thixotropic properties.[67] Thixotropy markedly influences barium flow, settling, x-ray absorption, and concentration.

Dispersions may possess dilatant flow (Fig. B-6) in which the viscosity increases with an increasing shear rate. Dilatant flow is the reverse of pseudoplastic flow and is seen in dispersions of powders at high concentrations. Because the particles are packed together closely, there is little intermediate fluid to act as a lubricant. Thus, the fluid flows readily under small stress but resists rapid change. Fischer stirred barium sulfate into water and found the onset of dilatant flow when the concentration of the powder reached 39% w/v.[68] Commercial barium sulfate pastes marketed for esophageal examinations often have dilatant flow. These pastes are useful in the detection of varices in a collapsed esophagus, but they are not useful for double-contrast esophagrams. Some pastes squeeze out water on standing, a process called syneresis. It is probably caused by agglomeration into denser aggregates that contain less fluid than the freshly formed product.

The quantitative determination of the flow properties of barium sulfate suspensions is not simple. Single viscosity values are not informative and may be misleading. Published reports often disregard the importance of pseudoplasticity in the behavior of barium suspensions. Thus, the apparent viscosity of barium sulfate suspensions varies widely with the rate of flow. Different readings are obtained with a Brookfield viscometer operating at different spindle rotation speeds.[6,45] In general, the apparent viscosity decreases at faster rotation speeds for many but not all commercial barium products.[6] Such a change in viscosity with flow rate can be clinically significant; a commercial product that is flowing through the esophagus under the force of gravity can exhibit a different viscosity than the same product that is barely flowing on the small bowel or colon mucosa.

The viscosity of commercial barium sulfate preparations varies considerably.[19] Some suspensions must flow readily through tubing. Other suspensions must not form clumps and must adhere to mucosa in a relatively thick coating. If the coating is too thin, there may not be enough radiographic contrast for adequate definition of small lesions.

Radiographic contrast is enhanced by a thick mucosal coating. The greatest factor in achieving a thick coating is the suspension's viscosity. Viscosity can have a linear relationship to the concentration. With an increase in viscosity, the suspension becomes stiffer and the yield value increases. Viscosity can be increased only to a point before the suspension starts to form a paste. Larger size barium sulfate particles help produce a thicker coating. Yet even with particle size in the 4-μm range, viscosity is still the major factor.[18]

Viscosity depends upon the system ionization. Discharging the particles by the addition of electrolytes decreases viscosity significantly. These salts both affect hydration and influence charge and thereby also affect viscosity.

The viscosity of a barium preparation is decreased by adding spherocolloid additives, such as 1 to 2% sodium citrate.[69] The viscosity of spherocolloids is essentially independent of particle size. If the particles are very small, the resultant increased surface area may lead to greater viscosity. If there are also increased amounts of adsorbed air cushions on the smaller particles, the viscosity may increase more than expected. In general, viscosity increases linearly with concentration until relatively high concentrations are reached.

The situation is even more complicated for linear colloid additives carrying a charge; charged particles can be considerably more viscous than similar uncharged particles. Thus, linear colloid additives, such as carboxymethylcellulose, may produce extremely viscous solutions even if the concentration of the colloid is only 0.5 to 1.0%. The viscosity is markedly dependent on the length of the particles. Different linear colloid additives behave quite differently even if they are of the same length, because of differences in the spatial configuration of the thread-like molecules. Thus, branching and coiling of these linear colloids profoundly influence viscosity, and the more bulky the molecules, the higher is the viscosity.[70] As an example, carboxymethylcellulose or its sodium salt in water produce a clear colloidal suspension. Its structure is

$$[C_6H_{10}O_5(CH_2COONa)_x]_n$$

Viscosity varies directly with the value of n. A 1% suspension produces viscosities ranging from 2 to 20,000 centipoises simply by increasing the value of n.[71] An increase in viscosity does not necessarily lead to better coating. Low viscosity allows a higher barium density; a material having a viscosity of less than 50 centipoises (at 2% w/w) is used in a commercial barium sulfate preparation.[72]

Salts and acids with their ions affect the viscosity of charged linear colloids not only by a direct reduction in electrical potential but also by changing the shape of the thread-like molecules. Salts cause a coiling of the long and flexible chains. A discharged chain is more likely to coil than a charged one because the negatively charged segments of a charged chain repel each other and tend to stretch out the chain.[73,74] The viscosity of some commercial barium products changes significantly with pH.[75] Some products have a marked increase in viscosity at the pH encountered in the stomach.

Sedimentation and viscosity with additives that contain thread-like molecules also vary markedly with concentration because of both mechanical entanglement and solvation phenomena. The results differ according to their structure. Cellulose and cellulose derivatives are probably unbranched, but such other polysaccharides as starch, gum arabic, and dextran have a high degree of branching. In extreme cases, branching ends in a complex structure. Modified branching is also possible through cross-links between the linear molecules.

In suspensions of mixtures with polymolecular additives where sedimentation is influenced by concentration, there may also be an anomalous behavior because of boundary effects. A slower moving substance may show an increased rate of sedimentation as the faster moving substance passes its boundary, and vice versa.

HYDRATION

Heat is produced when a highly insoluble, finely divided crystalline powder is immersed in a liquid. The heat is the result of a change in the interface energies between the solid and liquid and is directly proportional to the area of the solid. For barium sulfate immersed in water, the energy is 490 ergs per cm^2.[76]

Barium sulfate crystals can contain water in vacuoles within the crystal. The trapped water content can be up to 2.5% by weight.[77] If desired, dehydration of the barium crystals can be achieved by heating, with total removal of water occurring at 320°C.[77]

It is believed that the initially adsorbed "nonliquid" water differs from the excess water. The water adsorbed on a particle surface is more solid than liquid. The transition zone is gradual; the change in the state of the water changes rapidly with distance away from the surface.

A reduction in viscosity can be accomplished by adding substances, the ions of which tend to cause deflocculation when they are adsorbed on the barium sulfate particles. These substances can be added either to the dry powder or to the suspension. For example, when sodium phosphate is used, the phosphate ions *replace* the hydroxyl ions usually adsorbed on the barium sulfate particle surface, thus squeezing out *bound* water. These hydroxyl ions may be bound to the broken crystal faces by dipolar bonds or to another additive already adsorbed on the barium sulfate crystal. The hydroxyl ions that are adsorbed carry with them the water molecules (hydration) that make up the most important part of the water hulls. However, the particle size of these *micelles* must be very small to bring the attraction and repulsion forces into significant play. Similar effects may be achieved by use of more complex colloidal electrolytes, such as alkyl piridinium chloride or ethyltrimethyl ammonium bromide. In the case of barium sulfate contrast media, one is limited to nontoxic "dispersing" agents.

ADSORPTION

Adsorption is controlled by a number of chemical, physical, and colloidal forces. The adsorbed substance may have been added intentionally or be a contaminant. Some authors use the terms *deposition* and *adsorption* interchangeably.

Langmuir evolved a "surface active theory" in which the activity of a solid surface is dependent upon the spacing and arrangement of the atoms or ions comprising the surface layer.[78] The theory has been expanded to include the residual valences of the ions in the surface of the crystal lattice as being responsible for the adsorption of foreign ions.[79] The effectiveness of the surface ions is determined by their position. Those at the edges of the crystal are more effective than others. The crystal surface, instead of being a plane, has irregular submicroscopic "pits and humps" that become smoother with aging. Adsorption takes place first at "active centers" of the surface at the edges and at extra-lattice atoms, because they have a relatively high degree of unsaturation.[80,81] The valency forces at these active centers are only partly satisfied by the underlying lattice. Adsorption confirms the nonuniformity of the surface. Adsorbed particles first cover corners, then edges, and lastly the smoother surfaces. Initially, adsorption is in a monomolecular layer that may eventually progress into a multimolecular layer. The adsorption rate declines very rapidly as particles begin to accumulate in a monolayer on the substrate, so that an incomplete monolayer is present when the substrate is already saturated.[82] Furthermore, multilayer adsorption, as does monolayer adsorption, may exist in islands on the surface of the barium sulfate crystal. At times there is no interaction or only a weak interaction unless the particle to be adsorbed is brought together with the surface on which it is to stay with great force and energy. A strong chemical bond is then formed. In general, weakly bonded ions form ion-pair complexes that retain their primary hydration sphere upon adsorption, whereas strongly bonding ions form inner surface complexes.[83]

It is common to find contamination of precipitates by adsorption of substances dissolved in the solution from which they separate. Finely divided precipitates exhibit greater adsorption than coarse precipitates. The adsorbed substances may be organic or inorganic. For example, barium sulfate precipitated in a crystalline form from a hot, dilute hydrochloric solution containing potassium chloride adsorbs potassium quite strongly. The tendency of barium sulfate to precipitate potassium salts characterizes its behavior toward a large number of cation substances.[84]

Barium sulfate can adsorb anions. The nature of the ion and its valence determine the adsorption by a given adsorbing agent. With ions of the same general character, the specificity of adsorption should not be pronounced, and valency factors predominate. With ions that are not similar and yet have the same valence, the specificity of adsorption may be quite pronounced. The fundamental rule is that the adsorption is specific both as regards the adsorbing substance and the ion adsorbed. In practice, anion adsorption is still a poorly understood phenomenon, although computer models have been developed to study some of these processes.[85]

The adsorption of a substance in mixtures may increase or decrease. Inorganic salts included in a mixture increase the adsorption of acetic, propionic, and butyric acids; ethyl alcohol; and acetone. Adsorption has been studied extensively in the dye industry.[37,86–89] At a given pH, the bivalent barium ion decreases the adsorption of methylene blue more than the less strongly adsorbed sodium ion, and bivalent sulfate ion decreases the adsorption of lake scarlet ruby more than the less strongly adsorbed chloride ion. Furthermore, because of the strong adsorption of the hydrogen and hydroxyl ions, the adsorption and displacement of other ions are influenced considerably by the solution pH.[90–93]

FOAM

A foam is an emulsion consisting of gas bubbles dispersed in a liquid. Foams are colloidal systems. A gas dispersed in a solid, as in some polyurethane products, can also be a foam. Bubbles are physical structures in which visible gas collections are separated by thin liquid layers. Aggregates of these bubbles are called foams. The thickness of foam lamellae ranges from 4 nm to over 1 μm.[94]

Foam and bubbles in a barium sulfate suspension can cause artifacts. The resultant defects and irregular coating make radiographic interpretation difficult. Foam also produces inaccurate hydrometer readings, making standardization difficult.

Foam stability can be studied by passing a constant gas stream through a liquid and measuring the resultant foam. An alternate method is to shake a known volume of liquid and gas and then measure the rate of foam collapse. A calibrated apparatus for measuring foam has been described in the radiologic literature.[95]

The physical factors involved in the spreading of a liquid into a thin film apply to foams. Surface tension is one of them. In a liquid, the molecules attract each other, thus tending to form a globule with the least possible surface area. The resulting force represents the surface tension. It varies with the type of liquid and temperature and can be measured by a capillary size, ring, falling drop, and several other methods. The liquid surface acts as an elastic membrane.

Foam stability depends on surface viscosity and the formation of a lamellar framework. Gravity plays a role in foam collapse.[96] The walls in a foam may have various thicknesses. The framework holding the air may also be composed of solid particles with a layer of colloid joining the particles together. Bubbles tend to become less stable as their walls become thinner, and they eventually rupture by Brownian motion.

Pure liquids and saturated solutions do not foam, although contaminants, even in minute quantities, can influence foaming characteristics markedly. Foams occur in solutions of surface-active substances, such as organic salts, acids, sugars, glycerins, alcohols, and esters. Under certain situations the substance concentration, solubility, and viscosity may overcome a tendency to foam.[97]

Gibbs established the principle that solutes that lower the surface tension of a liquid become more concentrated in the surface layer than in the body of the solution.[98] Many common substances tend to accumulate at the surface of an aqueous solution.[99]

Low surface tension combined with low vapor pressure characterizes a foam. Although barium sulfate suspensions having a low surface tension do stick to mucosa and mix with intestinal fluids, they have a tendency to form bubbles and artifacts. However, good mucosal adhesion and coating can be obtained with low surface tension suspensions without any bubbles because of other factors. For example, as the amount of glycerin in a solution is increased, the viscosity rises but the surface tension falls. The lower surface tension results in a better coating that adheres to the surface. At the same time, the increase in viscosity permits a thicker film to adhere to the surface (Table B-1). The higher fluidity of pure water does not make it coat surfaces better than the more viscous glycerin solutions with lower surface tensions. In fact, the reverse is true.

Table B-1 Comparison of Surface Tension and Viscosity of Glycerol Solutions

% Sol	*Surface tension, dynes/cm*	*Relative viscosity, centipoises*
5	72.9	1.125
10	72.9	1.29
20	72.4	1.73
30	72.0	2.45
50	70.0	—
60	—	10.66
85	66.0	—
88	—	147.2
100	63.0	1487

Source: Handbook of Chemistry and Physics, ed 66 (pp D-232, F-31) by RC Weast (Ed), CRC Press Inc, © 1986.

Surface Viscosity

Although the greatest amount of foam formation occurs at low surface tensions, there is no absolute correlation. Deviations occur because of differences in surface-active substances. Two solutions of equal surface tension may differ greatly in the amount and stability of foam produced when they are shaken or when air is bubbled through, as in an air contrast examination of the bowel.

The properties of a substance near an interface are different from those within. The same holds true for bubbles. A film can be considered to consist of an inner layer, which has a viscosity approaching that of the liquid, and an outer variable thickness layer, where the viscosity can differ. In some liquids the surface viscosity is greater than the internal viscosity.

The type of bubble wall formed is the result of cohesion and viscosity of the liquid. When the ratio between superficial viscosity and surface tension is larger than and not proportional to internal viscosity, the bubbles collapse. The longest-lived foams have the lowest surface tension and a high surface viscosity. Their films are non-Newtonian and possess a yield point. They show plasticity or an increase in viscosity as the rate of shear decreases. The presence of monomolecular surface films increases the surface viscosity.[100]

In barium sulfate suspensions the bubble film must be sufficiently viscous or must be stabilized to form a persistent foam. Many hydrophilic colloids, such as soap, proteins, and such glycoproteins as mucin, give stable foams with a high surface viscosity.

A quality separate from surface tension and surface viscosity is internal viscosity. Some highly viscous fluids have high surface tensions and do not form bubbles easily. As an example, some molten glasses have surface tensions of 215 to 320 dynes/cm and viscosities from 10^6 to 10^7 centipoises.

Low viscosity (high fluidity) suspensions often have low surface tensions. They form large and long-lasting bubbles with a relatively high surface viscosity. The persistent foam on beer is an example of such a suspension. The chances of rupture or collapse of a bubble are considerably less in these liquids with a low viscosity because the bubbles are larger.

High internal viscosity does not cause more stable bubbles. Thus, the highly viscous glycerin does not increase the foam-holding capacity of beer to any degree. The addition of mere traces of egg albumen, however, produces an extremely long-lasting foam because of its influence on the surface properties of the bubble walls. The colloidal matter in beer that goes into the interfaces produces the larger bubbles (there are extensive literature and numerous patents on the foaming of beer). Bartsch tried to increase the stability of foam by increasing viscosity.[97] He found that only very high concentrations (about 60% glycerin) substantially increased their stability.

Certain substances, such as albumen and glycerol, have a greater influence on the formation of soft and flexible skins than on viscosity. Many of the resultant elastic foams consisting of large-size bubbles are more stable than those composed of small-size bubbles.

Many liquids possessing high viscosity and low surface tension do not form sustained bubbles. On the other hand, many highly fluid aqueous solutions and suspensions with a surface tension nearly equal to water are capable of marked foaming. Some barium sulfate suspensions have this capacity as well (Table B-2).

A low vapor pressure, in conjunction with a low surface tension and a high surface viscosity, is essential to foam. This is apparently the case in a fairly permanent foam. Salt solutions can also foam. Because salts may increase the fluidity of water, a high surface viscosity is not essential for foam formation, except for the more stable ones. To produce ordinary surface bubbles (not cavitation), it is essential that there be a distinct surface film. The concentration in the surface layer, therefore, must differ perceptibly from that in the mass of the liquid. All colloidal solutions will foam either when the colloid concentrates in the interface or when it is driven away from the interface. If the bubbles are to be fairly permanent, this surface film must either be sufficiently viscous in itself or must be stabilized in some way. In the more stable bubbles, when the surface film cracks and the bubble bursts, the stabilizer in the film does not readily go back into the colloidal solution, but is likely to remain as a precipitate.

Some colloidal solutions foam on agitation. However, colloidal solubility does not always favor foaming. The formation of larger molecular aggregates interferes with foaming.

Because pure substances do not foam, any foam can be considered a system in which the "foam agent" is present as an electrolyte, as a protective colloid, or as a solid foam former. The degree depends upon the components in the system and whether they are lyophilic or lyophobic substances. Lyophobic substances render

Table B-2 Bubble Formation as Measured in Millimeters at Top of Suspension after Vigorous Shaking

*Product**	*Foam, mm*	*Product**	*Foam, mm*
Baridol	6	Large particle barium (av. 4 μm)	4
Bari-O-Meal	11	Large particle barium with additives	2
Barium sulfate compound	12	Liquibarine	0
Baroloid	11	Mallinckrodt USP barium	0
Barosperse	4	Micropaque	13
Barotrast	3	Stabarium	0-1
Basolac	8	Ultrapaque B	5
Gastriloid	10	Ultrapaque C	6
Gastropaque	8	Unibaryt C	9
Intropaque	2	Unibaryt rectal	7
I-X Barium	0–1	Veri-O-Pake	5

*All suspensions are 20% w/w in distilled water.

Source: Reprinted with permission from "Barium Sulfate Suspensions" by RE Miller in *Radiology* (1965; 84:241–251), Copyright © 1965, Radiological Society of North America Inc.

foams unstable and sensitive to electrolytes. The lyophilic systems can be made sensitive to the action of electrolytes by means of other surface-active substances. Gelatin solutions, for example, foam easily, but the stability of the films is greatly influenced by any salts present in solution.

Solids and Foam

Finely divided solids may stabilize foam. The amount of solid that concentrates in the foam's film and the stability of the resulting foam depend upon the nature of the solid particles and the suspension concentration. The size of the solid particles is important, with the major factor being the wettability of the particles, which is inversely related to size. Within broad limits, the smaller the particles, the more stable the foam. All other factors being equal, the smallest particle barium contrast media also foam the most.

The bubbles or foam formed in barium sulfate suspensions can be compared to the ore froth flotation process in which iron sulfide ore is separated from the worthless gangue.[101] By adding a few tenths of a percent of oil per ton of ore separated, a foam is formed. Corresponding to some of the additives in barium contrast media, these minute droplets of oil are coated by the ore. When air is passed through the mixture, bubbles form. Although the specific gravity of the gangue is lower than that of ore, the gangue is hydrophilic (i.e., well wetted) and remains in the liquid. The ore particles are poorly wetted and are lifted to the surface by the air bubbles. The oil additive does not have an affinity for the silicous matter that settles. If there is no ore present, air bubbles are formed and coated with a thin film of oil around which is the water in the outer phase, but the oil film is not viscous enough to form stable bubbles. If finely divided ore is introduced into the oil-water interface, stabilization of the film is achieved because the ore particles enter the interface and form a coating about the air bubble. The resultant foam can then be scraped off the surface.

Protective colloids at certain concentrations can completely displace the solid phase. Generally, the more colloid that is dissolved in the liquid of the film, the more stable the bubble. Many barium sulfate contrast media contain colloids that form bubbles because of their surface activity.

Charge

Electrostatic and capillary forces help maintain foam stability. Gas bubbles are charged negatively against water so that stability of the foam depends upon the valence and adsorption ability of the cations. The electrical charge from the double layer can reach the bubble and give it a negative charge.

The presence of small amounts of electrolyte can result in major changes. If a gas is bubbled through water, the gas takes a negative charge; if the same gas is bubbled through a solution containing electrolytes, it takes a positive charge and the water becomes negative. Therefore, electrolytes, acids, and surface and other active substances produce a change in the charge of the bubbles and can destroy the bubbles. In general, these electrical forces are weak compared with capillary forces. Gas bubbles can adhere to solid particles in spite of electrostatic repulsion of the components. A pH greater than neutral helps prevent foam.

Gas, Temperature, and Pressure

It has been claimed that some gases and a low temperature (5°C) of the barium sulfate suspension will increase bubbles.[102] However, bubbling various gases (carbon dioxide, air, and nitrogen) through barium sulfate suspensions does not result in detectable differences in bubble size and duration when the same barium is used. Wark and Cox,[103] studying flotation, showed that the nature of the gas did not cause any difference in the contact angles and the size of the bubbles formed. Furthermore, barium suspensions, once inside the bowel, rapidly assume body temperature. In fact, high pressure and high temperature reduce the tendency to foam. The thickness of films in soap bubbles decreases considerably as the temperature is increased, resulting in less stable bubbles.[104,105] This property is used by heat transfer engineers, but has limited application in barium sulfate examinations.

Antifoam Agents

The prevention of bubbles is generally desired. Such additives as proteins, mucin, gelatin, dextrine, casein, and sodium citrate are foaming agents. Antifoam agents consist of oily materials, such as castor oil, petroleum, fat, peanut oil, vegetable oils, oil of cloves, alcohols, fatty acids and their esters, capric acid, eucerin, and various silicones. Dissolved salts also act as antifoam agents. Some of these agents have been and still are being used in barium sulfate contrast media.[106] They are not commonly listed on the label.

The rate of foam collapse is a function of the initial distribution of bubble sizes.[107] Once bubbles are formed, they can be broken in barium sulfate suspensions by various means. Capillary-active substances destroy the semisolid framework by dissolving its constituents, and as a result, the bubbles collapse. These surface-active substances can also destroy protein-containing films by denaturing the protein, thus producing flocculation of membranes. Capillary-active compounds, with a low solubility, such as octyl alcohol, break bubbles by interfering with the surface forces acting in the framework.

Silicones are mixed inorganic-organic linear polymers. Silicone oils with extremely low surface tensions are highly effective at low concentrations (0.01 to 2.5%) in breaking foam. A tiny droplet of an antifoam substance touching a bubble sharply reduces the local surface tension, disturbs the surface equilibrium, and thus breaks the bubble.[108] As little as 0.04 to 0.4 g of Dow Corning Antifoam A per kg/barium sulfate powder is sufficient in preventing foaming in most localities.

In general, antifoam agents depend not only on their own nature but also on the nature of the foam-producing substance. Therefore, a given substance can be an antifoam agent toward one foam-producing additive but not necessarily toward another.

Ions may also be antifoam agents. Many synthetic surfactants are ionic and are used as antifoam agents in industry. As an example, oil passing through an oil well may foam, resulting in decreased pumping capacity. The addition of a surfactant causes the bubbles to coalesce and the amount of foam to decrease.[109]

One company producing a barium product for double-contrast studies recommends that the liquid suspension be prepared the night before use, claiming that no sedimentation occurs during the interval before use; however, the extra time allows bubbles to settle.

MINING

Barium sulfate occurs as a natural mineral called barite. It contains varying amounts of impurities. Barytes, the semipurified form of barium sulfate, is obtained from the mineral barite. Other names used for barytes are *heavy spar, heavy earth, terra ponderosa,* and *Schwerspat*. It is presumably the same material as the Bolognan stone (Lapis Boloniensis) of medieval times.[110]

In the United States barite is mined in Alaska, Arkansas, California, Georgia, Missouri, Nevada, and Tennessee.[111] In Canada, extensive deposits are present and mined in Nova Scotia. Extensive barite deposits are also found in Europe and Asia, where they tend to be located below ground and are mined similar to coal. Barite is found either in sediment deposits or in hydrothermal veins.

Barite ore is not found in solid masses but rather in isolated nodules (Fig. B-8). The pieces vary in weight from 30 g to 10 kg. Larger isolated lumps are occasionally found. Much of the barite ore is associated with a silicate of magnesia and alumina, called *chirt*, that looks like barite but is distinguished by a difference in weight. Workers soon become adept at separating chirt from barite ore. Barite is

Figure B-8 Barium sulfate ore nodule. This specimen measures 8 cm in greatest length and undoubtedly contains a number of impurities.

relatively soft and pliable and is usually contaminated with iron and silica. To free the barite from impurities, the raw material can be bleached by a sulfuric acid process. It is then washed and floated so that the resulting barytes is not contaminated with acid.

Barite can be floated from iron by fatty acids and soaps.[112–115] Water-dispersible petroleum sulfonate reagents have also been used to float barite from minerals, with sodium silicate as a depressant for the iron.

One supplier of barium sulfate uses a combined process. Shovels dig the crude product from deposits. The barite is washed and broken into small pieces. Chunks of chirt, silica, other rocks, and heavy metal ores are separated, and the semicrude barytes is shipped to a refining plant. The semicrude product is then ground to a particle size of less than 25 μm. Magnetic particles are removed with a magnetic separator. Soluble impurities are chemically dissolved, and the slurry is placed in washing tanks where the chemicals with their impurities are removed. The remaining slurry is pumped from the wash tanks to a rotary filter and dryer that converts the slurry into a cake and removes nearly all the water. The caked barytes is then ready for industrial use.

By far, the major market for barytes is in oil well drilling muds. It is used as an emulsifier and, because of its high specific gravity, as a sealing or "weighting" agent against the high pressures generated in oil wells. Other commercial applications include paints, special glass products, brake linings, and some food containers. In paints it is used in high grade industrial primers for automobiles, refinery equipment, and other applications requiring high corrosion resistance.[116] In the United States, approximately 80,000 tons of barite are processed each year for the eventual manufacture of barium chemicals.[116]

At present little or no barite mined in the United States is used directly as USP barium sulfate for radiographic contrast agents. Some of the USP product comes from Europe, although deposits are also mined in Nova Scotia, Canada. There are only limited sources of bleached and ground barytes that meet the USP specifications of most industrial nations. The major limitation is the presence of impurities in excess of the amounts permitted by specifications. As a result, the majority of medical-grade barium sulfate is produced by a precipitation method, where the impurities and crystal size can be better controlled.

PRECIPITATION

Most fine particle powders, such as barium sulfate, are obtained by precipitation. The following reactions have been used commercially for precipitating barium sulfate: solutions of barium chloride and sodium sulfate or sulfuric acid; barium sulfide (black ash) solution and sodium sulfate; the peroxide of barium and sulfuric acid; dissolving barium carbonate in nitric or hydrochloric acid and then precipitating with sodium sulfate.

One typical method starts with ground barite ore mixed with coke or coal. In a rotary kiln it is reduced to barium sulfide. The soluble barium sulfide reacts with sodium carbonate to form barium carbonate. The barium carbonate is then treated directly with hot dilute sulfuric acid to form barium sulfate. In England, barium carbonate (Witherite) was mined commercially from natural deposits, but the last mine shut down in 1969.[117] (Witherite is named after William Withering who first identified barium carbonate. He is the same Withering who identified the medical uses of foxglove in 1785.[110])

In all these precipitation processes the final particle size usually ranges from 0.1 to 5 μm, with an average size of 0.5 to 2 μm. Through use of additives some barium contrast agents can be held to a smaller or larger particle size range (Fig. B-9).

The number of crystals formed and the rate of crystal growth depend on supersaturation, temperature, time of reaction, and additives present, with the major factor being the degree of supersaturation.[118] With increasing concentration, the size of the individual crystals initially increases and then starts to decrease. Usually larger crystals and a more uniform-size distribution are obtained

Figure B-9 Precipitated coarse barium sulfate crystals. The mean particle size is approximately 10 μm. This type is normally mixed with smaller particles and used for double-contrast stomach examinations. *Source*: Courtesy of Sachtleben Chemie, Duisburg, West Germany.

at lower concentrations of the reacting solutions. For example, 0.002-M solutions give larger crystals than 0.2-M solutions. In either case, precipitated barium sulfate particles consist of an irregular mosaic of numerous smaller rhomboid crystals.

The crystal growth and dissolution reactions are surface controlled. Little data are available on the effect of additives on the crystallization of barium sulfate. Polyelectrolytes and phosphonate derivatives seem to inhibit crystal growth.[119] Crystals prepared in the presence of sodium or potassium ions, such as sodium sulfate, have a rougher surface than when sulfuric acid is used.[15] Growth and dissolution of the crystal takes place at discrete, active surface sites. Adsorption of impurities, such as sodium, at these sites serves as centers for nucleation during crystallization. Dissolution of barium sulfate seems to be controlled by crystal surface action and not by passive diffusion.[15,120]

Von Weimarn,[25,121] investigating the dependence of particle size on the concentration of the components involved in the formation of colloidal particles, studied the formation of barium sulfate by this reaction:

$$MnSO_4 + Ba(CNS)_2 = BaSO_4 + Mn(CNS)_2$$

Because barium sulfate is sparingly soluble, the supersaturation even at concentrations of 0.001 to 0.0001 M is high, and barium sulfate will crystallize on many centers. However, as the total amount of ions in such extremely dilute solutions is limited, the nuclei cannot grow large and a sol is formed. If the concentration of the reagents is increased from 0.001 to 0.01 M or greater, the supersaturation and the number of nuclei remain the same, but because more material is available the particles can grow larger. Thus, at moderate concentrations of reagents, the particles are relatively coarse. Using 2 to 3 M solutions of manganous sulfate and barium thiocyanate, the particle size decreases again; the mixture viscosity increases and the rate of crystallization becomes impaired. If such a nearly saturated solution is mixed and shaken, the barium sulfate forms a semisolid gel. The higher the concentrations, the more transparent the gel because the particles are smaller. Thus, colloidal suspensions are formed when the concentrations of manganous sulfate and barium thiocyanate are either very high or very low.

Jellies are formed in chemical reactions of concentrated solutions if one product of the reaction is relatively insoluble and if the particles tend to form linear aggregates. Gelation is also promoted by solvation. If the solubility of the substance formed is low and a concentrated solution is used, no large crystals can grow because of impaired diffusion. Consequently, all of the insoluble material will crystallize out on billions of nuclei, with the resultant small crystals enclosing the associated liquid and forming a jelly. If a barium sulfate sol is to be stable, the solubility of the dispersed phase must not exceed 1 mg/L.[122] The reason for choosing $MnSO_4$ and $Ba(CNS)_2$ is that both these salts are extremely soluble, and hence it is possible to vary the concentrations over wide limits.

In precipitating barium sulfate, smaller crystals are obtained by high concentrations of the reacting solutions, low temperature, high viscosity, rapid addition of the precipitant, rapid mixing, and using solutions of lower acidity or those with a reduced solubility for barium sulfate. A stable supersaturated barium sulfate solution can be prepared by slowly mixing barium chloride and sulfuric acid.[15] One of the reagents must be added drop by drop, with constant stirring, to avoid an increase in local concentration that might induce nucleation. Extreme supersaturation and subsequent formation of exceedingly fine precipitates can be obtained by using a solvent in which the resulting salt is practically insoluble.[123] An example would be substances reacting to give barium sulfate in a 50% ethyl alcohol-water mixture. Both increased temperature and greater acidity decrease the supersaturation of barium sulfate, making the precipitate coarser. For example, when barium sulfate is precipitated hot, it does not pass through a filter as readily as when precipitated cold because the crystals are larger when precipitated hot.

Any substance that is adsorbed by another tends to disperse or deflocculate the latter. Crystal growth is prevented in a precipitate by the presence of a substance that is strongly adsorbed. It follows then that, other conditions being equal, barium sulfate is more finely divided when precipitated in the presence of those substances for which it has the greatest specific adsorption. The cations, hydrogen and barium, are strongly adsorbed. Thus, barium sulfate precipitates in a finely divided form in the presence of barium chloride or potassium sulfate.[31,124] It is more coarse in the presence of sulfuric acid or hydrochloric acid.[125] The solvent action of hydrochloric acid is also a factor in making a barium sulfate precipitate more coarse.[31,126] The peptizing action of sulfate ions in sulfuric acid is decreased by the tendency of the hydrogen ion to be adsorbed strongly. The selective nature of adsorption is shown even further by barium sulfate because it carries down much larger amounts of barium nitrate than barium chloride.[27]

The action of other substances in the solution depends upon their effect on supersaturation and whether they are adsorbed by the precipitate. The influence of foreign molecules can be studied when barium sulfate precipitates from aqueous reacting solutions to which some gelatin, sodium citrate, glucose, or agar has been added. These substances are adsorbed on the crystals and keep them from growing. Similarly, the addition of small amounts of sodium carboxymethylcellulose retards the growth rate of barium sulfate crystals, and the crystals are smaller in size.[127] Several substances can be adsorbed simultaneously. In addition, because of large adsorption forces, the solid surfaces become contaminated with at least a monolayer of water vapor.[128]

Other factors in the crystal growth and precipitation process include variations in solubility because of the size of the primary particles, the presence of dust particles, the extent of agitation or mixing, and the specific tendency for crystals to form and grow from nuclei. The rate of precipitation is not constant and follows second-order kinetics; at high supersaturation there is an initial surge, followed by a gradual slowing as the degree of supersaturation decreases.

The aggregation and orientation velocity of the reacting molecules influence the form of the precipitate.[129] If in precipitating the same substance the primary aggregation velocity predominates, the resulting precipitate is amorphous, but if the orientation velocity predominates, the precipitate is crystalline. The amorphous precipitate becomes crystalline, however, upon aging.

After precipitation, vacuum filters are frequently used for dewatering the barium sulfate. The thickened, partially settled suspension may enter at 50 to 70% moisture and be discharged at 35 to 50% moisture. Colloidal solids present in the suspension and a low temperature decrease the filtering rate and increase costs. After filtering, the dewatered product is fed to a dryer for removal of remaining water and is then packaged or processed further by the addition of surface-active agents. Precipitated barium sulfate is commonly known as blanc fixe.

MECHANICAL SIZE REDUCTION

Although most true sols are made by precipitation, the formation of sols by mechanical grinding can be obtained under certain circumstances. Pliny the Elder wrote in his *Natural History* that the Greeks made metallic inks by grinding with honey and carbon inks by grinding with gum. Sols can be made by grinding the dry solid to be dispersed with an indifferent solid that will dissolve on mixing with the dispersion medium. This method prevents the coalescence of extremely minute particles to form coarse aggregates. True sols of barium sulfate, silver, gold, and other substances have been obtained this way with glucose or lactose used as the neutral diluting agent.

Utzino ground barium sulfate with pure anhydrous glucose in an agate mortar at room temperature in a continuous series of three dilutions.[130] Each time he added nine parts of glucose to one part of the previous grinding that contained barium sulfate. The material was ground for a total of 4 hours, and a dispersion was obtained consisting of 0.038 μm particles in 55% ethyl alcohol. This sol contained 0.0025 g of barium sulfate/1 L and was a colorless transparent liquid that showed little sedimentation after 24 hours. In 5 months all of the particles had nearly settled and no longer showed any appreciable Tyndall effect.

Barium sulfate has a hardness rating of 3–3.5 on Moh's scale of hardness, which rates talc as 1 and diamond as 10. It has about the same hardness as marble. Lactose, because of its hardness, is often used as a grinding agent if mechanical particle size reduction is to be done.

Mechanical size reduction can be performed with various grinding machines, with a typical example being a ball mill. Such machines can produce particles of less than 44–46 μm in size.[131] Some hammer mills can result in average particles as small as 2 μm. Plastic deformation can increase the resistance to breakage of very small particles and limit the grinding size attainable.[132] In general, other means of size reduction must be used if particles smaller than 0.5 μm are required.

The mechanical size reduction devices produce particle sizes having a wide range and uneven surfaces. The mills are also useful in blending various additives. In general, mechanical grinding to relatively small particle sizes requires prolonged grinding times and high energy expenditure. Wet grinding tends to result in greater production rates than dry grinding.[132]

Commercially, barium sulfate particles in the fraction of a micrometer range (0.1 to 0.9) are obtained by chemical precipitation methods that use additives to control the speed of formation and size of the precipitating particles. Their agglomeration must be prevented by additives.

ADDITIVES

Additives and their size, shape, and charge markedly influence a final suspension's physical properties at least as much as the size and shape of the barium sulfate particles themselves. The major properties involved are surface tension, viscosity, density, and flocculation. These physical properties determine flow, clotting, adhesion, bubble formation, weak-spot resistance, ''alligatoring,'' flaking, settling, film thickness, and density. These factors, in turn, determine the suspension's ability to coat the intestinal mucosa.

The texture of barium sulfate preparations can be improved by the addition of a suspending agent, such as methylcellulose derivatives and various gums. Too great an amount, however, can change physical properties, such as viscosity, leading to subsequent poor mucosal coating. One method of introducing an additive is simply to mix a barium sulfate suspension with a hydrophilic colloid, such as a methylcellulose derivative. Such a colloid helps prevent the suspension from settling. The amount of additive adsorbed can be enhanced considerably by the presence of electrolytes.[133]

At very dilute concentrations, long-chain polymers can sensitize. Sensitization is the opposite of stabilization; it can be produced by an extremely small amount of additive. It can be produced by minute amounts of the same additives that act as stabilizing agents when used in larger amounts.[133] Polar additives may form a first layer with their hydrophobic part outward, thus sensitizing, and then a second layer may be oriented oppositely, leading to protection. With some long-chain additives, such as modified starches, flocculation occurs if two particles become joined by a long chain being adsorbed on both.[134] This keeps the two particles close together and favors more links by other chains. In dilute solutions the coverage of all particles by additives is incomplete so that a free-floating segment may be near another particle and can easily find a spot to which it can attach itself. On the other hand, in a concentrated solution the coverage is high so that the chances for multiple attachments are decreased. For example, a small amount of gelatin added to barium sulfate causes rapid flocculation and simplifies its filtration. This flocculation occurs only in a narrow range of concentration. At higher concentrations, there is a protective action.[135]

In general, in a colloid system anything that destroys the protector tends to cause the protected dispersion to flocculate. Anything that assists in making the protector more stable causes the dispersion to be more resistant to flocculation and changes in viscosity. Sometimes a blend of protecting agents is of benefit. For example, the addition of a germicidal agent in cases where casein is used as a protector offsets the bacterial action that would otherwise destroy the casein film around each particle. Some barium sulfate contrast agents have similar protective colloids that spoil on long standing.[18]

An example of a long-chain additive is pectin. The various pectins are complex hydrophilic linear polysaccharides containing carboxyl groups. Pectin depends on both charge and solvation for stability. Its charge comes from dissociation of atomic groups belonging to the pectin particles. The negative carboxyl groups in pectin repel anions and are sensitive to acids. Other examples of such substances in barium sulfate products are agar and various gums, such as gum arabic, that are polysaccharide chains that carry acid or ester groups. Still other additives are starch and the celluloses that are branched, and the straight-chain polysaccharides.

The long-chain polyelectrolyte additives thus interact with other additives to form complex systems to suspend and protect the barium sulfate particles from flocculation. The results can be subtle. As an example, in water sodium carboxymethylcellulose is adsorbed to barium sulfate as trains, whereas in the presence of electrolytes the molecules are in an extended configuration.[133] These complex systems, in turn, affect viscosity, thixotrophy, and other physical phenomena. They exert this effect primarily because of structure and change in structure with ionic charge, solvation, concentration, and different orientation at different flow rates. All of these factors in turn affect flow.

When lead pigments are used alone in paints, they tend to chalk and "alligator," whereas, zinc pigments, when used alone, often scale and flake. When lead and zinc are combined in the same paint, the overall tendency toward defect formation is reduced. Barium sulfate has similar properties. When a thin coating is applied to the mucosa, it will crack and flake. The addition of such compounds as bentonite (sodium montmorillonite) can prevent these defects.

Drying and flaking are minimized by nonabsorbable additives, such as sorbitol, mannitol, and sodium carboxymethylcellulose. These additives act as stabilizers by making the film formed by the suspension softer, more flexible, and less susceptible to cracking by the dehydrating action of the colon.

Anionic low polymer colloids, such as some polysaccharide derivatives, when adsorbed to barium sulfate, intensify a negative charge. As a result, they can inhibit agglomeration and flocculation. The suspension's viscosity may also be decreased.

Surface adhesion is often improved by additives that have active polar groups, such as carboxyl and hydroxyl moieties. Mucosal adhesion is also enhanced by an additive interlayer, the molecules of which are so oriented that one side adheres to the mucosal surface and the other side to the barium sulfate particles.

Many different additives have been used in barium sulfate contrast media. These additives were used in the past and are used currently to reduce flocculation; improve suspension properties, patient acceptance, viscosity, mixing, bubble formation, and coating; and as preservatives. The strong adsorption property of barium sulfate influences many of the processes used in commercial barium sulfate production. The aqueous slurries of precipitated particles of barium sulfate almost always contain adsorbed soluble salts and additives for controlling particle size; they may not be completely removed by washing. There may also be traces of other compounds that have been added for flocculation control. These agents are added for easier decantation or filtration or for a softer cake to make postdrying pulverization easier and more economical.

Great secrecy has surrounded many of these additives. Even now, only a limited number have been disclosed in the medical and patent literature. Few are listed on the product labels, although the manufacturers are becoming more open. Table B-3 lists some of the additives that have been used in the past or are used currently. Although incomplete, the table shows the complexity of the problem and indicates the almost endless number of possibilities when various substances are incorporated in various proportions and combinations.

Various food products were added in the early years of barium sulfate development. These ranged from eggs[136] to bread, butter, and coffee.[137] Chocolate milkshakes containing barium sulfate have been described.[138] One medium contained protein, fat, carbohydrate, and barium sulfate;[139] it was stored in powder form and was advocated for bowel motility studies.[140] Barium sulfate pellets have also been used to study motility.[141] Similar radiopaque markers became commercially available in 1987. A mixture of hamburger meat and barium sulfate measured gastric retention.[142] Another combination contained barium sulfate, gelatin, alcohol, and bicarbonate of soda.[143] Barium sulfate has been incorporated into ice cream (barium sulfate 30%, butterfat 8.5%, sugar 8.9%, defatted milk protein 7 to 8%, plus Tween emulsifier).

Sweetening agents have included saccharin and various sugars. Sugar should be minimized because of its effect on diabetic patients. Similarly, chocolate and chocolate-based syrup are probably best avoided because of possible allergy in an occasional patient. Flavoring helps minimize the unpleasant taste of barium sulfate. A mixture of flavors seems to be better tolerated than a single flavor.[144]

Table B-3 Additives in Barium Sulfate Contrast Media

Acacia
Agar
Alcohol
Alevaire*
Alginic acid
Alkali salts of polysaccharide-sulfuric acid esters
Aluminum hydroxide
Aluminum silicate
Avicel*
Bentonite (sodium montmorillonite)
Buttermilk
Carbopols
Carboxymethylcellulose
Carboxymethyldextran
Carrageenan
Cellulose sulfate
Cereals (processed)
Chocolate milk
Citric acid
Cocoa
Colloidal aluminum hydroxide
Dextrose
Diatomaceous earth
Dimethylpolysiloxane (simethicone)
Dioctyl sodium sulfosuccinate
Egg yolk
Ethyl polysilicate
Ethyl vanillin
Eucarin (wool fat and petrolatum)
Flavoring agents of many kinds
Gelatin
Glycerin
Gum arabic
Gum ghatti
Gum karaya
Gum tragacanth
Honey
Hydroxybenzoates
Koalin
Lecithin
Levulose
Lignosulfates
Magnesium aluminum silicate
Magnesium montmorillonite
Magnesium trisilicate
Malted milk
Maltose
Mannitol
Methylcellulose
Methylparaben
Methyl polysilicon
Mineral oil
Mono-oleate of sorbitan polyethylene glycol
Mucilaginous gels
Mucin
Myrj 45*
Oil of lemon
Oil of orange
Palatone* (Dow)
Pectin
Polyoxethylene derivatives
Polystyrene
Polyvinyl pyrrolidone
Potassium sorbate
Propantheline
Propyl paraben
Simethicone (dimethylpolysiloxane)
Sodium alginate
Sodium bicarbonate
Sodium carbonate
Sodium carboxymethylcellulose
Sodium carrageen (Irish moss)
Sodium cellulose acetate sulfate
Sodium chloride (saline)
Sodium chondroitin sulfate
Sodium citrate
Sodium cyclamate
Sodium dextran sulfate
Sodium hydroxyethylcellulose
Sodium lauryl sulfate
Sodium montmorillonite (bentonite)
Sodium saccharin
Sodium salt of polyalkylnaphthalene sulfonic acid
Sorbitol
Spiritus methyli-*p*-oxybenzoatis
Starch
Sucaryl*
Sucrose
Sugar
Tannic acid
Tartaric acid
Tragacanth
Tween-80*
Vanillin
Veegum*

*Trade names.

One mixture found acceptable contained sweet cherry, wild cherry, raspberry, orange, and vanilla. A sharp flavor, such as peppermint, lime, apricot, or citric acid, can also be useful.[144]

An antifoaming agent, such as dimethylpolysiloxane (simethicone), is commonly added to barium products to decrease bubble formation. Simethicone is a physiologically inert substance that is not absorbed.

Many commercial preparations contain a preservative. The alkyl-*p*-hydroxybenzoates, methylparaben, and other preservatives have been used. Because of a possible hypersensitivity reaction, methylparaben has been replaced by other, more innocuous preservatives in a number of current products. Although barium sulfate by itself does not support bacterial growth, many of the additives are organic compounds and do support such growth. It is possible that the preservatives may not inhibit growth completely, and any freshly mixed suspensions probably should be refrigerated if kept for an extended period of time. A change in odor of a barium preparation should be viewed with suspicion. Contamination with a number of organisms has been reported.[145]

Whether placed purposely for process control, left by accident, remaining because removal is costly, or used later as an additional and separate final processing step, the various additives influence markedly the coating characteristics of the final barium sulfate product.

MIXING

Dry Blending

It has been long known that fine powders adhere to coarser particles when they are blended.[146,147] When a coarse red powder is mixed with a fine white powder the resultant product appears white. If the red powder is fine and the white powder coarse, the blend appears red. Some red paints contain up to 90% barium sulfate; the white barium sulfate does not show up because of the fineness of the red powder. Thus, the color of a coarser powder may be masked completely by a comparatively small quantity of fine powder if it is fine enough to result in adequate coating. As a further example, a white nonconducting powder, such as thoria, when mixed with a black conducting powder, such as tungsten, may be either black and an electrical conductor or white and a nonconductor, depending entirely upon the relative coarseness of the two powders.[147] The finer powder in each case determines the insulating properties.

The blending of dry barium sulfate with dry additives may be done in a number of blenders or mixers. These machines can mix thoroughly powders of various densities, such as bentonite, sugars, flavorings, and barium sulfate. Such dry mixing, however, does not impart any significant surface treatment to the particles, such as coating them with an additive. In general, dry blending requires precise formulation with little further room for correction. Powdered particles already coated with additives, however, can be adequately dry mixed. Such dry mixing can be done in hospital pharmacies and manufacturing plants as a ''pre-mixing'' step before a surface treatment process.

The adsorbed air on each individual particle may make it difficult to wet some powders. This is often found with barium sulfate products when they are in a finely divided form. It may not be possible to release the adsorbed air from each particle; the adsorbed air is displaced slowly, even with vigorous mixing. Thus, mixing time affects the resultant coating properties of the product.

In liquids, large particles settle to the bottom first. The opposite is seen with powders. Because of gravity, the smaller, smoother, and more spherical particles tend to filter to the bottom.[148] If a container of powder is repeatedly handled, a large particle may move sufficiently so that smaller particles settle beneath it. The reverse, where a number of small particles are simultaneously displaced and a large opening is created, is not common. The larger particles thus migrate to the top of the container. As a result, a container of originally well-mixed barium sulfate powder may be nonhomogeneous when it reaches the user. The difference in subsequent coating may be noticeable between a suspension produced from the top of the container and one from the bottom.

Wet Mixing

In wet mixing all of the ingredients are in a suspension or solution. The water can eventually be removed by a number of drying processes. The finished dry product is then reconstituted with water just before use. The advantages of wet mixing include ease of final reconstitution and a decrease in the ''gritty'' feeling voiced by some patients.[149] Because of the need for the extra drying process, wet mixing tends to be more expensive than dry blending.

Because of an affinity for adsorbing air upon their surfaces, some fine particles may not come in close contact even after the powder is agitated in the presence of a liquid. One such example occurs when finely divided calcium carbonate is added to water and agitated. A thick paste is obtained after adding about 25% of the calcium carbonate. It is possible to shear off the adsorbed air cushions and obtain proper water-powder contact by passing this paste through a colloid mill. The resultant suspension is fluid. An additional 25% of calcium carbonate can then be added to the suspension and the operation repeated. The resultant suspension contains 50% calcium carbonate and is still quite fluid.

The importance of particle size on the amount of adsorbed air and the weight per volume can be illustrated by lampblack. With some of the finer grades, only 1.8 or

2.3 kg can be packed into a barrel, whereas with the coarser types, 9 to 14 kg may occupy the same space.

At times it is not possible to achieve adequate wetting by simple mixing; additional mechanical treatment may be necessary to remove the adsorbed air. Once the particles are thoroughly wetted, they will remain in suspension as long as there are no ions of opposite charge or some other condition to induce flocculation.

The final properties of a barium sulfate preparation are markedly affected by the type of equipment and method used to mix the suspensions. Different mixing processes can change the action of additives in barium sulfate preparations, which in turn can change the viscosity, settling, flocculation, and mucosal coating. For example, in a high shear mixer considerable adsorbed air can be sheared off and replaced by water. If the adsorbed water is primarily on the additives, the suspension may assume a gelatinous and highly viscous form. If the mixing time is then prolonged further, any long-chain additives may be broken and the viscosity will decrease. Thus, simply changing the length of time a suspension is in a mill or mixer may result in different final product properties. Likewise, whether a product is hand stirred or machine blended can affect the resultant viscosity.[150]

In one commercial manufacturing process, purified water, various additives, and barium sulfate are custom mixed to produce a suspension having known properties. The number of steps used, type of ingredients, and sequence of addition are proprietary. The resultant barium suspension is then packed in cans, the cans inspected for pinhole leaks, weighed, and packaged for eventual sale.

Granulation

Some barium sulfate processors granulate the fine powder into larger, more easily wettable, but still readily dispersible aggregates. Another procedure, applied to household flour, is to impart a strong electrostatic charge to the particles so that they repel each other vigorously and thus prevent clumping when they are first added to a liquid.

Granulation refers to the formation of 0.1–2.0 mm aggregates produced by moistened powders.[151] Granulation of barium sulfate is performed to debulk the material to prevent the formation of large cakes or lumps, to make it more easily mixable with water, and to enable more convenient storage and shipment.

A number of granulating methods have been described in the pharmaceutical literature.[152] As an intermediate pathway in forming granules from the dry powder, barium sulfate may be moistened with water, solvents, or binding solutions. Granulation formation requires a liquid-binding solution. A powder consisting of nonuniform-size individual particles tends to form stronger granules than one with uniform-size particles.

Granulation can be achieved with a slurry of barium sulfate in a spray drying process (Fig. B-10). The spray-dried suspensions vary in particle size and density depending upon the type of nozzle, drying temperatures, and efficiency and size of the dryer. The granule growth rate is dependent on the amount of liquid present and in some mixers on the droplet size.[151]

Various additives may be introduced, removed, or retained during the granulation process. Some of the binding materials that have been used for granulation are sugars, gelatin, dextrins, gums, starch, flour, molasses, and sodium carboxymethylcellulose. Among binders introduced dry and later activated are pulverized sugar, spray-dried glucose, malt extract, or any of the previously mentioned binders. More binder is required when it is introduced dry than when it is introduced in solution. A dry binder can result in smaller size granules and formation of lumps because of inhomogeneous binder distribution.[151] Most binders are therefore dissolved in the liquid before addition to the granulation material. After the damp mixture is produced, it is dried and reground in one or several stages. It can be processed through special granulation machines to produce the required granular product. Granulation results in a large reduction in bulk, but permits the small particle size to be retained when the powder is eventually mixed into an aqueous suspension ready for use.

Wetting Agents

The barium sulfate particles possess such a large surface area that interfacial forces among the particles, water, and the additives produce considerable effect both in the suspension and in the barium film deposited on the mucosa. Considerable air is trapped between particles in voids. This air must be removed before adequate wetting is achieved. Incomplete wetting can result from poor mixing. Unless thorough wetting is accomplished, inordinate settling and weak spots may form in the coating. Either a thin envelope of tightly bound air or an extremely tenacious moisture film may surround the barium sulfate or additive particles and retard good wetting.

Wetting agents form a monomolecular layer adsorbed over the particles. For example, an oriented layer with carboxyl groups directed toward the barium sulfate particles acts as a lubricant and prevents the particles from agglomerating, thus making them more easily wettable. Agents that lower the surface tension of a liquid are useful in wetting fine powders. The dispersion or wetting efficiency in some fatty acid additives increases with an increase in chain length until it reaches an optimal level in the 8-carbon atom molecule.[153]

During manufacture the barium sulfate particles are exposed successively to several additives. If an additive is strongly bound to the particle by adsorption, successive additives possessing weaker adsorption forces will not displace it.

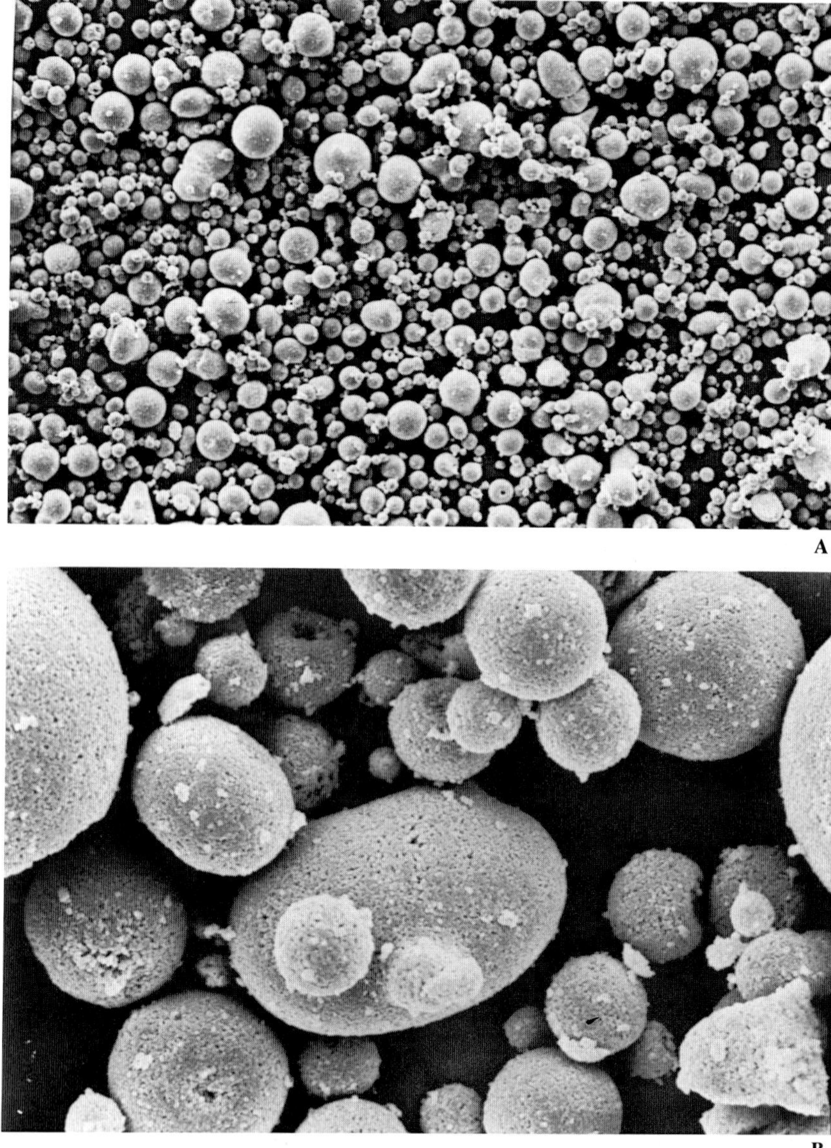

Figure B-10 Barium sulfate granulation. The product was granulated during a spray-drying process. The primary barium sulfate crystals are approximately 0.6 μm in size and were sprayed with a sodium citrate solution before being dried. Additives stabilize the suspension. In water these granules disperse rapidly into individual barium sulfate particles. (**A**) Approximate 100:1 magnification. (**B**) Further magnification to approximately 1000:1. *Source*: Courtesy of Sachtleben Chemie, Duisburg, West Germany.

Thus, the order of additive addition during manufacture influences the coating properties of the final product. Knowledge of all additives in a product is not enough to predict a product's quality. The sequence of manufacturing steps is generally considered a "trade secret."

Product Mixtures

Most barium sulfate suspensions are associated with electrical charges. The individual particle charges help keep them in suspension. If two products of similar charge are mixed together, the resultant suspending properties may be unchanged. If a product having an opposite charge is added, however, flocculation can occur and the suspending and coating properties will deteriorate.

In a mixture of oppositely charged sols, the degree of flocculation and sedimentation varies with the purity of the sols and with the presence of any electrolyte or additive not removed previously.[154] Particle-coating colloids can prevent such flocculation by keeping the particles separated. Some "colloidal" barium preparations contain such hydrophilic protective colloids and their mixtures can be compatible to a varying extent, whereas other preparations flocculate when mixed together. In some systems there is adsorption of one colloid by another, and at times two colloids of the same charge can even be adsorbed on each other.[154]

Some manufacturers pasteurize their liquid barium preparations; others do not, believing that the elevated temperatures required for pasteurization change the stability of the formulation.

A test of over 20 commercial barium sulfate products found that the products differed from each other in their additives, coating ability, grain size, foaming, settling, viscosity, and other characteristics.[18] Compatible preparations are those in which the different additives are compatible with one another. At times a combination of two preparations will improve the suspension properties of the final mixture. The combination may also have an effect on flow rate, mucosal coating, and foam formation.

WATER SUPPLY

The same brand of coffee might taste excellent in one location but be inferior in another, with the only difference being the water used. Similarly, a barium sulfate product can have different coating properties when mixed in different locations.

The hardness of water is a measure of the amount of soap needed to produce suds. Hardness is caused principally by calcium and magnesium ions; other alkaline earths, such as barium and strontium, free acid, and heavy-metal ions, also contribute to hardness. Hardness is expressed in milligrams per liter as calcium carbonate equivalent to the calcium and magnesium present in solution. The hardness range is classified as 0 to 60, soft; 61 to 120, moderately hard; 121 to 180, hard; and more than 180, very hard.[155] In both the treated and untreated water supplies of the 100 largest cities in the United States, the minimum and maximum degree of hardness ranges from 0 to 738.[155]

A change in salt concentration results in marked variability in the amount of foam. This partly explains why various barium sulfate products can have different foaming characteristics from one radiologist to another when there are differences in the amount and type of dissolved ions in the local water supplies. Water-insoluble soaps are formed readily from the replacement of sodium or potassium ions by the magnesium, calcium, iron, and other ions found in hard water. These calcium and magnesium salts can form insoluble compounds with the additives present in commercial barium sulfate products. In the absence of certain additives, low surface tension agents do not enter the gas-fluid interface in the foam and stabilize the gas; as a result, bubbles collapse because of lack of stability. Thus, some barium sulfate suspensions do not foam and will perform satisfactorily in one radiology department but foam in another. For example, Unibaryt C does not foam in Malmö, Sweden, where the water is very hard. Yet, the same barium foams markedly in Göteborg, Sweden, where the water is soft. It also foams in distilled water. Some barium products coat best when mixed with distilled water. In some localities, however, tap water gives best results.

Water hardness can differ between the hot and cold water supplies. Some hospitals have installed water softeners in the hot water supply to decrease boiler scale formation. As a result, a barium preparation mixed with the hot water supply can result in different coating characteristics than the same preparation mixed with cold water.

High calcium ion levels in the water supply are associated with poor mucosal coating. The loss in coating is believed to be proportional to the water calcium level.[156]

If sodium and potassium salts are present in a concentration greater than 50 mg/L and there is suspended matter present, foaming may result.[155] Of interest is that all 100 major water supplies in the United States contain barium ranging from 1.7 to 380 μg/L. Drinking 2 L/day would provide 0.3 to 30% of the total daily barium intake.[157]

SUMMARY

No true colloidal barium sulfate contrast media are commercially available. If available, they would be clear, and either extremely dilute, extremely viscous, or suspended in undesirable liquids. Thus, true colloidal barium sulfate is unsuitable for conventional radiologic contrast examinations.

All barium sulfate preparations contain additives that, through surface action, affect surface tension, hydration, particle charge, and pH. These physical properties influence particle size, flocculation, suspension, and viscosity. These properties, in turn, determine the suspension's ability to settle, flow, form blood clots, and produce artifacts; to coat and stick to the mucosa, form various film thicknesses, achieve and maintain various degrees of density; and to form an image on radiographs.

REFERENCES

1. Weast RC (ed): *CRC Handbook of Chemistry and Physics* ed 66. Boca Raton, FL, CRC Press, 1986, p B-9.

2. Weast RC (ed): *CRC Handbook of Chemistry and Physics*, ed 66. Boca Raton, FL, CRC Press, 1986, p B-76.

3. Weast RC (ed): *CRC Handbook of Chemistry and Physics*, ed 66. Boca Raton, FL, CRC Press, 1986, p B-213.

4. James AM, Goddard GH: A study of barium sulfate preparations used as x-ray opaque media. I. Particle size and particle charge. *Pharm Acta Helv* 1971;46:708–720.

5. Knoefel PK: *Radiopaque Diagnostic Agents*. Springfield, IL, Charles C Thomas, 1961, p 20.

6. Lotz W, Liebenow S: Erweiterte diagnostische Möglichkeiten der röntgenologischen Magenuntersuchung mit verbessertem Kontrastmittel. *ROEFO* 1979;131:157–165.

7. Virkkunen P, Loumatmaa K: On the differences between the $BaSO_4$ particles and additives in media for the double contrast examination of the stomach. *ROEFO* 1980;133:542–545.

8. Anderson W, Harthill JE, James WB, et al: Barium sulphate preparations for use in double contrast examination of the upper gastrointestinal tract. *Br J Radiol* 1980;53:1150–1159.

9. Bechhold H, Hebler F: Nephelometrie gefärbter Hydrosole. *Kolloid-Z* 1922;31:7–12.

10. Brown GR: High-density barium sulfate suspensions: improved diagnostic medium. *Radiology* 1963;81:839–846.

11. Parfitt GD: *Dispersion of Powders in Liquids*. New York, John Wiley and Sons, 1973, p 377.

12. Brindley GW: The interpretation of broadened x-ray reflections with special reference to clay minerals. *Faraday Soc Disc (London)* 1951;11:75–82.

13. Knoefel PK, Davis LA, Pilla LA: Agglomeration of barium sulfate and Roentgen visualization of the gastric mucosa. *Radiology* 1956;67:87–91.

14. Garret MD, Hess WM: Dispersion analysis of paint films. *J Paint Technol* 1968;40:367–378.

15. Nankollas GH, Liu ST: Crystal growth and dissolution of barium sulfate. *Soc Petrol Eng J* 1975;15:509–516.

16. Kralj-Lenardic S, Kozjek F: Physical stability of barium sulfate suspensions. *Farm Vestnik (Ljublijana)* 1981;32:27–32.

17. Oden S: Die automatisch registrierende sedimentiervorrichtung und ihre anwendung auf einige kolloidchemische probleme. *Kolloid-Z* 1919;26:100–121.

18. Miller RE: Barium sulfate suspensions. *Radiology* 1965;84:241–251.

19. Gelfand DW, Ott DJ: Barium sulfate suspensions. An evaluation of available products. *AJR* 1982;138:935–941.

20. Tyndall J: On the action of rays of high refrangibility upon gaseous matter. *Roy Soc Phil Trans (London)* 1870;160:333–365.

21. Vold RD, Vold MJ: *Colloid and Interface Chemistry*. London, Addison-Wesley Publishing Co, 1983, p XXI.

22. Embring G, Mattsson O: Barium contrast agents. *Acta Radiol Diagn* 1968;7:245–256.

23. Brown R: A brief account of microscopical observations made in the months of June, July, and August, 1827, on the particles contained in the pollen of plants; and on the general existence of active molecules in organic and inorganic bodies. *Phil Mag* 1828;4:161–173.

24. Perrin J: Mouvement brownian et realité moleculaire. *Ann Chim Phys* 1909;Ser 8;18:5-114.

25. Von Weimarn PP: Über die Darstellung sogenannter kolloidamorpher Bildungen gut kristallisierbarer und gut wasserlöslicher Salze der erdalkalischen Metalle. *Kolloid-Z* 1908;3:89–91.

26. Von Weimarn PP: *Zur Lehre von den Zuständen der Materie*. Leipzig, T Steinkopff, 1914.

27. Weiser HB, Sherrick JL: Adsorption by precipitates. I. Adsorption of anions by precipitated barium sulphate. *J Phys Chem* 1919;23:205–252.

28. Jirgensons B, Straumanis ME: *A Short Textbook of Colloid Chemistry* ed 2. Oxford, Pergamon Press, 1962, p 305.

29. Sheludko A: *Colloid Chemistry*. New York, Elsevier Publishing, 1966, p 208.

30. Kato Y: Studies on colloidal barium sulfate. *Mem Coll Sci Kyoto Imperial Univ* 1909; 2:187–215.

31. Weiser HB: The effect of adsorption on the physical character of precipitated barium sulfate. *J Phys Chem* 1917;21:314–333.

32. Sanders HL: Preparation of colloidal barium sulfate. US Patent, 2 597 384, 1952.

33. Lenher V: Some properties of selenium oxychloride. *J Am Chem Soc* 1921;43:29–35.

34. Gengou O: Beiträge zum Studium der molekularen Adhäsion und ihres Einflusses bei verschiedenen biologischen Erscheinungen. *Kolloid-Z* 1911;9:88–92.

35. Soxhlet F: Milk-globules, and a new theory of churning. *J Chem Soc* 1876;30:537–538.

36. Taylor WE: The precipitation of aluminum hydroxide in the granular form. *Chem News* 1911;103:169.

37. Bancroft WD: The theory of dyeing. III. *J Phys Chem* 1914;18:385–437.

38. Overbeck JTG: Birth, life and death of colloids, in Eicke H-F (ed): *Modern Trends of Colloid Science in Chemistry and Biology*. Basel, Birkhäuser Verlag, 1985, p 12.

39. Novosel B: Precipitation system of barium sulfate in aqueous medium and some mixed media of different dielectric constants. *Colloid Polymer Sci* 1976;254:650–655.

40. Morimoto T: The electrokinetic potential of sparingly soluble salts. *Bull Chem Soc Japan* 1964;37:386–392.

41. Hirtzel CS, Rajagopalan R: *Colloidal Phenomena*. Park Ridge, NJ, Noyes Publications, 1985, p 30.

42. Sato T, Ruch R: *Stabilization of Colloidal Dispersions by Polymer Adsorption*. New York, Marcel Dekker, 1980, pp 41–51.

43. Kuhn W: Über die Gestalt fadenförmiger Moleküle in Lösungen. *Kolloid-Z* 1934;68:2–15.

44. Illum L, Jacobsen LO, Müller RH, et al: Surface characteristics and the interaction of colloidal particles with mouse peritoneal macrophages. *Biomaterials* 1987;8:113–117.

45. Schwartz SE, Fischer HW, House AJS: Studies in adherence of contrast media to mucosal surfaces. *Radiology* 1974;112:727–731.

46. Coffey MD, Lauzon RV: A particle electrophoresis study of barium sulfate inhibition, in *International Symposium on Oilfield Chemistry*. New York, Society of Petroleum Engineering of AIME, 1975, pp 93–99.

47. Simmonds RJ, James AM: An *in vitro* investigation of barium meals. *Cytobios* 1976; 15:191–200.

48. Sato T, Ruch R: *Stabilization of Colloidal Dispersions by Polymer Adsorption*. New York, Marcel Dekker, 1980, p 106.

49. Thompson MW: Types of polymerisation, in Buscall R, Corner T, Stageman JF (eds): *Polymer Colloids*. London, Elsevier Applied Science Publishers, 1985, pp 13–14.

50. Buscall R, Ottewill RH: The stability of polymer latices, in: Buscall R, Corner T, Stageman JF (eds): *Polymer Colloids*. London, Elsevier Applied Science Publishers, 1985, p 158.

51. Williams PA, Harrop R, Phillips GO, et al: Effect of electrolyte on the stability of barium sulfate dispersions in the presence of sodium (carboxymethyl) cellulose. *Ind Eng Chem Prod Res Div* 1982;21:349–352.

52. Astley R, French JM: The small intestine pattern in normal children and in coeliac disease: its relationship to the nature of the opaque medium. *Br J Radiol* 1951;24:321–330.

53. Anderson W: A new deflocculant and protective colloid for barium sulphate. *J Pharm Pharmacol* 1962;14:64.

54. De Carvalho A, Madsen B: Prevention of crackle in double contrast examinations of the colon. *Acta Radiol Diagn* 1981;22:63–66.

55. Shirakabe H: *Atlas of X-Ray Diagnosis of Early Gastric Cancer*. Philadelphia, JB Lippincott, 1966.

56. Ford TF, Loomis AG, Fidham JF: The colloidal behavior of clays as related to their crystal structure. *J Phys Chem* 1940;44:1–12.

57. Hamilton JB: The use of tannic acid in barium enemas. *Am J Roentgenol* 1946;56:101–103.

58. Gelfand DW: Complications of gastrointestinal radiologic procedures: I. Complications of routine fluoroscopic studies. *Gastrointest Radiol* 1980;5:293–315.

59. McAlister WH, Anderson MS, Bloomberg GR, et al: Lethal effects of tannic acid in the barium enema: report of three fatalities and experimental studies. *Radiology* 1963;80:765–773.

60. Lucke HH, Hodge KE, Patt NL: Fatal liver damage after barium enema containing tannic acid. *Can Med Assoc J* 1963;89:1111–1114.

61. Rambo ON, Zboralske FF, Harris PA, et al: Toxicity studies of tannic acid administered by enema. I. Effects of the enema-administered tannic acid on the colon and liver of rats. *Am J Roentgenol* 1966;96:488–497.

62. Harris PA, Zboralske FF, Rambo ON, et al: Toxicity studies of tannic acid administered by enema. II. The colonic absorption and intraperitoneal toxicity of tannic acid and its hydrolytic products in rats. *Am J Roentgenol* 1966;96:498–504.

63. Zboralske FF, Harris PA, Riegelman S, et al: Toxicity studies of tannic acid administered by enema. III. Studies on the retention of enemas in humans. IV. Review and conclusions. *Am J Roentgenol* 1966;96:505–509.

64. Eshchar J, Friedman G: Acute hepatotoxicity of tannic acid added to barium enemas. *Am J Dig Dis* 1974;19:825–829.

65. Korolyuk IP, Shinkin VM, Kuzmina LI: The effect of structural and mechanical characteristics of barium sulphate suspension on the quality of x-ray image. *Farmatsiya (Moscow)* 1985;34:51–56.

66. Pawlaczyk J, Gorecki M: Rheometric characteristics of aqueous and stabilized suspensions of barium sulfate. *Acta Pol Pharm* 1978;35:87–92.

67. Wang HC, Hu XL: Effects of suspending agents on the properties of barium sulfate suspensions. *Acta Pharm Sinica* 1981;16:610–617.

68. Fischer EK: *Colloidal Dispersions*. New York, John Wiley and Sons, 1950, p 197.

69. O'Reilly GVA, Bryan G: The double contrast barium meal—a simplification. *Br J Radiol* 1974;47:482–483.

70. Vollmert B: *Polymer Chemistry*. New York, Springer-Verlag, 1973, p 514.

71. Kotzmann R: A technique of double contrast barium enema examination using sodium carboxymethyl cellulose as an additive. *Aust Radiol* 1969;13:178–188.

72. Brown GR: Barium sulfate and low viscosity monosaccharide polymer x-ray contrast media. United States Patent, 3 236 735, 1966.

73. Alexander P, Hitch SF: A comparative study of the anomalous viscosity of a high molecular weight polyelectrolyte and thymus nucleic acid. *Biochim Biophys Acta* 1952;9:229–236.

74. Markovitz H, Kimball GE: The effect of salts on the viscosity of solutions of polyacrylic acid. *J Colloid Sci* 1950;5:115–138.

75. Hirai K, Suezawa Y, Sugawara T: High-density barium sulfate as contrast media for mucosal examination of the stomach. *Sakura X-Ray Photograph Studies* 1975;26:1–18.

76. Harkins WD, Boyd GE: The binding energy between a crystalline solid and a liquid: the energy of adhesion and emersion. Energy of emersion of crystalline powders. *J Am Chem Soc* 1942; 64:1195–1204.

77. Melikhov IV, Sychev YN, Prokof MA: Mechanism for water removal during drying of a crystal with trapped inclusions. *Theoret Foundat Chem Eng* 1978;12:411–418.

78. Langmuir I: The constitution and fundamental properties of solids and liquids. *J Am Chem Soc* 1916;38:2221–2295.

79. Kolthoff IM, MacNevin WM: The adsorption of barium salts on barium sulfate from solutions in 50% ethanol. *J Am Chem Soc* 1936;58:1543–1546.

80. Taylor HS, Liang SC: The heterogeneity of catalyst surfaces. *J Am Chem Soc* 1947; 69:2989–2991.

81. Taylor HS, Liang SC: The heterogeneity of catalyst surfaces for chemisorption. *J Am Chem Soc* 1947;69:1306–1312.

82. Hirtzel CS, Rajagopalan R: *Colloidal Phenomena*. Park Ridge, NJ, Noyes Publications, 1985, p 124.

83. Hayes KF, Roe AL, Brown GE Jr, et al: In situ x-ray absorption study of surface complexes: selenium oxyanions on α-FeOOH. *Science* 1987;238:783–786.

84. Weiser HB: *Colloid Chemistry*, ed 2. New York, John Wiley & Sons, 1958.

85. Davis JA, Leckie JO: Surface ionization and complexation at the oxide/water interface. 3. Adsorption of anions. *J Colloid Interface Sci* 1980;74:32–43.

86. Bancroft WD: The theory of dyeing. I. *J Phys Chem* 1914;18:1–25.

87. Bancroft WD: The theory of dyeing. II. *J Phys Chem* 1914;18:118–151.

88. Briggs TR, Bull AW: The physical chemistry of dyeing: acid and basic dyes. *J Phys Chem* 1922;26:845–875.

89. Shenai VA, Patil VK, Churi RY: Use of barium sulfate in thin-layer chromatography of dyes. *Textile Research J* 1975;45:601–605.

90. Weiser HB: The physical chemistry of color lake formations. V. Hydrous oxide-alizarin lakes. *J Phys Chem* 1929;33:1713–1723.

91. Weiser HB, Porter EE: The physical chemistry of color lake formation. I. General principles. *J Phys Chem* 1927;31:1383–1399.

92. Weiser HB, Porter EE: The physical chemistry of color lake formation. II. Adsorption of typical dyes by basic mordants. *J Phys Chem* 1927;31:1704–1715.

93. Weiser HB, Porter EE: The physical chemistry of color lake formation. III. Alizarin lakes. *J Phys Chem 1927*;31:1824–1839.

94. Overbeek JTG: Colloids, a fascinating subject, in Goodwin JW (ed): *Colloidal Dispersions*. London, Royal Society of Chemistry, 1982, p 3.

95. Bagnall RD, Galloway RW, Annis JAD: Double contrast preparations: an *in vitro* study of some antifoaming agents. *Br J Radiol* 1977;50:546–550.

96. Marmur A: The effect of gravity on thin fluid films. *J Colloid Sci* 1984;100:407–413.

97. Bartsch O: Über Schaumbildungsfähigkeit und Oberflächenspannung. *Kolloid-Z* 1926; 38:177–179.

98. Gibbs JW: *The Collected Works of J.W. Gibbs*, vol 1, *Thermodynamics*. New York, Longmans, Green and Co, 1928.

99. Vold RD, Vold MJ: *Colloid and Interface Chemistry*. London, Addison-Wesley Publishing Co, 1983, p 37.

100. Hühnerfuss H: Surface viscosity measurements—a revival of a nearly forgotten surface chemical method? *J Colloid Interface Sci* 1985;107:84–95.

101. Bikerman JJ: *Foams*. New York, Springer-Verlag, 1973, p 263.

102. Levene G, Kaufman SA: An improved technic for double contrast examination of the colon by the use of compressed carbon dioxide. *Radiology* 1957;68:83–85.

103. Wark IW, Cox AB: Principles of flotation *AIME Trans* 1934;112:223.

104. Johonnott ES: Thickness of the black spot in liquid films. *Phil Mag* 1899;47:501–522.

105. Johonnott ES: The black spot in thin liquid films. *Phil Mag* 1906;11:746–753.

106. Miller RE: Antifoam barium sulfate suspensions. United States Patent, 3 201 317, 1965.

107. Monsalve A, Schechter RS: The stability of foams: dependence of observation on the bubble size distribution. *J Colloid Sci* 1984;97:327–335.

108. Robinson JV, Woods WW: A method of selecting foam inhibitors. *J Soc Chem Ind (London)* 1948;67:361–365.

109. Bikerman JJ: *Foams*. New York, Springer-Verlag, 1973, p 250.

110. Schott GD: Some observations on the history of the use of barium salts in medicine. *Med Hist* 1974;18:9–21.

111. Reznik RB, Toy HD Jr: Source assessment: Major barium chemicals. United States Environmental Agency, Environmental Protection Technology Series EPA-600/2-78-004b, 1978;1-138.

112. Borcherdt WO: Concentration of barite. United States Patents, 1 662 633 and 1 662 634, 1928.

113. Gieseke EW: Flotation of barite ores. United States Patent, 2 483 970, 1949.

114. Hoag EH: Flotation reagent. United States Patent, 2 371 292, 1945.

115. Hoag EH: Flotation process. United States Patent, 2 378 552, 1945.

116. Wells GA: Barytes market-filler-extender and chemical uses. Soc of Min Eng of AIME 1980 (presented at AIME meeting, Feb 24-28, 1980, Las Vegas, Nev.); Preprint number 80-29: 5 pages.

117. *Kirk-Othmer Encyclopedia of Chemical Technology*, ed 3., vol 3. New York, John Wiley & Sons, 1978, p 466.

118. Vetter OJG: How barium sulfate is formed: an interpretation. *J Petrol Technol* 1975; 27:1515–1524.

119. Vetter CJ: An evaluation of scale inhibitors. *J Petrol Technol* 1972;24:997–1006.

120. Liu ST, Nancollas GH: The crystal growth and dissolution of barium sulfate in the presence of additives. *J Colloid Interface Sci* 1975;52:582–592.

121. Von Weimarn PP: Zur lehre den Zuständen der Materie. *Kolloid-Z.* 1908;2:LII-LXI, 275-284, 301-307, 326-335.

122. Taylor WW: *The Chemistry of Colloids*, ed 2. London, Edward Arnold, 1921, p 132.

123. Byk H: Verfahren zur Herstellung von Erdalkalisalzen in kolloidaler und gelatinöser form. German patent, 178763, 1905. Summarized in Technisches Repertorium. *Z Electrochem* 1907;13:38–39.

124. Foulk CW. The effect of an excess of reagent in the precipitation of barium sulfate. *J Am Chem Soc* 1896;18:793–807.

125. Mar FW: On certain points in the estimation of barium as the sulfate. *Am J Sci* 1891; 41:288–295.

126. Osborne JL: The filtration of barium sulfate. *J Phys Chem* 1913;17:629–631.

127. Otsuka A, Sunada H, Bo M: Effect of sodium carboxymethylcellulose on the crystallization of barium sulfate. *Yakuzaigaku (Arch Pract Pharm)* 1971;31:97–103.

128. Boyd GE, Harkins WD: The energy of immersion of crystalline powders in water and organic liquids. Part 1. *J Am Chem Soc* 1942;64:1190–1194.

129. Hauser EA: Ueber die Thixotropie von Dispersionen geringer Konzentration. *Kolloid-Z* 1929;48:57–62.

130. Utzino S: *Quantitative Researches on Mechanical Dispersoid Synthesis: Colloid Chemistry*, vol 1. New York, Chemical Catalogue Co, 1926, p 664.

131. Perry JH, Chilton CH: *Chemical Engineers' Handbook*, ed 5. New York, McGraw-Hill, 1973, pp 8-48.

132. Snow RH, Kaye BH, Capes CE, et al: Size reduction and size enlargement, in Perry RH, Green DW, Maloney JO (eds): *Perry's Chemical Engineer's Handbook*, ed 6. New York, McGraw-Hill, 1984, pp 8–1 to 8–17.

133. Williams PA, Harrop R, Phillips GO, et al: Effect of electrolyte on the stability of barium sulfate dispersions in the presence of sodium (carboxymethyl) cellulose. *Ind Eng Chem Prod Res Dev* 1982;21:349–352.

134. La Mer VK, Smellie RH Jr: Flocculation, subsidence, and filtration of phosphate slimes. *J Colloid Sci* 1956;11:704–731.

135. Shaw DJ: *Introduction to Colloid and Surface Chemistry*. ed 2. London, Butterworths, 1970, p 185.

136. Kolta E, Scholtz A: Röntgenuntersuchungen des Magens mit Eiweiss und Fett enthaltenden Kontrastmitteln. *Med Klin* 1931;36:1313–1315.

137. Abbot WE, Krieger H, Levey S: Technical surgical factors which enhance or minimize postgastrectomy abnormalities. *Ann Surg* 1958;148:567–593.

138. Abbot WE, Krieger H, Bradshaw JS: The etiology and management of the dumping syndrome following a gastroenterostomy or subtotal gastrectomy. *Bull Soc Int Chir* 1961;20:40–52.

139. Embring PG, Mattsson O: Stabilized barium sulphate x-ray compositions. United States Patent, 3 216 900, 1965.

140. Embring G, Mattsson O: An improved physiologic contrast medium for the alimentary tract. *Acta Radiol Diagn* 1966;4:105–109.

141. Toyama T, Blickman JR: Serial roentgenographic observations on the passage of foods through the human stomach. *Radiol Clin Biol* 1971;40:29–38.

142. Raskin HF: Barium-burger roentgen study for unrecognized, clinically significant, gastric retention. *South Med J* 1971;64:1227–1235.

143. Abel MS: A barium-gelatine mixture for the x-ray examination of the digestive tract. *Radiology* 1944;43:175–180.

144. Miller RE: Flavoring barium sulfate. *Am J Roentgenol* 1966;96:484–487.

145. Amberg JR, Unger JD: Contamination of barium sulfate suspension. *Radiology* 1970; 97:182–183.

146. Briggs TR: The tinting strength of pigments. *J Phys Chem* 1918;22:216–230.

147. Fink CG: Chemical composition versus electrical conductivity. *J Phys Chem* 1917;21:32–36.

148. Rosato A, Prinz F, Standburg KJ, et al: Monte Carlo simulation of particulate matter segregation. *Powder Technol* 1986;49:59–69.

149. McKee MW, Jurgens RW Jr: Barium sulfate products for roentgenographic examination of the gastrointestinal tract. *Am J Hosp Pharm* 1986;43:145–148.

150. Moss AA, Beneventano TC, Gohel V, et al: The current status of upper gastrointestinal radiology. *Invest Radiol* 1980;15:92–102.

151. Kristensen HG, Schaefer T: Granulation. A review on pharmaceutical wet-granulation. *Drug Develop Industr Pharm* 1987;13:803–872.

152. Das S, Jarowski CI: Effect of granulating method on particle size distribution of granules and disintegrated tablets. *Drug Develop Industr Pharm* 1979;5:479–488.

153. Shaw DJ: *Introduction to Colloid and Surface Chemistry*, ed 2. London, Butterworths, 1970, p 125.

154. Weiser HB, Chapman TS: The mechanism of the mutual coagulation process. *J Phys Chem* 1931;35:543–556.

155. Durfor CN, Becker E: Public water supplies of the 100 largest cities in the United States. Geological Survey Water-Supply Paper 1812. US Government Printing Office, 1964.

156. James WB: Double contrast radiology in the gastrointestinal tract. *Clin Gastroenterol* 1978;7:397–430.

157. Schroeder HA, Tipton IH, Nason AP: Trace metals in man: strontium and barium. *J Chron Dis* 1972;25:491–517.

Index

I

K

L

M

R

S

T

W

X

Z